Student Solutions Manual

Multivariable Calculus

ELEVENTH EDITION

Ron Larson
The Pennsylvania State University, The Behrend College

Bruce Edwards
University of Florida

Prepared by

Ron Larson
The Pennsylvania State University, The Behrend College

Australia • Brazil • Mexico • Singapore • United Kingdom • United States

For product information and technology assistance, contact us at **Cengage Learning Customer & Sales Support, 1-800-354-9706**.

For permission to use material from this text or product, submit all requests online at **www.cengage.com/permissions** Further permissions questions can be emailed to **permissionrequest@cengage.com**.

ISBN: 978-1-337-27539-2

Cengage Learning
20 Channel Center Street
Boston, MA 02210
USA

Cengage Learning is a leading provider of customized learning solutions with office locations around the globe, including Singapore, the United Kingdom, Australia, Mexico, Brazil, and Japan. Locate your local office at: **www.cengage.com/global**.

Cengage Learning products are represented in Canada by Nelson Education, Ltd.

To learn more about Cengage Learning Solutions, visit **www.cengage.com**.

Purchase any of our products at your local college store or at our preferred online store **www.cengagebrain.com**.

Printed in the United States of America
Print Number: 01   Print Year: 2017

CONTENTS

CONTENTS

Multivariable Calculus

C H A P T E R   1 1
Vectors and the Geometry of Space

CHAPTER 11
Vectors and the Geometry of Space

Section 11.1 Vectors in the Plane

1. Answers will vary. *Sample answer*: A scalar is a real number such as 2. A vector is represented by a directed line segment. A vector has both magnitude and direction. For example $\langle \sqrt{3}, 1 \rangle$ has direction $\frac{\pi}{6}$ and a magnitude of 2.

3. (a) $\mathbf{v} = \langle 5 - 1, 4 - 2 \rangle = \langle 4, 2 \rangle$

   (b)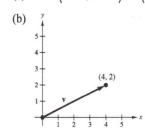

5. $\mathbf{u} = \langle 5 - 3, 6 - 2 \rangle = \langle 2, 4 \rangle$

   $\mathbf{v} = \langle 3 - 1, 8 - 4 \rangle = \langle 2, 4 \rangle$

   $\mathbf{u} = \mathbf{v}$

7. $\mathbf{u} = \langle 6 - 0, -2 - 3 \rangle = \langle 6, -5 \rangle$

   $\mathbf{v} = \langle 9 - 3, 5 - 10 \rangle = \langle 6, -5 \rangle$

   $\mathbf{u} = \mathbf{v}$

9. (b) $\mathbf{v} = \langle 5 - 2, 5 - 0 \rangle = \langle 3, 5 \rangle$

   (c) $\mathbf{v} = 3\mathbf{i} + 5\mathbf{j}$

   (a), (d)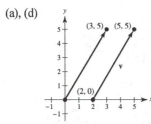

11. (b) $\mathbf{v} = \langle 6 - 8, -1 - 3 \rangle = \langle -2, -4 \rangle$

    (c) $\mathbf{v} = -2\mathbf{i} - 4\mathbf{j}$

    (a), (d)

13. (b) $\mathbf{v} = \langle 6 - 6, 6 - 2 \rangle = \langle 0, 4 \rangle$

    (c) $\mathbf{v} = 4\mathbf{j}$

    (a) and (d).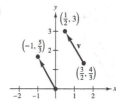

15. (b) $\mathbf{v} = \langle \frac{1}{2} - \frac{3}{2}, 3 - \frac{4}{3} \rangle = \langle -1, \frac{5}{3} \rangle$

    (c) $\mathbf{v} = -\mathbf{i} + \frac{5}{3}\mathbf{j}$

    (a) and (d)

17. $u_1 - 4 = -1$ $u_1 = 3$

    $u_2 - 2 = 3$ $u_2 = 5$

    $Q = (3, 5)$ Terminal point

19. $\mathbf{v} = 4i$

    $\|\mathbf{v}\| = \sqrt{4^2} = 4$

21. $\mathbf{v} = \langle 8, 15 \rangle$

    $\|\mathbf{v}\| = \sqrt{8^2 + 15^2} = 17$

23. $\mathbf{v} = -\mathbf{i} - 5\mathbf{j}$

    $\|\mathbf{v}\| = \sqrt{(-1)^2 + (-5)^2} = \sqrt{26}$

25. (a) $2\mathbf{v} = 2\langle 3, 5 \rangle = \langle 6, 10 \rangle$

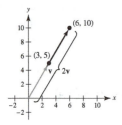

(b) $-3\mathbf{v} = \langle -9, -15 \rangle$

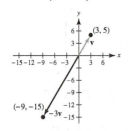

(c) $\frac{7}{2}\mathbf{v} = \left\langle \frac{21}{2}, \frac{35}{2} \right\rangle$

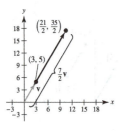

(d) $\frac{2}{3}\mathbf{v} = \left\langle 2, \frac{10}{3} \right\rangle$

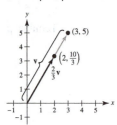

27. $\mathbf{u} = \langle 4, 9 \rangle$, $\mathbf{v} = \langle 2, -5 \rangle$

(a) $\frac{2}{3}\mathbf{u} = \frac{2}{3}\langle 4, 9 \rangle = \left\langle \frac{8}{3}, 6 \right\rangle$

(b) $3\mathbf{v} = 3\langle 2, -5 \rangle = \langle 6, -15 \rangle$

(c) $\mathbf{v} - \mathbf{u} = \langle 2, -5 \rangle - \langle 4, 9 \rangle = \langle -2, -14 \rangle$

(d) $2\mathbf{u} + 5\mathbf{v} = 2\langle 4, 9 \rangle + 5\langle 2, -5 \rangle$

$\qquad = \langle 8, 18 \rangle + \langle 10, -25 \rangle$

$\qquad = \langle 18, -7 \rangle$

29.

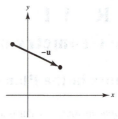

31.

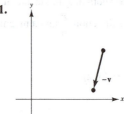

33.

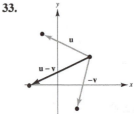

35. $\mathbf{v} = \langle 3, 12 \rangle$

$\|\mathbf{v}\| = \sqrt{3^2 + 12^2} = \sqrt{153}$

$\mathbf{u} = \dfrac{\mathbf{v}}{\|\mathbf{v}\|} = \dfrac{\langle 3, 12 \rangle}{\sqrt{153}} = \left\langle \dfrac{3}{\sqrt{153}}, \dfrac{12}{\sqrt{153}} \right\rangle$

$\qquad = \left\langle \dfrac{\sqrt{17}}{17}, \dfrac{4\sqrt{17}}{17} \right\rangle$ unit vector

37. $\mathbf{v} = \left\langle \dfrac{3}{2}, \dfrac{5}{2} \right\rangle$

$\|\mathbf{v}\| = \sqrt{\left(\dfrac{3}{2}\right)^2 + \left(\dfrac{5}{2}\right)^2} = \dfrac{\sqrt{34}}{2}$

$\mathbf{u} = \dfrac{\mathbf{v}}{\|\mathbf{v}\|} = \dfrac{\left\langle \left(\dfrac{3}{2}\right), \left(\dfrac{5}{2}\right) \right\rangle}{\dfrac{\sqrt{34}}{2}} = \left\langle \dfrac{3}{\sqrt{34}}, \dfrac{5}{\sqrt{34}} \right\rangle$

$\qquad = \left\langle \dfrac{3\sqrt{34}}{34}, \dfrac{5\sqrt{34}}{34} \right\rangle$ unit vector

39. $\mathbf{u} = \langle 1, -1 \rangle$, $\mathbf{v} = \langle -1, 2 \rangle$

(a) $\|\mathbf{u}\| = \sqrt{1+1} = \sqrt{2}$

(b) $\|\mathbf{v}\| = \sqrt{1+4} = \sqrt{5}$

(c) $\mathbf{u} + \mathbf{v} = \langle 0, 1 \rangle$

$\|\mathbf{u} + \mathbf{v}\| = \sqrt{0+1} = 1$

(d) $\dfrac{\mathbf{u}}{\|\mathbf{u}\|} = \dfrac{1}{\sqrt{2}}\langle 1, -1 \rangle$

$\left\| \dfrac{\mathbf{u}}{\|\mathbf{u}\|} \right\| = 1$

(e) $\dfrac{\mathbf{v}}{\|\mathbf{v}\|} = \dfrac{1}{\sqrt{5}}\langle -1, 2 \rangle$

$\left\| \dfrac{\mathbf{v}}{\|\mathbf{v}\|} \right\| = 1$

(f) $\dfrac{\mathbf{u}+\mathbf{v}}{\|\mathbf{u}+\mathbf{v}\|} = \langle 0, 1 \rangle$

$\left\| \dfrac{\mathbf{u}+\mathbf{v}}{\|\mathbf{u}+\mathbf{v}\|} \right\| = 1$

41. $\mathbf{u} = \left\langle 1, \dfrac{1}{2} \right\rangle$, $\mathbf{v} = \langle 2, 3 \rangle$

(a) $\|\mathbf{u}\| = \sqrt{1+\dfrac{1}{4}} = \dfrac{\sqrt{5}}{2}$

(b) $\|\mathbf{v}\| = \sqrt{4+9} = \sqrt{13}$

(c) $\mathbf{u} + \mathbf{v} = \left\langle 3, \dfrac{7}{2} \right\rangle$

$\|\mathbf{u} + \mathbf{v}\| = \sqrt{9+\dfrac{49}{4}} = \dfrac{\sqrt{85}}{2}$

(d) $\dfrac{\mathbf{u}}{\|\mathbf{u}\|} = \dfrac{2}{\sqrt{5}}\left\langle 1, \dfrac{1}{2} \right\rangle$

$\left\| \dfrac{\mathbf{u}}{\|\mathbf{u}\|} \right\| = 1$

(e) $\dfrac{\mathbf{v}}{\|\mathbf{v}\|} = \dfrac{1}{\sqrt{13}}\langle 2, 3 \rangle$

$\left\| \dfrac{\mathbf{v}}{\|\mathbf{v}\|} \right\| = 1$

(f) $\dfrac{\mathbf{u}+\mathbf{v}}{\|\mathbf{u}+\mathbf{v}\|} = \dfrac{2}{\sqrt{85}}\left\langle 3, \dfrac{7}{2} \right\rangle$

$\left\| \dfrac{\mathbf{u}+\mathbf{v}}{\|\mathbf{u}+\mathbf{v}\|} \right\| = 1$

43.
$\mathbf{u} = \langle 2, 1 \rangle$

$\|\mathbf{u}\| = \sqrt{5} \approx 2.236$

$\mathbf{v} = \langle 5, 4 \rangle$

$\|\mathbf{v}\| = \sqrt{41} \approx 6.403$

$\mathbf{u} + \mathbf{v} = \langle 7, 5 \rangle$

$\|\mathbf{u} + \mathbf{v}\| = \sqrt{74} \approx 8.602$

$\|\mathbf{u} + \mathbf{v}\| \le \|\mathbf{u}\| + \|\mathbf{v}\|$

$\sqrt{74} \le \sqrt{5} + \sqrt{41}$

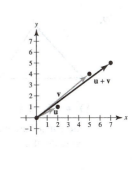

45. $\dfrac{\mathbf{u}}{\|\mathbf{u}\|} = \dfrac{1}{3}\langle 0, 3 \rangle = \langle 0, 1 \rangle$

$6\left(\dfrac{\mathbf{u}}{\|\mathbf{u}\|} \right) = 6\langle 0, 1 \rangle = \langle 0, 6 \rangle$

$\mathbf{v} = \langle 0, 6 \rangle$

47. $\dfrac{\mathbf{u}}{\|\mathbf{u}\|} = \dfrac{1}{\sqrt{5}}\langle -1, 2 \rangle = \left\langle -\dfrac{1}{\sqrt{5}}, \dfrac{2}{\sqrt{5}} \right\rangle$

$5\left(\dfrac{\mathbf{u}}{\|\mathbf{u}\|} \right) = 5\left\langle -\dfrac{1}{\sqrt{5}}, \dfrac{2}{\sqrt{5}} \right\rangle = \langle -\sqrt{5}, 2\sqrt{5} \rangle$

$\mathbf{v} = \langle -\sqrt{5}, 2\sqrt{5} \rangle$

49. $\mathbf{v} = 3\left[(\cos 0°)\mathbf{i} + (\sin 0°)\mathbf{j} \right] = 3\mathbf{i} = \langle 3, 0 \rangle$

51. $\mathbf{v} = 2\left[(\cos 150°)\mathbf{i} + (\sin 150°)\mathbf{j} \right]$

$= -\sqrt{3}\mathbf{i} + \mathbf{j} = \langle -\sqrt{3}, 1 \rangle$

53.
$\mathbf{u} = (\cos 0°)\mathbf{i} + (\sin 0°)\mathbf{j} = \mathbf{i}$

$\mathbf{v} = 3(\cos 45°)\mathbf{i} + 3(\sin 45°)\mathbf{j} = \dfrac{3\sqrt{2}}{2}\mathbf{i} + \dfrac{3\sqrt{2}}{2}\mathbf{j}$

$\mathbf{u} + \mathbf{v} = \left(\dfrac{2+3\sqrt{2}}{2} \right)\mathbf{i} + \dfrac{3\sqrt{2}}{2}\mathbf{j} = \left\langle \dfrac{2+3\sqrt{2}}{2}, \dfrac{3\sqrt{2}}{2} \right\rangle$

55.
$\mathbf{u} = 2(\cos 4)\mathbf{i} + 2(\sin 4)\mathbf{j}$

$\mathbf{v} = (\cos 2)\mathbf{i} + (\sin 2)\mathbf{j}$

$\mathbf{u} + \mathbf{v} = (2\cos 4 + \cos 2)\mathbf{i} + (2\sin 4 + \sin 2)\mathbf{j}$

$= \langle 2\cos 4 + \cos 2, \ 2\sin 4 + \sin 2 \rangle$

57. The forces act along the same direction. $\theta = 0°$.

59.

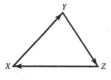

$\overrightarrow{XY} + \overrightarrow{YZ} + \overrightarrow{ZX} = 0.$

Vectors that start and end at the same point have a magnitude of 0.

61. $\mathbf{v} = \langle 4, 5 \rangle = a\langle 1, 2 \rangle + b\langle 1, -1 \rangle$

$4 = a + b$

$5 = 2a - b$

Adding the equations, $9 = 3a \Rightarrow a = 3.$

Then you have $b = 4 - a = 4 - 3 = 1.$

$a = 3, b = 1$

63. $\mathbf{v} = \langle -6, 0 \rangle = a\langle 1, 2 \rangle + 6\langle 1, -1 \rangle$

$-6 = a + b$

$0 = 2a - b$

Adding the equations, $-6 = 3a \Rightarrow a = -2.$

Then you have $b = -6 - a = -6 - (-2) = -4.$

$a = -2, b = -4$

65. $\mathbf{v} = \langle 1, -3 \rangle = a\langle 1, 2 \rangle + b\langle 1, -1 \rangle$

$1 = a + b$

$-3 = 2a - b$

Adding the equations, $-2 = 3a \Rightarrow a = -\dfrac{2}{3}.$

Then you have $b = 1 - a = 1 - \left(-\dfrac{2}{3}\right) = \dfrac{5}{3}.$

$a = -\dfrac{2}{3}, b = \dfrac{5}{3}$

67. $f(x) = x^2, f'(x) = 2x, f'(3) = 6$

(a) $m = 6.$ Let $\mathbf{w} = \langle 1, 6 \rangle, \|\mathbf{w}\| = \sqrt{37},$ then $\pm\dfrac{\mathbf{w}}{\|\mathbf{w}\|} = \pm\dfrac{1}{\sqrt{37}}\langle 1, 6 \rangle.$

(b) $m = -\dfrac{1}{6}.$ Let $\mathbf{w} = \langle -6, 1 \rangle, \|\mathbf{w}\| = \sqrt{37},$ then $\pm\dfrac{\mathbf{w}}{\|\mathbf{w}\|} = \pm\dfrac{1}{\sqrt{37}}\langle -6, 1 \rangle.$

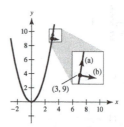

69. $f(x) = x^3, f'(x) = 3x^2 = 3$ at $x = 1.$

(a) $m = 3.$ Let $\mathbf{w} = \langle 1, 3 \rangle, \|\mathbf{w}\| = \sqrt{10},$ then $\dfrac{\mathbf{w}}{\|\mathbf{w}\|} = \pm\dfrac{1}{\sqrt{10}}\langle 1, 3 \rangle.$

(b) $m = -\dfrac{1}{3}.$ Let $\mathbf{w} = \langle 3, -1 \rangle, \|\mathbf{w}\| = \sqrt{10},$ then $\dfrac{\mathbf{w}}{\|\mathbf{w}\|} = \pm\dfrac{1}{\sqrt{10}}\langle 3, -1 \rangle.$

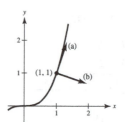

71. $f(x) = \sqrt{25 - x^2}$

$f'(x) = \dfrac{-x}{\sqrt{25 - x^2}} = \dfrac{-3}{4}$ at $x = 3.$

(a) $m = -\dfrac{3}{4}.$ Let $\mathbf{w} = \langle -4, 3 \rangle, \|\mathbf{w}\| = 5,$ then $\dfrac{\mathbf{w}}{\|\mathbf{w}\|} = \pm\dfrac{1}{5}\langle -4, 3 \rangle.$

(b) $m = \dfrac{4}{3}.$ Let $\mathbf{w} = \langle 3, 4 \rangle, \|\mathbf{w}\| = 5,$ then $\dfrac{\mathbf{w}}{\|\mathbf{w}\|} = \pm\dfrac{1}{5}\langle 3, 4 \rangle.$

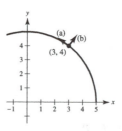

73. $\mathbf{u} = \dfrac{\sqrt{2}}{2}\mathbf{i} + \dfrac{\sqrt{2}}{2}\mathbf{j}$

$\mathbf{u} + \mathbf{v} = \sqrt{2}\mathbf{j}$

$\mathbf{v} = (\mathbf{u} + \mathbf{v}) - \mathbf{u} = -\dfrac{\sqrt{2}}{2}\mathbf{i} + \dfrac{\sqrt{2}}{2}\mathbf{j} = \left\langle -\dfrac{\sqrt{2}}{2}, \dfrac{\sqrt{2}}{2} \right\rangle$

75. $F_1 + F_2 = (500 \cos 30°i + 500 \sin 30°j) + (200 \cos(-45°)i + 200 \sin(-45°)j) = \left(250\sqrt{3} + 100\sqrt{2}\right)i + \left(250 - 100\sqrt{2}\right)j$

$\|F_1 + F_2\| = \sqrt{\left(250\sqrt{3} + 100\sqrt{2}\right)^2 + \left(250 - 100\sqrt{2}\right)^2} \approx 584.6$ lb

$\tan \theta = \dfrac{250 - 100\sqrt{2}}{250\sqrt{3} + 100\sqrt{2}} \Rightarrow \theta \approx 10.7°$

77. $F_1 + F_2 + F_3 = (75 \cos 30°i + 75 \sin 30°j) + (100 \cos 45°i + 100 \sin 45°j) + (125 \cos 120°i + 125 \sin 120°j)$

$= \left(\frac{75}{2}\sqrt{3} + 50\sqrt{2} - \frac{125}{2}\right)i + \left(\frac{75}{2} + 50\sqrt{2} + \frac{125}{2}\sqrt{3}\right)j$

$\|R\| = \|F_1 + F_2 + F_3\| \approx 228.5$ lb

$\theta_R = \theta_{F_1 + F_2 + F_3} \approx 71.3°$

79. $u = \overrightarrow{CB} = \|u\|(\cos 30°i + \sin 30°j)$

$v = \overrightarrow{CA} = \|v\|(\cos 130°i + \sin 130°j)$

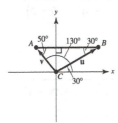

Vertical components: $\|u\| \sin 30° + \|v\| \sin 130° = 3000$

Horizontal components: $\|u\| \cos 30° + \|v\| \cos 130° = 0$

Solving this system, you obtain

$\|u\| \approx 1958.1$ pounds,

$\|v\| \approx 2638.2$ pounds.

81. Horizontal component $= \|v\| \cos \theta = 1200 \cos 6° \approx 1193.43$ ft/sec

Vertical component $= \|v\| \sin \theta = 1200 \sin 6° \approx 125.43$ ft/sec

83. $u = 900(\cos 148°i + \sin 148°j)$

$v = 100(\cos 45°i + \sin 45°j)$

$u + v = (900 \cos 148° + 100 \cos 45°)i + (900 \sin 148° + 100 \sin 45°)j$

$\approx -692.53i + 547.64j$

$\theta \approx \arctan\left(\dfrac{547.64}{-692.53}\right) \approx -38.34°; 38.34°$ North of West

$\|u + v\| \approx \sqrt{(-692.53)^2 + (547.64)^2} \approx 882.9$ km/h

85. False. Weight has direction.

87. True

89. True

91. True

93. False

$\|ai + bj\| = \sqrt{2}|a|$

95. $\|u\| = \sqrt{\cos^2 \theta + \sin^2 \theta} = 1,$

$\|v\| = \sqrt{\sin^2 \theta + \cos^2 \theta} = 1$

97. Let **u** and **v** be the vectors that determine the parallelogram, as indicated in the figure. The two diagonals are $\mathbf{u} + \mathbf{v}$ and $\mathbf{v} - \mathbf{u}$. So, $\mathbf{r} = x(\mathbf{u} + \mathbf{v})$, $\mathbf{s} = 4(\mathbf{v} - \mathbf{u})$. But,

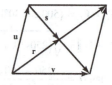

$$\mathbf{u} = \mathbf{r} - \mathbf{s} = x(\mathbf{u} + \mathbf{v}) - y(\mathbf{v} - \mathbf{u}) = (x + y)\mathbf{u} + (x - y)\mathbf{v}.$$

So, $x + y = 1$ and $x - y = 0$. Solving you have $x = y = \frac{1}{2}$.

99. The set is a circle of radius 5, centered at the origin.

$$\|\mathbf{u}\| = \|\langle x, y \rangle\| = \sqrt{x^2 + y^2} = 5 \Rightarrow x^2 + y^2 = 25$$

Section 11.2 Space Coordinates and Vectors in Space

1. x_0 is directed distance to yz-plane.

   y_0 is directed distance to xz-plane.

   z_0 is directed distance to xy-plane.

3. (a) $x = 4$ is a point on the number line.

   (b) $x = 4$ is a vertical line in the plane.

   (c) $x = 4$ is a plane in space.

5.

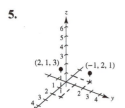

7.

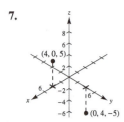

9. $x = -3, y = 4, z = 5$: $(-3, 4, 5)$

11. $y = z = 0, x = 12$: $(12, 0, 0)$

13. The point is 1 unit above the xy-plane.

15. The point is on the plane parallel to the yz-plane that passes through $x = -3$.

17. The point is to the left of the xz-plane.

19. The point is on or between the planes $y = 3$ and $y = -3$.

21. The point (x, y, z) is 3 units below the xy-plane, and below either quadrant I or III.

23. The point could be above the xy-plane and so above quadrants II or IV, or below the xy-plane, and so below quadrants I or III.

25. $d = \sqrt{(8-4)^2 + (2-1)^2 + (6-5)^2}$

   $= \sqrt{16 + 1 + 1}$

   $= \sqrt{18} = 3\sqrt{2}$

27. $d = \sqrt{(3-0)^2 + (2-2)^2 + (8-4)^2}$

   $= \sqrt{9 + 0 + 16}$

   $= \sqrt{25} = 5$

29. $A(0, 0, 4), B(2, 6, 7), C(6, 4, -8)$

   $|AB| = \sqrt{2^2 + 6^2 + 3^2} = \sqrt{49} = 7$

   $|AC| = \sqrt{6^2 + 4^2 + (-12)^2} = \sqrt{196} = 14$

   $|BC| = \sqrt{4^2 + (-2)^2 + (-15)^2} = \sqrt{245} = 7\sqrt{5}$

   $|BC|^2 = 245 = 49 + 196 = |AB|^2 + |AC|^2$

   Right triangle

31. $A(-1, 0, -2), B(-1, 5, 2), C(-3, -1, 1)$

   $|AB| = \sqrt{0 + 25 + 16} = \sqrt{41}$

   $|AC| = \sqrt{4 + 1 + 9} = \sqrt{14}$

   $|BC| = \sqrt{4 + 36 + 1} = \sqrt{41}$

   Because $|AB| = |BC|$, the triangle is isosceles.

33. $\left(\dfrac{4+8}{2}, \dfrac{0+8}{2}, \dfrac{-6+20}{2} \right) = (6, 4, 7)$

35. $\left(\dfrac{3+1}{2}, \dfrac{4+8}{2}, \dfrac{6+0}{2} \right) = (2, 6, 3)$

37. Center: $(7, 1, -2)$

   Radius: 1

   $(x - 7)^2 + (y - 1)^2 + (z + 2)^2 = 1$

39. Center is midpoint of diameter: $\left(\dfrac{2+1}{2}, \dfrac{1+3}{2}, \dfrac{3-1}{2}\right) = \left(\dfrac{3}{2}, 2, 1\right)$

Radius is distance from center to endpoint:

$$d = \sqrt{\left(\dfrac{3}{2}-1\right)^2 + (2-3)^2 + (1+1)^2} = \sqrt{\dfrac{1}{4}+1+4} = \dfrac{\sqrt{21}}{2}$$

$$\left(x - \dfrac{3}{2}\right)^2 + (y-2)^2 + (z-1)^2 = \dfrac{21}{4}$$

41. Center: $(-7, 7, 6)$

Tangent to *xy*-plane

Radius is *z*-coordinate, 6.

$$(x+7)^2 + (y-7)^2 + (z-6)^2 = 36$$

43.
$$x^2 + y^2 + z^2 - 2x + 6y + 8z + 1 = 0$$
$$\left(x^2 - 2x + 1\right) + \left(y^2 + 6y + 9\right) + \left(z^2 + 8z + 16\right) = -1 + 1 + 9 + 16$$
$$(x-1)^2 + (y+3)^2 + (z+4)^2 = 25$$

Center: $(1, -3, -4)$

Radius: 5

45.
$$9x^2 + 9y^2 + 9z^2 - 6x + 18y + 1 = 0$$
$$x^2 + y^2 + z^2 - \tfrac{2}{3}x + 2y + \tfrac{1}{9} = 0$$
$$\left(x^2 - \tfrac{2}{3}x + \tfrac{1}{9}\right) + \left(y^2 + 2y + 1\right) + z^2 = -\tfrac{1}{9} + \tfrac{1}{9} + 1$$
$$\left(x - \tfrac{1}{3}\right)^2 + (y+1)^2 + (z-0)^2 = 1$$

Center: $\left(\tfrac{1}{3}, -1, 0\right)$

Radius: 1

47. (a) $\mathbf{v} = \langle 2-4, 4-2, 3-1 \rangle = \langle -2, 2, 2 \rangle$

(b) $\mathbf{v} = -2\mathbf{i} + 2\mathbf{j} + 2\mathbf{k}$

(c)

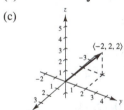

49. (b) $\mathbf{v} = \langle 3-(-1), 3-2, 4-3 \rangle = \langle 4, 1, 1 \rangle$

(c) $\mathbf{v} = 4\mathbf{i} + \mathbf{j} + \mathbf{k}$

(a), (d)

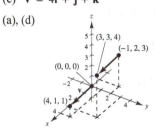

51. $\mathbf{v} = \langle 4-3, 1-2, 6-0 \rangle = \langle 1, -1, 6 \rangle$

$\|\mathbf{v}\| = \sqrt{1+1+36} = \sqrt{38}$

Unit vector: $\dfrac{\langle 1, -1, 6 \rangle}{\sqrt{38}} = \left\langle \dfrac{1}{\sqrt{38}}, \dfrac{-1}{\sqrt{38}}, \dfrac{6}{\sqrt{38}} \right\rangle$

53. $\mathbf{v} = \langle 0-4, 5-2, 2-0 \rangle = \langle -4, 3, 2 \rangle$

$\|\mathbf{v}\| = \sqrt{(-4)^2 + 3^2 + 2^2} = \sqrt{16+9+4} = \sqrt{29}$

Unit vector:

$$\dfrac{1}{\sqrt{29}}\langle -4, 3, 2 \rangle = \left\langle -\dfrac{4}{\sqrt{29}}, \dfrac{3}{\sqrt{29}}, \dfrac{2}{\sqrt{29}} \right\rangle$$

55. $(q_1, q_2, q_3) - (0, 6, 2) = (3, -5, 6)$

$Q = (3, 1, 8)$

57. (a) $2\mathbf{v} = \langle 2, 4, 4 \rangle$

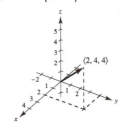

(b) $-\mathbf{v} = \langle -1, -2, -2 \rangle$

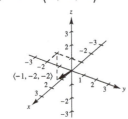

(c) $\frac{3}{2}\mathbf{v} = \langle \frac{3}{2}, 3, 3 \rangle$

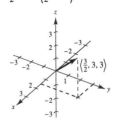

(d) $0\mathbf{v} = \langle 0, 0, 0 \rangle$

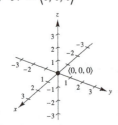

59. $\mathbf{z} = \mathbf{u} - \mathbf{v} + \mathbf{w}$

$= \langle 1, 2, 3 \rangle - \langle 2, 2, -1 \rangle + \langle 4, 0, -4 \rangle$

$= \langle 3, 0, 0 \rangle$

61. $\frac{1}{3}\mathbf{z} - 3\mathbf{u} = \mathbf{w}$

$\frac{1}{3}\mathbf{z} = 3\mathbf{u} + \mathbf{w}$

$\mathbf{z} = 9\mathbf{u} + 3\mathbf{w}$

$= 9\langle 1, 2, 3 \rangle + 3\langle 4, 0, -4 \rangle$

$= \langle 9, 18, 27 \rangle + \langle 12, 0, -12 \rangle$

$= \langle 21, 18, 15 \rangle$

63. (a) and (b) are parallel because

$\langle -6, -4, 10 \rangle = -2\langle 3, 2, -5 \rangle$ and

$\langle 2, \frac{4}{3}, -\frac{10}{3} \rangle = \frac{2}{3}\langle 3, 2, -5 \rangle$.

65. $\mathbf{z} = -3\mathbf{i} + 4\mathbf{j} + 2\mathbf{k}$

(a) is parallel because $-6\mathbf{i} + 8\mathbf{j} + 4\mathbf{k} = 2\mathbf{z}$.

67. $P(0, -2, -5), Q(3, 4, 4), R(2, 2, 1)$

$\overrightarrow{PQ} = \langle 3, 6, 9 \rangle$

$\overrightarrow{PR} = \langle 2, 4, 6 \rangle$

$\langle 3, 6, 9 \rangle = \frac{3}{2}\langle 2, 4, 6 \rangle$

So, $\overrightarrow{PQ}$ and $\overrightarrow{PR}$ are parallel, the points are collinear.

69. $P(1, 2, 4), Q(2, 5, 0), R(0, 1, 5)$

$\overrightarrow{PQ} = \langle 1, 3, -4 \rangle$

$\overrightarrow{PR} = \langle -1, -1, 1 \rangle$

Because $\overrightarrow{PQ}$ and $\overrightarrow{PR}$ are not parallel, the points are not collinear.

71. $A(2, 9, 1), B(3, 11, 4), C(0, 10, 2), D(1, 12, 5)$

$\overrightarrow{AB} = \langle 1, 2, 3 \rangle$

$\overrightarrow{CD} = \langle 1, 2, 3 \rangle$

$\overrightarrow{AC} = \langle -2, 1, 1 \rangle$

$\overrightarrow{BD} = \langle -2, 1, 1 \rangle$

Because $\overrightarrow{AB} = \overrightarrow{CD}$ and $\overrightarrow{AC} = \overrightarrow{BD}$, the given points form the vertices of a parallelogram.

73. $\|\mathbf{v}\| = \|\langle -1, 0, 1 \rangle\|$

$= \sqrt{(-1)^2 + 0^2 + 1^2}$

$= \sqrt{1 + 1} = \sqrt{2}$

75. $\mathbf{v} = 3\mathbf{j} - 5\mathbf{k} = \langle 0, 3, -5 \rangle$

$\|\mathbf{v}\| = \sqrt{0 + 9 + 25} = \sqrt{34}$

77. $\mathbf{v} = \mathbf{i} - 2\mathbf{j} - 3\mathbf{k} = \langle 1, -2, -3 \rangle$

$\|\mathbf{v}\| = \sqrt{1 + 4 + 9} = \sqrt{14}$

79. $\mathbf{v} = \langle 2, -1, 2 \rangle$

$\|\mathbf{v}\| = \sqrt{4 + 1 + 4} = 3$

(a) $\dfrac{\mathbf{v}}{\|\mathbf{v}\|} = \frac{1}{3}\langle 2, -1, 2 \rangle$

(b) $-\dfrac{\mathbf{v}}{\|\mathbf{v}\|} = -\frac{1}{3}\langle 2, -1, 2 \rangle$

81. $\mathbf{v} = 4\mathbf{i} - 5\mathbf{j} + 3\mathbf{k}$

$\|\mathbf{v}\| = \sqrt{16 + 25 + 9} = \sqrt{50} = 5\sqrt{2}$

(a) $\dfrac{\mathbf{v}}{\|\mathbf{v}\|} = \dfrac{1}{5\sqrt{2}}(4\mathbf{i} - 5\mathbf{j} + 3\mathbf{k}) = \dfrac{2\sqrt{2}}{5}\mathbf{i} - \dfrac{\sqrt{2}}{2}\mathbf{j} + \dfrac{3\sqrt{2}}{10}\mathbf{k}$

(b) $-\dfrac{\mathbf{v}}{\|\mathbf{v}\|} = -\dfrac{1}{5\sqrt{2}}(4\mathbf{i} - 5\mathbf{j} + 3\mathbf{k}) = -\dfrac{2\sqrt{2}}{5}\mathbf{i} + \dfrac{\sqrt{2}}{2}\mathbf{j} - \dfrac{3\sqrt{2}}{10}\mathbf{k}$

83. $\mathbf{v} = 10\dfrac{\mathbf{u}}{\|\mathbf{u}\|} = 10\dfrac{\langle 0, 3, 3 \rangle}{3\sqrt{2}} = 10\left\langle 0, \dfrac{1}{\sqrt{2}}, \dfrac{1}{\sqrt{2}} \right\rangle = \left\langle 0, \dfrac{10}{\sqrt{2}}, \dfrac{10}{\sqrt{2}} \right\rangle$

85. $\mathbf{v} = \dfrac{3}{2}\dfrac{\mathbf{u}}{\|\mathbf{u}\|} = \dfrac{3}{2}\dfrac{(2, -2, 1)}{3} = \dfrac{3}{2}\left\langle \dfrac{2}{3}, \dfrac{-2}{3}, \dfrac{1}{3} \right\rangle = \left\langle 1, -1, \dfrac{1}{2} \right\rangle$

87. $\mathbf{v} = 2\left[\cos(\pm 30°)\mathbf{j} + \sin(\pm 30°)\mathbf{k}\right] = \sqrt{3}\mathbf{j} \pm \mathbf{k} = \left\langle 0, \sqrt{3}, \pm 1 \right\rangle$

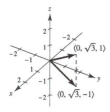

89.
$$\mathbf{v} = \langle -3, -6, 3 \rangle$$
$$\tfrac{2}{3}\mathbf{v} = \langle -2, -4, 2 \rangle$$
$$(4, 3, 0) + (-2, -4, 2) = (2, -1, 2)$$

91. A sphere of radius 4 centered at (x_1, y_1, z_1).

$\|\mathbf{v}\| = \|\langle x - x_2, y - y_1, z - z_1 \rangle\|$

$= \sqrt{(x - x_1)^2 + (y - y_1)^2 + (z - z_1)^2} = 4$

$(x - x_1)^2 + (y - y_1)^2 + (z - z_1)^2 = 16$

93. The set of all points (x, y, z) such that $\|\mathbf{r}\| > 1$ represent outside the sphere of radius 1 centered at the origin.

95. The terminal points of the vectors $t\mathbf{u}$, $\mathbf{u} + t\mathbf{v}$ and $s\mathbf{u} + t\mathbf{v}$ are collinear.

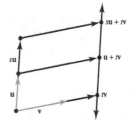

97. Let α be the angle between $\mathbf{v}$ and the coordinate axes.

$\mathbf{v} = (\cos \alpha)\mathbf{i} + (\cos \alpha)\mathbf{j} + (\cos \alpha)\mathbf{k}$

$\|\mathbf{v}\| = \sqrt{3}\cos \alpha = 1$

$\cos \alpha = \dfrac{1}{\sqrt{3}} = \dfrac{\sqrt{3}}{3}$

$\mathbf{v} = \dfrac{\sqrt{3}}{3}(\mathbf{i} + \mathbf{j} + \mathbf{k}) = \dfrac{\sqrt{3}}{3}\langle 1, 1, 1 \rangle$

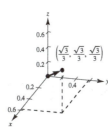

99. (a) The height of the right triangle is $h = \sqrt{L^2 - 18^2}$.

The vector $\overrightarrow{PQ}$ is given by $\overrightarrow{PQ} = \langle 0, -18, h \rangle$.

The tension vector $\mathbf{T}$ in each wire is $\mathbf{T} = c\langle 0, -18, h \rangle$ where $ch = \dfrac{24}{3} = 8$.

So, $\mathbf{T} = \dfrac{8}{h}\langle 0, -18, h \rangle$ and $T = \|\mathbf{T}\| = \dfrac{8}{h}\sqrt{18^2 + h^2} = \dfrac{8}{\sqrt{L^2 - 18^2}}\sqrt{18^2 + \left(L^2 - 18^2\right)} = \dfrac{8L}{\sqrt{L^2 - 18^2}},\ L > 18$.

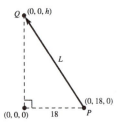

(b)

L	20	25	30	35	40	45	50
T	18.4	11.5	10	9.3	9.0	8.7	8.6

(c)

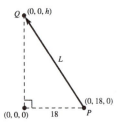

$x = 18$ is a vertical asymptote and $y = 8$ is a horizontal asymptote.

(d) $\displaystyle\lim_{L \to 18^+} \dfrac{8L}{\sqrt{L^2 - 18^2}} = \infty$

$\displaystyle\lim_{L \to \infty} \dfrac{8L}{\sqrt{L^2 - 18^2}} = \lim_{L \to \infty} \dfrac{8}{\sqrt{1 - (18/L)^2}} = 8$

(e) From the table, $T = 10$ implies $L = 30$ inches.

101. $\overrightarrow{AB} = \langle 0, 70, 115 \rangle,\ \mathbf{F_1} = C_1\langle 0, 70, 115 \rangle$

$\overrightarrow{AC} = \langle -60, 0, 115 \rangle,\ \mathbf{F_2} = C_2\langle -60, 0, 115 \rangle$

$\overrightarrow{AD} = \langle 45, -65, 115 \rangle,\ \mathbf{F_3} = C_3\langle 45, -65, 115 \rangle$

$\mathbf{F} = \mathbf{F_1} + \mathbf{F_2} + \mathbf{F_3} = \langle 0, 0, 500 \rangle$

So:

$-60C_2 + 45C_3 = 0$

$70C_1 \qquad\quad - 65C_3 = 0$

$115(C_1 + C_2 + C_3) = 500$

Solving this system yields $C_1 = \frac{104}{69}, C_2 = \frac{28}{23},$ and $C_3 = \frac{112}{69}$.

So: $\|\mathbf{F_1}\| \approx 202.919\,\text{N}$

$\|\mathbf{F_2}\| \approx 157.909\,\text{N}$

$\|\mathbf{F_3}\| \approx 226.521\,\text{N}$

103. $d(AP) = 2d(BP)$

$$\sqrt{x^2 + (y+1)^2 + (z-1)^2} = 2\sqrt{(x-1)^2 + (y-2)^2 + z^2}$$

$$x^2 + y^2 + z^2 + 2y - 2z + 2 = 4(x^2 + y^2 + z^2 - 2x - 4y + 5)$$

$$0 = 3x^2 + 3y^2 + 3z^2 - 8x - 18y + 2z + 18$$

$$-6 + \frac{16}{9} + 9 + \frac{1}{9} = \left(x^2 - \frac{8}{3}x + \frac{16}{9}\right) + \left(y^2 - 6y + 9\right) + \left(z^2 + \frac{2}{3}z + \frac{1}{9}\right)$$

$$\frac{44}{9} = \left(x - \frac{4}{3}\right)^2 + (y-3)^2 + \left(z + \frac{1}{3}\right)^2$$

Sphere; center: $\left(\frac{4}{3}, 3, -\frac{1}{3}\right)$, radius: $\dfrac{2\sqrt{11}}{3}$

Section 11.3   The Dot Product of Two Vectors

1. The vectors are orthogonal (perpendicular) if the dot product of the vectors is zero.

3. $\mathbf{u} = \langle 3, 4 \rangle$, $\mathbf{v} = \langle -1, 5 \rangle$

 (a) $\mathbf{u} \cdot \mathbf{v} = 3(-1) + 4(5) = 17$

 (b) $\mathbf{u} \cdot \mathbf{u} = 3(3) + 4(4) = 25$

 (c) $\|\mathbf{v}\|^2 = (-1)^2 + 5^2 = 26$

 (d) $(\mathbf{u} \cdot \mathbf{v})\mathbf{v} = 17\langle -1, 5 \rangle = \langle -17, 85 \rangle$

 (e) $\mathbf{u} \cdot (3\mathbf{v}) = 3(\mathbf{u} \cdot \mathbf{v}) = 3(17) = 51$

5. $\mathbf{u} = \langle 6, -4 \rangle$, $\mathbf{v} = \langle -3, 2 \rangle$

 (a) $\mathbf{u} \cdot \mathbf{v} = 6(-3) + (-4)(2) = -26$

 (b) $\mathbf{u} \cdot \mathbf{u} = 6(6) + (-4)(-4) = 52$

 (c) $\|\mathbf{v}\|^2 = (-3)^2 + 2^2 = 13$

 (d) $(\mathbf{u} \cdot \mathbf{v})\mathbf{v} = -26\langle -3, 2 \rangle = \langle 78, -52 \rangle$

 (e) $\mathbf{u} \cdot (3\mathbf{v}) = 3(\mathbf{u} \cdot \mathbf{v}) = 3(-26) = -78$

7. $\mathbf{u} = \langle 2, -3, 4 \rangle$, $\mathbf{v} = \langle 0, 6, 5 \rangle$

 (a) $\mathbf{u} \cdot \mathbf{v} = 2(0) + (-3)(6) + (4)(5) = 2$

 (b) $\mathbf{u} \cdot \mathbf{u} = 2(2) + (-3)(-3) + 4(4) = 29$

 (c) $\|\mathbf{v}\|^2 = 0^2 + 6^2 + 5^2 = 61$

 (d) $(\mathbf{u} \cdot \mathbf{v})\mathbf{v} = 2\langle 0, 6, 5 \rangle = \langle 0, 12, 10 \rangle$

 (e) $\mathbf{u} \cdot (3\mathbf{v}) = 3(\mathbf{u} \cdot \mathbf{v}) = 3(2) = 6$

9. $\mathbf{u} = 2\mathbf{i} - \mathbf{j} + \mathbf{k}$, $\mathbf{v} = \mathbf{i} - \mathbf{k}$

 (a) $\mathbf{u} \cdot \mathbf{v} = 2(1) + (-1)(0) + 1(-1) = 1$

 (b) $\mathbf{u} \cdot \mathbf{u} = 2(2) + (-1)(-1) + (1)(1) = 6$

 (c) $\|\mathbf{v}\|^2 = 1^2 + (-1)^2 = 2$

 (d) $(\mathbf{u} \cdot \mathbf{v})\mathbf{v} = \mathbf{v} = \mathbf{i} - \mathbf{k}$

 (e) $\mathbf{u} \cdot (3\mathbf{v}) = 3(\mathbf{u} \cdot \mathbf{v}) = 3(1) = 3$

11. $\mathbf{u} = \langle 1, 1 \rangle$, $\mathbf{v} = \langle 2, -2 \rangle$

$$\cos\theta = \frac{\mathbf{u} \cdot \mathbf{v}}{\|\mathbf{u}\|\|\mathbf{v}\|} = \frac{0}{\sqrt{2}\sqrt{8}} = 0$$

 (a) $\theta = \dfrac{\pi}{2}$ (b) $\theta = 90°$

13. $\mathbf{u} = 3\mathbf{i} + \mathbf{j}$, $\mathbf{v} = -2\mathbf{i} + 4\mathbf{j}$

$$\cos\theta = \frac{\mathbf{u} \cdot \mathbf{v}}{\|\mathbf{u}\|\|\mathbf{v}\|} = \frac{-2}{\sqrt{10}\sqrt{20}} = \frac{-1}{5\sqrt{2}}$$

 (a) $\theta = \arccos\left(-\dfrac{1}{5\sqrt{2}}\right) \approx 1.713$

 (b) $\theta \approx 98.1°$

15. $\mathbf{u} = \langle 1, 1, 1 \rangle$, $\mathbf{v} = \langle 2, 1, -1 \rangle$

$$\cos\theta = \frac{\mathbf{u} \cdot \mathbf{v}}{\|\mathbf{u}\|\|\mathbf{v}\|} = \frac{2}{\sqrt{3}\sqrt{6}} = \frac{\sqrt{2}}{3}$$

 (a) $\theta = \arccos\dfrac{\sqrt{2}}{3} \approx 1.080$

 (b) $\theta \approx 61.9°$

17. $u = 3i + 4j$, $v = -2j + 3k$

$$\cos \theta = \frac{u \cdot v}{\|u\| \|v\|} = \frac{-8}{5\sqrt{13}} = \frac{-8\sqrt{13}}{65}$$

(a) $\theta = \arccos\left(-\frac{8\sqrt{13}}{65}\right) \approx 2.031$

(b) $\theta \approx 116.3°$

19. $\dfrac{u \cdot v}{\|u\| \|v\|} = \cos \theta$

$u \cdot v = (8)(5) \cos \dfrac{\pi}{3} = 20$

21. $u = \langle 4, 3 \rangle$, $v = \left\langle \frac{1}{2}, -\frac{2}{3} \right\rangle$

$u \neq cv \Rightarrow$ not parallel

$u \cdot v = 0 \Rightarrow$ orthogonal

23. $u = j + 6k$, $v = i - 2j - k$

$u \neq cv \Rightarrow$ not parallel

$u \cdot v = -8 \neq 0 \Rightarrow$ not orthogonal

Neither

25. $u = \langle 2, -3, 1 \rangle$, $v = \langle -1, -1, -1 \rangle$

$u \neq cv \Rightarrow$ not parallel

$u \cdot v = 0 \Rightarrow$ orthogonal

27. The vector $\langle 1, 2, 0 \rangle$ joining $(1, 2, 0)$ and $(0, 0, 0)$ is perpendicular to the vector $\langle -2, 1, 0 \rangle$ joining $(-2, 1, 0)$ and $(0, 0, 0)$: $\langle 1, 2, 0 \rangle \cdot \langle -2, 1, 0 \rangle = 0$

The triangle has a right angle, so it is a right triangle.

29. $A(2, 0, 1)$, $B(0, 1, 2)$, $C\left(-\frac{1}{2}, \frac{3}{2}, 0\right)$

$\overline{AB} = \langle -2, 1, 1 \rangle \qquad \overline{BA} = \langle 2, -1, -1 \rangle$

$\overline{AC} = \left\langle -\frac{5}{2}, \frac{3}{2}, -1 \right\rangle \qquad \overline{CA} = \left\langle \frac{5}{2}, -\frac{3}{2}, 1 \right\rangle$

$\overline{BC} = \left\langle -\frac{1}{2}, \frac{1}{2}, -2 \right\rangle \qquad \overline{CB} = \left\langle \frac{1}{2}, -\frac{1}{2}, 2 \right\rangle$

$\overline{AB} \cdot \overline{AC} = 5 + \frac{3}{2} - 1 > 0$

$\overline{BA} \cdot \overline{BC} = -1 - \frac{1}{2} + 2 > 0$

$\overline{CA} \cdot \overline{CB} = \frac{5}{4} + \frac{3}{4} + 2 > 0$

The triangle has three acute angles, so it is an acute triangle.

31. $u = i + 2j + 2k$, $\|u\| = \sqrt{1 + 4 + 4} = 3$

$\cos \alpha = \frac{1}{3} \Rightarrow \alpha \approx 1.2310$ or $70.5°$

$\cos \beta = \frac{2}{3} \Rightarrow \beta \approx 0.8411$ or $48.2°$

$\cos \gamma = \frac{2}{3} \Rightarrow \gamma \approx 0.8411$ or $48.2°$

$\cos^2 \alpha + \cos^2 \beta + \cos^2 \gamma = \frac{1}{9} + \frac{4}{9} + \frac{4}{9} = 1$

33. $u = 7i + j - k$, $\|u\| = \sqrt{49 + 1 + 1} = \sqrt{51}$

$\cos \alpha = \frac{7}{\sqrt{51}} \Rightarrow \alpha \approx 11.4°$

$\cos \beta = \frac{1}{\sqrt{51}} \Rightarrow \beta \approx 82.0°$

$\cos \gamma = -\frac{1}{\sqrt{51}} \Rightarrow \gamma \approx 98.0°$

35. $u = \langle 0, 6, -4 \rangle$, $\|u\| = \sqrt{0 + 36 + 16} = \sqrt{52} = 2\sqrt{13}$

$\cos \alpha = 0 \Rightarrow \alpha = \frac{\pi}{2}$ or $90°$

$\cos \beta = \frac{3}{\sqrt{13}} \Rightarrow \beta \approx 0.5880$ or $33.7°$

$\cos \gamma = -\frac{2}{\sqrt{13}} \Rightarrow \gamma \approx 2.1588$ or $123.7°$

$\cos^2 \alpha + \cos^2 \beta + \cos^2 \gamma = 0 + \frac{9}{13} + \frac{4}{13} = 1$

37. $u = \langle 6, 7 \rangle$, $v = \langle 1, 4 \rangle$

(a) $w_1 = \text{proj}_v u = \left(\dfrac{u \cdot v}{\|v\|^2}\right) v$

$= \dfrac{6(1) + 7(4)}{1^2 + 4^2} \langle 1, 4 \rangle$

$= \dfrac{34}{17} \langle 1, 4 \rangle = \langle 2, 8 \rangle$

(b) $w_2 = u - w_1 = \langle 6, 7 \rangle - \langle 2, 8 \rangle = \langle 4, -1 \rangle$

39. $u = 2i + 3j = \langle 2, 3 \rangle$, $v = 5i + j = \langle 5, 1 \rangle$

(a) $w_1 = \text{proj}_v u = \left(\dfrac{u \cdot v}{\|v\|^2}\right) v$

$= \dfrac{2(5) + 3(1)}{5^2 + 1} \langle 5, 1 \rangle$

$= \dfrac{13}{26} \langle 5, 1 \rangle = \left\langle \frac{5}{2}, \frac{1}{2} \right\rangle$

(b) $w_2 = u - w_1 = \langle 2, 3 \rangle - \left\langle \frac{5}{2}, \frac{1}{2} \right\rangle = \left\langle -\frac{1}{2}, \frac{5}{2} \right\rangle$

41. $u = \langle 0, 3, 3 \rangle$, $v = \langle -1, 1, 1 \rangle$

(a) $w_1 = \text{proj}_v u = \left(\dfrac{u \cdot v}{\|v\|^2}\right) v$

$= \dfrac{0(-1) + 3(1) + 3(1)}{1 + 1 + 1} \langle -1, 1, 1 \rangle$

$= \dfrac{6}{3} \langle -1, 1, 1 \rangle = \langle -2, 2, 2 \rangle$

(b) $w_2 = u - w_1 = \langle 0, 3, 3 \rangle - \langle -2, 2, 2 \rangle = \langle 2, 1, 1 \rangle$

43. $\mathbf{u} = -9\mathbf{i} - 2\mathbf{j} - 4\mathbf{k}$, $\mathbf{v} = 4\mathbf{j} + 4\mathbf{k}$

  (a) $\mathbf{w}_1 = \text{proj}_\mathbf{v}\mathbf{u} = \left(\dfrac{\mathbf{u} \cdot \mathbf{v}}{\|\mathbf{v}\|^2}\right)\mathbf{v}$

  $= \left(\dfrac{(-2)(4) + (-4)(4)}{4^2 + 4^2}\right)\langle 0, 4, 4\rangle$

  $= -\dfrac{3}{4}\langle 0, 4, 4\rangle$

  $= \langle 0, -3, -3\rangle$

  (b) $\mathbf{w}_2 = \mathbf{u} - \mathbf{w}_1$

  $= \langle -9, -2, -4\rangle - \langle 0, -3, -3\rangle$

  $= \langle -9, 1, -1\rangle$

45. $\mathbf{u}$ is a vector and $\mathbf{v} \cdot \mathbf{w}$ is a scalar. You cannot add a vector and a scalar.

47. Yes, $\left\|\dfrac{\mathbf{u} \cdot \mathbf{v}}{\|\mathbf{v}\|^2}\mathbf{v}\right\| = \left\|\dfrac{\mathbf{v} \cdot \mathbf{u}}{\|\mathbf{u}\|^2}\mathbf{u}\right\|$

  $|\mathbf{u} \cdot \mathbf{v}|\dfrac{\|\mathbf{v}\|}{\|\mathbf{v}\|^2} = |\mathbf{v} \cdot \mathbf{u}|\dfrac{\|\mathbf{u}\|}{\|\mathbf{u}\|^2}$

  $\dfrac{1}{\|\mathbf{v}\|} = \dfrac{1}{\|\mathbf{u}\|}$

  $\|\mathbf{u}\| = \|\mathbf{v}\|$

49. $\mathbf{u} = \langle 3240, 1450, 2235\rangle$

  $\mathbf{v} = \langle 2.25, 2.95, 2.65\rangle$

  $\mathbf{u} \cdot \mathbf{v} = 3240(2.25) + 1450(2.95) + 2235(2.65)$

  $= \$17,490.25$

  This represents the total revenue the restaurant earned on its three products.

51. Answers will vary. *Sample answer:*

  $\mathbf{u} = -\dfrac{1}{4}\mathbf{i} + \dfrac{3}{2}\mathbf{j}$. Want $\mathbf{u} \cdot \mathbf{v} = 0$.

  $\mathbf{v} = 12\mathbf{i} + 2\mathbf{j}$ and $-\mathbf{v} = -12\mathbf{i} - 2\mathbf{j}$ are orthogonal to $\mathbf{u}$.

53. Answers will vary. *Sample answer:*

  $\mathbf{u} = \langle 3, 1, -2\rangle$. Want $\mathbf{u} \cdot \mathbf{v} = 0$.

  $\mathbf{v} = \langle 0, 2, 1\rangle$ and $-\mathbf{v} = \langle 0, -2, -1\rangle$ are orthogonal to $\mathbf{u}$.

55. Let $s = $ length of a side.

  $\mathbf{v} = \langle s, s, s\rangle$

  $\|\mathbf{v}\| = s\sqrt{3}$

  $\cos\alpha = \cos\beta = \cos\gamma = \dfrac{s}{s\sqrt{3}} = \dfrac{1}{\sqrt{3}}$

  $\alpha = \beta = \gamma = \arccos\left(\dfrac{1}{\sqrt{3}}\right) \approx 54.7°$

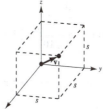

57. (a) Gravitational Force $\mathbf{F} = -48,000\mathbf{j}$

  $\mathbf{v} = \cos 10°\mathbf{i} + \sin 10°\mathbf{j}$

  $\mathbf{w}_1 = \dfrac{\mathbf{F} \cdot \mathbf{v}}{\|\mathbf{v}\|^2}\mathbf{v} = (\mathbf{F} \cdot \mathbf{v})\mathbf{v}$

  $= (-48,000)(\sin 10°)\mathbf{v}$

  $\approx -8335.1(\cos 10°\mathbf{i} + \sin 10°\mathbf{j})$

  $\|\mathbf{w}_1\| \approx 8335.1\,\text{lb}$

  (b) $\mathbf{w}_2 = \mathbf{F} - \mathbf{w}_1$

  $= -48,000\mathbf{j} + 8335.1(\cos 10°\mathbf{i} + \sin 10°\mathbf{j})$

  $= 8208.5\mathbf{i} - 46,552.6\mathbf{j}$

  $\|\mathbf{w}_2\| \approx 47,270.8\,\text{lb}$

59. $\mathbf{F} = 85\left(\dfrac{1}{2}\mathbf{i} + \dfrac{\sqrt{3}}{2}\mathbf{j}\right)$

  $\mathbf{v} = 10\mathbf{i}$

  $W = \mathbf{F} \cdot \mathbf{v} = 425\,\text{ft-lb}$

61. $\mathbf{F} = 1600(\cos 25°\,\mathbf{i} + \sin 25°\,\mathbf{j})$

  $\mathbf{v} = 2000\mathbf{i}$

  $W = \mathbf{F} \cdot \mathbf{v} = 1600(2000)\cos 25°$

  $\approx 2,900,184.9\,\text{Newton meters (Joules)}$

  $\approx 2900.2\quad\text{km-N}$

63. False.

  For example, let $\mathbf{u} = \langle 1, 1\rangle$, $\mathbf{v} = \langle 2, 3\rangle$ and $\mathbf{w} = \langle 1, 4\rangle$.

  Then $\mathbf{u} \cdot \mathbf{v} = 2 + 3 = 5$ and $\mathbf{u} \cdot \mathbf{w} = 1 + 4 = 5$.

65. (a) The graphs $y_1 = x^2$ and $y_2 = x^{1/3}$ intersect at $(0, 0)$ and $(1, 1)$.

(b) $y_1' = 2x$ and $y_2' = \dfrac{1}{3x^{2/3}}$.

At $(0, 0)$, $\pm\langle 1, 0\rangle$ is tangent to y_1 and $\pm\langle 0, 1\rangle$ is tangent to y_2.

At $(1, 1)$, $y_1' = 2$ and $y_2' = \dfrac{1}{3}$.

$\pm\dfrac{1}{\sqrt{5}}\langle 1, 2\rangle$ is tangent to y_1, $\pm\dfrac{1}{\sqrt{10}}\langle 3, 1\rangle$ is tangent to y_2.

(c) At $(0, 0)$, the vectors are perpendicular $(90°)$.

At $(1, 1)$,

$$\cos\theta = \frac{\dfrac{1}{\sqrt{5}}\langle 1, 2\rangle \cdot \dfrac{1}{\sqrt{10}}\langle 3, 1\rangle}{(1)(1)} = \frac{5}{\sqrt{50}} = \frac{1}{\sqrt{2}}$$

$$\theta = 45°$$

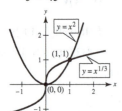

67. (a) The graphs of $y_1 = 1 - x^2$ and $y^2 = x^2 - 1$ intersect at $(1, 0)$ and $(-1, 0)$.

(b) $y_1' = -2x$ and $y_2' = 2x$.

At $(1, 0)$, $y_1' = -2$ and $y_2' = 2$. $\pm\dfrac{1}{\sqrt{5}}\langle 1, -2\rangle$ is tangent to y_1, $\pm\dfrac{1}{\sqrt{5}}\langle 1, 2\rangle$ is tangent to y_2.

At $(-1, 0)$, $y_1' = 2$ and $y_2' = -2$. $\pm\dfrac{1}{\sqrt{5}}\langle 1, 2\rangle$ is tangent to y_1, $\pm\dfrac{1}{\sqrt{5}}\langle 1, -2\rangle$ is tangent to y_2.

(c) At $(1, 0)$, $\cos\theta = \dfrac{1}{\sqrt{5}}\langle 1, -2\rangle \cdot \dfrac{-1}{\sqrt{5}}\langle 1, -2\rangle = \dfrac{3}{5}$.

$\theta \approx 0.9273$ or $53.13°$

By symmetry, the angle is the same at $(-1, 0)$.

69. In a rhombus, $\|\mathbf{u}\| = \|\mathbf{v}\|$. The diagonals are $\mathbf{u} + \mathbf{v}$ and $\mathbf{u} - \mathbf{v}$.

$$(\mathbf{u} + \mathbf{v}) \cdot (\mathbf{u} - \mathbf{v}) = (\mathbf{u} + \mathbf{v}) \cdot \mathbf{u} - (\mathbf{u} + \mathbf{v}) \cdot \mathbf{v}$$
$$= \mathbf{u} \cdot \mathbf{u} + \mathbf{v} \cdot \mathbf{u} - \mathbf{u} \cdot \mathbf{v} - \mathbf{v} \cdot \mathbf{v}$$
$$= \|\mathbf{u}\|^2 - \|\mathbf{v}\|^2 = 0$$

So, the diagonals are orthogonal.

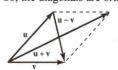

71. (a)

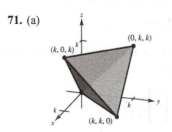

(b) Length of each edge: $\sqrt{k^2 + k^2 + 0^2} = k\sqrt{2}$

(c) $\cos\theta = \dfrac{k^2}{(k\sqrt{2})(k\sqrt{2})} = \dfrac{1}{2}$

$$\theta = \arccos\left(\frac{1}{2}\right) = 60°$$

(d) $\vec{r_1} = \langle k, k, 0\rangle - \left\langle \dfrac{k}{2}, \dfrac{k}{2}, \dfrac{k}{2}\right\rangle = \left\langle \dfrac{k}{2}, \dfrac{k}{2}, -\dfrac{k}{2}\right\rangle$

$\vec{r_2} = \langle 0, 0, 0\rangle - \left\langle \dfrac{k}{2}, \dfrac{k}{2}, \dfrac{k}{2}\right\rangle = \left\langle -\dfrac{k}{2}, -\dfrac{k}{2}, -\dfrac{k}{2}\right\rangle$

$$\cos\theta = \frac{-\dfrac{k^2}{4}}{\left(\dfrac{k}{2}\right)^2 \cdot 3} = -\frac{1}{3}$$

$$\theta = 109.5°$$

73. $\|\mathbf{u} + \mathbf{v}\|^2 = (\mathbf{u} + \mathbf{v}) \cdot (\mathbf{u} + \mathbf{v})$

$= (\mathbf{u} + \mathbf{v}) \cdot \mathbf{u} + (\mathbf{u} + \mathbf{v}) \cdot \mathbf{v}$

$= \mathbf{u} \cdot \mathbf{u} + \mathbf{v} \cdot \mathbf{u} + \mathbf{u} \cdot \mathbf{v} + \mathbf{v} \cdot \mathbf{v}$

$= \|\mathbf{u}\|^2 + 2\mathbf{u} \cdot \mathbf{v} + \|\mathbf{v}\|^2$

$\leq \|\mathbf{u}\|^2 + 2\|\mathbf{u}\|\|\mathbf{v}\| + \|\mathbf{v}\|^2 \leq (\|\mathbf{u}\| + \|\mathbf{v}\|)^2$

So, $\|\mathbf{u} + \mathbf{v}\| \leq \|\mathbf{u}\| + \|\mathbf{v}\|$.

75. $\mathbf{u} \cdot \mathbf{v} = \|\mathbf{u}\|\|\mathbf{v}\|\cos\theta$

$|\mathbf{u} \cdot \mathbf{v}| = |\|\mathbf{u}\|\|\mathbf{v}\|\cos\theta|$

$= \|\mathbf{u}\|\|\mathbf{v}\||\cos\theta|$

$\leq \|\mathbf{u}\|\|\mathbf{v}\|$ because $|\cos\theta| \leq 1$.

Section 11.4   The Cross Product of Two Vectors in Space

1. $\mathbf{u} \times \mathbf{v}$ is a vector that is perpendicular (orthogonal) to both $\mathbf{u}$ and $\mathbf{v}$.

3. $\mathbf{j} \times \mathbf{i} = \begin{vmatrix} \mathbf{i} & \mathbf{j} & \mathbf{k} \\ 0 & 1 & 0 \\ 1 & 0 & 0 \end{vmatrix} = -\mathbf{k}$

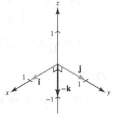

5. $\mathbf{i} \times \mathbf{k} = \begin{vmatrix} \mathbf{i} & \mathbf{j} & \mathbf{k} \\ 1 & 0 & 0 \\ 0 & 0 & 1 \end{vmatrix} = -\mathbf{j}$

7. (a) $\mathbf{u} \times \mathbf{v} = \begin{vmatrix} \mathbf{i} & \mathbf{j} & \mathbf{k} \\ -2 & 4 & 0 \\ 3 & 2 & 5 \end{vmatrix} = 20\mathbf{i} + 10\mathbf{j} - 16\mathbf{k}$

  (b) $\mathbf{v} \times \mathbf{u} = -(\mathbf{u} \times \mathbf{v}) = -20\mathbf{i} - 10\mathbf{j} + 16\mathbf{k}$

  (c) $\mathbf{v} \times \mathbf{v} = \mathbf{0}$

9. (a) $\mathbf{u} \times \mathbf{v} = \begin{vmatrix} \mathbf{i} & \mathbf{j} & \mathbf{k} \\ 7 & 3 & 2 \\ 1 & -1 & 5 \end{vmatrix} = 17\mathbf{i} - 33\mathbf{j} - 10\mathbf{k}$

  (b) $\mathbf{v} \times \mathbf{u} = -(\mathbf{u} \times \mathbf{v}) = -17\mathbf{i} + 33\mathbf{j} + 10\mathbf{k}$

  (c) $\mathbf{v} \times \mathbf{v} = \mathbf{0}$

11. $\mathbf{u} = \langle 4, -1, 0 \rangle$, $\mathbf{v} = \langle -6, 3, 0 \rangle$

$\mathbf{u} \times \mathbf{v} = \begin{vmatrix} \mathbf{i} & \mathbf{j} & \mathbf{k} \\ 4 & -1 & 0 \\ -6 & 3 & 0 \end{vmatrix} = 6\mathbf{k} = \langle 0, 0, 6 \rangle$

$\mathbf{u} \cdot (\mathbf{u} \times \mathbf{v}) = 4(0) + (-1)(0) + 0(6) = 0 \Rightarrow \mathbf{u} \perp \mathbf{u} \times \mathbf{v}$

$\mathbf{v} \cdot (\mathbf{u} \times \mathbf{v}) = -6(0) + 3(0) + 0(6) = 0 \Rightarrow \mathbf{v} \perp \mathbf{u} \times \mathbf{v}$

13. $\mathbf{u} = \mathbf{i} + \mathbf{j} + \mathbf{k}$, $\mathbf{v} = 2\mathbf{i} + \mathbf{j} - \mathbf{k}$

$\mathbf{u} \times \mathbf{v} = \begin{vmatrix} \mathbf{i} & \mathbf{j} & \mathbf{k} \\ 1 & 1 & 1 \\ 2 & 1 & -1 \end{vmatrix} = -2\mathbf{i} + 3\mathbf{j} - \mathbf{k} = \langle -2, 3, -1 \rangle$

$\mathbf{u} \cdot (\mathbf{u} \times \mathbf{v}) = 1(-2) + 1(3) + 1(-1) = 0 \Rightarrow \mathbf{u} \perp \mathbf{u} \times \mathbf{v}$

$\mathbf{v} \cdot (\mathbf{u} \times \mathbf{v}) = 2(-2) + 1(3) + (-1)(-1)$

$\qquad = 0 \Rightarrow \mathbf{v} \perp \mathbf{u} \times \mathbf{v}$

15. $\mathbf{u} = \langle 4, -3, 1 \rangle$

$\mathbf{v} = \langle 2, 5, 3 \rangle$

$\mathbf{u} \times \mathbf{v} = \begin{vmatrix} \mathbf{i} & \mathbf{j} & \mathbf{k} \\ 4 & -3 & 1 \\ 2 & 5 & 3 \end{vmatrix} = -14\mathbf{i} - 10\mathbf{j} + 26\mathbf{k}$

$\dfrac{\mathbf{u} \times \mathbf{v}}{\|\mathbf{u} \times \mathbf{v}\|} = \dfrac{1}{\sqrt{972}} \langle -14, -10, 26 \rangle$

$\qquad = \dfrac{1}{18\sqrt{3}} \langle -14, -10, 26 \rangle$

$\qquad = \left\langle -\dfrac{7}{9\sqrt{3}}, -\dfrac{5}{9\sqrt{3}}, \dfrac{13}{9\sqrt{3}} \right\rangle$

17. $\mathbf{u} = -3\mathbf{i} + 2\mathbf{j} - 5\mathbf{k}$

$\mathbf{v} = \mathbf{i} - \mathbf{j} + 4\mathbf{k}$

$\mathbf{u} \times \mathbf{v} = \begin{vmatrix} \mathbf{i} & \mathbf{j} & \mathbf{k} \\ -3 & 2 & -5 \\ 1 & -1 & -4 \end{vmatrix} = 3\mathbf{i} + 7\mathbf{j} + \mathbf{k}$

$\dfrac{\mathbf{u} \times \mathbf{v}}{\|\mathbf{u} \times \mathbf{v}\|} = \dfrac{1}{\sqrt{59}} \langle 3, 7, 1 \rangle$

$\qquad = \left\langle \dfrac{3}{\sqrt{59}}, \dfrac{7}{\sqrt{59}}, \dfrac{1}{\sqrt{59}} \right\rangle$

19. $\mathbf{u} = \mathbf{j}$

$\mathbf{v} = \mathbf{j} + \mathbf{k}$

$\mathbf{u} \times \mathbf{v} = \begin{vmatrix} \mathbf{i} & \mathbf{j} & \mathbf{k} \\ 0 & 1 & 0 \\ 0 & 1 & 1 \end{vmatrix} = \mathbf{i}$

$A = \|\mathbf{u} \times \mathbf{v}\| = \|\mathbf{i}\| = 1$

21.
$$\mathbf{u} = \langle 3, 2, -1 \rangle$$
$$\mathbf{v} = \langle 1, 2, 3 \rangle$$
$$\mathbf{u} \times \mathbf{v} = \begin{vmatrix} \mathbf{i} & \mathbf{j} & \mathbf{k} \\ 3 & 2 & -1 \\ 1 & 2 & 3 \end{vmatrix} = \langle 8, -10, 4 \rangle$$
$$A = \|\mathbf{u} \times \mathbf{v}\| = \|\langle 8, -10, 4 \rangle\| = \sqrt{180} = 6\sqrt{5}$$

23. $A(0, 3, 2), B(1, 5, 5), C(6, 9, 5), D(5, 7, 2)$

$$\overline{AB} = \langle 1, 2, 3 \rangle$$
$$\overline{DC} = \langle 1, 2, 3 \rangle$$
$$\overline{BC} = \langle 5, 4, 0 \rangle$$
$$\overline{AD} = \langle 5, 4, 0 \rangle$$

Because $\overline{AB} = \overline{DC}$ and $\overline{BC} = \overline{AD}$, the figure $ABCD$ is a parallelogram.

$\overline{AB}$ and $\overline{AD}$ are adjacent sides

$$\overline{AB} \times \overline{AD} = \begin{vmatrix} \mathbf{i} & \mathbf{j} & \mathbf{k} \\ 1 & 2 & 3 \\ 5 & 4 & 0 \end{vmatrix} = \langle -12, 15, -6 \rangle$$

$$A = \|\overline{AB} \times \overline{AD}\| = \sqrt{144 + 225 + 36} = 9\sqrt{5}$$

25. $A(0, 0, 0), B(1, 0, 3), C(-3, 2, 0)$

$$\overline{AB} = \langle 1, 0, 3 \rangle, \ \overline{AC} = \langle -3, 2, 0 \rangle$$

$$\overline{AB} \times \overline{AC} = \begin{vmatrix} \mathbf{i} & \mathbf{j} & \mathbf{k} \\ 1 & 0 & 3 \\ -3 & 2 & 0 \end{vmatrix} = \langle -6, -9, 2 \rangle$$

$$A = \tfrac{1}{2}\|\overline{AB} \times \overline{AC}\| = \tfrac{1}{2}\sqrt{36 + 81 + 4} = \tfrac{11}{2}$$

27. $\mathbf{F} = -20\mathbf{k}$

$$\overline{PQ} = \tfrac{1}{2}(\cos 40°\mathbf{j} + \sin 40°\mathbf{k})$$

$$\overline{PQ} \times \mathbf{F} = \begin{vmatrix} \mathbf{i} & \mathbf{j} & \mathbf{k} \\ 0 & \cos 40°/2 & \sin 40°/2 \\ 0 & 0 & -20 \end{vmatrix} = -10\cos 40°\mathbf{i}$$

$$\|\overline{PQ} \times \mathbf{F}\| = 10\cos 40° \approx 7.66 \text{ ft-lb}$$

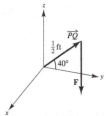

29. (a) $AC = 15 \text{ inches} = \dfrac{5}{4} \text{ feet}$

$BC = 12 \text{ inches} = 1 \text{ foot}$

$$\overline{AB} = -\tfrac{5}{4}\mathbf{j} + \mathbf{k}$$

$$\mathbf{F} = -180(\cos \theta\, \mathbf{j} + \sin \theta\, \mathbf{k})$$

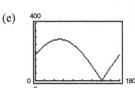

(b)
$$\overline{AB} \times \mathbf{F} = \begin{vmatrix} \mathbf{i} & \mathbf{j} & \mathbf{k} \\ 0 & -\tfrac{5}{4} & 1 \\ 0 & -180\cos\theta & -180\sin\theta \end{vmatrix} = (225\sin\theta + 180\cos\theta)\mathbf{i}$$

$$\|\overline{AB} \times \mathbf{F}\| = |225\sin\theta + 180\cos\theta|$$

(c) When $\theta = 30°$, $\|\overline{AB} \times \mathbf{F}\| = 225\left(\dfrac{1}{2}\right) + 180\left(\dfrac{\sqrt{3}}{2}\right) \approx 268.38$

(d) If $T = |225\sin\theta + 180\cos\theta|$, $T = 0$ for $225\sin\theta = -180\cos\theta \Rightarrow \tan\theta = -\dfrac{4}{5} \Rightarrow \theta \approx 141.34°$.

For $0 < \theta < 141.34$, $T'(\theta) = 225\cos\theta - 180\sin\theta = 0 \Rightarrow \tan\theta = \dfrac{5}{4} \Rightarrow \theta \approx 51.34°$. $\overline{AB}$ and $\mathbf{F}$ are perpendicular.

(e)

From part (d), the zero is $\theta \approx 141.34°$, when the vectors are parallel.

31. $\mathbf{u} \cdot (\mathbf{v} \times \mathbf{w}) = \begin{vmatrix} 1 & 0 & 0 \\ 0 & 1 & 0 \\ 0 & 0 & 1 \end{vmatrix} = 1$

33. $\mathbf{u} \cdot (\mathbf{v} \times \mathbf{w}) = \begin{vmatrix} 2 & 0 & 1 \\ 0 & 3 & 0 \\ 0 & 0 & 1 \end{vmatrix} = 6$

35. $\mathbf{u} \cdot (\mathbf{v} \times \mathbf{w}) = \begin{vmatrix} 1 & 1 & 0 \\ 0 & 1 & 1 \\ 1 & 0 & 1 \end{vmatrix} = 2$

$V = |\mathbf{u} \cdot (\mathbf{v} \times \mathbf{w})| = 2$

37. $\mathbf{u} = \langle 3, 0, 0 \rangle$

$\mathbf{v} = \langle 0, 5, 1 \rangle$

$\mathbf{w} = \langle 2, 0, 5 \rangle$

$\mathbf{u} \cdot (\mathbf{v} \times \mathbf{w}) = \begin{vmatrix} 3 & 0 & 0 \\ 0 & 5 & 1 \\ 2 & 0 & 5 \end{vmatrix} = 75$

$V = |\mathbf{u} \cdot (\mathbf{v} \times \mathbf{w})| = 75$

39. (a) $\mathbf{u} \cdot (\mathbf{v} \times \mathbf{w}) = (\mathbf{v} \times \mathbf{w}) \cdot \mathbf{u}$ (b)

$= \mathbf{w} \cdot (\mathbf{u} \times \mathbf{v}) = (\mathbf{u} \times \mathbf{v}) \cdot \mathbf{w}$ (c)

$= \mathbf{v} \cdot (\mathbf{w} \times \mathbf{u}) = (\mathbf{u}x - \mathbf{w}) \cdot \mathbf{v}$ (d)

$= \mathbf{v} \cdot (\mathbf{w} \times \mathbf{u}) = (\mathbf{w} \times \mathbf{u}) \cdot \mathbf{v}$ (h)

(e) $\mathbf{u} \cdot (\mathbf{w} \times \mathbf{v}) = \mathbf{w} \cdot (\mathbf{v} \times \mathbf{u})$ (f)

$= \mathbf{w} \cdot (\mathbf{v} \times \mathbf{u}) = (-\mathbf{u} \times \mathbf{v}) \cdot \mathbf{w}$ (g)

So, $a = b = c = d = h$ and $e = f = g$

41. The cross product is orthogonal to the two vectors, so it is orthogonal to the yz-plane. It lies on the x-axis, since it is of the form $\langle k, 0, 0 \rangle$.

43. False. If the vectors are ordered pairs, then the cross product does not exist.

45. False. Let $\mathbf{u} = \langle 1, 0, 0 \rangle$, $\mathbf{v} = \langle 1, 0, 0 \rangle$, $\mathbf{w} = \langle -1, 0, 0 \rangle$.

Then, $\mathbf{u} \times \mathbf{v} = \mathbf{u} \times \mathbf{w} = \mathbf{0}$, but $\mathbf{v} \neq \mathbf{w}$.

47. $\mathbf{u} = \langle u_1, u_2, u_3 \rangle$, $\mathbf{v} = \langle v_1, v_2, v_3 \rangle$, $\mathbf{w} = \langle w_1, w_2, w_3 \rangle$

$\mathbf{u} \times (\mathbf{v} + \mathbf{w}) = \begin{vmatrix} \mathbf{i} & \mathbf{j} & \mathbf{k} \\ u_1 & u_2 & u_3 \\ v_1 + w_1 & v_2 + w_2 & v_3 + w_3 \end{vmatrix}$

$= \left[u_2(v_3 + w_3) - u_3(v_2 + w_2) \right]\mathbf{i} - \left[u_1(v_3 + w_3) - u_3(v_1 + w_1) \right]\mathbf{j} + \left[u_1(v_2 + w_2) - u_2(v_1 + w_1) \right]\mathbf{k}$

$= (u_2v_3 - u_3v_2)\mathbf{i} - (u_1v_3 - u_3v_1)\mathbf{j} + (u_1v_2 - u_2v_1)\mathbf{k} + (u_2w_3 - u_3w_2)\mathbf{i} - (u_1w_3 - u_3w_1)\mathbf{j} + (u_1w_2 - u_2w_1)\mathbf{k}$

$= (\mathbf{u} \times \mathbf{v}) + (\mathbf{u} \times \mathbf{w})$

49. $\mathbf{u} = \langle u_1, u_2, u_3 \rangle$

$\mathbf{u} \times \mathbf{u} = \begin{vmatrix} \mathbf{i} & \mathbf{j} & \mathbf{k} \\ u_1 & u_2 & u_3 \\ u_1 & u_2 & u_3 \end{vmatrix} = (u_2u_3 - u_3u_2)\mathbf{i} - (u_1u_3 - u_3u_1)\mathbf{j} + (u_1u_2 - u_2u_1)\mathbf{k} = \mathbf{0}$

51. $\mathbf{u} \times \mathbf{v} = (u_2v_3 - u_3v_2)\mathbf{i} - (u_1v_3 - u_3v_1)\mathbf{j} + (u_1v_2 - u_2v_1)\mathbf{k}$

$(\mathbf{u} \times \mathbf{v}) \cdot \mathbf{u} = (u_2v_3 - u_3v_2)u_1 + (u_3v_1 - u_1v_3)u_2 + (u_1v_2 - u_2v_1)u_3 = 0$

$(\mathbf{u} \times \mathbf{v}) \cdot \mathbf{v} = (u_2v_3 - u_3v_2)v_1 + (u_3v_1 - u_1v_3)v_2 + (u_1v_2 - u_2v_1)v_3 = 0$

So, $\mathbf{u} \times \mathbf{v} \perp \mathbf{u}$ and $\mathbf{u} \times \mathbf{v} \perp \mathbf{v}$.

53. $\|\mathbf{u} \times \mathbf{v}\| = \|\mathbf{u}\| \|\mathbf{v}\| \sin \theta$

If $\mathbf{u}$ and $\mathbf{v}$ are orthogonal, $\theta = \pi/2$ and $\sin \theta = 1$. So, $\|\mathbf{u} \times \mathbf{v}\| = \|\mathbf{u}\| \|\mathbf{v}\|$.

55. $\mathbf{u} = u_1\mathbf{i} + u_2\mathbf{j} + u_3\mathbf{k}$, $\mathbf{v} = v_1\mathbf{i} + v_2\mathbf{j} + v_3\mathbf{k}$, $\mathbf{w} = w_1\mathbf{i} + w_2\mathbf{j} + w_3\mathbf{k}$

$$\mathbf{v} \times \mathbf{w} = \begin{vmatrix} \mathbf{i} & \mathbf{j} & \mathbf{k} \\ v_1 & v_2 & v_3 \\ w_1 & w_2 & w_3 \end{vmatrix} = \langle v_2 w_3 - w_2 v_3, -(v_1 w_3 - w_1 v_3), v_1 w_2 - w_1 v_2 \rangle$$

$$\mathbf{u} \times (\mathbf{v} \times \mathbf{w}) = \langle u_1, u_2, u_3 \rangle \cdot \langle v_2 w_3 - w_2 v_3, -(v_1 w_3 - w_1 v_3), v_1 w_2 - w_1 v_2 \rangle$$

$$= u_1 v_2 w_3 - u_1 v_3 w_2 - u_2 v_1 w_3 + u_2 v_3 w_1 + u_3 v_1 w_2 - u_3 v_2 w_1$$

$$= \begin{vmatrix} u_1 & u_2 & u_3 \\ v_1 & v_2 & v_3 \\ w_1 & w_2 & w_3 \end{vmatrix}$$

Section 11.5   Lines and Planes in Space

1. The parametric equations of a line L parallel to $\mathbf{v} = \langle a, b, c, \rangle$ and passing through the point $P(x_1, y_1, z_1)$ are

$x = x_1 + at$, $y = y_1 + bt$, $z = z_1 + ct$.

The symmetric equations are

$\dfrac{x - x_1}{a} = \dfrac{y - y_1}{b} = \dfrac{z - z_1}{c}$.

3. Answers will vary. Any plane that has a missing x-variable in its equation is parallel to the x-axis.

Sample answer: $3y - z = 5$

5. $x = -2 + t$, $y = 3t$, $z = 4 + t$

(a) $(0, 6, 6)$: For $x = 0 = -2 + t$, you have $t = 2$.

Then $y = 3(2) = 6$ and $z = 4 + 2 = 6$. Yes, $(0, 6, 6)$ lies on the line.

(b) $(2, 3, 5)$: For $x = 2 = -2 + t$, you have $t = 4$.

Then $y = 3(4) = 12 \neq 3$. No, $(2, 3, 5)$ does not lie on the line.

(c) $(-4, -6, 2)$: For $x = -4 = -2 + t$, you have $t = -2$.

Then $y = 3(-2) = -6$ and $z = 4 - 2 = 2$. Yes, $(-4, -6, 2)$ lies on the line.

7. Point: $(0, 0, 0)$

Direction vector: $\langle 3, 1, 5 \rangle$

Direction numbers: 3, 1, 5

(a) Parametric: $x = 3t$, $y = t$, $z = 5t$

(b) Symmetric: $\dfrac{x}{3} = y = \dfrac{z}{5}$

9. Point: $(-2, 0, 3)$

Direction vector: $\mathbf{v} = \langle 2, 4, -2 \rangle$

Direction numbers: 2, 4, −2

(a) Parametric: $x = -2 + 2t$, $y = 4t$, $z = 3 - 2t$

(b) Symmetric: $\dfrac{x + 2}{2} = \dfrac{y}{4} = \dfrac{z - 3}{-2}$

11. Point: $(1, 0, 1)$

Direction vector: $\mathbf{v} = 3\mathbf{i} - 2\mathbf{j} + \mathbf{k}$

Direction numbers: 3, −2, 1

(a) Parametric: $x = 1 + 3t$, $y = -2t$, $z = 1 + t$

(b) Symmetric: $\dfrac{x - 1}{3} = \dfrac{y}{-2} = \dfrac{z - 1}{1}$

13. Points: $(5, -3, -2)$, $\left(-\dfrac{2}{3}, \dfrac{2}{3}, 1\right)$

Direction vector: $\mathbf{v} = \dfrac{17}{3}\mathbf{i} - \dfrac{11}{3}\mathbf{j} - 3\mathbf{k}$

Direction numbers: 17, −11, −9

(a) Parametric:
$x = 5 + 17t$, $y = -3 - 11t$, $z = -2 - 9t$

(b) Symmetric: $\dfrac{x - 5}{17} = \dfrac{y + 3}{-11} = \dfrac{z + 2}{-9}$

15. Points: $(7, -2, 6)$, $(-3, 0, 6)$

Direction vector: $\langle -10, 2, 0 \rangle$

Direction numbers: −10, 2, 0

(a) Parametric: $x = 7 - 10t$, $y = -2 + 2t$, $z = 6$

(b) Symmetric: Not possible because the direction number for z is 0. But, you could describe the line as $\dfrac{x - 7}{10} = \dfrac{y + 2}{-2}$, $z = 6$.

17. Point: $(2, 3, 4)$

 Direction vector: $\mathbf{v} = \mathbf{k}$

 Direction numbers: $0, 0, 1$

 Parametric: $x = 2, y = 3, z = 4 + t$

19. Point: $(2, 3, 4)$

 Direction vector: $\mathbf{v} = 3\mathbf{i} + 2\mathbf{j} - \mathbf{k}$

 Direction numbers: $3, 2, -1$

 Parametric: $x = 2 + 3t, y = 3 + 2t, z = 4 - t$

21. Point: $(5, -3, -4)$

 Direction vector: $\mathbf{v} = \langle 2, -1, 3 \rangle$

 Direction numbers: $2, -1, 3$

 Parametric: $x = 5 + 2t, y = -3 - t, z = -4 + 3t$

23. Point: $(2, 1, 2)$

 Direction vector: $\langle -1, 1, 1 \rangle$

 Direction numbers: $-1, 1, 1$

 Parametric: $x = 2 - t, y = 1 + t, z = 2 + t$

25. Let $t = 0$: $P = (3, -1, -2)$ (other answers possible)

 $\mathbf{v} = \langle -1, 2, 0 \rangle$ (any nonzero multiple of $\mathbf{v}$ is correct)

27. Let each quantity equal 0:

 $P = (7, -6, -2)$ (other answers possible)

 $\mathbf{v} = \langle 4, 2, 1 \rangle$ (any nonzero multiple of $\mathbf{v}$ is correct)

29. L_1: $\mathbf{v}_1 = \langle -3, 2, 4 \rangle$ and $P = (6, -2, 5)$ on L_1

 L_2: $\mathbf{v}_2 = \langle 6, -4, -8 \rangle$ and $P = (6, -2, 5)$ on L_2

 The lines are identical.

31. L_1: $\mathbf{v}_1 = \langle 4, -2, 3 \rangle$ and $P = (8, -5, -9)$ on L_1

 L_2: $\mathbf{v}_2 = \langle -8, 4, -6 \rangle$ and $P = (8, -5, -9)$ on L_2

 The lines are identical.

33. At the point of intersection, the coordinates for one line equal the corresponding coordinates for the other line. So,

 (i) $4t + 2 = 2s + 2$, (ii) $3 = 2s + 3$, and

 (iii) $-t + 1 = s + 1$.

 From (ii), you find that $s = 0$ and consequently, from (iii), $t = 0$. Letting $s = t = 0$, you see that equation (i) is satisfied and so the two lines intersect. Substituting zero for s or for t, you obtain the point $(2, 3, 1)$.

 $\mathbf{u} = 4\mathbf{i} - \mathbf{k}$ \qquad (First line)

 $\mathbf{v} = 2\mathbf{i} + 2\mathbf{j} + \mathbf{k}$ \quad (Second line)

 $$\cos \theta = \frac{|\mathbf{u} \cdot \mathbf{v}|}{\|\mathbf{u}\|\|\mathbf{v}\|} = \frac{8 - 1}{\sqrt{17}\sqrt{9}} = \frac{7}{3\sqrt{17}} = \frac{7\sqrt{17}}{51}$$

 $\theta \approx 55.5°$

35. Writing the equations of the lines in parametric form you have

 $x = 3t$ \qquad $y = 2 - t$ \qquad $z = -1 + t$

 $x = 1 + 4s$ \qquad $y = -2 + s$ \qquad $z = -3 - 3s.$

 For the coordinates to be equal, $3t = 1 + 4s$ and $2 - t = -2 + s$. Solving this system yields $t = \frac{17}{7}$ and $s = \frac{11}{7}$. When using these values for s and t, the z coordinates are not equal. The lines do not intersect.

37. $x + 2y - 4z - 1 = 0$

 (a) $(-7, 2, -1)$: $(-7) + 2(2) - 4(-1) - 1 = 0$

 Point is in plane.

 (b) $(5, 2, 2)$: $5 + 2(2) - 4(2) - 1 = 0$

 Point is in plane.

 (c) $(-6, 1, -1)$: $-6 + 2(1) - 4(-1) - 1 = -1 \neq 0$

 Point is not in plane.

39. Point: $(1, 3, -7)$

 Normal vector: $\mathbf{n} = \mathbf{j} = \langle 0, 1, 0 \rangle$

 $0(x - 1) + 1(y - 3) + 0(z - (-7)) = 0$

 $y - 3 = 0$

41. Point: $(3, 2, 2)$

 Normal vector: $\mathbf{n} = 2\mathbf{i} + 3\mathbf{j} - \mathbf{k}$

 $2(x - 3) + 3(y - 2) - 1(z - 2) = 0$

 $2x + 3y - z - 10 = 0$

43. Point: $(-1, 4, 0)$

Normal vector: $\mathbf{v} = \langle 2, -1, -2 \rangle$

$$2(x + 1) - 1(y - 4) - 2(z - 0) = 0$$
$$2x - y - 2z + 6 = 0$$

45. Let $\mathbf{u}$ be the vector from $(0, 0, 0)$ to

$(2, 0, 3)$: $\mathbf{u} = \langle 2, 0, 3 \rangle$

Let $\mathbf{u}$ be the vector from $(0, 0, 0)$ to

$(-3, -1, 5)$: $\mathbf{v} = \langle -3, -1, 5 \rangle$

Normal vectors: $\mathbf{u} \times \mathbf{v} = \begin{vmatrix} \mathbf{i} & \mathbf{j} & \mathbf{k} \\ 2 & 0 & 3 \\ -3 & -1 & 5 \end{vmatrix} = \langle 3, -19, -2 \rangle$

$$3(x - 0) - 19(y - 0) - 2(z - 0) = 0$$
$$3x - 19y - 2z = 0$$

47. Let $\mathbf{u}$ be the vector from $(1, 2, 3)$ to

$(3, 2, 1)$: $\mathbf{u} = 2\mathbf{i} - 2\mathbf{k}$

Let $\mathbf{v}$ be the vector from $(1, 2, 3)$ to

$(-1, -2, 2)$: $\mathbf{v} = -2\mathbf{i} - 4\mathbf{j} - \mathbf{k}$

Normal vector:

$\left(\tfrac{1}{2}\mathbf{u}\right) \times (-\mathbf{v}) = \begin{vmatrix} \mathbf{i} & \mathbf{j} & \mathbf{k} \\ 1 & 0 & -1 \\ 2 & 4 & 1 \end{vmatrix} = 4\mathbf{i} - 3\mathbf{j} + 4\mathbf{k}$

$$4(x - 1) - 3(y - 2) + 4(z - 3) = 0$$
$$4x - 3y + 4z - 10 = 0$$

49. $(1, 2, 3)$, Normal vector:

$\mathbf{v} = \mathbf{k}$, $1(z - 3) = 0$, $z - 3 = 0$

51. The direction vectors for the lines are $\mathbf{u} = -2\mathbf{i} + \mathbf{j} + \mathbf{k}$,

$\mathbf{v} = -3\mathbf{i} + 4\mathbf{j} - \mathbf{k}$.

Normal vector: $\mathbf{u} \times \mathbf{v} = \begin{vmatrix} \mathbf{i} & \mathbf{j} & \mathbf{k} \\ -2 & 1 & 1 \\ -3 & 4 & -1 \end{vmatrix} = -5(\mathbf{i} + \mathbf{j} + \mathbf{k})$

Point of intersection of the lines: $(-1, 5, 1)$

$$(x + 1) + (y - 5) + (z - 1) = 0$$
$$x + y + z - 5 = 0$$

53. Let $\mathbf{v}$ be the vector from $(-1, 1, -1)$ to $(2, 2, 1)$:

$\mathbf{v} = 3\mathbf{i} + \mathbf{j} + 2\mathbf{k}$

Let $\mathbf{n}$ be a vector normal to the plane

$2x - 3y + z = 3$: $\mathbf{n} = 2\mathbf{i} - 3\mathbf{j} + \mathbf{k}$

Because $\mathbf{v}$ and $\mathbf{n}$ both lie in the plane P, the normal vector to P is

$\mathbf{v} \times \mathbf{n} = \begin{vmatrix} \mathbf{i} & \mathbf{j} & \mathbf{k} \\ 3 & 1 & 2 \\ 2 & -3 & 1 \end{vmatrix} = 7\mathbf{i} - \mathbf{j} - 11\mathbf{k}$

$$7(x - 2) + 1(y - 2) - 11(z - 1) = 0$$
$$7x + y - 11z - 5 = 0$$

55. Let $\mathbf{u} = \mathbf{i}$ and let v be the vector from $(1, -2, -1)$ to

$(2, 5, 6)$: $\mathbf{v} = \mathbf{i} + 7\mathbf{j} + 7\mathbf{k}$

Because $\mathbf{u}$ and $\mathbf{v}$ both lie in the plane P, the normal vector to P is:

$\mathbf{u} \times \mathbf{v} = \begin{vmatrix} \mathbf{i} & \mathbf{j} & \mathbf{k} \\ 1 & 0 & 0 \\ 1 & 7 & 7 \end{vmatrix} = -7\mathbf{j} + 7\mathbf{k} = -7(\mathbf{j} - \mathbf{k})$

$$\big[y - (-2)\big] - \big[z - (-1)\big] = 0$$
$$y - z + 1 = 0$$

57. Let (x, y, z) be equidistant from $(2, 2, 0)$ and $(0, 2, 2)$.

$$\sqrt{(x - 2)^2 + (y - 2)^2 + (z - 0)^2} = \sqrt{(x - 0)^2 + (y - 2)^2 + (z - 2)^2}$$
$$x^2 - 4x + 4 + y^2 - 4y + 4 + z^2 = x^2 + y^2 - 4y + 4 + z^2 - 4z + 4$$
$$-4x + 8 = -4z + 8$$
$$x - z = 0 \quad \text{Plane}$$

59. Let (x, y, z) be equidistant from $(-3, 1, 2)$ and $(6, -2, 4)$.

$$\sqrt{(x + 3)^2 + (y - 1)^2 + (z - 2)^2} = \sqrt{(x - 6)^2 + (y + 2)^2 + (z - 4)^2}$$
$$x^2 + 6x + 9 + y^2 - 2y + 1 + z^2 - 4z + 4 = x^2 - 12x + 36 + y^2 + 4y + 4 + z^2 - 8z + 16$$
$$6x - 2y - 4z + 14 = -12x + 4y - 8z + 56$$
$$18x - 6y + 4z - 42 = 0$$
$$9x - 3y + 2z - 21 = 0 \quad \text{Plane}$$

61. First plane: $\mathbf{n}_1 = \langle -5, 2, -8 \rangle$ and $P = (0, 3, 0)$ on plane

Second plane: $\mathbf{n}_2 = \langle 15, -6, 24 \rangle = -3\mathbf{n}_1$ and P not on plane

Parallel planes

(Note: The equations are not equivalent.)

63. First plane: $\mathbf{n}_1 = \langle 3, -2, 5 \rangle$ and $P = (0, 0, 2)$ on plane

Second plane: $\mathbf{n}_2 = \langle 75, -50, 125 \rangle = 25\mathbf{n}_1$ and P on plane

Planes are identical.

(Note: The equations are equivalent.)

65. (a) $\mathbf{n}_1 = 3\mathbf{i} + 2\mathbf{j} - \mathbf{k}$ and $\mathbf{n}_2 = \mathbf{i} - 4\mathbf{j} + 2\mathbf{k}$

$$\cos \theta = \frac{|\mathbf{n}_1 \cdot \mathbf{n}_2|}{\|\mathbf{n}_1\| \|\mathbf{n}_2\|} = \frac{|-7|}{\sqrt{14}\sqrt{21}} = \frac{\sqrt{6}}{6}$$

$$\Rightarrow \theta \approx 65.91°$$

(b) The direction vector for the line is

$$\mathbf{n}_2 \times \mathbf{n}_1 = \begin{vmatrix} \mathbf{i} & \mathbf{j} & \mathbf{k} \\ 1 & -4 & 2 \\ 3 & 2 & -1 \end{vmatrix} = 7(\mathbf{j} + 2\mathbf{k}).$$

Find a point of intersection of the planes.

$$\begin{array}{rcl} 6x + 4y - 2z &=& 14 \\ x - 4y + 2z &=& 0 \\ \hline 7x &=& 14 \\ x &=& 2 \end{array}$$

Substituting 2 for x in the second equation, you have $-4y + 2z = -2$ or $z = 2y - 1$. Letting $y = 1$,

a point of intersection is $(2, 1, 1)$.

$$x = 2, \ y = 1 + t, \ z = 1 + 2t$$

67. (a) $\mathbf{n}_1 = \langle 3, -1, 1 \rangle$ and $\mathbf{n}_2 = \langle 4, 6, 3 \rangle$

$$\cos \theta = \frac{|\mathbf{n}_1 \cdot \mathbf{n}_2|}{\|\mathbf{n}_1\| \|\mathbf{n}_2\|} = \frac{|9|}{\sqrt{11}\sqrt{61}} = \frac{9\sqrt{671}}{671}$$

$$\Rightarrow \theta \approx 69.67°$$

(b) The direction vector for the line is

$$\mathbf{n}_1 \times \mathbf{n}_2 = \begin{vmatrix} \mathbf{i} & \mathbf{j} & \mathbf{k} \\ 3 & -1 & 1 \\ 4 & 6 & 3 \end{vmatrix} = -9\mathbf{i} - 5\mathbf{j} + 22\mathbf{k}.$$

Find a point of intersection of the planes.

$$\begin{array}{rcl} 18x - 6y + 6z &=& 42 \\ 4x + 6y + 3z &=& 2 \\ \hline 22x \qquad + 9z &=& 44 \end{array}$$

Let $z = 0$, $22x = 44 \Rightarrow x = 2$ and

$3(2) - y + 0 = 7 \Rightarrow y = -1$.

A point of intersection is $(2, -1, 0)$.

$$x = 2 - 9t, \ y = -1 - 5t, \ z = 22t$$

69. The normal vectors to the planes are

$$\mathbf{n}_1 = \langle 5, -3, 1 \rangle, \ \mathbf{n}_2 = \langle 1, 4, 7 \rangle, \ \cos \theta = \frac{|\mathbf{n}_1 \cdot \mathbf{n}_2|}{\|\mathbf{n}_1\| \|\mathbf{n}_2\|} = 0.$$

So, $\theta = \pi/2$ and the planes are orthogonal.

71. The normal vectors to the planes are

$$\mathbf{n}_1 = \mathbf{i} - 3\mathbf{j} + 6\mathbf{k}, \ \mathbf{n}_2 = 5\mathbf{i} + \mathbf{j} - \mathbf{k},$$

$$\cos \theta = \frac{|\mathbf{n}_1 \cdot \mathbf{n}_2|}{\|\mathbf{n}_1\| \|\mathbf{n}_2\|} = \frac{|5 - 3 - 6|}{\sqrt{46}\sqrt{27}} = \frac{4\sqrt{138}}{414} = \frac{2\sqrt{138}}{207}.$$

So, $\theta = \arccos\left(\frac{2\sqrt{138}}{207}\right) \approx 83.5°$.

73. The normal vectors to the planes are $\mathbf{n}_1 = \langle 1, -5, -1 \rangle$ and $\mathbf{n}_2 = \langle 5, -25, -5 \rangle$. Because $\mathbf{n}_2 = 5\mathbf{n}_1$, the planes are parallel, but not equal.

75. $y \leq -2$

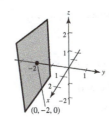

77. $x + z = 6$

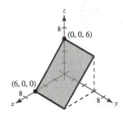

79. $4x + 2y + 6z = 12$

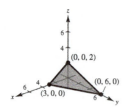

81. $2x - y + 3z = 4$

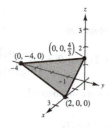

83. Writing the equation of the line in parametric form and substituting into the equation of the plane you have:

$x = -7 + 2t, \ y = 4 + t, \ z = -1 + 5t$

$(-7 + 2t) + 3(4 + t) - (-1 + 5t) = 6$

$$6 = 6$$

The equation is valid for all t.

The line lies in the plane.

85. Writing the equation of the line in parametric form and substituting into the equation of the plane you have:

$x = 1 + 3t, \ y = -1 - 2t, \ z = 3 + t$

$2(1 + 3t) + 3(-1 - 2t) = 10, \ -1 = 10$, contradiction

So, the line does not intersect the plane.

87. Point: $Q(0, 0, 0)$

Plane: $2x + 3y + z - 12 = 0$

Normal to plane: $\mathbf{n} = \langle 2, 3, 1 \rangle$

Point in plane: $P(6, 0, 0)$

Vector $\overrightarrow{PQ} = \langle -6, 0, 0 \rangle$

$$D = \frac{\left| \overrightarrow{PQ} \cdot \mathbf{n} \right|}{\|\mathbf{n}\|} = \frac{|-12|}{\sqrt{14}} = \frac{6\sqrt{14}}{7}$$

89. Point: $Q(2, 8, 4)$

Plane: $2x + y + z = 5$

Normal to plane: $\mathbf{n} = \langle 2, 1, 1 \rangle$

Point in plane: $P(0, 0, 5)$

Vector: $\overrightarrow{PQ} = \langle 2, 8, -1 \rangle$

$$D = \frac{\left| \overrightarrow{PQ} \cdot \mathbf{n} \right|}{\|\mathbf{n}\|} = \frac{11}{\sqrt{6}} = \frac{11\sqrt{6}}{6}$$

91. The normal vectors to the planes are $\mathbf{n}_1 = \langle 1, -3, 4 \rangle$ and $\mathbf{n}_2 = \langle 1, -3, 4 \rangle$. Because $\mathbf{n}_1 = \mathbf{n}_2$, the planes are parallel. Choose a point in each plane.

$P(10, 0, 0)$ is a point in $x - 3y + 4z = 10$.

$Q(6, 0, 0)$ is a point in $x - 3y + 4z = 6$.

$$\overrightarrow{PQ} = \langle -4, 0, 0 \rangle, \ D = \frac{\left| \overrightarrow{PQ} \cdot \mathbf{n}_1 \right|}{\|\mathbf{n}_1\|} = \frac{4}{\sqrt{26}} = \frac{2\sqrt{26}}{13}$$

93. The normal vectors to the planes are $\mathbf{n}_1 = \langle -3, 6, 7 \rangle$ and $\mathbf{n}_2 = \langle 6, -12, -14 \rangle$. Because $\mathbf{n}_2 = -2\mathbf{n}_1$, the planes are parallel. Choose a point in each plane.

$P(0, -1, 1)$ is a point in $-3x + 6y + 7z = 1$.

$Q\left(\dfrac{25}{6}, 0, 0 \right)$ is a point in $6x - 12y - 14z = 25$.

$\overrightarrow{PQ} = \left\langle \dfrac{25}{6}, 1, -1 \right\rangle$

$$D = \frac{\left| \overrightarrow{PQ} \cdot \mathbf{n}_1 \right|}{\|\mathbf{n}_1\|} = \frac{|-27/2|}{\sqrt{94}} = \frac{27}{2\sqrt{94}} = \frac{27\sqrt{94}}{188}$$

95. $\mathbf{u} = \langle 4, 0, -1 \rangle$ is the direction vector for the line.

$Q(1, 5, -2)$ is the given point, and $P(-2, 3, 1)$ is on the line.

$\overrightarrow{PQ} = \langle 3, 2, -3 \rangle$

$$\overrightarrow{PQ} \times \mathbf{u} = \begin{vmatrix} \mathbf{i} & \mathbf{j} & \mathbf{k} \\ 3 & 2 & -3 \\ 4 & 0 & -1 \end{vmatrix} = \langle -2, -9, -8 \rangle$$

$$D = \frac{\left\| \overrightarrow{PQ} \times \mathbf{u} \right\|}{\|\mathbf{u}\|} = \frac{\sqrt{149}}{\sqrt{17}} = \frac{\sqrt{2533}}{17}$$

97. $\mathbf{u} = \langle -1, 1, -2 \rangle$ is the direction vector for the line.

$Q(-2, 1, 3)$ is the given point, and $P(1, 2, 0)$ is on the line (let $t = 0$ in the parametric equations for the line).

$\overrightarrow{PQ} = \langle -3, -1, 3 \rangle$

$$\overrightarrow{PQ} \times \mathbf{u} = \begin{vmatrix} \mathbf{i} & \mathbf{j} & \mathbf{k} \\ -3 & -1 & 3 \\ -1 & 1 & -2 \end{vmatrix} = \langle -1, -9, -4 \rangle$$

$$D = \frac{\left\| \overrightarrow{PQ} \times \mathbf{u} \right\|}{\|\mathbf{u}\|} = \frac{\sqrt{1 + 81 + 16}}{\sqrt{1 + 1 + 4}} = \frac{\sqrt{98}}{\sqrt{6}} = \frac{7}{\sqrt{3}} = \frac{7\sqrt{3}}{3}$$

99. The direction vector for L_1 is $\mathbf{v}_1 = \langle -1, 2, 1 \rangle$.

The direction vector for L_2 is $\mathbf{v}_2 = \langle 3, -6, -3 \rangle$.

Because $\mathbf{v}_2 = -3\mathbf{v}_1$, the lines are parallel.

Let $Q(2, 3, 4)$ to be a point on L_1 and $P(0, 1, 4)$ a point on L_2. $\overrightarrow{PQ} = \langle 2, 2, 0 \rangle$. b

$\mathbf{u} = \mathbf{v}_2$ is the direction vector for L_2.

$$\overrightarrow{PQ} \times \mathbf{v}_2 = \begin{vmatrix} \mathbf{i} & \mathbf{j} & \mathbf{k} \\ 2 & 2 & 0 \\ 3 & -6 & -3 \end{vmatrix} = \langle -6, 6, -18 \rangle$$

$$D = \frac{\left\| \overrightarrow{PQ} \times \mathbf{v}_2 \right\|}{\left\| \mathbf{v}_2 \right\|} = \frac{\sqrt{36 + 36 + 324}}{\sqrt{9 + 36 + 9}} = \sqrt{\frac{396}{54}} = \sqrt{\frac{22}{3}} = \frac{\sqrt{66}}{3}$$

101. Exactly one plane contains the point and line. Select two points on the line and observe that three noncolinear points determine a unique plane.

103. Yes, Consider two points on one line, and a third distinct point on another line. Three distinct points determine a unique plane.

105. $z = 0.23x + 0.14y + 6.85$

(a)

Year	2009	2010	2011	2012	2013	2014
z(Approx)	18.93	19.46	20.31	21.10	21.58	22.62

The approximations are close to the actual values.

(b) If x and y both increase, then so does z.

107. L_1: $x_1 = 6 + t$; $y_1 = 8 - t$, $z_1 = 3 + t$

L_2: $x_2 = 1 + t$, $y_2 = 2 + t$, $z_2 = 2t$

(a) At $t = 0$, the first insect is at $P_1(6, 8, 3)$ and the second insect is at $P_2(1, 2, 0)$.

$$\text{Distance} = \sqrt{(6 - 1)^2 + (8 - 2)^2 + (3 - 0)^2}$$

$$= \sqrt{70}$$

$$\approx 8.37 \text{ inches.}$$

(b) $\text{Distance} = \sqrt{(x_1 - x_2)^2 + (y_1 - y_2)^2 + (z_1 - z_2)^2}$

$$= \sqrt{5^2 + (6 - 2t)^2 + (3 - t)^2}$$

$$= \sqrt{5t^2 - 30t + 70}, \ 0 \le t \le 10$$

(c) The distance is never zero.

(d) Using a graphing utility, the minimum distance is 5 inches when $t = 3$ minutes.

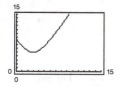

109. The direction vector $\mathbf{v}$ of the line is the normal to the plane, $\mathbf{v} = \langle 3, -1, 4 \rangle$.

The parametric equations of the line are $x = 5 + 3t$, $y = 4 - t$, $z = -3 + 4t$.

To find the point of intersection, solve for t in the following equation:

$$3(5 + 3t) - (4 - t) + 4(-3 + 4t) = 7$$

$$26t = 8$$

$$t = \tfrac{4}{13}$$

Point of intersection:

$$\left(5 + 3\left(\tfrac{4}{13}\right), 4 - \tfrac{4}{13}, -3 + 4\left(\tfrac{4}{13}\right) \right) = \left(\tfrac{77}{13}, \tfrac{48}{13}, -\tfrac{23}{13} \right)$$

111. $\mathbf{u} \times \mathbf{v} = \begin{vmatrix} \mathbf{i} & \mathbf{j} & \mathbf{k} \\ 2 & -5 & 1 \\ -3 & 1 & 4 \end{vmatrix} = -21\mathbf{i} - 11\mathbf{j} - 13\mathbf{k}$

Direction numbers: $21, 11, 13$

$x = 21t, \ y = 1 + 11t, \ z = 4 + 13t$

113. True

115. True

117. False. For example, planes $7x + y - 11z = 5$ and $5x + 2y - 4z = 1$ are both perpendicular to plane $2x - 3y + z = 3$, but are not parallel.

Section 11.6   Surfaces in Space

1. Quadric surfaces are the three-dimensional analogs of conic sections.

3. The trace of a surface is the intersection of the surface with a plane. You find a trace by setting one variable equal to a constant, such as $x = 0$ or $z = 2$.

5. Ellipsoid

Matches graph (c)

6. Hyperboloid of two sheets

Matches graph (e)

7. Hyperboloid of one sheet

Matches graph (f)

8. Elliptic cone

Matches graph (b)

9. Elliptic paraboloid

Matches graph (d)

10. Hyperbolic paraboloid

Matches graph (a)

11. $y^2 + z^2 = 9$

The x-coordinate is missing so you have a right circular cylinder with rulings parallel to the x-axis. The generating curve is a circle.

13. $4x^2 + y^2 = 4$

$$\frac{x^2}{1} + \frac{y^2}{4} = 1$$

The z-coordinate is missing so you have an elliptic cylinder with rulings parallel to the z-axis. The generating curve is an ellipse.

15. $4x^2 - y^2 - z^2 = 1$

Hyperboloid of two sheets

xy-trace: $4x^2 - y^2 = 1$ hyperbola

yz-trace: none

xz-trace: $4x^2 - z^2 = 1$ hyperbola

17. $16x^2 - y^2 + 16z^2 = 4$

$$4x^2 - \frac{y^2}{4} + 4z^2 = 1$$

Hyperboloid of one sheet

xy-trace: $4x^2 - \dfrac{y^2}{4} = 1$ hyperbola

xz-trace: $4(x^2 + z^2) = 1$ circle

yz-trace: $\dfrac{-y^2}{4} + 4z^2 = 1$ hyperbola

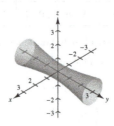

19. $\dfrac{x^2}{1} + \dfrac{y^2}{4} + \dfrac{z^2}{1} = 1$

Ellipsoid

xy-trace: $\dfrac{x^2}{1} + \dfrac{y^2}{4} = 1$ ellipse

xz-trace: $x^2 + z^2 = 1$ circle

yz-trace: $\dfrac{y^2}{4} + \dfrac{z^2}{1} = 1$ ellipse

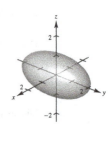

21. $z^2 = x^2 + \dfrac{y^2}{9}$

Elliptic cone

xy-trace: point $(0, 0, 0)$

xz-trace: $z = \pm x$

yz-trace: $z = \pm \dfrac{y}{3}$

When $z = \pm 1$, $x^2 + \dfrac{y^2}{9} = 1$ ellipse

23. $x^2 - y^2 + z = 0$

Hyperbolic paraboloid

xy-trace: $y = \pm x$

xz-trace: $z = -x^2$

yz-trace: $z = y^2$

$y = \pm 1$: $z = 1 - x^2$

25. $x^2 - y + z^2 = 0$

Elliptic paraboloid

xy-trace: $y = x^2$

xz-trace: $x^2 + z^2 = 0$,

point $(0, 0, 0)$

yz-trace: $y = z^2$

$y = 1$: $x^2 + z^2 = 1$

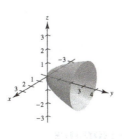

27. These have to be two minus signs in order to have a hyperboloid of two sheets. The number of sheets is the same as the number of minus signs.

29. No. See the table on pages 800 and 801.

31. $x^2 + z^2 = \left[r(y)\right]^2$ and $z = r(y) = 5y$, so

$x^2 + z^2 = 25y^2$.

33. $x^2 + y^2 = \left[r(z)\right]^2$ and $y = r(z) = 2z^{1/3}$, so

$x^2 + y^2 = 4z^{2/3}$.

35. $y^2 + z^2 = \left[r(x)\right]^2$ and $y = r(x) = \dfrac{2}{x}$, so

$y^2 + z^2 = \left(\dfrac{2}{x}\right)^2 \Rightarrow y^2 + z^2 = \dfrac{4}{x^2}$.

37. $x^2 + y^2 - 2z = 0$

$x^2 + y^2 = \left(\sqrt{2z}\right)^2$

Equation of generating curve: $y = \sqrt{2z}$ or $x = \sqrt{2z}$

39. $y^2 + z^2 = 5 - 8x^2 = \left(\sqrt{5 - 8x^2}\right)^2$

Equation of generating curve: $y = \sqrt{5 - 8x^2}$ or

$z = \sqrt{5 - 8x^2}$

41. $V = 2\pi\displaystyle\int_0^4 x\left(4x - x^2\right) dx = 2\pi\left[\dfrac{4x^3}{3} - \dfrac{x^4}{4}\right]_0^4 = \dfrac{218\pi}{3}$

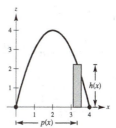

43. $z = \dfrac{x^2}{2} + \dfrac{y^2}{4}$

(a) When $z = 2$ we have $2 = \dfrac{x^2}{2} + \dfrac{y^2}{4}$, or

$1 = \dfrac{x^2}{4} + \dfrac{y^2}{8}$

Major axis: $2\sqrt{8} = 4\sqrt{2}$

Minor axis: $2\sqrt{4} = 4$

$c^2 = a^2 - b^2, c^2 = 4, c = 2$

Foci: $(0, \pm 2, 2)$

(b) When $z = 8$ we have $8 = \dfrac{x^2}{2} + \dfrac{y^2}{4}$, or

$1 = \dfrac{x^2}{16} + \dfrac{y^2}{32}$.

Major axis: $2\sqrt{32} = 8\sqrt{2}$

Minor axis: $2\sqrt{16} = 8$

$c^2 = 32 - 16 = 16, c = 4$

Foci: $(0, \pm 4, 8)$

45. If (x, y, z) is on the surface, then

$(y + 2)^2 = x^2 + (y - 2)^2 + z^2$

$y^2 + 4y + 4 = x^2 + y^2 - 4y + 4 + z^2$

$x^2 + z^2 = 8y$

Elliptic paraboloid

Traces parallel to xz-plane are circles.

47. $\dfrac{x^2}{3963^2} + \dfrac{y^2}{3963^2} + \dfrac{z^2}{3950^2} = 1$

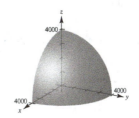

49. $z = \dfrac{y^2}{b^2} - \dfrac{x^2}{a^2}, \ z = bx + ay$

$$bx + ay = \frac{y^2}{b^2} - \frac{x^2}{a^2}$$

$$\frac{1}{a^2}\left(x^2 + a^2bx + \frac{a^4b^2}{4}\right) = \frac{1}{b^2}\left(y^2 - ab^2y + \frac{a^2b^4}{4}\right)$$

$$\frac{\left(x + \dfrac{a^2b}{2}\right)^2}{a^2} = \frac{\left(y - \dfrac{ab^2}{2}\right)^2}{b^2}$$

$$y = \pm\frac{b}{a}\left(x + \frac{a^2b}{2}\right) + \frac{ab^2}{2}$$

Letting $x = at$, you obtain the two intersecting lines

$x = at, \ y = -bt, \ z = 0$ and $x = at,$

$y = bt + ab^2, \ z = 2abt + a^2b^2.$

Section 11.7   Cylindrical and Spherical Coordinates

1. The cylindrical coordinate system is an extension of the polar coordinate system. In this system, a point P in space is represented by an ordered triple (r, θ, z). (r, θ) is a polar representation of the projection of P in the xy-plane, and z is the directed distance from (r, θ) to P.

3. $(-7, 0, 5)$, cylindrical

$x = r\cos\theta = -7\cos 0 = -7$

$y = r\sin\theta = -7\sin 0 = 0$

$z = 5$

$(-7, 0, 5)$, rectangular

5. $\left(3, \dfrac{\pi}{4}, 1\right)$, cylindrical

$x = r\cos\theta = 3\cos\dfrac{\pi}{4} = \dfrac{3\sqrt{2}}{2}$

$y = r\sin\theta = 3\sin\dfrac{\pi}{4} = \dfrac{3\sqrt{2}}{2}$

$z = 1$

$\left(\dfrac{3\sqrt{2}}{2}, \dfrac{3\sqrt{2}}{2}, 1\right)$, rectangular

7. $\left(4, \dfrac{7\pi}{6}, -3\right)$, cylindrical

$x = r\cos\theta = 4\cos\dfrac{7\pi}{6} = 4\left(-\dfrac{\sqrt{3}}{2}\right) = -2\sqrt{3}$

$y = r\sin\theta = 4\sin\dfrac{7\pi}{6} = 4\left(-\dfrac{1}{2}\right) = -2$

$z = -3$

$\left(-2\sqrt{3}, -2, -3\right)$, rectangular

9. $(0, 5, 1)$, rectangular

$r = \sqrt{(0)^2 + (5)^2} = 5$

$\tan\theta = \dfrac{5}{2} \Rightarrow \theta = \arctan\dfrac{5}{0} = \dfrac{\pi}{2}$

$z = 1$

$\left(5, \dfrac{\pi}{2}, 1\right)$, cylindrical

11. $(2, -2, -4)$, rectangular

$r = \sqrt{2^2 + (-2)^2} = 2\sqrt{2}$

$\tan\theta = \dfrac{-2}{2} \Rightarrow \theta = \arctan(-1) = -\dfrac{\pi}{4}$

$z = -4$

$\left(2\sqrt{2}, -\dfrac{\pi}{4}, -4\right)$, cylindrical

13. $\left(1, \sqrt{3}, 4\right)$, rectangular

$r = \sqrt{1^2 + \left(\sqrt{3}\right)^2} = 2$

$\tan\theta = \dfrac{\sqrt{3}}{1} \Rightarrow \theta = \arctan\sqrt{3} = \dfrac{\pi}{3}$

$z = 4$

$\left(2, \dfrac{\pi}{3}, 4\right)$, cylindrical

15. $z = 4$ is the equation in cylindrical coordinates. (plane)

51. The Klein bottle *does not* have both an "inside" and an "outside." It is formed by inserting the small open end through the side of the bottle and making it contiguous with the top of the bottle.

17. $x^2 + y^2 - 2z^2 = 5$, rectangular equation

$r^2 - 2z^2 = 5$, cylindrical equation

19. $y = x^2$, rectangular equation

$r \sin \theta = (r \cos \theta)^2$

$\sin \theta = r \cos^2 \theta$

$r = \sec \theta \cdot \tan \theta$, cylindrical equation

21. $y^2 = 10 - z^2$, rectangular equation

$(r \sin \theta)^2 = 10 - z^2$

$r^2 \sin^2 \theta + z^2 = 10$, cylindrical equation

23. $r = 3$, cylindrical equation

$\sqrt{x^2 + y^2} = 3$

$x^2 + y^2 = 9$, rectangular equation

25. $\theta = \dfrac{\pi}{6}$, cylindrical equation

$\tan \dfrac{\pi}{6} = \dfrac{y}{x}$

$\dfrac{1}{\sqrt{3}} = \dfrac{y}{x}$

$x = \sqrt{3}y$

$x - \sqrt{3}y = 0$, rectangular equation

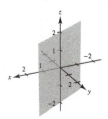

27. $r^2 + z^2 = 5$, cylindrical equation

$x^2 + y^2 + z^2 = 5$, rectangular equation

29. $r = 4 \sin \theta$, cylindrical equation

$r^2 = 4r \sin \theta$

$x^2 + y^2 = 4y$

$x^2 + y^2 - 4y + 4 = 4$

$x^2 + (y - 2)^2 = 4$, rectangular equation

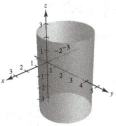

31. $(4, 0, 0)$, rectangular

$\rho = \sqrt{4^2 + 0^2 + 0^2} = 4$

$\tan \theta = \dfrac{y}{x} = 0 \Rightarrow \theta = 0$

$\phi = \arccos 0 = \dfrac{\pi}{2}$

$\left(4, 0, \dfrac{\pi}{2}\right)$, spherical

33. $\left(-2, 2\sqrt{3}, 4\right)$, rectangular

$\rho = \sqrt{(-2)^2 + \left(2\sqrt{3}\right)^2 + 4^2} = 4\sqrt{2}$

$\tan \theta = \dfrac{y}{x} = \dfrac{2\sqrt{3}}{-2} = -\sqrt{3}$

$\theta = \dfrac{2\pi}{3}$

$\phi = \arccos \dfrac{z}{\rho} = \arccos \dfrac{1}{\sqrt{2}} = \dfrac{\pi}{4}$

$\left(4\sqrt{2}, \dfrac{2\pi}{3}, \dfrac{\pi}{4}\right)$, spherical

35. $\left(\sqrt{3}, 1, 2\sqrt{3}\right)$, rectangular

$\rho = \sqrt{3 + 1 + 12} = 4$

$\tan \theta = \dfrac{y}{x} = \dfrac{1}{\sqrt{3}}$

$\theta = \dfrac{\pi}{6}$

$\phi = \arccos \dfrac{z}{\rho} = \arccos \dfrac{\sqrt{3}}{2} = \dfrac{\pi}{6}$

$\left(4, \dfrac{\pi}{6}, \dfrac{\pi}{6}\right)$, spherical

37. $\left(4, \dfrac{\pi}{6}, \dfrac{\pi}{4}\right)$, spherical

$x = \rho \sin \phi \cos \theta = 4 \sin \dfrac{\pi}{4} \cos \dfrac{\pi}{6} = \sqrt{6}$

$y = \rho \sin \phi \sin \theta = 4 \sin \dfrac{\pi}{4} \sin \dfrac{\pi}{6} = \sqrt{2}$

$z = \rho \cos \phi = 4 \cos \dfrac{\pi}{4} = 2\sqrt{2}$

$\left(\sqrt{6}, \sqrt{2}, 2\sqrt{2}\right)$, rectangular

39. $\left(12, -\dfrac{\pi}{4}, 0\right)$, spherical

$x = \rho \sin \phi \cos \theta = 12 \sin 0 \cos\left(-\dfrac{\pi}{4}\right) = 0$

$y = \rho \sin \phi \sin \theta = 12 \sin 0 \sin\left(-\dfrac{\pi}{4}\right) = 0$

$z = \rho \cos \phi = 12 \cos 0 = 12$

$(0, 0, 12)$, rectangular

41. $\left(5, \dfrac{\pi}{4}, \dfrac{\pi}{12}\right)$, spherical

$x = \rho \sin \phi \cos \theta = 5 \sin \dfrac{\pi}{12} \cos \dfrac{\pi}{4} \approx 0.915$

$y = \rho \sin \phi \sin \theta = 5 \sin \dfrac{\pi}{12} \sin \dfrac{\pi}{4} \approx 0.915$

$z = \rho \cos \theta = 5 \cos \dfrac{\pi}{12} \approx 4.830$

$(0.915, 0.915, 4.830)$, rectangular

43. $y = 2$, rectangular equation

$\rho \sin \phi \sin \theta = 2$

$\rho = 2 \csc \phi \csc \theta$, spherical equation

45. $x^2 + y^2 + z^2 = 49$, rectangular equation

$\rho^2 = 49$

$\rho = 7$, spherical equation

47. $x^2 + y^2 = 16$, rectangular equation

$\rho^2 \sin^2 \phi \sin^2 \theta + \rho^2 \sin^2 \phi \cos^2 \theta = 16$

$\rho^2 \sin^2 \phi \left(\sin^2 \theta + \cos^2 \theta\right) = 16$

$\rho^2 \sin^2 \phi = 16$

$\rho \sin \phi = 4$

$\rho = 4 \csc \phi$, spherical equation

49. $x^2 + y^2 = 2z^2$, rectangular equation

$\rho^2 \sin^2 \phi \cos^2 \theta + \rho^2 \sin^2 \phi \sin^2 \theta = 2\rho^2 \cos^2 \phi$

$\rho^2 \sin^2 \phi \left[\cos^2 \theta + \sin^2 \theta\right] = 2\rho^2 \cos^2 \phi$

$\rho^2 \sin^2 \phi = 2\rho^2 \cos^2 \theta$

$\dfrac{\sin^2 \phi}{\cos^2 \phi} = 2$

$\tan^2 \phi = 2$

$\tan \phi = \pm\sqrt{2}$, spherical equation

51. $\rho = 1$, spherical equation

$x^2 + y^2 + z^1 = 1$, rectangular equation

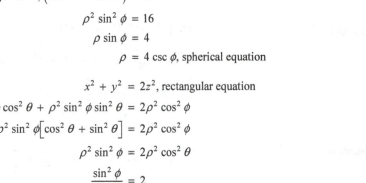

53. $\phi = \dfrac{\pi}{6}$, spherical equation

$$\cos\phi = \frac{z}{\sqrt{x^2 + y^2 + z^2}}$$

$$\frac{\sqrt{3}}{2} = \frac{z}{\sqrt{x^2 + y^2 + z^2}}$$

$$\frac{3}{4} = \frac{z^2}{x^2 + y^2 + z^2}$$

$3x^2 + 3y^2 - z^2 = 0,\ z \geq 0$, rectangular equation

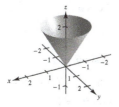

55. $\rho = 4\cos\phi$, spherical equation

$$\sqrt{x^2 + y^2 + z^2} = \frac{4z}{\sqrt{x^2 + y^2 + z^2}}$$

$$x^2 + y^2 + z^2 - 4z = 0$$

$x^2 + y^2 + (z-2)^2 = 4,\ z \geq 0$, rectangular equation

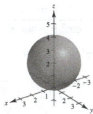

57. $\rho = \csc\phi$, spherical equation

$$\rho\sin\phi = 1$$

$$\sqrt{x^2 + y^2} = 1$$

$x^2 + y^2 = 1$, rectangular equation

59. $\left(4, \dfrac{\pi}{4}, 0\right)$, cylindrical

$$\rho = \sqrt{4^2 + 0^2} = 4$$

$$\theta = \frac{\pi}{4}$$

$$\phi = \arccos\frac{z}{\sqrt{r^2 + z^2}} = \arccos 0 = \frac{\pi}{2}$$

$\left(4, \dfrac{\pi}{4}, \dfrac{\pi}{2}\right)$, spherical

61. $\left(6, \dfrac{\pi}{2}, -6\right)$, cylindrical

$$\rho = \sqrt{6^2 + (-6)^2} = \sqrt{72} = 6\sqrt{2}$$

$$\theta = \frac{\pi}{2}$$

$$\phi = \arccos\frac{z}{\sqrt{r^2 + z^2}}$$

$$= \arccos\left(\frac{-6}{6\sqrt{2}}\right)$$

$$= \arccos\left(\frac{-1}{\sqrt{2}}\right)$$

$$= \frac{3\pi}{4}$$

$\left(6\sqrt{2}, \dfrac{\pi}{2}, \dfrac{3\pi}{4}\right)$, spherical

63. $(12, \pi, 5)$, cylindrical

$$\rho = \sqrt{12^2 + 5^2} = 13$$
$$\theta = \pi$$

$$\phi = \arccos\frac{z}{\sqrt{r^2 + z^2}} = \arccos\frac{5}{13}$$

$\left(13,\ \pi,\ \arccos\dfrac{5}{13}\right)$, spherical

65. $\left(10, \dfrac{\pi}{6}, \dfrac{\pi}{2}\right)$, spherical

$$r = 10\sin\frac{\pi}{2} = 10$$

$$\theta = \frac{\pi}{6}$$

$$z = 10\cos\frac{\pi}{2} = 0$$

$\left(10, \dfrac{\pi}{6}, 0\right)$, cylindrical

67. $\left(6, -\dfrac{\pi}{6}, \dfrac{\pi}{3}\right)$, spherical

$$r = 6 \sin \dfrac{\pi}{3} = 3\sqrt{3}$$

$$\theta = -\dfrac{\pi}{6}$$

$$z = 6 \cos \dfrac{\pi}{3} = 3$$

$\left(3\sqrt{3}, -\dfrac{\pi}{6}, 3\right)$, cylindrical

69. $\left(8, \dfrac{7\pi}{6}, \dfrac{\pi}{6}\right)$, spherical

$$r = 8 \sin \dfrac{\pi}{6} = 4$$

$$\theta = \dfrac{7\pi}{6}$$

$$z = 8 \cos \dfrac{\pi}{6} = \dfrac{8\sqrt{3}}{2}$$

$\left(4, \dfrac{7\pi}{6}, 4\sqrt{3}\right)$, cylindrical

71. $r = 5$

Cylinder

Matches graph (d)

72. $\theta = \dfrac{\pi}{4}$

Plane

Matches graph (e)

73. $\rho = 5$

Sphere

Matches graph (c)

74. $\phi = \dfrac{\pi}{4}$

Cone

Matches graph (a)

75. $r^2 = z, x^2 + y^2 = z$

Paraboloid

Matches graph (f)

76. $\rho = 4 \sec \phi, z = \rho \cos \phi = 4$

Plane

Matches graph (b)

77. $\theta = c$ is a half-plane because of the restriction $r \geq 0$.

79. $x^2 + y^2 + z^2 = 27$

(a) $r^2 + z^2 = 27$

(b) $\rho^2 = 27 \Rightarrow \rho = 3\sqrt{3}$

81. $x^2 + y^2 + z^2 - 2z = 0$

(a) $r^2 + z^2 - 2z = 0 \Rightarrow r^2 + (z-1)^2 = 1$

(b) $\rho^2 - 2\rho \cos \phi = 0$

$\rho(\rho - 2\cos \phi) = 0$

$\rho = 2 \cos \phi$

83. $x^2 + y^2 = 4y$

(a) $r^2 = 4r \sin \theta, r = 4 \sin \theta$

(b) $\rho^2 \sin^2 \phi = 4\rho \sin \phi \sin \theta$

$\rho \sin \phi (\rho \sin \phi - 4 \sin \theta) = 0$

$$\rho = \dfrac{4 \sin \theta}{\sin \phi}$$

$\rho = 4 \sin \theta \csc \phi$

85. $x^2 - y^2 = 9$

(a) $r^2 \cos^2 \theta - r^2 \sin^2 \theta = 9$

$$r^2 = \dfrac{9}{\cos^2 \theta - \sin^2 \theta}$$

(b) $\rho^2 \sin^2 \phi \cos^2 \theta - \rho^2 \sin^2 \phi \sin^2 \theta = 9$

$$\rho^2 \sin^2 \phi = \dfrac{9}{\cos^2 \theta - \sin^2 \theta}$$

$$\rho^2 = \dfrac{9 \csc^2 \phi}{\cos^2 \theta - \sin^2 \theta}$$

87. $0 \leq \theta \leq \dfrac{\pi}{2}$

$0 \leq r \leq 2$

$0 \leq z \leq 4$

89. $0 \leq \theta \leq 2\pi$

$0 \leq r \leq a$

$r \leq z \leq a$

91. $0 \le \theta \le 2\pi$

$0 \le \phi \le \dfrac{\pi}{6}$

$0 \le \rho \le a \sec \phi$

93. $0 \le \theta \le \dfrac{\pi}{2}$

$0 \le \phi \le \dfrac{\pi}{2}$

$0 \le \rho \le 2$

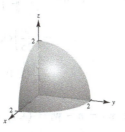

95. Rectangular

$0 \le x \le 10$

$0 \le y \le 10$

$0 \le z \le 10$

97. Spherical

$4 \le \rho \le 6$

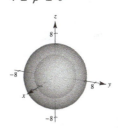

99. Cylindrical coordinates:

$r^2 + z^2 \le 9,$

$r \le 3 \cos \theta, 0 \le \theta \le \pi$

101. False. $(r, \theta, z) = (0, 0, 1)$ and $(r, \theta, z) = (0, \pi, 1)$ represent the same point $(x, y, z) = (0, 0, 1)$.

103. $z = \sin \theta, r = 1$

$z = \sin \theta = \dfrac{y}{r} = \dfrac{y}{1} = y$

The curve of intersection is the ellipse formed by the intersection of the plane $z = y$ and the cylinder $r = 1$.

Review Exercises for Chapter 11

1. $P = (1, 2), Q = (4, 1), R = (5, 4)$

(a) $\mathbf{u} = \overrightarrow{PQ} = \langle 4 - 1, 1 - 2 \rangle = \langle 3, -1 \rangle$

$\mathbf{v} = \overrightarrow{PR} = \langle 5 - 1, 4 - 2 \rangle = \langle 4, 2 \rangle$

(b) $\mathbf{u} = 3\mathbf{i} - \mathbf{j}, \mathbf{v} = 4\mathbf{i} + 2\mathbf{j}$

(c) $\|\mathbf{u}\| = \sqrt{3^2 + (-1)^2} = \sqrt{10} \quad \|\mathbf{v}\| = \sqrt{4^2 + 2^2} = \sqrt{20} = 2\sqrt{5}$

(d) $-3\mathbf{u} + \mathbf{v} = -3\langle 3, -1 \rangle + \langle 4, 2 \rangle = \langle -5, 5 \rangle$

3. $\mathbf{v} = \|\mathbf{v}\|(\cos \theta\, \mathbf{i} + \sin \theta\, \mathbf{j})$

$= 8(\cos 60° \, \mathbf{i} + \sin 60° \, \mathbf{j})$

$= 8\left(\dfrac{1}{2}\mathbf{i} + \dfrac{\sqrt{3}}{2}\mathbf{j}\right) = 4\mathbf{i} + 4\sqrt{3}\, \mathbf{j} = \langle 4, 4\sqrt{3} \rangle$

5. $z = 0, y = 4, x = -5: (-5, 4, 0)$

7. $d = \sqrt{(-2 - 1)^2 + (3 - 6)^2 + (5 - 3)^2}$

$= \sqrt{9 + 9 + 4} = \sqrt{22}$

9. $(x - 3)^2 + (y + 2)^2 + (z - 6)^2 = 4^2$

$(x - 3)^2 + (y + 2)^2 + (z - 6)^2 = 16$

11. $(x^2 - 4x + 4) + (y^2 - 6y + 9) + z^2 = -4 + 4 + 9$

$(x - 2)^2 + (y - 3)^3 + z^2 = 9$

Center: $(2, 3, 0)$

Radius: 3

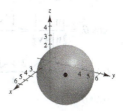

13. (a), (d)

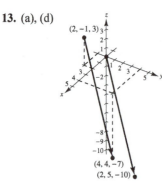

(b) $\mathbf{v} = \langle 4 - 2, 4 - (-1), -7 - 3 \rangle = \langle 2, 5, -10 \rangle$

(c) $\mathbf{v} = 2\mathbf{i} + 5\mathbf{j} - 10\mathbf{k}$

15. $z = -\mathbf{u} + 3\mathbf{v} + \dfrac{1}{2}\mathbf{w}$

$\qquad = -\langle 5, -2, 3 \rangle + 3\langle 0, 2, 1 \rangle + \dfrac{1}{2}\langle -6, -6, 2 \rangle$

$\qquad = \langle -5, 2, -3 \rangle + \langle 0, 6, 3 \rangle + \langle -3, -3, 1 \rangle$

$\qquad = \langle -8, 5, 1 \rangle$

17. $\mathbf{v} = \langle -1 - 3, 6 - 4, 9 + 1 \rangle = \langle -4, 2, 10 \rangle$

$\quad \mathbf{w} = \langle 5 - 3, 3 - 4, -6 + 1 \rangle = \langle 2, -1, -5 \rangle$

Because $-2\mathbf{w} = \mathbf{v}$, the points lie in a straight line.

19. Unit vector: $\dfrac{\mathbf{u}}{\|\mathbf{u}\|} = \left\langle \dfrac{2, 3, 5}{\sqrt{38}} \right\rangle = \left\langle \dfrac{2}{\sqrt{38}}, \dfrac{3}{\sqrt{38}}, \dfrac{5}{\sqrt{38}} \right\rangle$

21. $P = \langle 5, 0, 0 \rangle, Q = \langle 4, 4, 0 \rangle, R = \langle 2, 0, 6 \rangle$

(a) $\mathbf{u} = \overrightarrow{PQ} = \langle -1, 4, 0 \rangle$

$\qquad \mathbf{v} = \overrightarrow{PR} = \langle -3, 0, 6 \rangle$

(b) $\mathbf{u} \cdot \mathbf{v} = (-1)(-3) + 4(0) + 0(6) = 3$

(c) $\mathbf{v} \cdot \mathbf{v} = 9 + 36 = 45$

23. $\mathbf{u} = 5\left(\cos \dfrac{3\pi}{4}\mathbf{i} + \sin \dfrac{3\pi}{4}\mathbf{j} \right) = \dfrac{5\sqrt{2}}{2}[-\mathbf{i} + \mathbf{j}]$

$\quad \mathbf{v} = 2\left(\cos \dfrac{2\pi}{3}\mathbf{i} + \sin \dfrac{2\pi}{3}\mathbf{j} \right) = -\mathbf{i} + \sqrt{3}\,\mathbf{j}$

$\quad \mathbf{u} \cdot \mathbf{v} = \dfrac{5\sqrt{2}}{2}\left(1 + \sqrt{3} \right)$

$\quad \|\mathbf{u}\| = \sqrt{\dfrac{25}{2} + \dfrac{25}{2}} = 5 \qquad \|\mathbf{v}\| = \sqrt{1 + 3} = 2$

$\quad \cos \theta = \dfrac{|\mathbf{u} \cdot \mathbf{v}|}{\|\mathbf{u}\|\|\mathbf{v}\|} = \dfrac{\left(5\sqrt{2}/2 \right)\left(1 + \sqrt{3} \right)}{5(2)} = \dfrac{\sqrt{2} + \sqrt{6}}{4}$

(a) $\theta = \arccos \dfrac{\sqrt{2} + \sqrt{6}}{4} = \dfrac{\pi}{12} \approx 0.262$

(b) $\theta \approx 15°$

25. $\mathbf{u} = \langle 7, -2, 3 \rangle, \mathbf{v} = \langle -1, 4, 5 \rangle$

Because $\mathbf{u} \cdot \mathbf{v} = 0$, the vectors are orthogonal.

27. $\mathbf{u} = \langle 4, 2 \rangle, \mathbf{v} = \langle 3, 4 \rangle$

(a) $\mathbf{w}_1 = \text{proj}_{\mathbf{v}}\mathbf{u} = \left(\dfrac{\mathbf{u} \cdot \mathbf{v}}{\|\mathbf{v}\|^2} \right)\mathbf{v}$

$\qquad\quad = \left(\dfrac{20}{25} \right)\langle 3, 4 \rangle$

$\qquad\quad = \dfrac{4}{5}\langle 3, 4 \rangle = \left\langle \dfrac{12}{5}, \dfrac{16}{5} \right\rangle$

(b) $\mathbf{w}_2 = \mathbf{u} - \mathbf{w}_1 = \langle 4, 2 \rangle - \left\langle \dfrac{12}{5}, \dfrac{16}{5} \right\rangle = \left\langle \dfrac{8}{5}, -\dfrac{6}{5} \right\rangle$

29. There are many correct answers.

For example: $\mathbf{v} = \pm\langle 6, -5, 0 \rangle$.

31. (a) $\mathbf{u} \times \mathbf{v} = \begin{vmatrix} \mathbf{i} & \mathbf{j} & \mathbf{k} \\ 4 & 3 & 6 \\ 5 & 2 & 1 \end{vmatrix} = -9\mathbf{i} + 26\mathbf{j} - 7\mathbf{k}$

(b) $\mathbf{v} \times \mathbf{u} = -(\mathbf{u} \times \mathbf{v}) = 9\mathbf{i} - 26\mathbf{j} + 7\mathbf{k}$

(c) $\mathbf{v} \times \mathbf{v} = \mathbf{0}$

33. $\mathbf{u} \times \mathbf{v} = \begin{vmatrix} \mathbf{i} & \mathbf{j} & \mathbf{k} \\ 2 & -10 & 8 \\ 4 & 6 & -8 \end{vmatrix} = 32\mathbf{i} + 48\mathbf{j} + 52\mathbf{k}$

$\|\mathbf{u} \times \mathbf{v}\| = \sqrt{6032} = 4\sqrt{377}$

Unit vector: $\dfrac{1}{\sqrt{377}}\langle 8, 12, 13 \rangle$

35. $\mathbf{F} = -40\mathbf{k} \qquad\qquad \left(9 \text{ in.} = \frac{3}{4} \text{ ft} \right)$

$\overrightarrow{PQ} = \dfrac{3}{4}(\cos 60°\mathbf{j} + \sin 60°\mathbf{k}) = \dfrac{3}{8}\mathbf{j} + \dfrac{3\sqrt{3}}{8}\mathbf{k}$

The moment of $\mathbf{F}$ about P is

$M = \overrightarrow{PQ} \times \mathbf{F} = \begin{vmatrix} \mathbf{i} & \mathbf{j} & \mathbf{k} \\ 0 & \dfrac{3}{8} & \dfrac{3\sqrt{3}}{8} \\ 0 & 0 & -40 \end{vmatrix} = -15\mathbf{i}$

Torque $= 15$ ft-lb

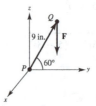

37. $\mathbf{v} = \langle 9 - 3, 11 - 0, 6 - 2 \rangle = \langle 6, 11, 4 \rangle$

(a) Parametric equations:

$x = 3 + 6t, \ y = 11t, \ z = 2 + 4t$

(b) Symmetric equations: $\dfrac{x-3}{6} = \dfrac{y}{11} = \dfrac{z-2}{4}$

39. $P = (-6, -8, 2)$

$\mathbf{v} = \mathbf{j} = \langle 0, 1, 0 \rangle$

$x = -6, \ y = -8 + t, \ z = 2$

41. $P = (-3, -4, 2), \ Q = (-3, 4, 1), \ R = (1, 1, -2)$

$\overrightarrow{PQ} = \langle 0, 8, -1 \rangle, \ \overrightarrow{PR} = [4, 5, -4]$

$\mathbf{n} = \overrightarrow{PQ} \times \overrightarrow{PR} = \begin{vmatrix} \mathbf{i} & \mathbf{j} & \mathbf{k} \\ 0 & 8 & -1 \\ 4 & 5 & -4 \end{vmatrix} = -27\mathbf{i} - 4\mathbf{j} - 32\mathbf{k}$

$-27(x + 3) - 4(y + 4) - 32(z - 2) = 0$

$27x + 4y + 32z = -33$

43. The two lines are parallel as they have the same direction numbers, $-2, 1, 1$. Therefore, a vector parallel to the plane is $\mathbf{v} = -2\mathbf{i} + \mathbf{j} + \mathbf{k}$. A point on the first line is $(1, 0, -1)$ and a point on the second line is $(-1, 1, 2)$.

The vector $\mathbf{u} = 2\mathbf{i} - \mathbf{j} - 3\mathbf{k}$ connecting these two points is also parallel to the plane. Therefore, a normal to the plane is

$\mathbf{v} \times \mathbf{u} = \begin{vmatrix} \mathbf{i} & \mathbf{j} & \mathbf{k} \\ -2 & 1 & 1 \\ 2 & -1 & -3 \end{vmatrix} = -2\mathbf{i} - 4\mathbf{j} = -2(\mathbf{i} + 2\mathbf{j}).$

Equation of the plane: $(x - 1) + 2y = 0$

$x + 2y = 1$

45. $Q(1, 0, 2)$ point

$2x - 3y + 6z = 6$

A point P on the plane is $(3, 0, 0)$.

$\overrightarrow{PQ} = \langle -2, 0, 2 \rangle$

$\mathbf{n} = \langle 2, -3, 6 \rangle$ normal to plane

$D = \dfrac{|\overrightarrow{PQ} \cdot \mathbf{n}|}{\|\mathbf{n}\|} = \dfrac{8}{7}$

47. The normal vectors to the planes are the same,

$\mathbf{n} = (5, -3, 1)$.

Choose a point in the first plane $P(0, 0, 2)$. Choose a point in the second plane, $Q(0, 0, -3)$.

$\overrightarrow{PQ} = \langle 0, 0, -5 \rangle$

$D = \dfrac{|\overrightarrow{PQ} \cdot \mathbf{n}|}{\|\mathbf{n}\|} = \dfrac{|-5|}{\sqrt{35}} = \dfrac{5}{\sqrt{35}} = \dfrac{\sqrt{35}}{7}$

49. $x + 2y + 3z = 6$

Plane

Intercepts: $(6, 0, 0), (0, 3, 0), (0, 0, 2),$

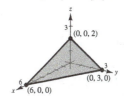

51. $y = \dfrac{1}{2}z$

Plane with rulings parallel to the x-axis.

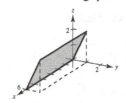

53. $\dfrac{x^2}{16} + \dfrac{y^2}{9} + z^2 = 1$

Ellipsoid

xy-trace: $\dfrac{x^2}{16} + \dfrac{y^2}{9} = 1$

xz-trace: $\dfrac{x^2}{16} + z^2 = 1$

yz-trace: $\dfrac{y^2}{9} + z^2 = 1$

55. $\dfrac{x^2}{16} - \dfrac{y^2}{9} + z^2 = -1$

$\dfrac{y^2}{9} - \dfrac{x^2}{16} - z^2 = 1$

Hyperboloid of two sheets

xy-trace: $\dfrac{y^2}{9} - \dfrac{x^2}{16} = 1$

xz-trace: None

yz-trace: $\dfrac{y^2}{9} - z^2 = 1$

57. $x^2 + z^2 = 4$.

Cylinder of radius 2 about *y*-axis

59. $z^2 = 2y$ revolved about *y*-axis

$$z = \pm\sqrt{2y}$$

$$x^2 + z^2 = \left[r(y) \right]^2 = 2y$$

$$x^2 + z^2 = 2y$$

61. $\left(-\sqrt{3}, 3, -5 \right)$, rectangular

(a) $r = \sqrt{\left(-\sqrt{3} \right)^2 + 3^2} = \sqrt{12} = 2\sqrt{3}$

$\tan \theta = \dfrac{-3}{\sqrt{3}} \Rightarrow \theta = -\dfrac{\pi}{3}$

$z = -5$

$\left(2\sqrt{3}, -\dfrac{\pi}{3}, -5 \right)$, cylindrical

(b) $\rho = \sqrt{\left(-\sqrt{3} \right)^2 + 3^2 + (-5)^2} = \sqrt{37}$

$\tan \theta = -\dfrac{3}{\sqrt{3}} \Rightarrow \theta = -\dfrac{\pi}{3}$

$\phi = \arccos \dfrac{z}{\rho} = \arccos\left(\dfrac{-5}{\sqrt{37}} \right)$

$\left(\sqrt{37}, -\dfrac{\pi}{3}, \arccos\left(-\dfrac{5\sqrt{37}}{37} \right) \right)$, spherical

63. $(5, \pi, 1)$, cylindrical

$x = r \cos \theta = 5 \cos \pi = -5$

$y = r \sin \theta = 5 \sin \pi = 0$

$z = 1$

$(-5, 0, 1)$, rectangular

65. $\left(4, \pi, \dfrac{\pi}{4} \right)$, spherical

$x = \rho \sin \phi \cos \theta = 4 \sin \dfrac{\pi}{4} \cos \pi = -2\sqrt{2}$

$y = \rho \sin \phi \sin \theta = 4 \sin \dfrac{\pi}{4} \sin \pi = 0$

$z = \rho \cos \phi = 4 \cos \dfrac{\pi}{4} = 2\sqrt{2}$

$\left(-2\sqrt{2}, 0, 2\sqrt{2} \right)$, rectangular

67. $x^2 - y^2 = 2z$

(a) Cylindrical:

$r^2 \cos^2 \theta - r^2 \sin^2 \theta = 2z \Rightarrow r^2 \cos 2\theta = 2z$

(b) Spherical:

$\rho^2 \sin^2 \phi \cos^2 \theta - \rho^2 \sin^2 \phi \sin^2 \theta = 2\rho \cos \phi$

$\rho \sin^2 \phi \cos 2\theta - 2 \cos \phi = 0$

$\rho = 2 \sec 2\theta \cos \phi \csc^2 \phi$

69. $z = r^2 \sin^2 \theta + 3r \cos \theta$, cylindrical equation

$z = y^2 + 3x$, rectangular equation

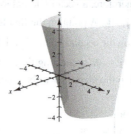

71. $\phi = \dfrac{\pi}{4}$, spherical equation

$$\phi = \arccos \dfrac{z}{\sqrt{x^2 + y^2 + z^2}} = \dfrac{\pi}{4}$$

$$\dfrac{z}{\sqrt{x^2 + y^2 + z^2}} = \cos \dfrac{\pi}{4} = \dfrac{\sqrt{2}}{2}$$

$$z^2 = \dfrac{1}{2}\left(x^2 + y^2 + z^2 \right)$$

$$2z^2 = x^2 + y^2 + z^2$$

$$x^2 + y^2 - z^2 = 0, \text{ rectangular equation}$$

Problem Solving for Chapter 11

1.
$$\mathbf{a} + \mathbf{b} + \mathbf{c} = \mathbf{0}$$
$$\mathbf{b} \times (\mathbf{a} + \mathbf{b} + \mathbf{c}) = \mathbf{0}$$
$$(\mathbf{b} \times \mathbf{a}) + (\mathbf{b} \times \mathbf{c}) = \mathbf{0}$$
$$\|\mathbf{a} \times \mathbf{b}\| = \|\mathbf{b} \times \mathbf{c}\|$$
$$\|\mathbf{b} \times \mathbf{c}\| = \|\mathbf{b}\|\|\mathbf{c}\| \sin A$$
$$\|\mathbf{a} \times \mathbf{b}\| = \|\mathbf{a}\|\|\mathbf{b}\| \sin C$$

Then,
$$\frac{\sin A}{\|\mathbf{a}\|} = \frac{\|\mathbf{b} \times \mathbf{c}\|}{\|\mathbf{a}\|\|\mathbf{b}\|\|\mathbf{c}\|}$$
$$= \frac{\|\mathbf{a} \times \mathbf{b}\|}{\|\mathbf{a}\|\|\mathbf{b}\|\|\mathbf{c}\|}$$
$$= \frac{\sin C}{\|\mathbf{c}\|}.$$

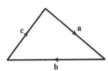

The other case, $\dfrac{\sin A}{\|\mathbf{a}\|} = \dfrac{\sin B}{\|\mathbf{b}\|}$ is similar.

3. Label the figure as indicated.
From the figure, you see that

$$\overrightarrow{SP} = \frac{1}{2}\mathbf{a} - \frac{1}{2}\mathbf{b} = \overrightarrow{RQ} \text{ and } \overrightarrow{SR} = \frac{1}{2}\mathbf{a} + \frac{1}{2}\mathbf{b} = \overrightarrow{PQ}.$$

Because $\overrightarrow{SP} = \overrightarrow{RQ}$ and $\overrightarrow{SR} = \overrightarrow{PQ}$,
$PSRQ$ is a parallelogram.

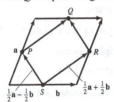

11. (a) $\rho = 2 \sin \phi$ (b) $\rho = 2 \cos \phi$

    Torus   Sphere

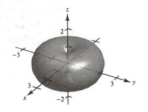 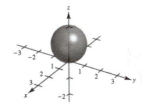

5. (a) $\mathbf{u} = \langle 0, 1, 1 \rangle$ is the direction vector of the line determined by P_1 and P_2.

$$D = \frac{\|\overrightarrow{P_1Q} \times \mathbf{u}\|}{\|\mathbf{u}\|}$$
$$= \frac{\|\langle 2, 0, -1 \rangle \times \langle 0, 1, 1 \rangle\|}{\sqrt{2}}$$
$$= \frac{\|\langle 1, -2, 2 \rangle\|}{\sqrt{2}} = \frac{3}{\sqrt{2}} = \frac{3\sqrt{2}}{2}$$

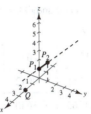

(b) The shortest distance to the line **segment** is
$$\|\overrightarrow{P_1Q}\| = \|\langle 2, 0, -1 \rangle\| = \sqrt{5}.$$

7. (a) $V = \pi \int_0^1 (\sqrt{z})^2 \, dz = \left[\pi \frac{z^2}{2}\right]_0^1 = \frac{1}{2}\pi$

    Note: $\dfrac{1}{2}$ (base)(altitude) $= \dfrac{1}{2}\pi(1) = \dfrac{1}{2}\pi$

(b) $\dfrac{x^2}{a^2} + \dfrac{y^2}{b^2} = z$: (slice at $z = c$)

$$\frac{x^2}{\left(\sqrt{ca}\right)^2} + \frac{y^2}{\left(\sqrt{cb}\right)^2} = 1$$

At $z = c$, figure is ellipse of area
$$\pi\left(\sqrt{ca}\right)\left(\sqrt{cb}\right) = \pi abc.$$

$$V = \int_0^k \pi abc \cdot dc = \left[\frac{\pi abc^2}{2}\right]_0^k = \frac{\pi abk^2}{2}$$

(c) $V = \dfrac{1}{2}(\pi abk)k = \dfrac{1}{2}$ (area of base)(height)

9. From Exercise 54, Section 11.4,
$$(\mathbf{u} \times \mathbf{v}) \times (\mathbf{w} \times \mathbf{z}) = \left[(\mathbf{u} \times \mathbf{v}) \cdot \mathbf{z}\right]\mathbf{w} - \left[(\mathbf{u} \times \mathbf{v}) \cdot \mathbf{w}\right]\mathbf{z}.$$

13. (a) $\mathbf{u} = \|\mathbf{u}\|(\cos 0\, \mathbf{i} + \sin 0\, \mathbf{j}) = \|\mathbf{u}\|\mathbf{i}$

Downward force $\mathbf{w} = -\mathbf{j}$

$\mathbf{T} = \|\mathbf{T}\|(\cos(90° + \theta)\mathbf{i} + \sin(90° + \theta)\mathbf{j})$

$\quad = \|\mathbf{T}\|(-\sin\theta\, \mathbf{i} + \cos\theta\, \mathbf{j})$

$\mathbf{0} = \mathbf{u} + \mathbf{w} + \mathbf{T} = \|\mathbf{u}\|\mathbf{i} - \mathbf{j} + \|\mathbf{T}\|(-\sin\theta\, \mathbf{i} + \cos\theta\, \mathbf{j})$

$\|\mathbf{u}\| = \sin\theta\|\mathbf{T}\|$

$\quad 1 = \cos\theta\|\mathbf{T}\|$

If $\theta = 30°, \|\mathbf{u}\| = (1/2)\|\mathbf{T}\|$ and $1 = (\sqrt{3}/2)\|\mathbf{T}\| \Rightarrow \|\mathbf{T}\| = \dfrac{2}{\sqrt{3}} \approx 1.1547$ lb and $\|\mathbf{u}\| = \dfrac{1}{2}\left(\dfrac{2}{\sqrt{3}}\right) \approx 0.5774$ lb

(b) From part (a), $\|\mathbf{u}\| = \tan\theta$ and $\|\mathbf{T}\| = \sec\theta$.

Domain: $0 \le \theta \le 90°$

(c)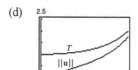

θ	0°	10°	20°	30°	40°	50°	60°
T	1	1.0154	1.0642	1.1547	1.3054	1.5557	2
$\|\mathbf{u}\|$	0	0.1763	0.3640	0.5774	0.8391	1.1918	1.7321

(d)

2.5

T

$\|\mathbf{u}\|$

0

0   60

(e) Both are increasing functions.

(f) $\displaystyle\lim_{\theta\to\pi/2^-} T = \infty$ and $\displaystyle\lim_{\theta\to\pi/2^-} \|\mathbf{u}\| = \infty.$

Yes. As θ increases, both T and $\|\mathbf{u}\|$ increase.

15. Let $\theta = \alpha - \beta$, the angle between **u** and **v**. Then

$\sin(\alpha - \beta) = \dfrac{\|\mathbf{u}\times\mathbf{v}\|}{\|\mathbf{u}\|\|\mathbf{v}\|} = \dfrac{\|\mathbf{v}\times\mathbf{u}\|}{\|\mathbf{u}\|\|\mathbf{v}\|}.$

For $\mathbf{u} = \langle\cos\alpha, \sin\alpha, 0\rangle$ and

$\mathbf{v} = \langle\cos\beta, \sin\beta, 0\rangle, \|\mathbf{u}\| = \|\mathbf{v}\| = 1$ and

$\mathbf{v}\times\mathbf{u} = \begin{vmatrix} \mathbf{i} & \mathbf{j} & \mathbf{k} \\ \cos\beta & \sin\beta & 0 \\ \cos\alpha & \sin\alpha & 0 \end{vmatrix}$

$\quad = (\sin\alpha\cos\beta - \cos\alpha\sin\beta)\mathbf{k}.$

So, $\sin(\alpha - \beta) = \|\mathbf{v}\times\mathbf{u}\| = \sin\alpha\cos\beta - \cos\alpha\sin\beta.$

17. From Theorem 11.13 and Theorem 11.7 (6) you have

$D = \dfrac{\left|\overrightarrow{PQ}\cdot\mathbf{n}\right|}{\|\mathbf{n}\|}$

$\quad = \dfrac{\left|\mathbf{w}\cdot(\mathbf{u}\times\mathbf{v})\right|}{\|\mathbf{u}\times\mathbf{v}\|} = \dfrac{\left|(\mathbf{u}\times\mathbf{v})\cdot\mathbf{w}\right|}{\|\mathbf{u}\times\mathbf{v}\|} = \dfrac{\left|\mathbf{u}\cdot(\mathbf{v}\times\mathbf{w})\right|}{\|\mathbf{u}\times\mathbf{v}\|}.$

19. $x^2 + y^2 = 1$ cylinder

$\quad z = 2y$ plane

Introduce a coordinate system in the plane $z = 2y$.

The new u-axis is the original x-axis.

The new v-axis is the line $z = 2y, x = 0$.

Then the intersection of the cylinder and plane satisfies the equation of an ellipse:

$x^2 + y^2 = 1$

$x^2 + \left(\dfrac{z}{2}\right)^2 = 1$

$x^2 + \dfrac{z^2}{4} = 1$ ellipse

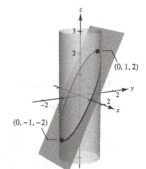

CHAPTER 12
Vector-Valued Functions

CHAPTER 12
Vector-Valued Functions

Section 12.1 Vector-Valued Functions

1. You can use a vector-valued function to trace the graph of a curve. Recall that the terminal point of the position vector $\mathbf{r}(t)$ coincides with a point on the curve.

3. $\mathbf{r}(t) = \dfrac{1}{t+1}\mathbf{i} + \dfrac{t}{2}\mathbf{j} - 3t\mathbf{k}$

   Component functions: $f(t) = \dfrac{1}{t+1}$

   $g(t) = \dfrac{t}{2}$

   $h(t) = -3t$

   Domain: $(-\infty, -1) \cup (-1, \infty)$

5. $\mathbf{r}(t) = \ln t\mathbf{i} - e^t\mathbf{j} - t\mathbf{k}$

   Component functions: $f(t) = \ln t$

   $g(t) = -e^t$

   $h(t) = -t$

   Domain: $(0, \infty)$

7. $\mathbf{r}(t) = \mathbf{F}(t) + \mathbf{G}(t)$

   $= \left(\cos t\mathbf{i} - \sin t\mathbf{j} + \sqrt{t}\mathbf{k}\right) + (\cos t\mathbf{i} + \sin t\mathbf{j})$

   $= 2\cos t\mathbf{i} + \sqrt{t}\mathbf{k}$

   Domain: $[0, \infty)$

9. $\mathbf{r}(t) = \mathbf{F}(t) \times \mathbf{G}(t) = \begin{vmatrix} \mathbf{i} & \mathbf{j} & \mathbf{k} \\ \sin t & \cos t & 0 \\ 0 & \sin t & \cos t \end{vmatrix} = \cos^2 t\mathbf{i} - \sin t \cos t\mathbf{j} + \sin^2 t\mathbf{k}$

   Domain: $(-\infty, \infty)$

11. $\mathbf{r}(t) = \frac{1}{2}t^2\mathbf{i} - (t-1)\mathbf{j}$

   (a) $\mathbf{r}(1) = \frac{1}{2}\mathbf{i}$

   (b) $\mathbf{r}(0) = \mathbf{j}$

   (c) $\mathbf{r}(s+1) = \frac{1}{2}(s+1)^2\mathbf{i} - (s+1-1)\mathbf{j} = \frac{1}{2}(s+1)^2\mathbf{i} - s\mathbf{j}$

   (d) $\mathbf{r}(2+\Delta t) - \mathbf{r}(2) = \frac{1}{2}(2+\Delta t)^2\mathbf{i} - (2+\Delta t - 1)\mathbf{j} - (2\mathbf{i} - \mathbf{j}) = \left(2 + 2\Delta t + \frac{1}{2}(\Delta t)^2\right)\mathbf{i} - (1+\Delta t)\mathbf{j} - 2\mathbf{i} + \mathbf{j}$

   $= \left(2\Delta t + \frac{1}{2}(\Delta t)^2\right)\mathbf{i} - (\Delta t)\mathbf{j} = \frac{1}{2}\Delta t(\Delta t + 4)\mathbf{i} - \Delta t\mathbf{j}$

13. $P(0, 0, 0),\ Q(5, 2, 2)$

   $\mathbf{v} = \overrightarrow{PQ} = \langle 5, 2, 2 \rangle$

   $\mathbf{r}(t) = 5t\mathbf{i} + 2t\mathbf{j} + 2t\mathbf{k},\ 0 \le t \le 1$

   $x = 5t,\ y = 2t,\ z = 2t,\ 0 \le t \le 1,$ Parametric equation

   (Answers may vary.)

15. $P(-3, -6, -1),\ Q(-1, -9, -8)$

   $\mathbf{v} = \overrightarrow{PQ} = \langle -1 + 3, -9 + 6, -8 + 1 \rangle = \langle 2, -3, -7 \rangle$

   $\mathbf{r}(t) = (-3 + 2t)\mathbf{i} + (-6 - 3t)\mathbf{j} + (-1 - 7t)\mathbf{k},\ 0 \le t \le 1$

   $x = -3 + 2t,\ y = -6 - 3t,\ z = -1 - 7t,\ 0 \le t \le 1,$ Parametric equation

   (Answers may vary.)

17. $\mathbf{r}(t) \cdot \mathbf{u}(t) = (3t - 1)(t^2) + \left(\frac{1}{4}t^3\right)(-8) + 4(t^3) = 3t^3 - t^2 - 2t^3 + 4t^3 = 5t^3 - t^2$, a scalar.

No, the dot product is a scalar-valued function.

19. $\mathbf{r}(t) = t\mathbf{i} + 2t\mathbf{j} + t^2\mathbf{k}, \ -2 \le t \le 2$

$x = t, \ y = 2t, \ z = t^2$

So, $z = x^2$. Matches (b)

20. $\mathbf{r}(t) = \cos(\pi t)\mathbf{i} + \sin(\pi t)\mathbf{j} + t^2\mathbf{k}, \ -1 \le t \le 1$

$x = \cos(\pi t), \ y = \sin(\pi t), \ z = t^2$

So, $x^2 + y^2 = 1$. Matches (c)

21. $\mathbf{r}(t) = t\mathbf{i} + t^2\mathbf{j} + e^{0.75t}\mathbf{k}, \ -2 \le t \le 2$

$x = t, \ y = t^2, \ z = e^{0.75t}$

So, $y = x^2$. Matches (d)

22. $\mathbf{r}(t) = t\mathbf{i} + \ln t\mathbf{j} + \dfrac{2t}{3}\mathbf{k}, \ 0.1 \le t \le 5$

$x = t, \ y = \ln t, \ z = \dfrac{2t}{3}$

So, $z = \frac{2}{3}x$ and $y = \ln x$. Matches (a)

23. $x = \dfrac{t}{4} \Rightarrow t = 4x$

$y = t - 1$

$y = 4x - 1$

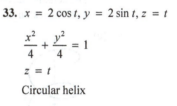

25. $x = t^3, \ y = t^2$

$y = x^{2/3}$

27. $x = \cos\theta, \ y = 3\sin\theta$

$x^2 + \dfrac{y^2}{9} = 1$, Ellipse

29. $x = 3\sec\theta, \ y = 2\tan\theta$

$\dfrac{x^2}{9} = \dfrac{y^2}{4} + 1$, Hyperbola

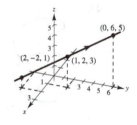

31. $x = -t + 1$

$y = 4t + 2$

$z = 2t + 3$

Line passing through
the points: $(0, 6, 5), (1, 2, 3)$

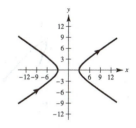

33. $x = 2\cos t, \ y = 2\sin t, \ z = t$

$\dfrac{x^2}{4} + \dfrac{y^2}{4} = 1$

$z = t$

Circular helix

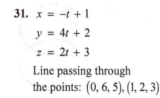

35. $x = 2\sin t, \ y = 2\cos t, \ z = e^{-t}$

$x^2 + y^2 = 4$

$z = e^{-t}$

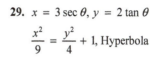

37. $x = t, \ y = t^2, \ z = \frac{2}{3}t^3$

$y = x^2, \ z = \frac{2}{3}x^3$

t	-2	-1	0	1	2
x	-2	-1	0	1	2
y	4	1	0	1	4
z	$-\frac{16}{3}$	$-\frac{2}{3}$	0	$\frac{2}{3}$	$\frac{16}{3}$

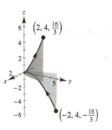

39. $\mathbf{r}(t) = -\dfrac{1}{2}t^2\mathbf{i} + t\mathbf{j} - \dfrac{\sqrt{3}}{2}t^2\mathbf{k}$

Parabola

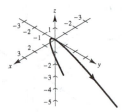

41.

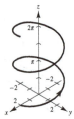

(a)

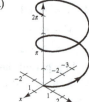

The helix is translated
2 units back on the x-axis.

(b)

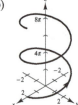

The height of the helix
increases at a faster rate.

(c)

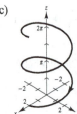

The orientation of the
helix is reversed.

(d)

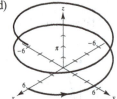

The radius of the helix is
increased from 2 to 6.

43. $\mathbf{u}(t) = \mathbf{r}(t) + 2\mathbf{k} = 3t^2\mathbf{i} + (t-1)\mathbf{j} + (t+2)\mathbf{k}$

45. $\mathbf{u}(t) = 3t^2\mathbf{i} + 2(t-1)\mathbf{j} + t\mathbf{k}$

47. $y = x + 5$

Let $x = t$, then $y = t + 5$

$\mathbf{r}(t) = t\mathbf{i} + (t+5)\mathbf{j}$

49. $y = (x-2)^2$

Let $x = t$, then $y = (t-2)^2$.

$\mathbf{r}(t) = t\mathbf{i} + (t-2)^2\mathbf{j}$

51. $x^2 + y^2 = 25$

Let $x = 5\cos t$, then $y = 5\sin t$.

$\mathbf{r}(t) = 5\cos t\mathbf{i} + 5\sin t\mathbf{j}$

53. $\dfrac{x^2}{16} - \dfrac{y^2}{4} = 1$

Let $x = 4\sec t$, $y = 2\tan t$.

$\mathbf{r}(t) = 4\sec t\mathbf{i} + 2\tan t\mathbf{j}$

55. $z = x^2 + y^2$, $x + y = 0$

Let $x = t$, then $y = -x = -t$

and $z = x^2 + y^2 = 2t^2$.

So, $x = t$, $y = -t$, $z = 2t^2$.

$\mathbf{r}(t) = t\mathbf{i} - t\mathbf{j} + 2t^2\mathbf{k}$

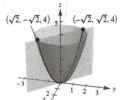

57. $x^2 + y^2 = 4$, $z = x^2$

$x = 2 \sin t$, $y = 2 \cos t$

$z = x^2 = 4 \sin^2 t$

t	0	$\dfrac{\pi}{6}$	$\dfrac{\pi}{4}$	$\dfrac{\pi}{2}$	$\dfrac{3\pi}{4}$	π
x	0	1	$\sqrt{2}$	2	$\sqrt{2}$	0
y	2	$\sqrt{3}$	$\sqrt{2}$	0	$-\sqrt{2}$	-2
z	0	1	2	4	2	0

$\mathbf{r}(t) = 2 \sin t \mathbf{i} + 2 \cos t \mathbf{j} + 4 \sin^2 t \mathbf{k}$

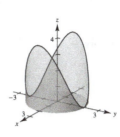

59. $x^2 + y^2 + z^2 = 4$, $x + z = 2$

Let $x = 1 + \sin t$, then $z = 2 - x = 1 - \sin t$ and $x^2 + y^2 + z^2 = 4$.

$(1 + \sin t)^2 + y^2 + (1 - \sin t)^2 = 2 + 2 \sin^2 t + y^2 = 4$

$y^2 = 2 \cos^2 t$, $y = \pm\sqrt{2} \cos t$

$x = 1 + \sin t$, $y = \pm\sqrt{2} \cos t$

$z = 1 - \sin t$

$\mathbf{r}(t) = (1 + \sin t)\mathbf{i} + \sqrt{2} \cos t \mathbf{j} - (1 - \sin t)\mathbf{k}$ and

$\mathbf{r}(t) = (1 + \sin t)\mathbf{i} - \sqrt{2} \cos t \mathbf{j} + (1 - \sin t)\mathbf{k}$

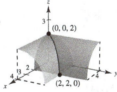

t	$-\dfrac{\pi}{2}$	$-\dfrac{\pi}{6}$	0	$\dfrac{\pi}{6}$	$\dfrac{\pi}{2}$
x	0	$\dfrac{1}{2}$	1	$\dfrac{3}{2}$	2
y	0	$\pm\dfrac{\sqrt{6}}{2}$	$\pm\sqrt{2}$	$\pm\dfrac{\sqrt{6}}{2}$	0
z	2	$\dfrac{3}{2}$	1	$\dfrac{1}{2}$	0

61. $x^2 + z^2 = 4$, $y^2 + z^2 = 4$

Subtracting, you have $x^2 - y^2 = 0$ or $y = \pm x$.

So, in the first octant, if you let $x = t$,

then $x = t$, $y = t$, $z = \sqrt{4 - t^2}$.

$\mathbf{r}(t) = t\mathbf{i} + t\mathbf{j} + \sqrt{4 - t^2}\mathbf{k}$

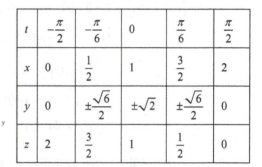

63. $y^2 + z^2 = (2t \cos t)^2 + (2t \sin t)^2 = 4t^2 = 4x^2$

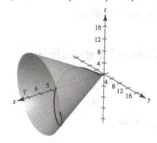

65. $\displaystyle\lim_{t \to \pi} (t\mathbf{i} + \cos t\mathbf{j} + \sin t\mathbf{k}) = \pi\mathbf{i} - \mathbf{j}$

67. $\displaystyle\lim_{t \to 0}\left[t^2\mathbf{i} + 3t\mathbf{j} + \frac{1 - \cos t}{t}\mathbf{k}\right] = 0$

because $\displaystyle\lim_{t \to 0}\frac{1 - \cos t}{t} = \lim_{t \to 0}\frac{\sin t}{1} = 0$.

(L'Hôpital's Rule)

69. $\displaystyle\lim_{t \to 0}\left[e^t\mathbf{i} + \frac{\sin t}{t}\mathbf{j} + e^{-t}\mathbf{k}\right] = \mathbf{i} + \mathbf{j} + \mathbf{k}$

because $\displaystyle\lim_{t \to 0}\frac{\sin t}{t} = \lim_{t \to 0}\frac{\cos t}{1} = 1$ (L'Hôpital's Rule)

71. $\mathbf{r}(t) = \dfrac{1}{2t + 1}\mathbf{i} + \dfrac{1}{t}\mathbf{j}$

The function is not continuous at $t = -\dfrac{1}{2}$ and $t = 0$. So,

$\mathbf{r}(t)$ is continuous on the intervals $\left(-\infty, -\dfrac{1}{2}\right), \left(-\dfrac{1}{2}, 0\right)$,

and $(0, \infty)$.

73. $\mathbf{r}(t) = t\mathbf{i} + \arcsin t\mathbf{j} + (t - 1)\mathbf{k}$

Continuous on $[-1, 1]$

75. $\mathbf{r}(t) = \left\langle e^{-t}, t^2, \tan t \right\rangle$

Discontinuous at $t = \dfrac{\pi}{2} + n\pi$

Continuous on $\left(-\dfrac{\pi}{2} + n\pi, \dfrac{\pi}{2} + n\pi \right)$

77. Yes. $\mathbf{r}(t)$ represents a line. Answers will vary.

79. Answers will vary. *Sample answer*:

$$\mathbf{r}(t) = \begin{cases} \mathbf{i} + \mathbf{j}, \ t \geq 3 \\ -\mathbf{i} + \mathbf{j}, \ t < 3 \end{cases}$$

81. $\mathbf{r}(t) = \cos t\mathbf{i} + \sin t\mathbf{j} + \dfrac{1}{\pi}t\mathbf{k}, \ 0 \leq t \leq 4\pi$

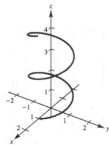

83. Let $\mathbf{r}(t) = x_1(t)\mathbf{i} + y_1(t)\mathbf{j} + z_1(t)\mathbf{k}$ and $\mathbf{u}(t) = x_2(t)\mathbf{i} + y_2(t)\mathbf{j} + z_2(t)\mathbf{k}$. Then:

$$\lim_{t \to c}\left[\mathbf{r}(t) \times \mathbf{u}(t)\right] = \lim_{t \to c}\left\{\left[y_1(t)z_2(t) - y_2(t)z_1(t)\right]\mathbf{i} - \left[x_1(t)z_2(t) - x_2(t)z_1(t)\right]\mathbf{j} + \left[x_1(t)y_2(t) - x_2(t)y_1(t)\right]\mathbf{k}\right\}$$

$$= \left[\lim_{t \to c} y_1(t) \lim_{t \to c} z_2(t) - \lim_{t \to c} y_2(t) \lim_{t \to c} z_1(t)\right]\mathbf{i} - \left[\lim_{t \to c} x_1(t) \lim_{t \to c} z_2(t) - \lim_{t \to c} x_2(t) \lim_{t \to c} z_1(t)\right]\mathbf{j}$$

$$+ \left[\lim_{t \to c} x_1(t) \lim_{t \to c} y_2(t) - \lim_{t \to c} x_2(t) \lim_{t \to c} y_1(t)\right]\mathbf{k}$$

$$= \left[\lim_{t \to c} x_1(t)\mathbf{i} + \lim_{t \to c} y_1(t)\mathbf{j} + \lim_{t \to c} z_1(t)\mathbf{k}\right] \times \left[\lim_{t \to c} x_2(t)\mathbf{i} + \lim_{t \to c} y_2(t)\mathbf{j} + \lim_{t \to c} z_2(t)\mathbf{k}\right]$$

$$= \lim_{t \to c} \mathbf{r}(t) \times \lim_{t \to c} \mathbf{u}(t)$$

85. Let $\mathbf{r}(t) = x(t)\mathbf{i} + y(t)\mathbf{j} + z(t)\mathbf{k}$. Because $\mathbf{r}$ is

continuous at $t = c$, then $\lim\limits_{t \to c} \mathbf{r}(t) = \mathbf{r}(c)$.

$\mathbf{r}(c) = x(c)\mathbf{i} + y(c)\mathbf{j} + z(c)\mathbf{k} \Rightarrow x(c), y(c), z(c)$

are defined at c.

$$\|\mathbf{r}\| = \sqrt{\left(x(t)\right)^2 + \left(y(t)\right)^2 + \left(z(t)\right)^2}$$

$$\lim_{t \to c} \|\mathbf{r}\| = \sqrt{\left(x(c)\right)^2 + \left(y(c)\right)^2 + \left(z(c)\right)^2} = \|\mathbf{r}(c)\|$$

So, $\|\mathbf{r}\|$ is continuous at c.

87. No, not necessarily. See Exercise 90.

89. $\mathbf{r}(t) = t^2\mathbf{i} + (9t - 20)\mathbf{j} + t^2\mathbf{k}$

$\mathbf{u}(s) = (3s + 4)\mathbf{i} + s^2\mathbf{j} + (5s - 4)\mathbf{k}$.

Equating components:

$$t^2 = 3s + 4$$
$$9t - 20 = s^2$$
$$t^2 = 5s - 4$$

So, $3s + 4 = 5s - 4 \Rightarrow s = 4$

$9t - 20 = s^2 = 16 \Rightarrow t = 4$.

The paths intersect at the same time $t = 4$ at the point $(16, 16, 16)$. The particles collide.

Section 12.2   Differentiation and Integration of Vector-Valued Functions

1. $\mathbf{r}'(t_0)$ represents the vector that is tangent to the curve represented by $\mathbf{r}(t)$ at the point t_0.

3. $\mathbf{r}(t) = (1 - t^2)\mathbf{i} + t\mathbf{j}, \ t_0 = 3$

$x(t) = 1 - t^2, \ y(t) = t$

$x = 1 - y^2$

$\mathbf{r}(3) = -8\mathbf{i} + 3\mathbf{j}$

$\mathbf{r}'(t) = -2t\mathbf{i} + \mathbf{j}$

$\mathbf{r}'(3) = -6\mathbf{i} + \mathbf{j}$

$\mathbf{r}'(t_0)$ is tangent to the curve at $t_0 = 3$.

5. $\mathbf{r}(t) = \cos t\mathbf{i} + \sin t\mathbf{j}, \ t_0 = \dfrac{\pi}{2}$

$x(t) = \cos t, \ y(t) = \sin t$

$x^2 + y^2 = 1$

$\mathbf{r}\left(\dfrac{\pi}{2}\right) = \mathbf{j}$

$\mathbf{r}'(t) = -\sin t\mathbf{i} + \cos t\mathbf{j}$

$\mathbf{r}'\left(\dfrac{\pi}{2}\right) = -\mathbf{i}$

$\mathbf{r}'(t_0)$ is tangent to the curve at t_0.

7. $\mathbf{r}(t) = \langle e^t, e^{2t} \rangle$, $t_0 = 0$

$x(t) = e^t$, $y(t) = e^{2t} = (e^t)^2$

$y = x^2$, $x > 0$

$\mathbf{r}(0) = \langle 1, 1 \rangle$

$\mathbf{r}'(t) = \langle e^t, 2e^{2t} \rangle$

$\mathbf{r}'(0) = \langle 1, 2 \rangle$

$\mathbf{r}'(t_0)$ is tangent to the curve at t_0.

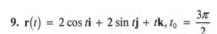

9. $\mathbf{r}(t) = 2\cos t\mathbf{i} + 2\sin t\mathbf{j} + t\mathbf{k}$, $t_0 = \dfrac{3\pi}{2}$

$x^2 + y^2 = 4$, $z = t$

$\mathbf{r}'(t) = -2\sin t\mathbf{i} + 2\cos t\mathbf{j} + \mathbf{k}$

$\mathbf{r}\left(\dfrac{3\pi}{2}\right) = -2\mathbf{j} + \dfrac{3\pi}{2}\mathbf{k}$

$\mathbf{r}'\left(\dfrac{3\pi}{2}\right) = 2\mathbf{i} + \mathbf{k}$

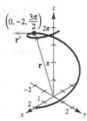

11. $\mathbf{r}(t) = t^4\mathbf{i} - 5t\mathbf{j}$

$\mathbf{r}'(t) = 4t^3\mathbf{i} - 5\mathbf{j}$

13. $\mathbf{r}(t) = 3\cos^3 t\mathbf{i} + 2\sin^3 t\mathbf{j} + \mathbf{k}$

$\mathbf{r}'(t) = 9\cos^2 t(-\sin t)\mathbf{i} + 6\sin^2 t(\cos t)\mathbf{j}$

$= -9\sin t\cos^2 t\mathbf{i} + 6\sin^2 t\cos t\mathbf{j}$

15. $\mathbf{r}(t) = e^{-t}\mathbf{i} + 4\mathbf{j} + 5te^t\mathbf{k}$

$\mathbf{r}'(t) = -e^{-t}\mathbf{i} + (5e^t + 5te^t)\mathbf{k}$

17. $\mathbf{r}(t) = \langle t\sin t, t\cos t, t \rangle$

$\mathbf{r}'(t) = \langle \sin t + t\cos t, \cos t - t\sin t, 1 \rangle$

19. $\mathbf{r}(t) = t^3\mathbf{i} + \dfrac{1}{2}t^2\mathbf{j}$

  (a) $\mathbf{r}'(t) = 3t^2\mathbf{i} + t\mathbf{j}$

  (b) $\mathbf{r}''(t) = 6t\mathbf{i} + \mathbf{j}$

  (c) $\mathbf{r}'(t) \cdot \mathbf{r}''(t) = 3t^2(6t) + t = 18t^3 + t$

21. $\mathbf{r}(t) = 4\cos t\mathbf{i} + 4\sin t\mathbf{j}$

  (a) $\mathbf{r}'(t) = -4\sin t\mathbf{i} + 4\cos t\mathbf{j}$

  (b) $\mathbf{r}''(t) = -4\cos t\mathbf{i} - 4\sin t\mathbf{j}$

  (c) $\mathbf{r}'(t) \cdot \mathbf{r}''(t) = (-4\sin t)(-4\cos t) + 4\cos t(-4\sin t) = 0$

23. $\mathbf{r}(t) = \dfrac{1}{2}t^2\mathbf{i} - t\mathbf{j} + \dfrac{1}{6}t^3\mathbf{k}$

  (a) $\mathbf{r}'(t) = t\mathbf{i} - \mathbf{j} + \dfrac{1}{2}t^2\mathbf{k}$

  (b) $\mathbf{r}''(t) = \mathbf{i} + t\mathbf{k}$

  (c) $\mathbf{r}'(t) \cdot \mathbf{r}''(t) = t(1) + (-1)(0) + \dfrac{1}{2}t^2(t) = t + \dfrac{1}{2}t^3$

  (d) $\mathbf{r}'(t) \times \mathbf{r}''(t) = \begin{vmatrix} \mathbf{i} & \mathbf{j} & \mathbf{k} \\ t & -1 & \dfrac{1}{2}t^2 \\ 1 & 0 & t \end{vmatrix} = (-t)\mathbf{i} - \left(t^2 - \dfrac{1}{2}t^2\right)\mathbf{j} + \mathbf{k} = -t\mathbf{i} - \dfrac{1}{2}t^2\mathbf{j} + \mathbf{k}$

25. $\mathbf{r}(t) = \langle \cos t + t\sin t, \sin t - t\cos t, t \rangle$

  (a) $\mathbf{r}'(t) = \langle -\sin t + \sin t + t\cos t, \cos t - \cos t + t\sin t, 1 \rangle = \langle t\cos t, t\sin t, 1 \rangle$

  (b) $\mathbf{r}''(t) = \langle \cos t - t\sin t, \sin t + t\cos t, 0 \rangle$

  (c) $\mathbf{r}'(t) \cdot \mathbf{r}''(t) = (t\cos t)(\cos t - t\sin t) + t\sin t(\sin t + t\cos t) + 1(0)$

  $= t\cos^2 t - t^2\cos t\sin t + t\sin^2 t + t^2\sin t\cos t$

  $= t(\cos^2 t + \sin^2 t) = t$

(d) $\mathbf{r}'(t) \times \mathbf{r}''(t) = \begin{bmatrix} \mathbf{i} & \mathbf{j} & \mathbf{k} \\ t\cos t & t\sin t & 1 \\ \cos t - t\sin t & \sin t + t\cos t & 0 \end{bmatrix}$

$\qquad = (-\sin t - t\cos t)\mathbf{i} + (\cos t - t\sin t)\mathbf{j} + (t\cos t\sin t + t^2\cos^2 t - t\sin t\cos t + t^2\sin^2 t)\mathbf{k}$

$\qquad = (-\sin t - t\cos t)\mathbf{i} + (\cos t - t\sin t)\mathbf{j} + t^2\mathbf{k}$

$\qquad = \langle -\sin t - t\cos t, \cos t - t\sin t, t^2 \rangle$

27. $\mathbf{r}(t) = t^2\mathbf{i} + t^3\mathbf{j}$

$\mathbf{r}'(t) = 2t\mathbf{i} + 3t^2\mathbf{j}$

$\mathbf{r}'(0) = \mathbf{0}$

Smooth on $(-\infty, 0), (0, \infty)$

(Note: $t = 0$ corresponds to a cusp at $(x, y) = (0, 0)$.)

29. $\mathbf{r}(\theta) = 2\cos^3\theta\mathbf{i} + 3\sin^3\theta\mathbf{j}$

$\mathbf{r}'(\theta) = -6\cos^2\theta\sin\theta\mathbf{i} + 9\sin^2\theta\cos\theta\mathbf{j}$

$\mathbf{r}'\left(\dfrac{n\pi}{2}\right) = \mathbf{0}$

Smooth on $\left(\dfrac{n\pi}{2}, \dfrac{(n+1)\pi}{2}\right)$, n any integer.

31. $\mathbf{r}(t) = \dfrac{2t}{8+t^3}\mathbf{i} + \dfrac{2t^2}{8+t^3}\mathbf{j}$

$\mathbf{r}'(t) = \dfrac{16 - 4t^3}{(t^3+8)^2}\mathbf{i} + \dfrac{32t - 2t^4}{(t^3+8)^2}\mathbf{j}$

$\mathbf{r}'(t) \neq \mathbf{0}$ for any value of t.

$\mathbf{r}$ is not continuous when $t = -2$.

Smooth on $(-\infty, -2), (-2, \infty)$

33. $\mathbf{r}(t) = t\mathbf{i} - 3t\mathbf{j} + \tan t\mathbf{k}$

$\mathbf{r}'(t) = \mathbf{i} - 3\mathbf{j} + \sec^2 t\mathbf{k} \neq \mathbf{0}$

$\mathbf{r}$ is smooth for all $t \neq \dfrac{\pi}{2} + n\pi = \dfrac{2n+1}{2}\pi$.

Smooth on intervals of form $\left(-\dfrac{\pi}{2} + n\pi, \dfrac{\pi}{2} + n\pi\right)$, n is an integer.

35. $\mathbf{r}(t) = t\mathbf{i} + 3t\mathbf{j} + t^2\mathbf{k}$, $\mathbf{u}(t) = 4t\mathbf{i} + t^2\mathbf{j} + t^3\mathbf{k}$

$\mathbf{r}'(t) = \mathbf{i} + 3\mathbf{j} + 2t\mathbf{k}$, $\mathbf{u}'(t) = 4\mathbf{i} + 2t\mathbf{j} + 3t^2\mathbf{k}$

(a) $\mathbf{r}'(t) = \mathbf{i} + 3\mathbf{j} + 2t\mathbf{k}$

(b) $\dfrac{d}{dt}\left[3\mathbf{r}(t) - \mathbf{u}(t)\right] = 3\mathbf{r}'(t) - \mathbf{u}'(t) = 3(\mathbf{i} + 3\mathbf{j} + 2t\mathbf{k}) - (4\mathbf{i} + 2t\mathbf{j} + 3t^2\mathbf{k})$

$\qquad = (3 - 4)\mathbf{i} + (9 - 2t)\mathbf{j} + (6t - 3t^2)\mathbf{k}$

$\qquad = -\mathbf{i} + (9 - 2t)\mathbf{j} + (6t - 3t^2)\mathbf{k}$

(c) $\dfrac{d}{dt}(5t)\mathbf{u}(t) = (5t)\mathbf{u}'(t) + 5\mathbf{u}(t) = 5t(4\mathbf{i} + 2t\mathbf{j} + 3t^2\mathbf{k}) + 5(4t\mathbf{i} + t^2\mathbf{j} + t^3\mathbf{k})$

$\qquad = (20t + 20t)\mathbf{i} + (10t^2 + 5t^2)\mathbf{j} + (15t^3 + 5t^3)\mathbf{k}$

$\qquad = 40t\mathbf{i} + 15t^2\mathbf{j} + 20t^3\mathbf{k}$

(d) $\dfrac{d}{dt}\left[\mathbf{r}(t) \cdot \mathbf{u}(t)\right] = \mathbf{r}(t) \cdot \mathbf{u}'(t) + \mathbf{r}'(t) \cdot \mathbf{u}(t) = \left[(t)(4) + (3t)(2t) + (t^2)(3t^2)\right] + \left[(1)(4t) + (3)(t^2) + (2t)(t^3)\right]$

$\qquad = (4t + 6t^2 + 3t^4) + (4t + 3t^2 + 2t^4)$

$\qquad = 8t + 9t^2 + 5t^4$

(e) $\dfrac{d}{dt}\left[\mathbf{r}(t) \times \mathbf{u}(t)\right] = \mathbf{r}(t) \times \mathbf{u}'(t) + \mathbf{r}'(t) \times \mathbf{u}(t) = \left[7t^3\mathbf{i} + (4t^2 - 3t^3)\mathbf{j} + (2t^2 - 12t)\mathbf{k}\right] + \left[t^3\mathbf{i} + (8t^2 - t^3)\mathbf{j} + (t^2 - 12t)\mathbf{k}\right]$

$\qquad = 8t^3\mathbf{i} + (12t^2 - 4t^3)\mathbf{j} + (3t^2 - 24t)\mathbf{k}$

(f) $\dfrac{d}{dt}\mathbf{r}(2t) = 2\mathbf{r}'(2t) = 2\left[\mathbf{i} + 3\mathbf{j} + 2(2t)\mathbf{k}\right] = 2\mathbf{i} + 6\mathbf{j} + 8t\mathbf{k}$

37. $\mathbf{r}(t) = t\mathbf{i} + 2t^2\mathbf{j} + t^3\mathbf{k}, \mathbf{u}(t) = t^4\mathbf{k}$

   (a) $\mathbf{r}(t) \cdot \mathbf{u}(t) = t^7$

      (i) $D_t\big[\mathbf{r}(t) \cdot \mathbf{u}(t)\big] = 7t^6$

      (ii) Alternate Solution:

$$D_t\big[\mathbf{r}(t) \cdot \mathbf{u}(t)\big] = \mathbf{r}(t) \cdot \mathbf{u}'(t) + \mathbf{r}'(t) \cdot \mathbf{u}(t) = \big(t\mathbf{i} + 2t^2\mathbf{j} + t^3\mathbf{k}\big) \cdot \big(4t^3\mathbf{k}\big) + \big(\mathbf{i} + 4t\mathbf{j} + 3t^2\mathbf{k}\big) \cdot \big(t^4\mathbf{k}\big) = 4t^6 + 3t^6 = 7t^6$$

   (b) $\mathbf{r}(t) \times \mathbf{u}(t) = \begin{vmatrix} \mathbf{i} & \mathbf{j} & \mathbf{k} \\ t & 2t^2 & t^3 \\ 0 & 0 & t^4 \end{vmatrix} = 2t^6\mathbf{i} - t^5\mathbf{j}$

      (i) $D_t\big[\mathbf{r}(t) \times \mathbf{u}(t)\big] = 12t^5\mathbf{i} - 5t^4\mathbf{j}$

      (ii) Alternate Solution: $D_t\big[\mathbf{r}(t) \times \mathbf{u}(t)\big] = \mathbf{r}(t) \times \mathbf{u}'(t) \times \mathbf{r}'(t) \times \mathbf{u}(t) = \begin{vmatrix} \mathbf{i} & \mathbf{j} & \mathbf{k} \\ t & 2t^2 & t^3 \\ 0 & 0 & 4t^3 \end{vmatrix} + \begin{vmatrix} \mathbf{i} & \mathbf{j} & \mathbf{k} \\ 1 & 4t & 3t^2 \\ 0 & 0 & t^4 \end{vmatrix} = 12t^5\mathbf{i} - 5t^4\mathbf{j}$

39. $\int (2t\mathbf{i} + \mathbf{j} + 9\mathbf{k})\,dt = t^2\mathbf{i} + t\mathbf{j} + 9t\mathbf{k} + \mathbf{C}$

41. $\int \left(\dfrac{1}{t}\mathbf{i} + \mathbf{j} - t^{3/2}\mathbf{k}\right) dt = \ln|t|\mathbf{i} + t\mathbf{j} - \dfrac{2}{5}t^{5/2}\mathbf{k} + \mathbf{C}$

43. $\int \big(\mathbf{i} + 4t^3\mathbf{j} + 5^t\mathbf{k}\big) dt = t\mathbf{i} + t^4\mathbf{j} + \dfrac{5^t}{\ln 5}\mathbf{k} + \mathbf{C}$

45. $\int \big(e^t\mathbf{i} + \mathbf{j} + t\cos t\mathbf{k}\big) dt = e^t\mathbf{i} + t\mathbf{j} + (\cos t + t\sin t)\mathbf{k} + \mathbf{C}$

   (Note: Integration by parts needed for $\int t\cos t\, dt$.)

47. $\int_0^1 (8t\mathbf{i} + t\mathbf{j} - \mathbf{k})\, dt = \Big[4t^2\mathbf{i}\Big]_0^1 + \left[\dfrac{t^2}{2}\mathbf{j}\right]_0^1 - \big[t\mathbf{k}\big]_0^1 = 4\mathbf{i} + \dfrac{1}{2}\mathbf{j} - \mathbf{k}$

49. $\int_0^{\pi/2} (5\cos t\mathbf{i} + 6\sin t\mathbf{j} + \mathbf{k})\, dt = \big[5\sin t\mathbf{i} - 6\cos t\mathbf{j} + t\mathbf{k}\big]_0^{\pi/2} = \left(5\mathbf{i} + \dfrac{\pi}{2}\mathbf{k}\right) - (-6\mathbf{j}) = 5\mathbf{i} + 6\mathbf{j} + \dfrac{\pi}{2}\mathbf{k}$

51. $\int_0^2 \big(t\mathbf{i} + e^t\mathbf{j} - te^t\mathbf{k}\big) dt = \left[\dfrac{t^2}{2}\mathbf{i}\right]_0^2 + \Big[e^t\mathbf{j}\Big]_0^2 - \Big[(t-1)e^t\mathbf{k}\Big]_0^2 = 2\mathbf{i} + (e^2 - 1)\mathbf{j} - (e^2 + 1)\mathbf{k}$

53. $\mathbf{r}(t) = \int \big(4e^{2t}\mathbf{i} + 3e^t\mathbf{j}\big) dt = 2e^{2t}\mathbf{i} + 3e^t\mathbf{j} + \mathbf{C}$

   $\mathbf{r}(0) = 2\mathbf{i} + 3\mathbf{j} + \mathbf{C} = 2\mathbf{i} \Rightarrow \mathbf{C} = -3\mathbf{j}$

   $\mathbf{r}(t) = 2e^{2t}\mathbf{i} + 3(e^t - 1)\mathbf{j}$

55. $\mathbf{r}'(t) = \int -32\mathbf{j}\, dt = -32t\mathbf{j} + \mathbf{C}_1$

   $\mathbf{r}'(0) = \mathbf{C}_1 = 600\sqrt{3}\mathbf{i} + 600\mathbf{j}$

   $\mathbf{r}'(t) = 600\sqrt{3}\mathbf{i} + (600 - 32t)\mathbf{j}$

   $\mathbf{r}(t) = \int\big[600\sqrt{3}\mathbf{i} + (600 - 32t)\mathbf{j}\big] dt$

   $= 600\sqrt{3}t\mathbf{i} + \big(600t - 16t^2\big)\mathbf{j} + \mathbf{C}$

   $\mathbf{r}(0) = \mathbf{C} = 0$

   $\mathbf{r}(t) = 600\sqrt{3}t\mathbf{i} + \big(600t - 16t^2\big)\mathbf{j}$

57. $\mathbf{r}(t) = \int\big(te^{-t^2}\mathbf{i} - e^{-t}\mathbf{j} + \mathbf{k}\big) dt = -\dfrac{1}{2}e^{-t^2}\mathbf{i} + e^{-t}\mathbf{j} + t\mathbf{k} + \mathbf{C}$

   $\mathbf{r}(0) = -\dfrac{1}{2}\mathbf{i} + \mathbf{j} + \mathbf{C} = \dfrac{1}{2}\mathbf{i} - \mathbf{j} + \mathbf{k} \Rightarrow \mathbf{C} = \mathbf{i} - 2\mathbf{j} + \mathbf{k}$

   $\mathbf{r}(t) = \left(1 - \dfrac{1}{2}e^{-t^2}\right)\mathbf{i} + (e^{-t} - 2)\mathbf{j} + (t + 1)\mathbf{k}$

   $= \left(\dfrac{2 - e^{-t^2}}{2}\right)\mathbf{i} + (e^{-t} - 2)\mathbf{j} + (t + 1)\mathbf{k}$

59. At $t = t_0$, the graph of $\mathbf{u}(t)$ is increasing in the x, y, and z directions simultaneously.

61. Let $\mathbf{r}(t) = x(t)\mathbf{i} + y(t)\mathbf{j} + z(t)\mathbf{k}$. Then $c\mathbf{r}(t) = cx(t)\mathbf{i} + cy(t)\mathbf{j} + cz(t)\mathbf{k}$ and

$$\frac{d}{dt}\big[c\mathbf{r}(t)\big] = cx'(t)\mathbf{i} + cy'(t)\mathbf{j} + cz'(t)\mathbf{k} = c\big[x'(t)\mathbf{i} + y'(t)\mathbf{j} + z'(t)\mathbf{k}\big] = c\mathbf{r}'(t).$$

63. Let $\mathbf{r}(t) = x(t)\mathbf{i} + y(t)\mathbf{j} + z(t)\mathbf{k}$, then $w(t)\mathbf{r}(t) = w(t)x(t)\mathbf{i} + w(t)y(t)\mathbf{j} + w(t)z(t)\mathbf{k}$.

$$\frac{d}{dt}\big[w(t)\mathbf{r}(t)\big] = \big[w(t)x'(t) + w'(t)x(t)\big]\mathbf{i} + \big[w(t)y'(t) + w'(t)y(t)\big]\mathbf{j} + \big[w(t)z'(t) + w'(t)z(t)\big]\mathbf{k}$$

$$= w(t)\big[x'(t)\mathbf{i} + y'(t)\mathbf{j} + z'(t)\mathbf{k}\big] + w'(t)\big[x(t)\mathbf{i} + y(t)\mathbf{j} + z(t)\mathbf{k}\big] = w(t)\mathbf{r}'(t) + w'(t)\mathbf{r}(t)$$

65. Let $\mathbf{r}(t) = x(t)\mathbf{i} + y(t)\mathbf{j} + z(t)\mathbf{k}$. Then $\mathbf{r}(w(t)) = x(w(t))\mathbf{i} + y(w(t))\mathbf{j} + z(w(t))\mathbf{k}$ and

$$\frac{d}{dt}\big[\mathbf{r}(w(t))\big] = x'(w(t))w'(t)\mathbf{i} + y'(w(t))w'(t)\mathbf{j} + z'(w(t))w'(t)\mathbf{k} \quad \text{(Chain Rule)}$$

$$= w'(t)\big[x'(w(t))\mathbf{i} + y'(w(t))\mathbf{j} + z'(w(t))\mathbf{k}\big] = w'(t)\mathbf{r}'(w(t)).$$

67. Let $\mathbf{r}(t) = x_1(t)\mathbf{i} + y_1(t)\mathbf{j} + z_1(t)\mathbf{k}$, $\mathbf{u}(t) = x_2(t)\mathbf{i} + y_2(t)\mathbf{j} + z_2(t)\mathbf{k}$, and $\mathbf{v}(t) = x_3(t)\mathbf{i} + y_3(t)\mathbf{j} + z_3(t)\mathbf{k}$. Then:

$$\mathbf{r}(t) \cdot \big[\mathbf{u}(t) \times \mathbf{v}(t)\big] = x_1(t)\big[y_2(t)z_3(t) - z_2(t)y_3(t)\big] - y_1(t)\big[x_2(t)z_3(t) - z_2(t)x_3(t)\big] + z_1(t)\big[x_2(t)y_3(t) - y_2(t)x_3(t)\big]$$

$$\frac{d}{dt}\big[\mathbf{r}(t) \cdot (\mathbf{u}(t) \times \mathbf{v}(t))\big] = x_1(t)y_2(t)z_3'(t) + x_1(t)y_2'(t)z_3(t) + x_1'(t)y_2(t)z_3(t) - x_1(t)y_3(t)z_2'(t)$$

$$- x_1(t)y_3'(t)z_2(t) - x_1'(t)y_3(t)z_2(t) - y_1(t)x_2(t)z_3'(t) - y_1(t)x_2'(t)z_3(t) - y_1'(t)x_2(t)z_3(t)$$

$$+ y_1(t)z_2(t)x_3'(t) + y_1(t)z_2'(t)x_3(t) + y_1'(t)z_2(t)x_3(t) + z_1(t)x_2(t)y_3'(t) + z_1(t)x_2'(t)y_3(t)$$

$$+ z_1'(t)x_2(t)y_3(t) - z_1(t)y_2(t)x_3'(t) - z_1(t)y_2'(t)x_3(t) - z_1'(t)y_2(t)x_3(t)$$

$$= \Big\{x_1'(t)\big[y_2(t)z_3(t) - y_3(t)z_2(t)\big] + y_1'(t)\big[-x_2(t)z_3(t) + z_2(t)x_3(t)\big] + z_1'(t)\big[x_2(t)y_3(t) - y_2(t)x_3(t)\big]\Big\}$$

$$+ \Big\{x_1(t)\big[y_2'(t)z_3(t) - y_3(t)z_2'(t)\big] + y_1(t)\big[-x_2'(t)z_3(t) + z_2'(t)x_3(t)\big] + z_1(t)\big[x_2'(t)y_3(t) - y_2'(t)x_3(t)\big]\Big\}$$

$$+ \Big\{x_1(t)\big[y_2(t)z_3'(t) - y_3'(t)z_2(t)\big] + y_1(t)\big[-x_2(t)z_3'(t) + z_2(t)x_3'(t)\big] + z_1(t)\big[x_2(t)y_3'(t) - y_2(t)x_3'(t)\big]\Big\}$$

$$= \mathbf{r}'(t) \cdot \big[\mathbf{u}(t) \times \mathbf{v}(t)\big] + \mathbf{r}(t) \cdot \big[\mathbf{u}'(t) \times \mathbf{v}(t)\big] + \mathbf{r}(t) \cdot \big[\mathbf{u}(t) \times \mathbf{v}'(t)\big]$$

69. $\mathbf{r}(t) = (t - \sin t)\mathbf{i} + (1 - \cos t)\mathbf{j}$

(a)
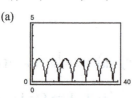

The curve is a cycloid.

(b) $\mathbf{r}'(t) = (1 - \cos t)\mathbf{i} + \sin t\,\mathbf{j}$

$\mathbf{r}''(t) = \sin t\,\mathbf{i} + \cos t\,\mathbf{j}$

$\|\mathbf{r}'(t)\| = \sqrt{1 - 2\cos t + \cos^2 t + \sin^2 t} = \sqrt{2 - 2\cos t}$

Minimum of $\|\mathbf{r}'(t)\|$ is 0, $(t = 0)$.

Maximum of $\|\mathbf{r}'(t)\|$ is 2, $(t = \pi)$.

$\|\mathbf{r}''(t)\| = \sqrt{\sin^2 t + \cos^2 t} = 1$

Minimum and maximum of $\|\mathbf{r}''(t)\|$ is 1.

71. $\mathbf{r}(t) = e^t \sin t\mathbf{i} + e^t \cos t\mathbf{j}$

$\mathbf{r}'(t) = \left(e^t \cos t + e^t \sin t\right)\mathbf{i} + \left(e^t \cos t - e^t \sin t\right)\mathbf{j}$

$\mathbf{r}''(t) = \left(-e^t \sin t + e^t \cos t + e^t \sin t + e^t \cos t\right)\mathbf{i} + \left(e^t \cos t - e^t \sin t - e^t \sin t - e^t \cos t\right)\mathbf{j} = 2e^t \cos t\mathbf{i} - 2e^t \sin t\mathbf{j}$

$\mathbf{r}(t) \cdot \mathbf{r}''(t) = 2e^{2t} \sin t \cos t - 2e^{2t} \sin t \cos t = 0$

So, $\mathbf{r}(t)$ is always perpendicular to $\mathbf{r}''(t)$.

73. True

75. False. Let $\mathbf{r}(t) = \cos t\mathbf{i} + \sin t\mathbf{j} + \mathbf{k}$.

$$\|\mathbf{r}(t)\| = \sqrt{2}$$

$$\frac{d}{dt}\big[\|\mathbf{r}(t)\|\big] = 0$$

$$\mathbf{r}'(t) = -\sin t\mathbf{i} + \cos t\mathbf{j}$$

$$\|\mathbf{r}'(t)\| = 1$$

Section 12.3   Velocity and Acceleration

1. The direction of the velocity vector provides the direction of motion at time t. The magnitude of the velocity vector provides the speed of the object.

3. $\mathbf{r}(t) = 3t\mathbf{i} + (t - 1)\mathbf{j}, \ (3, 0)$

   (a) $\mathbf{v}(t) = \mathbf{r}'(t) = 3\mathbf{i} + \mathbf{j}$

       Speed $= \|\mathbf{v}(t)\| = \sqrt{3^2 + 1^2} = \sqrt{10}$

       $\mathbf{a}(t) = \mathbf{r}''(t) = \mathbf{0}$

   (b) At $(3, 0), t = 1$.

       $\mathbf{v}(1) = 3\mathbf{i} + \mathbf{j}, \mathbf{a}(1) = \mathbf{0}$

   (c) $x = 3t, \ y = t - 1$

      $y = \dfrac{x}{3} - 1$, line

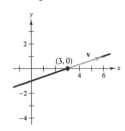

5. $\mathbf{r}(t) = t^2\mathbf{i} + t\mathbf{j}, \ (4, 2)$

   (a) $\mathbf{v}(t) = \mathbf{r}'(t) = 2t\mathbf{i} + \mathbf{j}$

       Speed $= \|\mathbf{v}(t)\| = \sqrt{4t^2 + 1}$

       $\mathbf{a}(t) = \mathbf{r}''(t) = 2\mathbf{i}$

   (b) At $(4, 2), t = 2$.

       $\mathbf{v}(2) = 4\mathbf{i} + \mathbf{j}, \mathbf{a}(2) = 2\mathbf{i}$

   (c) $x = t^2, \ y = t$

      $x = y^2$, parabola

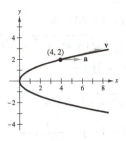

7. $\mathbf{r}(t) = 2 \cos t \mathbf{i} + 2 \sin t \mathbf{j}, \; \left(\sqrt{2}, \sqrt{2}\right)$

(a) $\mathbf{v}(t) = \mathbf{r}'(t) = -2 \sin t \mathbf{i} + 2 \cos t \mathbf{j}$

$\text{Speed} = \|\mathbf{v}(t)\| = \sqrt{4 \sin^2 t + 4 \cos^2 t} = 2$

$\mathbf{a}(t) = -2 \cos t \mathbf{i} - 2 \sin t \, \mathbf{j}$

(b) At $\left(\sqrt{2}, \sqrt{2}\right), t = \dfrac{\pi}{4}.$

$\mathbf{v}\!\left(\dfrac{\pi}{4}\right) = -\sqrt{2}\mathbf{i} + \sqrt{2}\mathbf{j}$

$\mathbf{a}\!\left(\dfrac{\pi}{4}\right) = -\sqrt{2}\mathbf{i} - \sqrt{2}\mathbf{j}$

(c) $x = 2 \cos t, \; y = 2 \sin t$

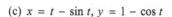

$x^2 + y^2 = 4$, circle

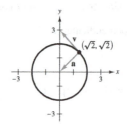

9. $\mathbf{r}(t) = \langle t - \sin t, 1 - \cos t \rangle, \; (\pi, 2)$

(a) $\mathbf{v}(t) = \mathbf{v}'(t) = \langle 1 - \cos t, \sin t \rangle$

$\text{Speed} = \|\mathbf{v}(t)\| = \sqrt{(1 - \cos t)^2 + \sin^2 t} = \sqrt{2 - 2 \cos t}$

$\mathbf{a}(t) = \langle \sin t, \cos t \rangle$

(b) At $(\pi, 2), t = \pi.$

$\mathbf{v}(\pi) = \langle 2, 0 \rangle, \; \mathbf{a}(\pi) = \langle 0, -1 \rangle$

(c) $x = t - \sin t, \; y = 1 - \cos t$

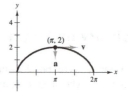

11. $\mathbf{r}(t) = t\mathbf{i} + 5t\mathbf{j} + 3t\mathbf{k}, \; t = 1$

(a) $\mathbf{v}(t) = \mathbf{r}'(t) = \mathbf{i} + 5\mathbf{j} + 3\mathbf{k}$

$\text{Speed} = \|\mathbf{v}(t)\| = \sqrt{1^2 + 5^2 + 3^2} = \sqrt{35}$

$\mathbf{a}(t) = \mathbf{r}''(t) = \mathbf{0}$

(b) $\mathbf{v}(1) = \mathbf{i} + 5\mathbf{j} + 3\mathbf{k}$

$\mathbf{a}(1) = \mathbf{0}$

13. $\mathbf{r}(t) = t\mathbf{i} + t^2\mathbf{j} + \dfrac{1}{2}t^2\mathbf{k}, \; t = 4$

(a) $\mathbf{v}(t) = \mathbf{r}'(t) = \mathbf{i} + 2t\mathbf{j} + t\mathbf{k}$

$\text{Speed} = \|\mathbf{v}(t)\| = \sqrt{1 + 4t^2 + t^2} = \sqrt{1 + 5t^2}$

$\mathbf{a}(t) = \mathbf{r}''(t) = 2\mathbf{j} + \mathbf{k}$

(b) $\mathbf{v}(4) = \mathbf{i} + 8\mathbf{j} + 4\mathbf{k}$

$\mathbf{a}(4) = 2\mathbf{j} + \mathbf{k}$

15. $\mathbf{r}(t) = t\mathbf{i} - t\mathbf{j} + \sqrt{9 - t^2}\mathbf{k}, \; t = 0$

(a) $\mathbf{v}(t) = \mathbf{r}'(t) = \mathbf{i} - \mathbf{j} - \dfrac{t}{\sqrt{9 - t^2}}\mathbf{k}$

$\text{Speed} = \|\mathbf{v}(t)\| = \sqrt{1 + 1 + \dfrac{t^2}{9 - t^2}} = \sqrt{\dfrac{18 - t^2}{9 - t^2}}$

$\mathbf{a}(t) = \mathbf{r}''(t) = -\dfrac{9}{\left(9 - t^2\right)^{3/2}}\mathbf{k}$

(b) $\mathbf{v}(0) = \mathbf{i} - \mathbf{j}$

$\mathbf{a}(0) = -\dfrac{9}{9^{3/2}}\mathbf{k} = -\dfrac{1}{3}\mathbf{k}$

17. $\mathbf{r}(t) = \langle 4t, 3\cos t, 3\sin t\rangle, \ t = \pi$

   (a) $\mathbf{v}(t) = \mathbf{r}'(t) = \langle 4, -3\sin t, 3\cos t\rangle$

        Speed $= \|\mathbf{v}(t)\| = \sqrt{4^2 + (-3\sin t)^2 + (3\cos t)^2} = \sqrt{16 + 9} = 5$

        $\mathbf{a}(t) = \mathbf{r}''(t) = \langle 0, -3\cos t, -3\sin t\rangle$

   (b) $\mathbf{v}(\pi) = \langle 4, 0, -3\rangle$

       $\mathbf{a}(\pi) = \langle 0, 3, 0\rangle$

19. $\mathbf{r}(t) = \langle e^t\cos t, e^t\sin t, e^t\rangle, \ t = 0$

   (a) $\mathbf{v}(t) = \mathbf{r}'(t) = \langle e^t\cos t - e^t\sin t, e^t\sin t + e^t\cos t, e^t\rangle$

       Speed $= \|\mathbf{v}(t)\| = \sqrt{e^{2t}(\cos t - \sin t)^2 + e^{2t}(\cos t + \sin t)^2 + e^{2t}} = e^t\sqrt{3}$

       $\mathbf{a}(t) = \mathbf{r}''(t) = \langle e^t\cos t - e^t\sin t - e^t\sin t - e^t\cos t, e^t\sin t + e^t\cos t + e^t\cos t - e^t\sin t, e^t\rangle$

          $= \langle -2e^t\sin t, 2e^t\cos t, e^t\rangle$

   (b) $\mathbf{v}(0) = \langle 1, 1, 1\rangle$

       $\mathbf{a}(0) = \langle 0, 2, 1\rangle$

21. $\mathbf{a}(t) = \mathbf{i} + \mathbf{j} + \mathbf{k}, \ \mathbf{v}(0) = 0, \mathbf{r}(0) = 0$

   $\mathbf{v}(t) = \int(\mathbf{i} + \mathbf{j} + \mathbf{k})\,dt = t\mathbf{i} + t\mathbf{j} + t\mathbf{k} + \mathbf{C}$

   $\mathbf{v}(0) = \mathbf{C} = 0, \mathbf{v}(t) = t\mathbf{i} + t\mathbf{j} + t\mathbf{k}, \mathbf{v}(t) = t(\mathbf{i} + \mathbf{j} + \mathbf{k})$

   $\mathbf{r}(t) = \int(t\mathbf{i} + t\mathbf{j} + t\mathbf{k})\,dt = \dfrac{t^2}{2}(\mathbf{i} + \mathbf{j} + \mathbf{k}) + \mathbf{C}$

   $\mathbf{r}(0) = \mathbf{C} = 0, \mathbf{r}(t) = \dfrac{t^2}{2}(\mathbf{i} + \mathbf{j} + \mathbf{k}),$

   $\mathbf{r}(2) = 2(\mathbf{i} + \mathbf{j} + \mathbf{k}) = 2\mathbf{i} + 2\mathbf{j} + 2\mathbf{k}$

23. $\mathbf{a}(t) = t\mathbf{j} + t\mathbf{k}, \ \mathbf{v}(1) = 5\mathbf{j}, \mathbf{r}(1) = 0$

   $\mathbf{v}(t) = \int(t\mathbf{j} + t\mathbf{k})\,dt = \dfrac{t^2}{2}\mathbf{j} + \dfrac{t^2}{2}\mathbf{k} + \mathbf{C}$

   $\mathbf{v}(1) = \dfrac{1}{2}\mathbf{j} + \dfrac{1}{2}\mathbf{k} + \mathbf{C} = 5\mathbf{j} \Rightarrow \mathbf{C} = \dfrac{9}{2}\mathbf{j} - \dfrac{1}{2}\mathbf{k}$

   $\mathbf{v}(t) = \left(\dfrac{t^2}{2} + \dfrac{9}{2}\right)\mathbf{j} + \left(\dfrac{t^2}{2} - \dfrac{1}{2}\right)\mathbf{k}$

   $\mathbf{r}(t) = \int\left[\left(\dfrac{t^2}{2} + \dfrac{9}{2}\right)\mathbf{j} + \left(\dfrac{t^2}{2} - \dfrac{1}{2}\right)\mathbf{k}\right]dt = \left(\dfrac{t^3}{6} + \dfrac{9}{2}t\right)\mathbf{j} + \left(\dfrac{t^3}{6} - \dfrac{1}{2}t\right)\mathbf{k} + \mathbf{C}$

   $\mathbf{r}(1) = \dfrac{14}{3}\mathbf{j} - \dfrac{1}{3}\mathbf{k} + \mathbf{C} = 0 \Rightarrow \mathbf{C} = -\dfrac{14}{3}\mathbf{j} + \dfrac{1}{3}\mathbf{k}$

   $\mathbf{r}(t) = \left(\dfrac{t^3}{6} + \dfrac{9}{2}t - \dfrac{14}{3}\right)\mathbf{j} + \left(\dfrac{t^3}{6} - \dfrac{1}{2}t + \dfrac{1}{3}\right)\mathbf{k}$

   $\mathbf{r}(2) = \dfrac{17}{3}\mathbf{j} + \dfrac{2}{3}\mathbf{k}$

25. $\mathbf{a}(t) = -\cos t\mathbf{i} - \sin t\mathbf{j}, \mathbf{v}(0) = \mathbf{j} + \mathbf{k}, \mathbf{r}(0) = \mathbf{i}$

$\mathbf{v}(t) = \int(-\cos t\mathbf{i} - \sin t\mathbf{j})\, dt = -\sin t\mathbf{i} + \cos t\mathbf{j} + \mathbf{C}$

$\mathbf{v}(0) = \mathbf{j} + \mathbf{C} = \mathbf{j} + \mathbf{k} \Rightarrow \mathbf{C} = \mathbf{k}$

$\mathbf{v}(t) = -\sin t\mathbf{i} + \cos t\mathbf{j} + \mathbf{k}$

$\mathbf{r}(t) = \int(-\sin t\mathbf{i} + \cos t\mathbf{j} + \mathbf{k})\, dt = \cos t\mathbf{i} + \sin t\mathbf{j} + t\mathbf{k} + \mathbf{C}$

$\mathbf{r}(0) = \mathbf{i} + \mathbf{C} = \mathbf{i} \Rightarrow \mathbf{C} = \mathbf{0}$

$\mathbf{r}(t) = \cos t\mathbf{i} + \sin t\mathbf{j} + t\mathbf{k}$

$\mathbf{r}(2) = (\cos 2)\mathbf{i} + (\sin 2)\mathbf{j} + 2\mathbf{k}$

27. $\mathbf{r}(t) = 140(\cos 22°)t\mathbf{i} + (2.5 + 140(\sin 22°)t - 16t^2)\mathbf{j}$

$\mathbf{v}(t) = \mathbf{r}'(t) = 140(\cos 22°)\mathbf{i} + (140(\sin 22°) - 32t)\mathbf{j}$

The maximum height occurs when

$y'(t) = 140(\sin 22°) - 32t = 0 \Rightarrow t = \dfrac{140 \sin 22°}{32} = \dfrac{35}{8}\sin 22° \approx 1.639.$

The maximum height is

$y = 2.5 + 140(\sin 22°)\left(\dfrac{35}{8}\sin 22°\right) - 16\left(\dfrac{35}{8}\sin 22°\right)^2 \approx 45.5 \text{ feet.}$

When $x = 375, t = \dfrac{375}{140 \cos 22°} \approx 2.889.$

For this value of $t, y \approx 20.47$ feet.

So the ball clears the 10-foot fence.

29. $\mathbf{r}(t) = (v_0 \cos \theta)t\mathbf{i} + \left[h + (v_0 \sin \theta)t - \dfrac{1}{2}gt^2\right]\mathbf{j} = \dfrac{v_0}{\sqrt{2}}t\mathbf{i} + \left(3 + \dfrac{v_0}{\sqrt{2}}t - 16t^2\right)\mathbf{j}$

$\dfrac{v_0}{\sqrt{2}}t = 300$ when $3 + \dfrac{v_0}{\sqrt{2}}t - 16t^2 = 3.$

$t = \dfrac{300\sqrt{2}}{v_0}, \dfrac{v_0}{\sqrt{2}}\left(\dfrac{300\sqrt{2}}{v_0}\right) - 16\left(\dfrac{300\sqrt{2}}{v_0}\right)^2 = 0, \ 300 - \dfrac{300^2(32)}{v_0^2} = 0$

$v_0^2 = 300(32), v_0 = \sqrt{9600} = 40\sqrt{6}, v_0 = 40\sqrt{6} \approx 97.98 \text{ ft/sec}$

The maximum height is reached when the derivative of the vertical component is zero.

$y(t) = 3 + \dfrac{tv_0}{\sqrt{2}} - 16t^2 = 3 + \dfrac{40\sqrt{6}}{\sqrt{2}}t - 16t^2 = 3 + 40\sqrt{3}t - 16t^2$

$y'(t) = 40\sqrt{3} - 32t = 0$

$t = \dfrac{40\sqrt{3}}{32} = \dfrac{5\sqrt{3}}{4}$

Maximum height: $y\left(\dfrac{5\sqrt{3}}{4}\right) = 3 + 40\sqrt{3}\left(\dfrac{5\sqrt{3}}{4}\right) - 16\left(\dfrac{5\sqrt{3}}{4}\right)^2 = 78 \text{ feet}$

31. $x(t) = t(v_0 \cos \theta)$ or $t = \dfrac{x}{v_0 \cos \theta}$

$y(t) = t(v_0 \sin \theta) - 16t^2 + h$

$y = \dfrac{x}{v_0 \cos \theta}(v_0 \sin \theta) - 16\left(\dfrac{x^2}{v_0^2 \cos^2 \theta}\right) + h = (\tan \theta)x - \left(\dfrac{16}{v_0^2}\sec^2 \theta\right)x^2 + h$

33. $100 \text{ mi/h} = \left(100\dfrac{\text{mi}}{\text{hr}}\right)\left(5280\dfrac{\text{ft}}{\text{mi}}\right)\bigg/\left(3600\dfrac{\text{sec}}{\text{hr}}\right) = \dfrac{440}{3}\text{ft/sec}$

(a) $\mathbf{r}(t) = \left(\dfrac{440}{3}\cos \theta_0\right)t\mathbf{i} + \left[3 + \left(\dfrac{440}{3}\sin \theta_0\right)t - 16t^2\right]\mathbf{j}$

Graphing these curves together with $y = 10$ shows that $\theta_0 = 20°$.

(b)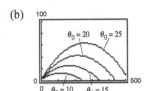

(c) You want

$x(t) = \left(\dfrac{440}{3}\cos \theta\right)t \geq 400$ and $y(t) = 3 + \left(\dfrac{440}{3}\sin \theta\right)t - 16t^2 \geq 10.$

From $x(t)$, the minimum angle occurs when $t = 30/(11\cos \theta)$. Substituting this for t in $y(t)$ yields:

$3 + \left(\dfrac{440}{3}\sin \theta\right)\left(\dfrac{30}{11\cos \theta}\right) - 16\left(\dfrac{30}{11\cos \theta}\right)^2 = 10$

$400 \tan \theta - \dfrac{14,400}{121}\sec^2 \theta = 7$

$\dfrac{14,400}{121}(1 + \tan^2 \theta) - 400\tan \theta + 7 = 0$

$14,400 \tan^2 \theta - 48,400 \tan \theta + 15,247 = 0$

$\tan \theta = \dfrac{48,400 \pm \sqrt{48,400^2 - 4(14,400)(15,247)}}{2(14,400)}$

$\theta = \tan^{-1}\left(\dfrac{48,400 - \sqrt{1,464,332,800}}{28,800}\right) \approx 19.38°$

35. $\mathbf{r}(t) = (v \cos \theta)t\mathbf{i} + \left[(v \sin \theta)t - 16t^2\right]\mathbf{j}$

(a) You want to find the minimum initial speed v as a function of the angle θ. Because the bale must be thrown to the position $(16, 8)$, you have

$$16 = (v \cos \theta)t$$

$$8 = (v \sin \theta)t - 16t^2.$$

$t = 16/(v \cos \theta)$ from the first equation. Substituting into the second equation and solving for v, you obtain:

$$8 = (v \sin \theta)\left(\frac{16}{v \cos \theta}\right) - 16\left(\frac{16}{v \cos \theta}\right)^2$$

$$1 = 2\left(\frac{\sin \theta}{\cos \theta}\right) - 512\left(\frac{1}{v^2 \cos^2 \theta}\right)$$

$$512\left(\frac{1}{v^2 \cos^2 \theta}\right) = 2\left(\frac{\sin \theta}{\cos \theta}\right) - 1$$

$$\frac{1}{v^2} = \left(2\frac{\sin \theta}{\cos \theta} - 1\right)\frac{\cos^2 \theta}{512} = \frac{2 \sin \theta \cos \theta - \cos^2 \theta}{512}$$

$$v^2 = \frac{512}{2 \sin \theta \cos \theta - \cos^2 \theta}$$

You minimize $f(\theta) = \dfrac{512}{2 \sin \theta \cos \theta - \cos^2 \theta}$.

$$f'(\theta) = -512\left[\frac{2\cos^2 \theta - 2 \sin^2 \theta + 2 \sin \theta \cos \theta}{\left(2 \sin \theta \cos \theta - \cos^2 \theta\right)^2}\right]$$

$$f'(\theta) = 0 \Rightarrow 2 \cos (2\theta) + \sin(2\theta) = 0$$

$$\tan(2\theta) = -2$$

$$\theta \approx 1.01722 \approx 58.28°$$

Substituting into the equation for v, $v \approx 28.78$ ft/sec.

(b) If $\theta = 45°$,

$$16 = (v \cos \theta)t = v\frac{\sqrt{2}}{2}t$$

$$8 = (v \sin \theta)t - 16t^2 = v\frac{\sqrt{2}}{2}t - 16t^2$$

From part (a), $v^2 = \dfrac{512}{2\left(\sqrt{2}/2\right)\left(\sqrt{2}/2\right) - \left(\sqrt{2}/2\right)^2} = \dfrac{512}{1/2} = 1024 \Rightarrow v = 32$ ft/sec.

37. $\mathbf{r}(t) = (v_0 \cos \theta)t\mathbf{i} + \left[(v_0 \sin \theta)t - 16t^2\right]\mathbf{j}$

$(v_0 \sin \theta)t - 16t^2 = 0$ when $t = 0$ and $t = \dfrac{v_0 \sin \theta}{16}$.

The range is $x = (v_0 \cos \theta)t = (v_0 \cos \theta)\dfrac{v_0 \sin \theta}{16} = \dfrac{v_0^2}{32} \sin 2\theta$.

So, $x = \dfrac{1200^2}{32}\sin(2\theta) = 3000 \Rightarrow \sin 2\theta = \dfrac{1}{15} \Rightarrow \theta \approx 1.91°$.

39. (a) $\theta = 10°$, $v_0 = 66$ ft/sec

$$\mathbf{r}(t) = (66\cos 10°)t\mathbf{i} + \left[0 + (66\sin 10°)t - 16t^2\right]\mathbf{j}$$

$$\mathbf{r}(t) \approx (65t)\mathbf{i} + \left(11.46t - 16t^2\right)\mathbf{j}$$

Maximum height: 2.052 feet

Range: 46.557 feet

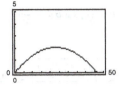

(b) $\theta = 10°$, $v_0 = 146$ ft/sec

$$\mathbf{r}(t) = (146\cos 10°)t\mathbf{i} + \left[0 + (146\sin 10°)t - 16t^2\right]\mathbf{j}$$

$$\mathbf{r}(t) \approx (143.78t)\mathbf{i} + \left(25.35t - 16t^2\right)\mathbf{j}$$

Maximum height: 10.043 feet

Range: 227.828 feet

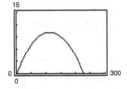

(c) $\theta = 45°$, $v_0 = 66$ ft/sec

$$\mathbf{r}(t) = (66\cos 45°)t\mathbf{i} + \left[0 + (66\sin 45°)t - 16t^2\right]\mathbf{j}$$

$$\mathbf{r}(t) \approx (46.67t)\mathbf{i} + \left(46.67t - 16t^2\right)\mathbf{j}$$

Maximum height: 34.031 feet

Range: 136.125 feet

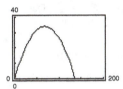

(d) $\theta = 45°$, $v_0 = 146$ ft/sec

$$\mathbf{r}(t) = (146\cos 45°)t\mathbf{i} + \left[0 + (146\sin 45°)t - 16t^2\right]\mathbf{j}$$

$$\mathbf{r}(t) \approx (103.24t)\mathbf{i} + \left(103.24t - 16t^2\right)\mathbf{j}$$

Maximum height: 166.531 feet

Range: 666.125 feet

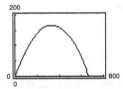

(e) $\theta = 60°$, $v_0 = 66$ ft/sec

$$\mathbf{r}(t) = (66\cos 60°)t\mathbf{i} + \left[0 + (66\sin 60°)t - 16t^2\right]\mathbf{j}$$

$$\mathbf{r}(t) \approx (33t)\mathbf{i} + \left(57.16t - 16t^2\right)\mathbf{j}$$

Maximum height: 51.047 feet

Range: 117.888 feet

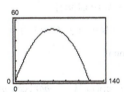

(f) $\theta = 60°$, $v_0 = 146$ ft/sec

$$\mathbf{r}(t) = (146\cos 60°)t\mathbf{i} + \left[0 + (146\sin 60°)t - 16t^2\right]\mathbf{j}$$

$$\mathbf{r}(t) \approx (73t)\mathbf{i} + \left(126.44t - 16t^2\right)\mathbf{j}$$

Maximum height: 249.797 feet

Range: 576.881 feet

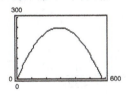

41. $\mathbf{r}(t) = (v_0\cos\theta)t\mathbf{i} + \left[h + (v_0\sin\theta)t - 4.9t^2\right]\mathbf{j} = (100\cos 30°)t\mathbf{i} + \left[1.5 + (100\sin 30°)t - 4.9t^2\right]\mathbf{j}$

The projectile hits the ground when $-4.9t^2 + 100\left(\frac{1}{2}\right)t + 1.5 = 0 \Rightarrow t \approx 10.234$ seconds.

So the range is $(100\cos 30°)(10.234) \approx 886.3$ meters.

The maximum height occurs when $dy/dt = 0$.

$100\sin 30 = 9.8t \Rightarrow t \approx 5.102$ sec

The maximum height is $y = 1.5 + (100\sin 30°)(5.102) - 4.9(5.102)^2 \approx 129.1$ meters.

43. To find the range, set $y(t) = h + (v_0 \sin \theta)t - \frac{1}{2}gt^2 = 0$ then $0 = \left(\frac{1}{2}g\right)t^2 - (v_0 \sin \theta)t - h$. By the Quadratic Formula, (discount the negative value)

$$t = \frac{v_0 \sin \theta + \sqrt{(-v_0 \sin \theta)^2 - 4[(1/2)g](-h)}}{2[(1/2)g]} = \frac{v_0 \sin \theta + \sqrt{v_0^2 \sin^2 \theta + 2gh}}{g} \text{ second.}$$

At this time,

$$x(t) = v_0 \cos \theta \left(\frac{v_0 \sin \theta + \sqrt{v_0^2 \sin^2 \theta + 2gh}}{g} \right)$$

$$= \frac{v_0 \cos \theta}{g}\left(v_0 \sin \theta + \sqrt{v_0^2 \left(\sin^2 \theta + \frac{2gh}{v_0^2} \right)} \right)$$

$$= \frac{v_0^2 \cos \theta}{g}\left(\sin \theta + \sqrt{\sin^2 \theta + \frac{2gh}{v_0^2}} \right) \text{ feet.}$$

45. $\mathbf{r}(t) = b(\omega t - \sin \omega t)\mathbf{i} + b(1 - \cos \omega t)\mathbf{j}$

$\mathbf{v}(t) = b(\omega - \omega \cos \omega t)\mathbf{i} + b\omega \sin \omega t\mathbf{j} = b\omega(1 - \cos \omega t)\mathbf{i} + b\omega \sin \omega t\mathbf{j}$

$\mathbf{a}(t) = \left(b\omega^2 \sin \omega t\right)\mathbf{i} + \left(b\omega^2 \cos \omega t\right)\mathbf{j} = b\omega^2\left[\sin(\omega t)\mathbf{i} + \cos(\omega t)\mathbf{j}\right]$

$\|\mathbf{v}(t)\| = \sqrt{2}b\omega\sqrt{1 - \cos(\omega t)}$

$\|\mathbf{a}(t)\| = b\omega^2$

(a) $\|\mathbf{v}(t)\| = 0$ when $\omega t = 0, 2\pi, 4\pi, \dots$.

(b) $\|\mathbf{v}(t)\|$ is maximum when $\omega t = \pi, 3\pi, \dots$, then $\|\mathbf{v}(t)\| = 2b\omega$.

47. $\quad\mathbf{v}(t) = -b\omega \sin(\omega t)\mathbf{i} + b\omega \cos(\omega t)\mathbf{j}$

$\mathbf{r}(t) \cdot \mathbf{v}(t) = -b^2\omega \sin(\omega t) \cos(\omega t) + b^2\omega \sin(vt) \cos(\omega t) = 0$

So, $\mathbf{r}(t)$ and $\mathbf{v}(t)$ are orthogonal.

49. $\mathbf{a}(t) = -b\omega^2 \cos(\omega t)\mathbf{i} - b\omega^2 \sin(\omega t)\mathbf{j} = -b\omega^2\left[\cos(\omega t)\mathbf{i} + \sin(\omega t)\mathbf{j}\right] = -\omega^2\mathbf{r}(t)$

$\mathbf{a}(t)$ is a negative multiple of a unit vector from $(0, 0)$ to $(\cos \omega t, \sin \omega t)$ and so $\mathbf{a}(t)$ is directed toward the origin.

51. $\|\mathbf{a}(t)\| = \omega^2 b, b = \frac{1}{2}$

$$\mathbf{F} = m\mathbf{a} \Rightarrow \frac{1}{4} = m(32) \Rightarrow m = \frac{1}{4(32)}$$

$$2 = m(\omega^2 b) = \frac{1}{4(32)}\left(\frac{1}{2}\omega^2\right) \Rightarrow \omega^2 = 512 \Rightarrow \omega = 16\sqrt{2}$$

$$\|\mathbf{v}(t)\| = b\omega = \frac{1}{2}(16\sqrt{2}) = 8\sqrt{2} \text{ ft/sec}$$

53. The particle could be changing direction. For example, let $\mathbf{r}(t) = \cos t\mathbf{i} + \sin t\mathbf{j}$. Then the speed is $\|\mathbf{r}(t)\| = 1$, but $\mathbf{a}(t)$ is not constant.

55. No. This is true for uniform circular motion only.

57. $\mathbf{a}(t) = \sin t\mathbf{i} - \cos t\mathbf{j}$

$\mathbf{v}(t) = \int \mathbf{a}(t)\, dt = -\cos t\mathbf{i} - \sin t\mathbf{j} + \mathbf{C}_1$

$\mathbf{v}(0) = -\mathbf{i} = -\mathbf{i} + \mathbf{C}_1 \Rightarrow \mathbf{C}_1 = \mathbf{0}$

$\mathbf{v}(t) = -\cos t\mathbf{i} - \sin t\mathbf{j}$

$\mathbf{r}(t) = \int \mathbf{v}(t)\, dt = -\sin t\mathbf{i} + \cos t\mathbf{j} + \mathbf{C}_2$

$\mathbf{r}(0) = \mathbf{j} = \mathbf{j} + \mathbf{C}_2 \Rightarrow \mathbf{C}_2 = \mathbf{0}$

$\mathbf{r}(t) = -\sin t\mathbf{i} + \cos t\mathbf{j}$

The path is a circle.

59. $\mathbf{r}(t) = x(t)\mathbf{i} + y(t)\mathbf{j} + z(t)\mathbf{k}$ Position vector

$\mathbf{v}(t) = x'(t)\mathbf{i} + y'(t)\mathbf{j} + z'(t)\mathbf{k}$ Velocity vector

$\mathbf{a}(t) = x''(t)\mathbf{i} + y''(t)\mathbf{j} + z''(t)\mathbf{k}$ Acceleration vector

$\text{Speed} = \|\mathbf{v}(t)\| = \sqrt{x'(t)^2 + y'(t)^2 + z'(t)^2}$

$\qquad\qquad = C, C \text{ is a constant.}$

$\dfrac{d}{dt}\left[x'(t)^2 + y'(t)^2 + z'(t)^2 \right] = 0$

$2x'(t)x''(t) + 2y'(t)y''(t) + 2z'(t)z''(t) = 0$

$2\left[x'(t)x''(t) + y'(t)y''(t) + z'(t)z''(t) \right] = 0$

$\mathbf{v}(t) \cdot \mathbf{a}(t) = 0$

Orthogonal

61. True

63. False. Consider $\mathbf{r}(t) = \langle t^2, -t^2 \rangle$. Then $\mathbf{v}(t) = \langle 2t, -2t \rangle$ and $\|\mathbf{v}(t)\| = \sqrt{8t^2}$.

Section 12.4   Tangent Vectors and Normal Vectors

1. The unit tangent vector points in the direction of motion.

3. $\mathbf{r}(t) = t^2\mathbf{i} + 2t\mathbf{j},\ t = 1$

$\mathbf{r}'(t) = 2t\mathbf{i} + 2\mathbf{j},\ \|\mathbf{r}'(t)\| = \sqrt{4t^2 + 4} = 2\sqrt{t^2 + 1}$

$\mathbf{T}(1) = \dfrac{\mathbf{r}'(1)}{\|\mathbf{r}'(1)\|} = \dfrac{1}{\sqrt{2}}(\mathbf{i} + \mathbf{j}) = \dfrac{\sqrt{2}}{2}\mathbf{i} + \dfrac{\sqrt{2}}{2}\mathbf{j}$

5. $\mathbf{r}(t) = 5\cos t\mathbf{i} + 5\sin t\mathbf{j},\ t = \dfrac{\pi}{3}$

$\mathbf{r}'(t) = -5\sin t\mathbf{i} + 5\cos t\mathbf{j}$

$\|\mathbf{r}'(t)\| = \sqrt{25\sin^2 t + 25\cos^2 t} = 5$

$\mathbf{T}\left(\dfrac{\pi}{3}\right) = \dfrac{\mathbf{r}'\left(\dfrac{\pi}{3}\right)}{\left\|\mathbf{r}'\left(\dfrac{\pi}{3}\right)\right\|} = -\sin\dfrac{\pi}{3}\mathbf{i} + \cos\dfrac{\pi}{3}\mathbf{j} = -\dfrac{\sqrt{3}}{2}\mathbf{i} + \dfrac{1}{2}\mathbf{j}$

7. $\mathbf{r}(t) = 3t\mathbf{i} - \ln t\mathbf{j},\ t = e$

$\mathbf{r}'(t) = 3\mathbf{i} - \dfrac{1}{t}\mathbf{j}$

$\mathbf{r}'(e) = 3\mathbf{i} - \dfrac{1}{e}\mathbf{j}$

$\mathbf{T}(e) = \dfrac{\mathbf{r}'(e)}{\|\mathbf{r}'(e)\|}$

$= \dfrac{3\mathbf{i} - \dfrac{1}{e}\mathbf{j}}{\sqrt{9 + \dfrac{1}{e^2}}} = \dfrac{3e\mathbf{i} - \mathbf{j}}{\sqrt{9e^2 + 1}} \approx 0.9926\mathbf{i} - 0.1217\mathbf{j}$

9. $\mathbf{r}(t) = t\mathbf{i} + t^2\mathbf{j} + t\mathbf{k},\ P(0, 0, 0)$

$\mathbf{r}'(t) = \mathbf{i} + 2t\mathbf{j} + \mathbf{k}$

When $t = 0$, $\mathbf{r}'(0) = \mathbf{i} + \mathbf{k}$, $\left[t = 0 \text{ at } (0, 0, 0) \right]$.

$\mathbf{T}(0) = \dfrac{\mathbf{r}'(0)}{\|\mathbf{r}'(0)\|} = \dfrac{\sqrt{2}}{2}(\mathbf{i} + \mathbf{k})$

Direction numbers: $a = 1,\ b = 0,\ c = 1$

Parametric equations: $x = t,\ y = 0,\ z = t$

11. $\mathbf{r}(t) = \cos t\mathbf{i} + 3\sin t\mathbf{j} + (3t - 4)\mathbf{k},\ P(1, 0, -4)$

$\mathbf{r}'(t) = -\sin t\mathbf{i} + 3\cos t\mathbf{j} + 3\mathbf{k}$

$t = 0$ at $P(1, 0, -4)$

$\mathbf{r}'(0) = 3\mathbf{j} + 3\mathbf{k}$

$\|\mathbf{r}'(0)\| = \sqrt{9 + 9} = 3\sqrt{2}$

$\mathbf{T}(0) = \dfrac{1}{3\sqrt{2}}(3\mathbf{j} + 3\mathbf{k}) = \dfrac{\sqrt{2}}{2}(\mathbf{j} + \mathbf{k})$

Direction numbers: $a = 0,\ b = 1,\ c = 1$

Parametric equations: $x = 1,\ y = 3t,\ z = -4 + 3t$ or
$x = 1,\ y = t,\ z = -4 + t$

13. $\mathbf{r}(t) = \langle 2\cos t, 2\sin t, 4 \rangle, P(\sqrt{2}, \sqrt{2}, 4)$

$\mathbf{r}'(t) = \langle -2\sin t, 2\cos t, 0 \rangle$

When $t = \dfrac{\pi}{4}$, $\mathbf{r}'\left(\dfrac{\pi}{4}\right) = \langle -\sqrt{2}, \sqrt{2}, 0 \rangle$,

$\left[t = \dfrac{\pi}{4} \text{ at } \left(\sqrt{2}, \sqrt{2}, 4\right) \right].$

$\mathbf{T}\left(\dfrac{\pi}{4}\right) = \dfrac{\mathbf{r}'(\pi/4)}{\|\mathbf{r}'(\pi/4)\|} = \dfrac{1}{2}\langle -\sqrt{2}, \sqrt{2}, 0 \rangle$

Direction numbers: $a = -\sqrt{2}, b = \sqrt{2}, c = 0$

Parametric equations: $x = -\sqrt{2}\,t + \sqrt{2}$, $y = \sqrt{2}\,t + \sqrt{2}$, $z = 4$

15. $\mathbf{r}(t) = t\mathbf{i} + \dfrac{1}{2}t^2\mathbf{j}, t = 2$

$\mathbf{r}'(t) = \mathbf{i} + t\mathbf{j}$

$\mathbf{T}(t) = \dfrac{\mathbf{r}'(t)}{\|\mathbf{r}'(t)\|} = \dfrac{\mathbf{i} + t\mathbf{j}}{\sqrt{1 + t^2}}$

$\mathbf{T}'(t) = \dfrac{-t}{\left(t^2 + 1\right)^{3/2}}\mathbf{i} + \dfrac{1}{\left(t^2 + 1\right)^{3/2}}\mathbf{j}$

$\mathbf{T}'(2) = \dfrac{-2}{5^{3/2}}\mathbf{i} + \dfrac{1}{5^{3/2}}\mathbf{j}$

$\mathbf{N}(2) = \dfrac{\mathbf{T}'(2)}{\|\mathbf{T}'(2)\|} = \dfrac{1}{\sqrt{5}}(-2\mathbf{i} + \mathbf{j}) = \dfrac{-2\sqrt{5}}{5}\mathbf{i} + \dfrac{\sqrt{5}}{5}\mathbf{j}$

17. $\mathbf{r}(t) = t\mathbf{i} + t^2\mathbf{j} + \ln t\mathbf{k}, t = 1$

$\mathbf{r}'(t) = \mathbf{i} + 2t\mathbf{j} + \dfrac{1}{t}\mathbf{k}$

$\mathbf{T}(t) = \dfrac{\mathbf{r}'(t)}{\|\mathbf{r}'(t)\|} = \dfrac{\mathbf{i} + 2t\mathbf{j} + \dfrac{1}{t}\mathbf{k}}{\sqrt{1 + 4t^2 + \dfrac{1}{t^2}}} = \dfrac{t\mathbf{i} + 2t^2\mathbf{j} + \mathbf{k}}{\sqrt{4t^4 + t^2 + 1}}$

$\mathbf{T}'(t) = \dfrac{1 - 4t^4}{\left(4t^4 + t^2 + 1\right)^{3/2}}\mathbf{i} + \dfrac{2t^3 + 4t}{\left(4t^4 + t^2 + 1\right)^{3/2}}\mathbf{j} + \dfrac{-8t^3 - t}{\left(4t^4 + t^2 + 1\right)^{3/2}}\mathbf{k}$

$\mathbf{T}'(1) = \dfrac{-3}{6^{3/2}}\mathbf{i} + \dfrac{6}{6^{3/2}}\mathbf{j} + \dfrac{-9}{6^{3/2}}\mathbf{k} = \dfrac{3}{6^{3/2}}[-\mathbf{i} + 2\mathbf{j} - 3\mathbf{k}]$

$\mathbf{N}(1) = \dfrac{-\mathbf{i} + 2\mathbf{j} - 3\mathbf{k}}{\sqrt{14}} = \dfrac{-\sqrt{14}}{14}\mathbf{i} + \dfrac{2\sqrt{14}}{14}\mathbf{j} - \dfrac{3\sqrt{14}}{14}\mathbf{k}$

19. $\mathbf{r}(t) = 6\cos t\mathbf{i} + 6\sin t\mathbf{j} + \mathbf{k}, t = \dfrac{3\pi}{4}$

$\mathbf{r}'(t) = -6\sin t\mathbf{i} + 6\cos t\mathbf{j}$

$\mathbf{T}(t) = \dfrac{\mathbf{r}'(t)}{\|\mathbf{r}'(t)\|} = -\sin t\mathbf{i} + \cos t\mathbf{j}$

$\mathbf{T}'(t) = -\cos t\mathbf{i} - \sin t\mathbf{j}, \|\mathbf{T}'(t)\| = 1$

$\mathbf{N}\left(\dfrac{3\pi}{4}\right) = \dfrac{\mathbf{T}'(3\pi/4)}{\|\mathbf{T}'(3\pi/4)\|} = \dfrac{\sqrt{2}}{2}\mathbf{i} - \dfrac{\sqrt{2}}{2}\mathbf{j}$

21. $\mathbf{r}(t) = t\mathbf{i} + \dfrac{1}{t}\mathbf{j}, t_0 = 2$

$x = t, y = \dfrac{1}{t} \Rightarrow xy = 1$

$\mathbf{r}'(t) = \mathbf{i} - \dfrac{1}{t^2}\mathbf{j}$

$\mathbf{T}(t) = \dfrac{t^2\mathbf{i} - \mathbf{j}}{\sqrt{t^4 + 1}}$

$\mathbf{N}(t) = \dfrac{\mathbf{i} + t^2\mathbf{j}}{\sqrt{t^4 + 1}}$

$\mathbf{r}(2) = 2\mathbf{i} + \dfrac{1}{2}\mathbf{j}$

$\mathbf{T}(2) = \dfrac{\sqrt{17}}{17}(4\mathbf{i} - \mathbf{j})$

$\mathbf{N}(2) = \dfrac{\sqrt{17}}{17}(\mathbf{i} + 4\mathbf{j})$

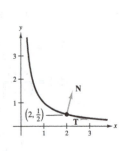

23. $\mathbf{r}(t) = (2t + 1)\mathbf{i} - t^2\mathbf{j}, t_0 = 2$

$x = 2t + 1,$

$y = -t^2 = -\left(\dfrac{x - 1}{2}\right)^2$

$\mathbf{r}(2) = 5\mathbf{i} - 4\mathbf{j}$

$\mathbf{r}'(t) = 2\mathbf{i} - 2t\mathbf{j}$

$\mathbf{T}(t) = \dfrac{2\mathbf{i} - 2t\mathbf{j}}{\sqrt{4 + 4t^2}} = \dfrac{\mathbf{i} - t\mathbf{j}}{\sqrt{1 + t^2}}$

$\mathbf{T}(2) = \dfrac{\mathbf{i} - 2\mathbf{j}}{\sqrt{5}}$

$\mathbf{N}(2) = \dfrac{-2\mathbf{i} - \mathbf{j}}{\sqrt{5}}$, perpendicular to $\mathbf{T}(2)$

25. $\mathbf{r}(t) = t\mathbf{i} + \dfrac{1}{t}\mathbf{j}, t = 1$

$\mathbf{v}(t) = \mathbf{i} - \dfrac{1}{t^2}\mathbf{j}, \mathbf{v}(1) = \mathbf{i} - \mathbf{j},$

$\mathbf{a}(t) = \dfrac{2}{t^3}\mathbf{j}, \mathbf{a}(1) = 2\mathbf{j}$

$\mathbf{T}(t) = \dfrac{\mathbf{v}(t)}{\|\mathbf{v}(t)\|} = \dfrac{t^2}{\sqrt{t^4 + 1}}\left(\mathbf{i} - \dfrac{1}{t^2}\mathbf{j}\right) = \dfrac{1}{\sqrt{t^4 + 1}}(t^2\mathbf{i} - \mathbf{j})$

$\mathbf{T}(1) = \dfrac{1}{\sqrt{2}}(\mathbf{i} - \mathbf{j}) = \dfrac{\sqrt{2}}{2}(\mathbf{i} - \mathbf{j})$

$\mathbf{N}(t) = \dfrac{\mathbf{T}'(t)}{\|\mathbf{T}'(t)\|} = \dfrac{\dfrac{2t}{(t^4 + 1)^{3/2}}\mathbf{i} + \dfrac{2t^3}{(t^4 + 1)^{3/2}}\mathbf{j}}{\dfrac{2t}{(t^4 + 1)}}$

$\qquad = \dfrac{1}{\sqrt{t^4 + 1}}(\mathbf{i} + t^2\mathbf{j})$

$\mathbf{N}(1) = \dfrac{1}{\sqrt{2}}(\mathbf{i} + \mathbf{j}) = \dfrac{\sqrt{2}}{2}(\mathbf{i} + \mathbf{j})$

$a_\mathbf{T} = \mathbf{a} \cdot \mathbf{T} = -\sqrt{2}$

$a_\mathbf{N} = \mathbf{a} \cdot \mathbf{N} = \sqrt{2}$

27. $\mathbf{r}(t) = e^t\mathbf{i} + e^{-2t}\mathbf{j}, t = 0$

$\mathbf{v}(t) = e^t\mathbf{i} - 2e^{-2t}\mathbf{j}, \mathbf{v}(0) = \mathbf{i} - 2\mathbf{j}$

$\mathbf{a}(t) = e^t\mathbf{i} + 4e^{-2t}\mathbf{j}, \mathbf{a}(0) = \mathbf{i} + 4\mathbf{j}$

$\mathbf{T}(t) = \dfrac{\mathbf{v}(t)}{\|\mathbf{v}(t)\|} = \dfrac{e^t\mathbf{i} - 2e^{-2t}\mathbf{j}}{\sqrt{4e^{-4t} + e^{2t}}}$

$\mathbf{T}(0) = \dfrac{\mathbf{i} - 2\mathbf{j}}{\sqrt{5}}$

$\mathbf{N}(0) = \dfrac{2\mathbf{i} + \mathbf{j}}{\sqrt{5}}$

$a_\mathbf{T} = \mathbf{a} \cdot \mathbf{T} = \dfrac{1}{\sqrt{5}}(1 - 8) = \dfrac{-7\sqrt{5}}{5}$

$a_\mathbf{N} = \mathbf{a} \cdot \mathbf{N} = \dfrac{1}{\sqrt{5}}(2 + 4) = \dfrac{6\sqrt{5}}{5}$

29. $\mathbf{r}(t) = (e^t \cos t)\mathbf{i} + (e^t \sin t)\mathbf{j}, t = \dfrac{\pi}{2}$

$\mathbf{v}(t) = e^t(\cos t - \sin t)\mathbf{i} + e^t(\cos t + \sin t)\mathbf{j}$

$\mathbf{a}(t) = e^t(-2 \sin t)\mathbf{i} + e^t(2 \cos t)\mathbf{j}$

At $t = \dfrac{\pi}{2}, \mathbf{T} = \dfrac{\mathbf{v}}{\|\mathbf{v}\|} = \dfrac{1}{\sqrt{2}}(-\mathbf{i} + \mathbf{j}) = \dfrac{\sqrt{2}}{2}(-\mathbf{i} + \mathbf{j}).$

Motion along $\mathbf{r}$ is counterclockwise. So,

$\mathbf{N} = \dfrac{1}{\sqrt{2}}(-\mathbf{i} - \mathbf{j}) = -\dfrac{\sqrt{2}}{2}(\mathbf{i} + \mathbf{j}).$

$a_\mathbf{T} = \mathbf{a} \cdot \mathbf{T} = \sqrt{2}e^{\pi/2}$

$a_\mathbf{N} = \mathbf{a} \cdot \mathbf{N} = \sqrt{2}e^{\pi/2}$

31. $\mathbf{r}(t) = a \cos \omega t\mathbf{i} + a \sin \omega t\mathbf{j}$

$\mathbf{v}(t) = -a\omega \sin \omega t\mathbf{i} + a\omega \cos \omega t\mathbf{j}$

$\mathbf{a}(t) = -a\omega^2 \cos \omega t\mathbf{i} - a\omega^2 \sin \omega t\mathbf{j}$

$\mathbf{T}(t) = \dfrac{\mathbf{v}(t)}{\|\mathbf{v}(t)\|} = -\sin \omega t\mathbf{i} + \cos \omega t\mathbf{j}$

$\mathbf{N}(t) = \dfrac{\mathbf{T}'(t)}{\|\mathbf{T}'(t)\|} = -\cos \omega t\mathbf{i} - \sin \omega t\mathbf{j}$

$a_\mathbf{T} = \mathbf{a} \cdot \mathbf{T} = 0$

$a_\mathbf{N} = \mathbf{a} \cdot \mathbf{N} = a\omega^2$

33. Speed: $\|\mathbf{v}(t)\| = a\omega$

The speed is constant because $a_\mathbf{T} = 0$.

35. $\mathbf{r}(t) = t\mathbf{i} + 2t\mathbf{j} - 3t\mathbf{k}, t = 1$

$\mathbf{v}(t) = \mathbf{i} + 2\mathbf{j} - 3\mathbf{k}$

$\mathbf{a}(t) = \mathbf{0}$

$\mathbf{T}(t) = \dfrac{\mathbf{v}}{\|\mathbf{v}\|}$

$\qquad = \dfrac{1}{\sqrt{14}}(\mathbf{i} + 2\mathbf{j} - 3\mathbf{k})$

$\qquad = \dfrac{\sqrt{14}}{14}(\mathbf{i} + 2\mathbf{j} - 3\mathbf{k}) = \mathbf{T}(1)$

$\mathbf{N}(t) = \dfrac{\mathbf{T}'}{\|\mathbf{T}'\|}$ is undefined.

$a_{\mathbf{T}}, a_{\mathbf{N}}$ are not defined.

37. $\mathbf{r}(t) = t\mathbf{i} + t^2\mathbf{j} + \dfrac{t^2}{2}\mathbf{k}, t = 1$

$\mathbf{v}(t) = \mathbf{i} + 2t\mathbf{j} + t\mathbf{k}$

$\mathbf{v}(1) = \mathbf{i} + 2\mathbf{j} + \mathbf{k}$

$\mathbf{a}(t) = 2\mathbf{j} + \mathbf{k}$

$\mathbf{T}(t) = \dfrac{\mathbf{v}}{\|\mathbf{v}\|} = \dfrac{1}{\sqrt{1 + 5t^2}}(\mathbf{i} + 2t\mathbf{j} + t\mathbf{k})$

$\mathbf{T}(1) = \dfrac{\sqrt{6}}{6}(\mathbf{i} + 2\mathbf{j} + \mathbf{k})$

$\mathbf{N}(t) = \dfrac{\mathbf{T}'}{\|\mathbf{T}'\|} = \dfrac{\dfrac{-5t\mathbf{i} + 2\mathbf{j} + \mathbf{k}}{\left(1 + 5t^2\right)^{3/2}}}{\dfrac{\sqrt{5}}{1 + 5t^2}} = \dfrac{-5t\mathbf{i} + 2\mathbf{j} + \mathbf{k}}{\sqrt{5}\sqrt{1 + 5t^2}}$

$\mathbf{N}(1) = \dfrac{\sqrt{30}}{30}(-5\mathbf{i} + 2\mathbf{j} + \mathbf{k})$

$a_{\mathbf{T}} = \mathbf{a} \cdot \mathbf{T} = \dfrac{5\sqrt{6}}{6}$

$a_{\mathbf{N}} = \mathbf{a} \cdot \mathbf{N} = \dfrac{\sqrt{30}}{6}$

39. $\mathbf{r}(t) = e^t \sin t\,\mathbf{i} + e^t \cos t\,\mathbf{j} + e^t\mathbf{k}, t = 0$

$\mathbf{v}(t) = \left(e^t \cos t + e^t \sin t\right)\mathbf{i} + \left(-e^t \sin t + e^t \cos t\right)\mathbf{j} + e^t\mathbf{k}$

$\mathbf{v}(0) = \mathbf{i} + \mathbf{j} + \mathbf{k}$

$\mathbf{a}(t) = 2e^t \cos t\,\mathbf{i} - 2e^t \sin t\,\mathbf{j} + e^t\mathbf{k}$

$\mathbf{a}(0) = 2\mathbf{i} + \mathbf{k}$

$\mathbf{T}(t) = \dfrac{\mathbf{v}}{\|\mathbf{v}\|}$

$\qquad = \dfrac{1}{\sqrt{3}}\left[(\cos t + \sin t)\mathbf{i} + (-\sin t + \cos t)\mathbf{j} + \mathbf{k}\right]$

$\mathbf{T}(0) = \dfrac{1}{\sqrt{3}}[\mathbf{i} + \mathbf{j} + \mathbf{k}]$

$\mathbf{N}(t) = \dfrac{1}{\sqrt{2}}\left[(-\sin t + \cos t)\mathbf{i} + (-\cos t - \sin t)\mathbf{j}\right]$

$\mathbf{N}(0) = \dfrac{\sqrt{2}}{2}\mathbf{i} - \dfrac{\sqrt{2}}{2}\mathbf{j}$

$a_{\mathbf{T}} = \mathbf{a} \cdot \mathbf{T} = \sqrt{3}$

$a_{\mathbf{N}} = \mathbf{a} \cdot \mathbf{N} = \sqrt{2}$

41. If $a_{\mathbf{N}} = 0$, then the motion is in a straight line.

43. $\mathbf{r}(t) = 3t\mathbf{i} + 4t\mathbf{j}$

$\mathbf{v}(t) = \mathbf{r}'(t) = 3\mathbf{i} + 4\mathbf{j}, \|\mathbf{v}(t)\| = \sqrt{9 + 16} = 5$

$\mathbf{a}(t) = \mathbf{v}'(t) = \mathbf{0}$

$\mathbf{T}(t) = \dfrac{\mathbf{v}(t)}{\|\mathbf{v}(t)\|} = \dfrac{3}{5}\mathbf{i} + \dfrac{4}{5}\mathbf{j}$

$\mathbf{T}'(t) = \mathbf{0} \Rightarrow \mathbf{N}(t)$ does not exist.

The path is a line. The speed is constant (5).

45. $\mathbf{r}(t) = \langle \pi t - \sin \pi t, 1 - \cos \pi t \rangle$

The graph is a cycloid.

(a) $\mathbf{r}(t) = \langle \pi t - \sin \pi t, 1 - \cos \pi t \rangle$

$\mathbf{v}(t) = \langle \pi - \pi \cos \pi t, \pi \sin \pi t \rangle$

$\mathbf{a}(t) = \langle \pi^2 \sin \pi t, \pi^2 \cos \pi t \rangle$

$\mathbf{T}(t) = \dfrac{\mathbf{v}(t)}{\|\mathbf{v}(t)\|} = \dfrac{1}{\sqrt{2(1 - \cos \pi t)}} \langle 1 - \cos \pi t, \sin \pi t \rangle$

$\mathbf{N}(t) = \dfrac{\mathbf{T}'(t)}{\|\mathbf{T}'(t)\|} = \dfrac{1}{\sqrt{2(1 - \cos \pi t)}} \langle \sin \pi t, -1 + \cos \pi t \rangle$

$a_{\mathbf{T}} = \mathbf{a} \cdot \mathbf{T} = \dfrac{1}{\sqrt{2(1 - \cos \pi t)}} \left[\pi^2 \sin \pi t (1 - \cos \pi t) + \pi^2 \cos \pi t \sin \pi t \right] = \dfrac{\pi^2 \sin \pi t}{\sqrt{2(1 - \cos \pi t)}}$

$a_{\mathbf{N}} = \mathbf{a} \cdot \mathbf{N} = \dfrac{1}{\sqrt{2(1 - \cos \pi t)}} \left[\pi^2 \sin^2 \pi t + \pi^2 \cos \pi t (-1 + \cos \pi t) \right] = \dfrac{\pi^2 (1 - \cos \pi t)}{\sqrt{2(1 - \cos \pi t)}} = \dfrac{\pi^2 \sqrt{2(1 - \cos \pi t)}}{2}$

When $t = \dfrac{1}{2}$: $a_{\mathbf{T}} = \dfrac{\pi^2}{\sqrt{2}} = \dfrac{\sqrt{2}\pi^2}{2}$, $a_{\mathbf{N}} = \dfrac{\sqrt{2}\pi^2}{2}$

When $t = 1$: $a_{\mathbf{T}} = 0$, $a_{\mathbf{N}} = \pi^2$

When $t = \dfrac{3}{2}$: $a_{\mathbf{T}} = -\dfrac{\sqrt{2}\pi^2}{2}$, $a_{\mathbf{N}} = \dfrac{\sqrt{2}\pi^2}{2}$

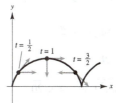

(b) Speed: $s = \|\mathbf{v}(t)\| = \pi\sqrt{2(1 - \cos \pi t)}$

$\dfrac{ds}{dt} = \dfrac{\pi^2 \sin \pi t}{\sqrt{2(1 - \cos \pi t)}} = a_{\mathbf{T}}$

When $t = \dfrac{1}{2}$: $a_{\mathbf{T}} = \dfrac{\sqrt{2}\pi^2}{2} > 0 \Rightarrow$ the speed in increasing.

When $t = 1$: $a_{\mathbf{T}} = 0 \Rightarrow$ the height is maximum.

When $t = \dfrac{3}{2}$: $a_{\mathbf{T}} = -\dfrac{\sqrt{2}\pi^2}{2} < 0 \Rightarrow$ the speed is decreasing.

47. $\mathbf{r}(t) = 2\cos t\,\mathbf{i} + 2\sin t\,\mathbf{j} + \dfrac{t}{2}\mathbf{k}$, $t_0 = \dfrac{\pi}{2}$

$\mathbf{r}'(t) = -2\sin t\,\mathbf{i} + 2\cos t\,\mathbf{j} + \dfrac{1}{2}\mathbf{k}$

$\mathbf{T}(t) = \dfrac{2\sqrt{17}}{17}\left(-2\sin t\,\mathbf{i} + 2\cos t\,\mathbf{j} + \dfrac{1}{2}\mathbf{k} \right)$

$\mathbf{N}(t) = -\cos t\,\mathbf{i} - \sin t\,\mathbf{j}$

$\mathbf{r}\left(\dfrac{\pi}{2}\right) = 2\mathbf{j} + \dfrac{\pi}{4}\mathbf{k}$

$\mathbf{T}\left(\dfrac{\pi}{2}\right) = \dfrac{2\sqrt{17}}{17}\left(-2\mathbf{i} + \dfrac{1}{2}\mathbf{k} \right) = \dfrac{\sqrt{17}}{17}(-4\mathbf{i} + \mathbf{k})$

$\mathbf{N}\left(\dfrac{\pi}{2}\right) = -\mathbf{j}$

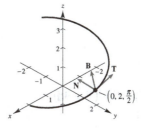

$\mathbf{B}\left(\dfrac{\pi}{2}\right) = \mathbf{T}\left(\dfrac{\pi}{2}\right) \times \mathbf{N}\left(\dfrac{\pi}{2}\right) = \begin{vmatrix} \mathbf{i} & \mathbf{j} & \mathbf{k} \\ -\dfrac{4\sqrt{17}}{17} & 0 & \dfrac{\sqrt{17}}{17} \\ 0 & -1 & 0 \end{vmatrix} = \dfrac{\sqrt{17}}{17}\mathbf{i} + \dfrac{4\sqrt{17}}{17}\mathbf{k} = \dfrac{\sqrt{17}}{17}(\mathbf{i} + 4\mathbf{k})$

49. $\mathbf{r}(t) = \mathbf{i} + \sin t\mathbf{j} + \cos t\mathbf{k}, \ t_0 = \dfrac{\pi}{4}$

$\mathbf{r}'(t) = \cos t\mathbf{j} - \sin t\mathbf{k},$

$\|\mathbf{r}'(t)\| = 1$

$\mathbf{r}'\left(\dfrac{\pi}{4}\right) = \mathbf{T}\left(\dfrac{\pi}{4}\right) = \dfrac{\sqrt{2}}{2}\mathbf{j} - \dfrac{\sqrt{2}}{2}\mathbf{k}$

$\mathbf{T}'(t) = -\sin t\mathbf{j} - \cos t\mathbf{k},$

$\mathbf{N}\left(\dfrac{\pi}{4}\right) = -\dfrac{\sqrt{2}}{2}\mathbf{j} - \dfrac{\sqrt{2}}{2}\mathbf{k}$

$\mathbf{B}\left(\dfrac{\pi}{4}\right) = \mathbf{T}\left(\dfrac{\pi}{4}\right) \times \mathbf{N}\left(\dfrac{\pi}{4}\right) = \begin{vmatrix} \mathbf{i} & \mathbf{j} & \mathbf{k} \\ 0 & \dfrac{\sqrt{2}}{2} & -\dfrac{\sqrt{2}}{2} \\ 0 & -\dfrac{\sqrt{2}}{2} & -\dfrac{\sqrt{2}}{2} \end{vmatrix} = -\mathbf{i}$

51. $\mathbf{r}(t) = 4\sin t\mathbf{i} + 4\cos t\mathbf{j} + 2t\mathbf{k}, \ t_0 = \dfrac{\pi}{3}$

$\mathbf{r}'(t) = 4\cos t\mathbf{i} - 4\sin t\mathbf{j} + 2\mathbf{k},$

$\|\mathbf{r}'(t)\| = \sqrt{16\cos^2 t + 16\sin^2 t + 4} = \sqrt{20} = 2\sqrt{5}$

$\mathbf{r}'\left(\dfrac{\pi}{3}\right) = 2\mathbf{i} - 2\sqrt{3}\mathbf{j} + 2\mathbf{k}$

$\mathbf{T}\left(\dfrac{\pi}{3}\right) = \dfrac{1}{2\sqrt{5}}\left(2\mathbf{i} - 2\sqrt{3}\mathbf{j} + 2\mathbf{k}\right) = \dfrac{\sqrt{5}}{5}\mathbf{i} - \dfrac{\sqrt{15}}{5}\mathbf{j} + \dfrac{\sqrt{5}}{5}\mathbf{k} = \dfrac{\sqrt{5}}{5}\left(\mathbf{i} - \sqrt{3}\mathbf{j} + \mathbf{k}\right)$

$\mathbf{T}'(t) = \dfrac{1}{2\sqrt{5}}\left(-4\sin t\mathbf{i} - 4\cos t\mathbf{j}\right)$

$\mathbf{N}\left(\dfrac{\pi}{3}\right) = -\dfrac{\sqrt{3}}{2}\mathbf{i} - \dfrac{1}{2}\mathbf{j}$

$\mathbf{B}\left(\dfrac{\pi}{3}\right) = \mathbf{T}\left(\dfrac{\pi}{3}\right) \times \mathbf{N}\left(\dfrac{\pi}{3}\right) = \begin{vmatrix} \mathbf{i} & \mathbf{j} & \mathbf{k} \\ \dfrac{\sqrt{5}}{5} & -\dfrac{\sqrt{15}}{5} & \dfrac{\sqrt{5}}{5} \\ -\dfrac{\sqrt{3}}{2} & -\dfrac{1}{2} & 0 \end{vmatrix} = \dfrac{\sqrt{5}}{10}\mathbf{i} - \dfrac{\sqrt{15}}{10}\mathbf{j} - \dfrac{4\sqrt{5}}{10}\mathbf{k} = \dfrac{\sqrt{5}}{10}\left(\mathbf{i} - \sqrt{3}\mathbf{j} - 4\mathbf{k}\right)$

53. $\mathbf{r}(t) = 3t\mathbf{i} + 2t^2\mathbf{j}$

$\mathbf{v}(t) = 3\mathbf{i} + 4t\mathbf{j}$

$\mathbf{a}(t) = 4\mathbf{j}$

$\mathbf{v} \cdot \mathbf{v} = 9 + 16t^2$

$\mathbf{v} \cdot \mathbf{a} = 16t$

$(\mathbf{v} \cdot \mathbf{v})\mathbf{a} - (\mathbf{v} \cdot \mathbf{a})\mathbf{v} = \left(9 + 16t^2\right)4\mathbf{j} - (16t)(3\mathbf{i} + 4t\mathbf{j})$

$\qquad\qquad = -48t\mathbf{i} + 36\mathbf{j}$

$\mathbf{N} = \dfrac{(\mathbf{v} \cdot \mathbf{v})\mathbf{a} - (\mathbf{v} \cdot \mathbf{a})\mathbf{v}}{\|(\mathbf{v} \cdot \mathbf{v})\mathbf{a} - (\mathbf{v} \cdot \mathbf{a})\mathbf{v}\|} = \dfrac{1}{\sqrt{9 + 16t^2}}(-4t\mathbf{i} + 3\mathbf{j})$

55. $\mathbf{r}(t) = 2t\mathbf{i} + 4t\mathbf{j} + t^2\mathbf{k}$

$\mathbf{v}(t) = 2\mathbf{i} + 4\mathbf{j} + 2t\mathbf{k}$

$\mathbf{a}(t) = 2\mathbf{k}$

$\mathbf{v} \cdot \mathbf{v} = 4 + 16 + 4t^2 = 20 + 4t^2$

$\mathbf{v} \cdot \mathbf{a} = 4t$

$(\mathbf{v} \cdot \mathbf{v})\mathbf{a} - (\mathbf{v} \cdot \mathbf{a})\mathbf{v} = \left(20 + 4t^2\right)2\mathbf{k} - 4t(2\mathbf{i} + 4\mathbf{j} + 2t\mathbf{k})$

$\qquad\qquad = -8t\mathbf{i} - 16t\mathbf{j} + 40\mathbf{k}$

$\mathbf{N} = \dfrac{(\mathbf{v} \cdot \mathbf{v})\mathbf{a} - (\mathbf{v} \cdot \mathbf{a})\mathbf{v}}{\|(\mathbf{v} \cdot \mathbf{v})\mathbf{a} - (\mathbf{v} \cdot \mathbf{a})\mathbf{v}\|} = \dfrac{1}{\sqrt{5t^2 + 25}}(-t\mathbf{i} - 2t\mathbf{j} + 5\mathbf{k})$

57. From Theorem 12.3 you have:

$$\mathbf{r}(t) = (v_0 t \cos \theta)\mathbf{i} + \left(h + v_0 t \sin \theta - 16t^2\right)\mathbf{j}$$

$$\mathbf{v}(t) = v_0 \cos \theta\mathbf{i} + (v_0 \sin \theta - 32t)\mathbf{j}$$

$$\mathbf{a}(t) = -32\mathbf{j}$$

$$\mathbf{T}(t) = \frac{(v_0 \cos \theta)\mathbf{i} + (v_0 \sin \theta - 32t)\mathbf{j}}{\sqrt{v_0^2 \cos^2 \theta + (v_0 \sin \theta - 32t)^2}}$$

$$\mathbf{N}(t) = \frac{(v_0 \sin \theta - 32t)\mathbf{i} - v_0 \cos \theta\mathbf{j}}{\sqrt{v_0^2 \cos^2 \theta + (v_0 \sin \theta - 32t)^2}} \quad \text{(Motion is clockwise.)}$$

$$a_\mathbf{T} = \mathbf{a} \cdot \mathbf{T} = \frac{-32(v_0 \sin \theta - 32t)}{\sqrt{v_0^2 \cos^2 \theta + (v_0 \sin \theta - 32t)^2}}$$

$$a_\mathbf{N} = \mathbf{a} \cdot \mathbf{N} = \frac{32v_0 \cos \theta}{\sqrt{v_0^2 \cos^2 \theta + (v_0 \sin \theta - 32t)^2}}$$

Maximum height when $v_0 \sin \theta - 32t = 0$; (vertical component of velocity)

At maximum height, $a_\mathbf{T} = 0$ and $a_\mathbf{N} = 32$.

59. (a) $\mathbf{r}(t) = (v_0 \cos \theta)t\mathbf{i} + \left[h + (v_0 \sin \theta)t - \frac{1}{2}gt^2\right]\mathbf{j}$

$$= (120 \cos 30°)t\mathbf{i} + \left[5 + (120 \sin 30°)t - 16t^2\right]\mathbf{j} = 60\sqrt{3}t\mathbf{i} + \left[5 + 60t - 16t^2\right]\mathbf{j}$$

(b)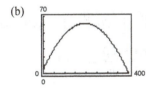

Maximum height ≈ 61.25 feet

range ≈ 398.2 feet

(c) $\mathbf{v}(t) = 60\sqrt{3}\mathbf{i} + (60 - 32t)\mathbf{j}$

$$\text{Speed} = \|\mathbf{v}(t)\| = \sqrt{3600(3) + (60 - 32t)^2} = 8\sqrt{16t^2 - 60t + 225}$$

$$\mathbf{a}(t) = -32\mathbf{j}$$

(d)

t	0.5	1.0	1.5	2.0	2.5	3.0
Speed	112.85	107.63	104.61	104.0	105.83	109.98

(e) From Exercise 61, using $v_0 = 120$ and $\theta = 30°$,

$$a_\mathbf{T} = \frac{-32(60 - 32t)}{\sqrt{\left(60\sqrt{3}\right)^2 + (60 - 32t)^2}}$$

$$a_\mathbf{N} = \frac{32\left(60\sqrt{3}\right)}{\sqrt{\left(60\sqrt{3}\right)^2 + (60 - 32t)^2}}$$

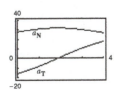

At $t = 1.875$, $a_\mathbf{T} = 0$ and the projectile is at its maximum height. When $a_\mathbf{T}$ and $a_\mathbf{N}$ have opposite signs, the speed is decreasing.

61. $\mathbf{r}(t) = \langle 10 \cos 10\pi t, 10 \sin 10\pi t, 4 + 4t \rangle, 0 \le t \le \frac{1}{20}$

   (a) $\mathbf{r}'(t) = \langle -100\pi \sin(10\pi t), 100\pi \cos(10\pi t), 4 \rangle$

$$\|\mathbf{r}'(t)\| = \sqrt{(100\pi)^2 \sin^2(10\pi t) + (100\pi)^2 \cos^2(10\pi t) + 16}$$

$$= \sqrt{(100\pi)^2 + 16} = 4\sqrt{625\pi^2 + 1} \approx 314 \text{ mi/h}$$

   (b) $a_{\mathbf{T}} = 0$ and $a_{\mathbf{N}} = 1000\pi^2$

       $a_{\mathbf{T}} = 0$ because the speed is constant

63. $\mathbf{r}(t) = (a \cos \omega t)\mathbf{i} + (a \sin \omega t)\mathbf{j}$

From Exercise 31, you know $\mathbf{a} \cdot \mathbf{T} = 0$ and
$\mathbf{a} \cdot \mathbf{N} = a\omega^2$.

   (a) Let $\omega_0 = 2\omega$. Then

$$\mathbf{a} \cdot \mathbf{N} = a\omega_0^2 = a(2\omega)^2 = 4a\omega^2$$

or the centripetal acceleration is increased by a factor
of 4 when the velocity is doubled.

   (b) Let $a_0 = a/2$. Then

$$\mathbf{a} \cdot \mathbf{N} = a_0 \omega^2 = \left(\frac{a}{2}\right)\omega^2 = \left(\frac{1}{2}\right)a\omega^2$$

or the centripetal acceleration is halved when the
radius is halved.

65. $\mathbf{v} = \sqrt{\dfrac{GM}{r}} = \sqrt{\dfrac{9.56 \times 10^4}{4000 + 255}} \approx 4.74 \text{ mi/sec}$

67. $v = \sqrt{\dfrac{9.56 \times 10^4}{4000 + 385}} \approx 4.67 \text{ mi/sec}$

69. False. These vectors are perpendicular for an object
traveling at a constant speed but not for an object
traveling at a variable speed.

71. (a) $\mathbf{r}(t) = \cosh(bt)\mathbf{i} + \sinh(bt)\mathbf{j}, b > 0$

      $x = \cosh(bt), y = \sinh(bt)$

      $x^2 - y^2 = \cosh^2(bt) - \sinh^2(bt) = 1$, hyperbola

  (b) $\mathbf{v}(t) = b \sinh(bt)\mathbf{i} + b \cosh(bt)\mathbf{j}$

      $\mathbf{a}(t) = b^2 \cosh(bt)\mathbf{i} + b^2 \sinh(bt)\mathbf{j} = b^2 \mathbf{r}(t)$

73. $\mathbf{r}(t) = x(t)\mathbf{i} + y(t)\mathbf{j}$

    $y(t) = m(x(t)) + b$, m and b are constants.

    $\mathbf{r}(t) = x(t)\mathbf{i} + \left[m(x(t)) + b \right]\mathbf{j}$

    $\mathbf{v}(t) = x'(t)\mathbf{i} + mx'(t)\mathbf{j}$

    $\|\mathbf{v}(t)\| = \sqrt{\left[x'(t)\right]^2 + \left[mx'(t)\right]^2} = |x'(t)|\sqrt{1 + m^2}$

    $\mathbf{T}(t) = \dfrac{\mathbf{v}(t)}{\|\mathbf{v}(t)\|} = \dfrac{\pm(\mathbf{i} + m\mathbf{j})}{\sqrt{1 + m^2}}$, constant

    So, $\mathbf{T}'(t) = \mathbf{0}$.

75. $\|\mathbf{a}\|^2 = \mathbf{a} \cdot \mathbf{a}$

$$= (a_{\mathbf{T}}\mathbf{T} + a_{\mathbf{N}}\mathbf{N}) \cdot (a_{\mathbf{T}}\mathbf{T} + a_{\mathbf{N}}\mathbf{N})$$

$$= a_{\mathbf{T}}^2 \|\mathbf{T}\|^2 + 2a_{\mathbf{T}}a_{\mathbf{N}}\mathbf{T} \cdot \mathbf{N} + a_{\mathbf{N}}^2 \|\mathbf{N}\|^2$$

$$= a_{\mathbf{T}}^2 + a_{\mathbf{N}}^2$$

$$a_{\mathbf{N}}^2 = \|\mathbf{a}\|^2 - a_{\mathbf{T}}^2$$

Because $a_{\mathbf{N}} > 0$, we have $a_{\mathbf{N}} = \sqrt{\|\mathbf{a}\|^2 - a_{\mathbf{T}}^2}$.

Section 12.5 Arc Length and Curvature

1. The curve bends more sharply at Q than at P.

3. $\mathbf{r}(t) = 3t\mathbf{i} - t\mathbf{j}, [0, 3]$

$$\frac{dx}{dt} = 3, \frac{dy}{dt} = -1, \frac{dz}{dt} = 0$$

$$s = \int_0^3 \sqrt{3^2 + (-1)^2}\, dt = \left[\sqrt{10}\,t\right]_0^3 = 3\sqrt{10}$$

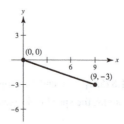

5. $\mathbf{r}(t) = t^3\mathbf{i} + t^2\mathbf{j}, [0, 1]$

$$\frac{dx}{dt} = 3t^2, \frac{dy}{dt} = 2t, \frac{dz}{dt} = 0$$

$$s = \int_0^1 \sqrt{9t^4 + 4t^2}\, dt = \int_0^1 \sqrt{9t^2 + 4}\, t\, dt$$

$$= \frac{1}{18}\int_0^1 (9t^2 + 4)^{1/2}(18t)\, dt = \frac{1}{27}\left[(9t^2 + 4)^{3/2}\right]_0^1 = \frac{1}{27}(13^{3/2} - 8) \approx 1.4397$$

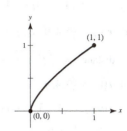

7. $\mathbf{r}(t) = a\cos^3 t\mathbf{i} + a\sin^3 t\mathbf{j}, [0, 2\pi]$

$$\frac{dx}{dt} = -3a\cos^2 t\sin t, \frac{dy}{dt} = 3a\sin^2 t\cos t$$

$$s = 4\int_0^{\pi/2}\sqrt{\left[-3a\cos^2 t\sin t\right]^2 + \left[3a\sin^2 t\cos t\right]^2}\, dt$$

$$= 12a\int_0^{\pi/2}\sin t\cos t\, dt = 3a\int_0^{\pi/2} 2\sin 2t\, dt = \left[-3a\cos 2t\right]_0^{\pi/2} = 6a$$

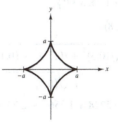

9. $\mathbf{r}(t) = 50\sqrt{2}t\mathbf{i} + \left[3 + 50\sqrt{2}t - 16t^2\right]\mathbf{j}$

$$\mathbf{v}(t) = 50\sqrt{2}\mathbf{i} + \left(50\sqrt{2} - 32t\right)\mathbf{j}$$

$$3 + 50\sqrt{2}t - 16t^2 = 0 \Rightarrow t \approx 4.4614$$

$$s = \int_0^{4.4614}\sqrt{\left(50\sqrt{2}\right)^2 + \left(50\sqrt{2} - 32t\right)^2}\, dt \approx 362.9\ \text{ft}$$

11. $\mathbf{r}(t) = -t\mathbf{i} + 4t\mathbf{j} + 3t\mathbf{k}, [0, 1]$

$$\frac{dx}{dt} = -1, \frac{dy}{dt} = 4, \frac{dz}{dt} = 3$$

$$s = \int_0^1\sqrt{1 + 16 + 9}\, dt = \left[\sqrt{26}t\right]_0^1 = \sqrt{26}$$

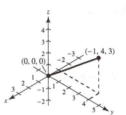

13. $\mathbf{r}(t) = \langle 4t, -\cos t, \sin t\rangle, \left[0, \dfrac{3\pi}{2}\right]$

$$\frac{dx}{dt} = 4, \frac{dy}{dt} = \sin t, \frac{dz}{dt} = \cos t$$

$$s = \int_0^{3\pi/2}\sqrt{16 + \sin^2 t + \cos^2 t}\, dt = \int_0^{3\pi/2}\sqrt{17}\, dt = \left[\sqrt{17}t\right]_0^{3\pi/2} = \frac{3\pi}{2}\sqrt{17}$$

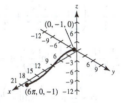

15. $\mathbf{r}(t) = a\cos t\mathbf{i} + a\sin t\mathbf{j} + bt\mathbf{k}, [0, 2\pi]$

$$\frac{dx}{dt} = -a\sin t, \frac{dy}{dt} = a\cos t, \frac{dz}{dt} = b$$

$$s = \int_0^{2\pi}\sqrt{a^2\sin^2 t + a^2\cos^2 t + b^2}\, dt$$

$$= \int_0^{2\pi}\sqrt{a^2 + b^2}\, dt = \left[\sqrt{a^2 + b^2}t\right]_0^{2\pi} = 2\pi\sqrt{a^2 + b^2}$$

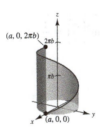

17. $\mathbf{r}(t) = t\mathbf{i} + (4 - t^2)\mathbf{j} + t^3\mathbf{k}, \ [0, 2]$

    (a) $\mathbf{r}(0) = \langle 0, 4, 0 \rangle, \mathbf{r}(2) = \langle 2, 0, 8 \rangle$

        distance $= \sqrt{2^2 + 4^2 + 8^2} = \sqrt{84} = 2\sqrt{21} \approx 9.165$

    (b) $\mathbf{r}(0) = \langle 0, 4, 0 \rangle$

        $\mathbf{r}(0.5) = \langle 0.5, 3.75, 0.125 \rangle$

        $\mathbf{r}(1) = \langle 1, 3, 1 \rangle$

        $\mathbf{r}(1.5) = \langle 1.5, 1.75, 3.375 \rangle$

        $\mathbf{r}(2) = \langle 2, 0, 8 \rangle$

        distance $\approx \sqrt{(0.5)^2 + (0.25)^2 + (0.125)^2} + \sqrt{(0.5)^2 + (0.75)^2 + (0.875)^2} + \sqrt{(0.5)^2 + (1.25)^2 + (2.375)^2}$

                    $+ \sqrt{(0.5)^2 + (1.75)^2 + (4.625)^2}$

          $\approx 0.5728 + 1.2562 + 2.7300 + 4.9702 \approx 9.529$

    (c) Increase the number of line segments.

    (d) Using a graphing utility, you obtain 9.57057.

19. $\mathbf{r}(s) = \left(1 + \dfrac{\sqrt{2}}{2}s\right)\mathbf{i} + \left(1 - \dfrac{\sqrt{2}}{2}s\right)\mathbf{j}$

    $\mathbf{r}'(s) = \dfrac{\sqrt{2}}{2}\mathbf{i} - \dfrac{\sqrt{2}}{2}\mathbf{j}$ and $\|\mathbf{r}'(s)\| = \sqrt{\dfrac{1}{2} + \dfrac{1}{2}} = 1$

    $\mathbf{T}(s) = \dfrac{\mathbf{r}'(s)}{\|\mathbf{r}'(s)\|} = \mathbf{r}'(s)$

    $\mathbf{T}'(s) = \mathbf{0} \Rightarrow K = \|\mathbf{T}'(s)\| = 0$ (The curve is a line.)

21. $\mathbf{r}(s) = \cos\left(\dfrac{s}{2}\right)\mathbf{i} + \dfrac{\sqrt{3}}{2}s\mathbf{j} + \sin\left(\dfrac{s}{2}\right)\mathbf{k}$

    $\mathbf{r}'(s) = -\dfrac{1}{2}\sin\left(\dfrac{s}{2}\right)\mathbf{i} + \dfrac{\sqrt{3}}{2}\mathbf{j} + \dfrac{1}{2}\cos\left(\dfrac{s}{2}\right)\mathbf{k}, \ \|\mathbf{r}'(s)\| = 1$

    $\mathbf{T}(s) = \dfrac{\mathbf{r}'(s)}{\|\mathbf{r}'(s)\|} = -\dfrac{1}{2}\sin\left(\dfrac{s}{2}\right)\mathbf{i} + \dfrac{\sqrt{3}}{2}\mathbf{j} + \dfrac{1}{2}\cos\left(\dfrac{s}{2}\right)\mathbf{k}$

    $\mathbf{T}'(s) = -\dfrac{1}{4}\cos\left(\dfrac{s}{2}\right)\mathbf{i} - \dfrac{1}{4}\sin\left(\dfrac{s}{2}\right)\mathbf{k}$

    $K = \|\mathbf{T}'(s)\| = \sqrt{\dfrac{1}{16}\cos^2\left(\dfrac{s}{2}\right) + \dfrac{1}{16}\sin^2\left(\dfrac{s}{2}\right)} = \dfrac{1}{4}$

23. $\mathbf{r}(t) = 4t\mathbf{i} - 2t\mathbf{j}, \ t = 1$

    $\mathbf{v}(t) = 4\mathbf{i} - 2\mathbf{j}$

    $\mathbf{T}(t) = \dfrac{1}{\sqrt{5}}(2\mathbf{i} - \mathbf{j})$

    $\mathbf{T}'(t) = \mathbf{0}$

    $K = \dfrac{\|\mathbf{T}'(t)\|}{\|\mathbf{r}'(t)\|} = 0$

    (The curve is a line.)

25. $\mathbf{r}(t) = t\mathbf{i} + \dfrac{1}{t}\mathbf{j}, \ t = 1$

    $\mathbf{v}(t) = \mathbf{i} - \dfrac{1}{t^2}\mathbf{j}$

    $\mathbf{v}(1) = \mathbf{i} - \mathbf{j}$

    $\mathbf{a}(t) = \dfrac{2}{t^3}\mathbf{j}$

    $\mathbf{a}(1) = 2\mathbf{j}$

    $\mathbf{T}(t) = \dfrac{t^2\mathbf{i} - \mathbf{j}}{\sqrt{t^4 + 1}}$

    $\mathbf{N}(t) = \dfrac{1}{(t^4 + 1)^{1/2}}(\mathbf{i} + t^2\mathbf{j})$

    $\mathbf{N}(1) = \dfrac{1}{\sqrt{2}}(\mathbf{i} + \mathbf{j})$

    $K = \dfrac{\mathbf{a} \cdot \mathbf{N}}{\|\mathbf{v}\|^2} = \dfrac{\sqrt{2}}{2}$

27. $\mathbf{r}(t) = \langle t, \sin t \rangle, t = \dfrac{\pi}{2}$

$\mathbf{r}'(t) = \langle 1, \cos t \rangle, \|\mathbf{r}'(t)\| = \sqrt{1 + \cos^2 t}$

$\mathbf{r}'\left(\dfrac{\pi}{2}\right) = \langle 1, 0 \rangle, \left\|\mathbf{r}\left(\dfrac{\pi}{2}\right)\right\| = 1$

$\mathbf{a}(t) = \langle 0, -\sin t \rangle, \mathbf{a}\left(\dfrac{\pi}{2}\right) = \langle 0, -1 \rangle$

$\mathbf{T}(t) = \dfrac{1}{\sqrt{1 + \cos^2 t}} \langle 1, \cos t \rangle$

$\mathbf{T}\left(\dfrac{\pi}{2}\right) = \langle 1, 0 \rangle$

$\mathbf{N}\left(\dfrac{\pi}{2}\right) = \langle 0, -1 \rangle$

$K = \dfrac{\mathbf{a} \cdot \mathbf{N}}{\|\mathbf{v}\|^2} = \dfrac{1}{1} = 1$

29. $\mathbf{r}(t) = 4\cos 2\pi t\,\mathbf{i} + 4\sin 2\pi t\,\mathbf{j}$

$\mathbf{r}'(t) = -8\pi \sin 2\pi t\,\mathbf{i} + 8\pi \cos 2\pi t\,\mathbf{j}$

$\mathbf{T}(t) = -\sin 2\pi t\,\mathbf{i} + \cos 2t\,\mathbf{j}$

$\mathbf{T}'(t) = -2\pi \cos 2\pi t\,\mathbf{i} - 2\pi \sin 2\pi t\,\mathbf{j}$

$K = \dfrac{\|\mathbf{T}'(t)\|}{\|\mathbf{r}'(t)\|} = \dfrac{2\pi}{8\pi} = \dfrac{1}{4}$

31. $\mathbf{r}(t) = a\cos \omega t\,\mathbf{i} + a\sin \omega t\,\mathbf{j}$

$\mathbf{r}'(t) = -a\omega \sin \omega t\,\mathbf{i} + a\omega \cos \omega t\,\mathbf{j}$

$\mathbf{T}(t) = -\sin \omega t\,\mathbf{i} + \cos \omega t\,\mathbf{j}$

$\mathbf{T}'(t) = -\omega \cos \omega t\,\mathbf{i} - \omega \sin \omega t\,\mathbf{j}$

$K = \dfrac{\|\mathbf{T}'(t)\|}{\|\mathbf{r}'(t)\|} = \dfrac{\omega}{a\omega} = \dfrac{1}{a}$

33. $\mathbf{r}(t) = t\mathbf{i} + t^2\mathbf{j} + \dfrac{t^2}{2}\mathbf{k}$

$\mathbf{r}'(t) = \mathbf{i} + 2t\mathbf{j} + t\mathbf{k}$

$\mathbf{T}(t) = \dfrac{\mathbf{i} + 2t\mathbf{j} + t\mathbf{k}}{\sqrt{1 + 5t^2}}$

$\mathbf{T}'(t) = \dfrac{-5t\mathbf{i} + 2\mathbf{j} + \mathbf{k}}{\left(1 + 5t^2\right)^{3/2}}$

$K = \dfrac{\|\mathbf{T}'(t)\|}{\|\mathbf{r}'(t)\|} = \dfrac{\dfrac{\sqrt{5}}{\left(1 + 5t^2\right)}}{\sqrt{1 + 5t^2}} = \dfrac{\sqrt{5}}{\left(1 + 5t^2\right)^{3/2}}$

35. $\mathbf{r}(t) = 4t\mathbf{i} + 3\cos t\mathbf{j} + 3\sin t\mathbf{k}$

$\mathbf{r}'(t) = 4\mathbf{i} - 3\sin t\mathbf{j} + 3\cos t\mathbf{k}$

$\mathbf{T}(t) = \dfrac{1}{5}[4\mathbf{i} - 3\sin t\mathbf{j} + 3\cos t\mathbf{k}]$

$\mathbf{T}'(t) = \dfrac{1}{5}[-3\cos t\mathbf{j} - 3\sin t\mathbf{k}]$

$K = \dfrac{\|\mathbf{T}'(t)\|}{\mathbf{r}'(t)} = \dfrac{3/5}{5} = \dfrac{3}{25}$

37. $\mathbf{r}(t) = 3t\mathbf{i} + 2t^2\mathbf{j}, \; P(-3, 2) \Rightarrow t = -1$

$x = 3t, \; x' = 3, \; x'' = 0$

$y = 2t^2, \; y' = 4t, \; y'' = 4$

$K = \dfrac{|x'y'' - y'x''|}{\left[(x')^2 + (y')^2\right]^{3/2}} = \dfrac{|3(4) - 0|}{\left[9 + (4t)^2\right]^{3/2}}$

At $t = -1$, $K = \dfrac{12}{(9 + 16)^{3/2}} = \dfrac{12}{125}$

39. $\mathbf{r}(t) = t\mathbf{i} + t^2\mathbf{j} + \dfrac{t^3}{4}\mathbf{k}, \; P(2, 4, 2) \Rightarrow t = 2$

$\mathbf{r}'(t) = \mathbf{i} + 2t\mathbf{j} + \dfrac{3}{4}t^2\mathbf{k}$

$\mathbf{r}'(2) = \mathbf{i} + 4\mathbf{j} + 3\mathbf{k}, \|\mathbf{r}'(2)\| = \sqrt{26}$

$\mathbf{r}''(t) = 2\mathbf{j} + \dfrac{3}{2}t\mathbf{k}$

$\mathbf{r}''(2) = 2\mathbf{j} + 3\mathbf{k}$

$\mathbf{r}'(2) \times \mathbf{r}''(2) = \begin{vmatrix} \mathbf{i} & \mathbf{j} & \mathbf{k} \\ 1 & 4 & 3 \\ 0 & 2 & 3 \end{vmatrix} = 6\mathbf{i} - 3\mathbf{j} + 2\mathbf{k}$

$\|\mathbf{r}'(2) \times \mathbf{r}''(2)\| = \sqrt{49} = 7$

$K = \dfrac{\|\mathbf{r}' \times \mathbf{r}''\|}{\|\mathbf{r}'\|^3} = \dfrac{7}{26^{3/2}} = \dfrac{7\sqrt{26}}{676}$

41. $y = 6x, x = 3$

This is a line, so the curvature is 0.

$K = 0, \dfrac{1}{K}$ is undefined.

43. $y = 5x^2 + 7, \ x = -1$

$y' = 10x, \ y'' = 10$

At $x = -1, \ y' = -10$ and $y'' = 10$.

$K = \dfrac{|y''|}{\left[1 + (y')^2\right]^{3/2}} = \dfrac{10}{\left[1 + 10^2\right]^{3/2}} = \dfrac{10}{101^{3/2}}$

$\dfrac{1}{K} = \dfrac{101^{3/2}}{10}$

45. $y = \sin 2x, \ x = \dfrac{\pi}{4}$

$y' = 2 \cos 2x, \ y'' = -4 \sin 2x$

At $x = \dfrac{\pi}{4}, \ y' = 0$ and $y'' = -4$.

$K = \dfrac{|y''|}{\left[1 + (y')^2\right]^{3/2}} = \dfrac{4}{1} = 4$

$\dfrac{1}{K} = \dfrac{1}{4}$

47. $y = x^3, \ x = 2$

$y' = 3x^2, \ y'' = 6x$

At $x = 2, \ y = 8, \ y' = 12, \ y'' = 12$

$K = \dfrac{12}{\left[1 + (12)^2\right]^{3/2}} = \dfrac{12}{(145)^{3/2}}$

$\dfrac{1}{K} = \dfrac{145\sqrt{145}}{12}$

49. $y = (x - 1)^2 + 3, \ y' = 2(x - 1), \ y'' = 2$

$K = \dfrac{2}{\left(1 + \left[2(x - 1)\right]^2\right)^{3/2}} = \dfrac{2}{\left[1 + 4(x - 1)^2\right]^{3/2}}$

$\dfrac{dK}{dx} = \dfrac{-24(x - 1)}{\left[1 + 4(x - 1)^2\right]^{5/2}}$

(a) K is maximum when $x = 1$ or at the vertex $(1, 3)$.

(b) $\lim\limits_{x \to \infty} K = 0$

51. $y = x^{2/3}, \ y' = \dfrac{2}{3}x^{-1/3}, \ y'' = -\dfrac{2}{9}x^{-4/3}$

$K = \left|\dfrac{(-2/9)x^{-4/3}}{\left[1 + (4/9)x^{-2/3}\right]^{3/2}}\right| = \left|\dfrac{6}{x^{1/3}\left(9x^{2/3} + 4\right)^{3/2}}\right|$

$\dfrac{dK}{dx} = \dfrac{-8\left(9x^{2/3} + 1\right)}{x^{4/3}\left(9x^{2/3} + 4\right)^{5/2}}$

(a) $K \to \infty$ as $x \to 0$. No maximum

(b) $\lim\limits_{x \to \infty} K = 0$

53. $y = \ln x, \ y' = \dfrac{1}{x}, \ y'' = -\dfrac{1}{x^2}$

$K = \left|\dfrac{-1/x^2}{\left[1 + (1/x)^2\right]^{3/2}}\right| = \dfrac{x}{\left(x^2 + 1\right)^{3/2}}$

$\dfrac{dK}{dx} = \dfrac{-2x^2 + 1}{\left(x^2 + 1\right)^{5/2}}$

(a) K has a maximum when $x = \dfrac{1}{\sqrt{2}}$.

(b) $\lim\limits_{x \to \infty} K = 0$

55. $y = 1 - x^4, \ y' = -4x^3, \ y'' = -12x^2$

$K = \dfrac{12x^2}{\left[1 + \left(-4x^3\right)^2\right]^{3/2}} = 0 \Rightarrow x = 0$

Curvature is 0 at $(0, 1)$.

57. $y = \cos \dfrac{x}{2}, \ y' = -\dfrac{1}{2} \sin \dfrac{x}{2}, \ y'' = -\dfrac{1}{4} \cos \dfrac{x}{2}$

$K = \dfrac{\left|-\dfrac{1}{4} \cos \dfrac{x}{2}\right|}{\left[1 + (y')^2\right]^{3/2}} = 0 \Rightarrow \cos \dfrac{x}{2} = 0$

So, $\dfrac{x}{2} = \dfrac{\pi}{2} + n\pi \Rightarrow x = \pi + 2n\pi$

Curvature is 0 at $(\pi + 2n\pi, 0)$.

59. $y = e^{cx}, \ y' = ce^{cx}, \ y'' = c^2 e^{cx}$

$K = \dfrac{c^2 e^{cx}}{\left[1 + \left(ce^{cx}\right)^2\right]^{3/2}}$

At $x = 0, \ K = \dfrac{c^2}{\left[1 + c^2\right]^{3/2}}$.

$\dfrac{dK}{dc} = \dfrac{c\left(2 - c^2\right)}{\left(c^2 + 1\right)^{5/2}}$

K is a maximum at $c = \pm\sqrt{2}$.

61. $f(x) = x^4 - x^2$

(a) $K = \dfrac{2\left|6x^2 - 1\right|}{\left|16x^6 - 16x^4 + 4x^2 + 1\right|^{3/2}}$

(b) For $x = 0$, $K = 2$. $f(0) = 0$. At $(0, 0)$, the circle

of curvature has radius $\dfrac{1}{2}$. Using the symmetry of the

graph of f, you obtain $x^2 + \left(y + \dfrac{1}{2}\right)^2 = \dfrac{1}{4}$.

For $x = 1$, $K = \left(2\sqrt{5}\right)/5$. $f(1) = 0$. At $(1, 0)$, the

circle of curvature has radius $\dfrac{\sqrt{5}}{2} = \dfrac{1}{K}$.

Using the graph of f, you see that the center of

curvature is $\left(0, \dfrac{1}{2}\right)$. So, $x^2 + \left(y - \dfrac{1}{2}\right)^2 = \dfrac{5}{4}$.

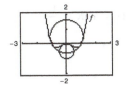

To graph these circles, use $y = -\dfrac{1}{2} \pm \sqrt{\dfrac{1}{4} - x^2}$

and $y = \dfrac{1}{2} \pm \sqrt{\dfrac{5}{4} - x^2}$.

(c) The curvature tends to be greatest near the extrema
of f, and K decreases as $x \to \pm\infty$. f and K,
however, do not have the same critical numbers.

Critical numbers of f: $x = 0$, $\pm\dfrac{\sqrt{2}}{2} \approx \pm0.7071$

Critical numbers of K: $x = 0$, ±0.7647, ±0.4082

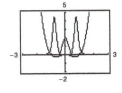

63. $y_1 = ax(b - x)$, $y_2 = \dfrac{x}{x + 2}$

You observe that $(0, 0)$ is a solution point to both
equations. So, the point P is origin.

$y_1 = ax(b - x)$, $y_1' = a(b - 2x)$, $y_1'' = -2a$

$y_2 = \dfrac{x}{x + 2}$, $y_2' = \dfrac{2}{(x + 2)^2}$, $y_2'' = \dfrac{-4}{(x + 2)^3}$

At P, $y_1'(0) = ab$ and $y_2'(0) = \dfrac{2}{(0 + 2)^2} = \dfrac{1}{2}$.

Because the curves have a common tangent at P,

$y_1'(0) = y_2'(0)$ or $ab = \dfrac{1}{2}$. So, $y_1'(0) = \dfrac{1}{2}$.

Because the curves have the same curvature at P,
$K_1(0) = K_2(0)$.

$K_1(0) = \left|\dfrac{y_1''(0)}{\left[1 + (y_1(0))^2\right]^{3/2}}\right| = \left|\dfrac{-2a}{\left[1 + (1/2)^2\right]^{3/2}}\right|$

$K_2(0) = \left|\dfrac{y_2''(0)}{\left[1 + (y_2(0))^2\right]^{3/2}}\right| = \left|\dfrac{-1/2}{\left[1 + (1/2)^2\right]^{3/2}}\right|$

So, $2a = \pm\dfrac{1}{2}$ or $a = \pm\dfrac{1}{4}$. In order that the curves

intersect at only one point, the parabola must be
concave downward. So,

$a = \dfrac{1}{4}$ and $b = \dfrac{1}{2a} = 2$.

$y_1 = \dfrac{1}{4}x(2 - x)$ and $y_2 = \dfrac{x}{x + 2}$

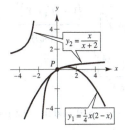

65. (a) Imagine dropping the circle $x^2 + (y - k)^2 = 16$ into the parabola $y = x^2$. The circle will drop to the point where the tangents to the circle and parabola are equal.

$y = x^2$ and $x^2 + (y - k)^2 = 16 \Rightarrow x^2 + (x^2 - k)^2 = 16$

Taking derivatives, $2x + 2(y - k)y' = 0$ and $y' = 2x$. So, $(y - k)y' = -x \Rightarrow y' = \dfrac{-x}{y - k}$.

So, $\dfrac{-x}{y - k} = 2x \Rightarrow -x = 2x(y - k) \Rightarrow -1 = 2(x^2 - k) \Rightarrow x^2 - k = -\dfrac{1}{2}$.

So, $x^2 + (x^2 - k)^2 = x^2 + \left(-\dfrac{1}{2}\right)^2 = 16 \Rightarrow x^2 = 15.75$.

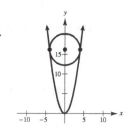

Finally, $k = x^2 + \dfrac{1}{2} = 16.25$, and the center of the circle is 16.25 units from the vertex of the parabola. Because the radius of the circle is 4, the circle is 12.25 units from the vertex.

(b) In 2-space, the parabola $z = y^2$ $\left(\text{or } z = x^2\right)$ has a curvature of $K = 2$ at $(0, 0)$. The radius of the largest sphere that will touch the vertex has radius $= 1/K = \dfrac{1}{2}$.

67. $P(x_0, y_0)$ point on curve $y = f(x)$. Let (α, β) be the center of curvature. The radius of curvature is $\dfrac{1}{K}$.

$y' = f'(x)$. Slope of normal line at (x_0, y_0) is $\dfrac{-1}{f'(x_0)}$.

Equation of normal line: $y - y_0 = \dfrac{-1}{f'(x_0)}(x - x_0)$

(α, β) is on the normal line: $-f'(x_0)(\beta - y_0) = \alpha - x_0$ Equation 1

(x_0, y_0) lies on the circle: $(x_0 - \alpha)^2 + (y_0 - \beta)^2 = \left(\dfrac{1}{K}\right)^2 = \left[\dfrac{\left(1 + f'(x_0)^2\right)^{3/2}}{|f''(x_0)|}\right]^2$ Equation 2

Substituting Equation 1 into Equation 2:

$\left[f'(x_0)(\beta - y_0)\right]^2 + (y_0 - \beta)^2 = \left(\dfrac{1}{K}\right)^2$

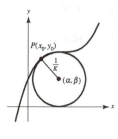

$(\beta - y_0)^2 + \left[1 + f'(x_0)^2\right] = \dfrac{\left(1 + f'(x_0)^2\right)^3}{\left(f''(x_0)\right)^2}$

$(\beta - y_0)^2 = \dfrac{\left[1 + f'(x_0)^2\right]^2}{f''(x_0)^2}$

When $f''(x_0) > 0$, $\beta - y_0 > 0$, and if $f''(x_0) < 0$, then $\beta - y_0 < 0$.

So $\beta - y_0 = \dfrac{1 + f'(x_0)^2}{f''(x_0)}$

$\beta = y_0 + \dfrac{1 + f'(x_0)^2}{f''(x_0)} = y_0 + z$

Similarly, $\alpha = x_0 - f'(x_0)z$.

69. $r(\theta) = r \cos \theta \mathbf{i} + r \sin \theta \mathbf{j} = f(\theta) \cos \theta \mathbf{i} + f(\theta) \sin \theta \mathbf{j}$

$x(\theta) = f(\theta) \cos \theta$

$y(\theta) = f(\theta) \sin \theta$

$x'(\theta) = -f(\theta) \sin \theta + f'(\theta) \cos \theta$

$y'(\theta) = f(\theta) \cos \theta + f'(\theta) \sin \theta$

$x''(\theta) = -f(\theta) \cos \theta - f'(\theta) \sin \theta - f'(\theta) \sin \theta + f''(\theta) \cos \theta = -f(\theta) \cos \theta - 2f'(\theta) \sin \theta + f''(\theta) \cos \theta$

$y''(\theta) = -f(\theta) \sin \theta + f'(\theta) \cos \theta + f'(\theta) \cos \theta + f''(\theta) \sin \theta = -f(\theta) \sin \theta + 2f'(\theta) \cos \theta + f''(\theta) \sin \theta$

$$K = \frac{|x'y'' - y'x''|}{\left[(x')^2 + (y')^2\right]^{3/2}} = \frac{\left|f^2(\theta) - f(\theta)f''(\theta) + 2\left(f'(\theta)\right)^2\right|}{\left[f^2(\theta) + \left(f'(\theta)\right)^2\right]^{3/2}} = \frac{\left|r^2 - rr'' + 2(r')^2\right|}{\left[r^2 + (r')^2\right]^{3/2}}$$

71. $r = e^{a\theta}, a > 0$

$r' = ae^{a\theta}$

$r'' = a^2 e^{a\theta}$

$$K = \frac{\left|2(r')^2 - rr'' + r^2\right|}{\left[(r')^2 + r^2\right]^{3/2}} = \frac{\left|2a^2 e^{2a\theta} - a^2 e^{2a\theta} + e^{2a\theta}\right|}{\left[a^2 e^{2a\theta} + e^{2a\theta}\right]^{3/2}} = \frac{1}{e^{a\theta}\sqrt{a^2 + 1}}$$

(a) As $\theta \to \infty$, $K \to 0$.

(b) As $a \to \infty$, $K \to 0$.

73. $r = 4 \sin 2\theta$

$r' = 8 \cos 2\theta$

At the pole: $K = \dfrac{2}{|r'(0)|} = \dfrac{2}{8} = \dfrac{1}{4}$

75. $x = f(t), y = g(t)$

$$y' = \frac{dy}{dx} = \frac{\dfrac{dy}{dt}}{\dfrac{dx}{dt}} = \frac{g'(t)}{f'(t)}$$

$$y'' = \frac{\dfrac{d}{dt}\left[\dfrac{g'(t)}{f'(t)}\right]}{\dfrac{dx}{dt}} = \frac{\dfrac{f'(t)g''(t) - g'(t)f''(t)}{\left[f'(t)\right]^2}}{f'(t)} = \frac{f'(t)g''(t) - g'(t)f''(t)}{\left[f'(t)\right]^3}$$

$$K = \frac{|y''|}{\left[1 + (y')^2\right]^{3/2}} = \frac{\left|\dfrac{f'(t)g''(t) - g'(t)f''(t)}{\left[f'(t)\right]^3}\right|}{\left[1 + \left(\dfrac{g'(t)}{f'(t)}\right)^2\right]^{3/2}} = \frac{\left|\dfrac{f'(t)g''(t) - g'(t)f''(t)}{\left[f'(t)\right]^3}\right|}{\sqrt{\left\{\dfrac{\left[f'(t)\right]^2 + \left[g'(t)\right]^2}{\left[f'(t)\right]^2}\right\}^3}} = \frac{|f'(t)g''(t) - g'(t)f''(t)|}{\left(\left[f'(t)\right]^2 + \left[g'(t)\right]^2\right)^{3/2}}$$

77. $x(\theta) = a(\theta - \sin\theta)$ $y(\theta) = a(1 - \cos\theta)$

$x'(\theta) = a(1 - \cos\theta)$ $y'(\theta) = a\sin\theta$

$x''(\theta) = a\sin\theta$ $y''(\theta) = a\cos\theta$

$$K = \frac{|x'(\theta)y''(\theta) - y'(\theta)x''(\theta)|}{\left[x'(\theta)^2 + y'(\theta)^2\right]^{3/2}}$$

$$= \frac{|a^2(1 - \cos\theta)\cos\theta - a^2\sin^2\theta|}{\left[a^2(1 - \cos\theta)^2 + a^2\sin^2\theta\right]^{3/2}}$$

$$= \frac{1}{a}\frac{|\cos\theta - 1|}{[2 - 2\cos\theta]^{3/2}}$$

$$= \frac{1}{a}\frac{1 - \cos\theta}{2\sqrt{2}[1 - \cos\theta]^{3/2}} \quad (1 - \cos \geq 0)$$

$$= \frac{1}{2a\sqrt{2 - 2\cos\theta}} = \frac{1}{4a}\csc\left(\frac{\theta}{2}\right)$$

Minimum: $\dfrac{1}{4a}$ $(\theta = \pi)$

Maximum: none $(K \to \infty \text{ as } \theta \to 0)$

79. $F = ma_N = mK\left(\dfrac{ds}{dt}\right)^2$

$$= \left(\frac{5500\text{ lb}}{32\text{ ft/sec}^2}\right)\left(\frac{1}{100\text{ ft}}\right)\left(\frac{30(5280)\text{ ft}}{3600\text{ sec}}\right)^2 = 3327.5\text{ lb}$$

81. $y = \cosh x = \dfrac{e^x + e^{-x}}{2}$

$y' = \dfrac{e^x - e^{-x}}{2} = \sinh x$

$y'' = \dfrac{e^x + e^{-x}}{2} = \cosh x$

$$K = \frac{|\cosh x|}{\left[1 + (\sinh x)^2\right]^{3/2}} = \frac{\cosh x}{(\cosh^2 x)^{3/2}} = \frac{1}{\cosh^2 x} = \frac{1}{y^2}$$

83. False. See Exploration on page 855.

85. True

87. Let $\mathbf{r} = x(t)\mathbf{i} + y(t)\mathbf{j} + z(t)\mathbf{k}$. Then $r = \|\mathbf{r}\| = \sqrt{\left[x(t)\right]^2 + \left[y(t)\right]^2 + \left[z(t)\right]^2}$ and $\mathbf{r}' = x'(t)\mathbf{i} + y'(t)\mathbf{j} + z'(t)\mathbf{k}$. Then,

$$r\left(\frac{dr}{dt}\right) = \sqrt{\left[x(t)\right]^2 + \left[y(t)\right]^2 + \left[z(t)\right]^2}\left[\frac{1}{2}\left\{\left[x(t)\right]^2 + \left[y(t)\right]^2 + \left[z(t)\right]^2\right\}^{-1/2} \cdot \left(2x(t)x'(t) + 2y(t)y'(t) + 2z(t)z'(t)\right)\right]$$

$$= x(t)x'(t) + y(t)y'(t) + z(t)z'(t) = \mathbf{r} \cdot \mathbf{r}'.$$

89. Let $\mathbf{r} = x\mathbf{i} + y\mathbf{j} + z\mathbf{k}$ where x, y, and z are function of t, and $r = \|\mathbf{r}\|$.

$$\frac{d}{dt}\left[\frac{\mathbf{r}}{r}\right] = \frac{r\mathbf{r}' - \mathbf{r}(dr/dt)}{r^2} = \frac{r\mathbf{r}' - \mathbf{r}\left[(\mathbf{r} \cdot \mathbf{r}')/r\right]}{r^2}$$

$$= \frac{r^2\mathbf{r}' - (\mathbf{r} \cdot \mathbf{r}')\mathbf{r}}{r^3} \text{ (using Exercise 105)}$$

$$= \frac{(x^2 + y^2 + z^2)(x'\mathbf{i} + y'\mathbf{j} + z'\mathbf{k}) - (xx' + yy' + zz')(x\mathbf{i} + y\mathbf{j} + z\mathbf{k})}{r^3}$$

$$= \frac{1}{r^3}\left[(x'y^2 + x'z^2 - xyy' - xzz')\mathbf{i} + (x^2y' + z^2y' - xx'y - zz'y)\mathbf{j} + (x^2z' + y^2z' - xx'z - yy'z)\mathbf{k}\right]$$

$$= \frac{1}{r^3}\begin{vmatrix} \mathbf{i} & \mathbf{j} & \mathbf{k} \\ yz' - y'z & -(xz' - x'z) & xy' - x'y \\ x & y & z \end{vmatrix} = \frac{1}{r^3}\{[\mathbf{r} \times \mathbf{r}'] \times \mathbf{r}\}$$

91. From Exercise 88, you have concluded that planetary motion is planar. Assume that the planet moves in the *xy*-plane with the sum at the origin. From Exercise 90, you have

$$\mathbf{r}' \times \mathbf{L} = GM\left(\frac{\mathbf{r}}{r} + \mathbf{e}\right).$$

Because $\mathbf{r}' \times \mathbf{L}$ and $\mathbf{r}$ are both perpendicular to $\mathbf{L}$, so is $\mathbf{e}$.

So, $\mathbf{e}$ lies in the *xy*-plane. Situate the coordinate system so that $\mathbf{e}$ lies along the positive *x*-axis and θ is the angle between $\mathbf{e}$ and $\mathbf{r}$. Let $e = \|\mathbf{e}\|$. Then $\mathbf{r} \cdot \mathbf{e} = \|\mathbf{r}\|\|\mathbf{e}\|\cos\theta = re\cos\theta$. Also,

$$
\begin{aligned}
\|\mathbf{L}\|^2 &= \mathbf{L} \cdot \mathbf{L}\\
&= (\mathbf{r} \times \mathbf{r}') \cdot \mathbf{L}\\
&= \mathbf{r} \cdot (\mathbf{r}' \times \mathbf{L})\\
&= \mathbf{r} \cdot \left[GM\left(\mathbf{e} + \frac{\mathbf{r}}{r}\right)\right]\\
&= GM\left[\mathbf{r} \cdot \mathbf{e} + \frac{\mathbf{r} \cdot \mathbf{r}}{r}\right]\\
&= GM[re\cos\theta + r].
\end{aligned}
$$

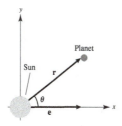

So, $\dfrac{\|\mathbf{L}\|^2/GM}{1 + e\cos\theta} = r$

and the planetary motion is a conic section. Because the planet returns to its initial position periodically, the conic is an ellipse.

93. $A = \dfrac{1}{2}\displaystyle\int_{\alpha}^{\beta} r^2\, d\theta$

So, $\dfrac{dA}{dt} = \dfrac{dA}{d\theta}\dfrac{d\theta}{dt} = \dfrac{1}{2}r^2\dfrac{d\theta}{dt} = \dfrac{1}{2}\|\mathbf{L}\|$ and $\mathbf{r}$ sweeps out area at a constant rate.

Review Exercises for Chapter 12

1. $\mathbf{r}(t) = \tan t\,\mathbf{i} + \mathbf{j} + t\,\mathbf{k}$

   (a) Domain: $t \neq \dfrac{\pi}{2} + n\pi$, *n* an integer

   (b) Continuous for all $t \neq \dfrac{\pi}{2} + n\pi$, *n* an integer

3. $\mathbf{r}(t) = \sqrt{t^2 - 9}\,\mathbf{i} - \mathbf{j} + \ln(t - 1)\mathbf{k}$

   (a) $t^2 - 9 \geq 0 \Rightarrow t^2 \geq 9 \Rightarrow t \leq -3, t \geq 3$

      $t - 1 > 0 \Rightarrow t > 1$

      Domain: $[3, \infty)$

   (b) Continuous for all $t \geq 3$

5. $\mathbf{r}(t) = (2t + 1)\mathbf{i} + t^2\mathbf{j} - \sqrt{t + 2}\,\mathbf{k}$

   (a) $\mathbf{r}(0) = \mathbf{i} - \sqrt{2}\,\mathbf{k}$

   (b) $\mathbf{r}(-2) = -3\mathbf{i} + 4\mathbf{j}$

   (c) $\mathbf{r}(c - 1) = (2c - 1)\mathbf{i} + (c - 1)^2\mathbf{j} - \sqrt{c + 1}\,\mathbf{k}$

   (d) $\mathbf{r}(1 + \Delta t) - \mathbf{r}(1) = \left[2(1 + \Delta t) + 1\right]\mathbf{i} + (1 + \Delta t)^2\mathbf{j} - \sqrt{3 + \Delta t}\,\mathbf{k} - \left(3\mathbf{i} + \mathbf{j} - \sqrt{3}\,\mathbf{k}\right)$

                      $= 2\Delta t\,\mathbf{i} + \left(\Delta t^2 + 2\Delta t\right)\mathbf{j} - \left(\sqrt{3 + \Delta t} - \sqrt{3}\right)\mathbf{k}$

7. $P(3, 0, 5),\ Q(2, -2, 3)$

   $\mathbf{v} = \overrightarrow{PQ} = \langle -1, -2, -2 \rangle$

   $\mathbf{r}(t) = (3 - t)\mathbf{i} - 2t\mathbf{j} + (5 - 2t)\mathbf{k},\ 0 \leq t \leq 1$

     $x = 3 - t,\ y = -2t,\ z = 5 - 2t,\ 0 \leq t \leq 1$

   (Answers may vary)

9. $\mathbf{r}(t) = \langle \pi\cos t, \pi\sin t \rangle$

   $x = \pi\cos t,\ y = \pi\sin t$

   $x^2 + y^2 = \pi^2$, circle

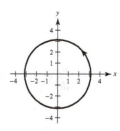

11. $\mathbf{r}(t) = (t + 1)\mathbf{i} + (3t - 1)\mathbf{j} + 2t\mathbf{k}$

$x = t + 1, \, y = 3t - 1, \, z = 2t$

This is a line passing through the points $(1, -1, 0)$ and $(2, 2, 2)$.

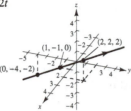

13. $3x + 4y - 12 = 0$

Let $x = t$, then $y = \dfrac{12 - 3t}{4}$.

$r(t) = t\mathbf{i} + \dfrac{12 - 3t}{4}\mathbf{j}$

Alternate solution: $x = 4t, \, y = 3 - 3t$

$r(t) = 4t\mathbf{i} + (3 - 3t)\mathbf{j}$

15. $z = x^2 + y^2, \, y = 2$

$z = x^2 + 4$

$x = t, \, y = 2, \, z = t^2 + 4$

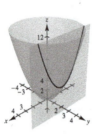

17. $\displaystyle\lim_{t \to 3}\left(\sqrt{3 - t}\,\mathbf{i} + \ln t\mathbf{j} - \frac{1}{t}\mathbf{k}\right) = \ln 3\mathbf{j} - \frac{1}{3}\mathbf{k}$

19. $\mathbf{r}(t) = (t^2 + 4t)\mathbf{i} - 3t^2\mathbf{j}$

   (a) $\mathbf{r}'(t) = (2t + 4)\mathbf{i} - 6t\mathbf{j}$

   (b) $\mathbf{r}''(t) = 2\mathbf{i} - 6\mathbf{j}$

   (c) $\mathbf{r}'(t) \cdot \mathbf{r}''(t) = (2t + 4)(2) + (-6t)(-6) = 40t + 8$

21. $\mathbf{r}(t) = 2t^3\mathbf{i} + 4t\mathbf{j} - t^2\mathbf{k}$

   (a) $\mathbf{r}'(t) = 6t^2\mathbf{i} + 4\mathbf{j} - 2t\mathbf{k}$

   (b) $\mathbf{r}''(t) = 12t\mathbf{i} - 2\mathbf{k}$

   (c) $\mathbf{r}'(t) \cdot \mathbf{r}''(t) = 6t^2(12t) + (-2t)(-2) = 72t^3 + 4t$

   (d) $\mathbf{r}'(t) \times \mathbf{r}''(t) = \begin{vmatrix} \mathbf{i} & \mathbf{j} & \mathbf{k} \\ 6t^2 & 4 & -2t \\ 12t & 0 & -2 \end{vmatrix}$

                $= -8\mathbf{i} - (-12t^2 + 24t^2)\mathbf{j} + (-48t)\mathbf{k}$

                $= -8\mathbf{i} - 12t^2\mathbf{j} - 48t\mathbf{k}$

23. $\mathbf{r}(t) = (t - 1)^3\mathbf{i} + (t - 1)^4\mathbf{j}$

    $\mathbf{r}'(t) = 3(t - 1)^2\mathbf{i} + 4(t - 1)^3\mathbf{j}$

    Not smooth at $t = 1$.

    Smooth on $(-\infty, 1)$ and $(1, \infty)$

25. $\mathbf{r}(t) = 3t\mathbf{i} + (t - 1)\mathbf{j}, \, \mathbf{u}(t) = t\mathbf{i} + t^2\mathbf{j} + \frac{2}{3}t^3\mathbf{k}$

    $\mathbf{r}'(t) = 3\mathbf{i} + \mathbf{j}, \, \mathbf{u}'(t) = \mathbf{i} + 2t\mathbf{j} + 2t^2\mathbf{k}$

(a) $\mathbf{r}'(t) = 3\mathbf{i} + \mathbf{j}$

(b) $\dfrac{d}{dt}\left[\mathbf{u}(t) - 2\mathbf{r}(t)\right] = \mathbf{u}'(t) - 2\mathbf{r}'(t) = \left(\mathbf{i} + 2t\mathbf{j} + 2t^2\mathbf{k}\right) - 2(3\mathbf{i} + \mathbf{j})$

                      $= (1 - 6)\mathbf{i} + (2t - 2)\mathbf{j} + 2t^2\mathbf{k} = -5\mathbf{i} + (2t - 2)\mathbf{j} + 2t^2\mathbf{k}$

(c) $\dfrac{d}{dt}(3t)\mathbf{r}(t) = (3t)\mathbf{r}'(t) + 3\mathbf{r}(t) = (3t)(3\mathbf{i} + \mathbf{j}) + 3\left[3t\mathbf{i} + (t - 1)\mathbf{j}\right] = 9t\mathbf{i} + 3t\mathbf{j} + 9t\mathbf{i} + (3t - 3)\mathbf{j} = 18t\mathbf{i} + (6t - 3)\mathbf{j}$

(d) $\dfrac{d}{dt}\left[\mathbf{r}(t) \cdot \mathbf{u}(t)\right] = \mathbf{r}(t) \cdot \mathbf{u}'(t) + \mathbf{r}'(t) \cdot \mathbf{u}(t) = \left[(3t)(1) + (t - 1)(2t) + (0)(2t^2)\right] + \left[(3)(t) + (1)(t^2) + (0)\left(\frac{2}{3}t^3\right)\right]$

                    $= \left(3t + 2t^2 - 2t\right) + \left(3t + t^2\right) = 4t + 3t^2$

(e) $\dfrac{d}{dt}\left[\mathbf{r}(t) \times \mathbf{u}(t)\right] = \mathbf{r}(t) \times \mathbf{u}'(t) + \mathbf{r}'(t) \times \mathbf{u}(t)$

                    $= \left[\left(2t^3 - 2t^2\right)\mathbf{i} - 6t^3\mathbf{j} + \left(6t^2 - t + 1\right)\mathbf{k}\right] + \left[\frac{2}{3}t^3\mathbf{i} - 2t^3\mathbf{j} + \left(3t^2 - t\right)\mathbf{k}\right]$

                    $= \left(\frac{8}{3}t^3 - 2t^2\right)\mathbf{i} - 8t^3\mathbf{j} + \left(9t^2 - 2t + 1\right)\mathbf{k}$

(f) $\dfrac{d}{dt}\left[\mathbf{u}(2t)\right] = 2\mathbf{u}'(2t) = 2\left[\mathbf{i} + 2(2t)\mathbf{j} + 2(2t)^2\mathbf{k}\right] = 2\mathbf{i} + 8t\mathbf{j} + 16t^2\mathbf{k}$

27. $\int (t^2\mathbf{i} + 5t\mathbf{j} + 8t^3\mathbf{k})\,dt = \dfrac{t^3}{3}\mathbf{i} + \dfrac{5}{2}t^2\mathbf{j} + 2t^4\mathbf{j} + \mathbf{C}$

29. $\int \left(3\sqrt{t}\,\mathbf{i} + \dfrac{2}{t}\mathbf{j} + \mathbf{k}\right) dt = 2t^{3/2}\mathbf{i} + 2\ln|t|\,\mathbf{j} + t\mathbf{k} + \mathbf{C}$

31. $\displaystyle\int_{-2}^{2} (3t\mathbf{i} + 2t^2\,\mathbf{j} - t^3\,\mathbf{k})\,dt = \left[\dfrac{3t^2}{2}\mathbf{i} + \dfrac{2t^3}{3}\mathbf{j} - \dfrac{t^4}{4}\mathbf{k}\right]_{-2}^{2}$

$\qquad\qquad = \dfrac{32}{3}\mathbf{j}$

33. $\displaystyle\int_{0}^{2} (e^{t/2}\mathbf{i} - 3t^2\mathbf{j} - \mathbf{k})\,dt = \left[2e^{t/2}\mathbf{i} - t^3\mathbf{j} - t\mathbf{k}\right]_{0}^{2}$

$\qquad\qquad = (2e - 2)\mathbf{i} - 8\mathbf{j} - 2\mathbf{k}$

35. $\mathbf{r}(t) = \int (2t\mathbf{i} + e^t\mathbf{j} + e^{-t}\mathbf{k})\,dt = t^2\mathbf{i} + e^t\mathbf{j} - e^{-t}\mathbf{k} + \mathbf{C}$

$\quad\ \mathbf{r}(0) = \mathbf{j} - \mathbf{k} + \mathbf{C} = \mathbf{i} + 3\mathbf{j} - 5\mathbf{k} \Rightarrow \mathbf{C} = \mathbf{i} + 2\mathbf{j} - 4\mathbf{k}$

$\quad\ \mathbf{r}(t) = (t^2 + 1)\mathbf{i} + (e^t + 2)\mathbf{j} - (e^{-t} + 4)\mathbf{k}$

37. $\mathbf{r}(t) = 4t\mathbf{i} + t^3\mathbf{j} - t\mathbf{k},\ t = 1$

$\quad$ (a) $\quad \mathbf{v}(t) = \mathbf{r}'(t) = 4\mathbf{i} + 3t^2\mathbf{j} - \mathbf{k}$

$\qquad\qquad$ Speed $= \|\mathbf{v}(t)\| = \sqrt{16 + 9t^4 + 1} = \sqrt{17 + 9t^4}$

$\qquad\qquad \mathbf{a}(t) = \mathbf{r}''(t) = 6t\mathbf{j}$

$\quad$ (b) $\mathbf{v}(1) = 4\mathbf{i} + 3\mathbf{j} - \mathbf{k}$

$\qquad\ \mathbf{a}(1) = 6\mathbf{j}$

39. $\mathbf{r}(t) = \langle \cos^3 t, \sin^3 t, 3t \rangle,\ t = \pi$

$\quad$ (a) $\mathbf{v}(t) = \mathbf{r}'(t) = \langle -3\cos^2 t \sin t, 3\sin^2 t \cos t, 3 \rangle$

$\qquad$ Speed $= \|\mathbf{v}(t)\| = \sqrt{9\cos^4 t \sin^2 t + 9\sin^4 t \cos^2 t + 9} = 3\sqrt{\cos^2 t \sin^2 t(\cos^2 t + \sin^2 t) + 1} = 3\sqrt{\cos^2 t \sin^2 t + 1}$

$\qquad \mathbf{a}(t) = \langle -6\cos t(-\sin^2 t) + (-3\cos^2 t)\cos t, 6\sin t\cos^2 t + 3\sin^2 t(-\sin t), 0 \rangle$

$\qquad\quad = \langle 3\cos t(2\sin^2 t - \cos^2 t), 3\sin t(2\cos^2 t - \sin^2 t), 0 \rangle$

$\quad$ (b) $\mathbf{v}(\pi) = \langle 0, 0, 3 \rangle$

$\qquad\ \mathbf{a}(\pi) = \langle 3, 0, 0 \rangle$

41. $\mathbf{r}(t) = (v_0 \cos\theta)t\mathbf{i} + \left[h + (v_0 \sin\theta)t - 16t^2\right]\mathbf{j} = (120\cos 30°)t\mathbf{i} + (3.5 + (120\sin 30°)t - 16t^2)\mathbf{j}$

$\qquad\ = 60\sqrt{3}\,t\mathbf{i} + (3.5 + 60t - 16t^2)\mathbf{j}$

$\quad\ \mathbf{r}'(t) = \mathbf{v}(t) = 60\sqrt{3}\,\mathbf{i} + (60 - 32t)\mathbf{j}$

$\quad$ Find the maximum height

$\quad\ y'(t) = 60 - 32t = 0 \Rightarrow t = \dfrac{60}{32} = \dfrac{15}{8} = 1.875$

$\quad\ y\left(\dfrac{15}{8}\right) = 3.5 + 60\left(\dfrac{15}{8}\right) - 16\left(\dfrac{15}{8}\right)^2 = 59.75$ ft, Maximum height

$\quad\ 375 = 60\sqrt{3}\,t \Rightarrow t = \dfrac{375}{60\sqrt{3}} = \dfrac{25}{4\sqrt{3}} = \dfrac{25\sqrt{3}}{12} \approx 3.608$

$\quad\ y\left(\dfrac{25\sqrt{3}}{12}\right) \approx 11.67$ ft

$\quad$ The baseball clears the 8-foot fence.

43. $\mathbf{r}(t) = 6t\mathbf{i} - t^2\mathbf{j},\ t = 2$

$\quad\ \mathbf{r}'(t) = 6\mathbf{i} - 2t\mathbf{j}$

$\quad\ \mathbf{r}'(2) = 6\mathbf{i} - 4\mathbf{j},$

$\quad\ \|\mathbf{r}'(2)\| = \sqrt{36 + 16} = 2\sqrt{13}$

$\quad\ \mathbf{T}(2) = \dfrac{1}{\sqrt{13}}(3\mathbf{i} - 2\mathbf{j})$

45. $\mathbf{r}(t) = e^{2t}\mathbf{i} + \cos t\mathbf{j} - \sin 3t\mathbf{k}$, $P(1, 1, 0)$

$\mathbf{r}'(t) = 2e^{2t}\mathbf{i} - \sin t\mathbf{j} - 3\cos 3t\mathbf{k}$

At $P(1, 1, 0)$, $t = 0$.

$\mathbf{r}(0) = \mathbf{i} + \mathbf{j}$

$\mathbf{r}'(0) = 2\mathbf{i} - 3\mathbf{k}$

$\|\mathbf{r}'(0)\| = \sqrt{4 + 9} = \sqrt{13}$

$\mathbf{T}(0) = \dfrac{1}{\sqrt{13}}(2\mathbf{i} - 3\mathbf{k})$

Direction numbers: $a = 2, b = 0, c = -3$

Parametric equations: $x = 1 + 2t, y = 1, z = -3t$

47. $\mathbf{r}(t) = 2t\mathbf{i} + 3t^2\mathbf{j}$, $t = 1$

$\mathbf{r}'(t) = 2\mathbf{i} + 6t\mathbf{j}, \mathbf{r}'(1) = 2\mathbf{i} + 6\mathbf{j}$

$\|\mathbf{r}'(t)\| = \sqrt{4 + 36t^2}, \|\mathbf{r}'(1)\| = \sqrt{40} = 2\sqrt{10}$

$\mathbf{T}(1) = \dfrac{\mathbf{r}'(1)}{\|\mathbf{r}'(1)\|} = \dfrac{2\mathbf{i} + 6\mathbf{j}}{2\sqrt{10}} = \dfrac{1}{\sqrt{10}}(\mathbf{i} + 3\mathbf{j})$

$\mathbf{N}(1)$ is orthogonal to $\mathbf{T}(1)$ and points towards the concave side. Hence,

$\mathbf{N}(1) = \dfrac{1}{\sqrt{10}}(-3\mathbf{i} + \mathbf{j})$.

49. $\mathbf{r}(t) = 3\cos 2t\mathbf{i} + 3\sin 2t\,\mathbf{j} + 3\mathbf{k}$, $t = \dfrac{\pi}{4}$

$\mathbf{r}'(t) = -6\sin 2t\mathbf{i} + 6\cos 2t\,\mathbf{j}, \|\mathbf{r}'(t)\| = 6$

$\mathbf{T}(t) = -\sin 2t\mathbf{i} + \cos 2t\,\mathbf{j}, \mathbf{T}(\pi/4) = -\mathbf{i}$

$\mathbf{T}'(t) = -2\cos 2t\mathbf{i} - 2\sin 2t\,\mathbf{j}, \|\mathbf{T}'(t)\| = 2$

$\mathbf{N}(t) = -\cos 2t\mathbf{i} - \sin 2t\,\mathbf{j}$

$\mathbf{N}(\pi/4) = -\mathbf{j}$

51. $\mathbf{r}(t) = \dfrac{3}{t}\mathbf{i} - 6t\,\mathbf{j}$, $t = 3$

$\mathbf{v}(t) = -\dfrac{3}{t^2}\mathbf{i} - 6\mathbf{j}, \mathbf{v}(3) = -\dfrac{1}{3}\mathbf{i} - 6\mathbf{j}$

$\mathbf{a}(t) = \dfrac{6}{t^3}\mathbf{i}, \mathbf{a}(3) = \dfrac{2}{9}\mathbf{i}$

$\mathbf{T}(t) = \dfrac{\mathbf{v}(t)}{\|\mathbf{v}(t)\|} = \dfrac{\left(-\dfrac{3}{t^2}\right)\mathbf{i} - 6\mathbf{j}}{\sqrt{\dfrac{9}{t^4} + 36}} = \dfrac{-3\mathbf{i} - 6t^2\mathbf{j}}{3\sqrt{1 + 4t^4}}$

$\mathbf{T}(3) = \dfrac{-3\mathbf{i} - 54\mathbf{j}}{3\sqrt{1 + 324}} = \dfrac{-\mathbf{i} - 18\mathbf{j}}{\sqrt{325}}$

$\quad = -\dfrac{\sqrt{13}}{65}\mathbf{i} - \dfrac{18\sqrt{13}}{65}\mathbf{j}$

$\mathbf{N}(3)$ is orthogonal to $\mathbf{T}(3)$, and points in the direction the curve is bending. Hence,

$\mathbf{N}(3) = \dfrac{18\mathbf{i} - \mathbf{j}}{\sqrt{325}} = \dfrac{18\sqrt{13}}{65}\mathbf{i} - \dfrac{\sqrt{13}}{65}\mathbf{j}$.

$a_{\mathbf{T}} = \mathbf{a} \cdot \mathbf{T} = -\dfrac{2}{9\sqrt{325}} = -\dfrac{2\sqrt{13}}{585}$

$a_{\mathbf{N}} = \mathbf{a} \cdot \mathbf{N} = \dfrac{4}{\sqrt{325}} = \dfrac{4\sqrt{13}}{65}$

53. $\mathbf{r}(t) = \sin t\mathbf{i} - 3t\mathbf{j} + \cos t\mathbf{k}$, $t = \dfrac{\pi}{6}$

$\mathbf{r}'(t) = \mathbf{v}(t) = \cos t\mathbf{i} - 3\mathbf{j} - \sin t\mathbf{k}$

$\mathbf{r}''(t) = \mathbf{a}(t) = -\sin t\mathbf{i} - \cos t\mathbf{k}$

$\mathbf{r}'\left(\dfrac{\pi}{6}\right) = \dfrac{\sqrt{3}}{2}\mathbf{i} - 3\mathbf{j} - \dfrac{1}{2}\mathbf{k}; \mathbf{r}''\left(\dfrac{\pi}{6}\right) = -\dfrac{1}{2}\mathbf{i} - \dfrac{\sqrt{3}}{2}\mathbf{k}$

$\left\|\mathbf{r}'\left(\dfrac{\pi}{6}\right)\right\| = \sqrt{\dfrac{3}{4} + 9 + \dfrac{1}{4}} = \sqrt{10}$

$\mathbf{T}\left(\dfrac{\pi}{6}\right) = \dfrac{1}{\sqrt{10}}\left(\dfrac{\sqrt{3}}{2}\mathbf{i} - 3\mathbf{j} - \dfrac{1}{2}\mathbf{k}\right)$

$a_{\mathbf{T}} = \mathbf{a} \cdot \mathbf{T} = \dfrac{1}{\sqrt{10}}\left(-\dfrac{1}{2}\mathbf{i} - \dfrac{\sqrt{3}}{2}\mathbf{k}\right) \cdot \left(\dfrac{\sqrt{3}}{2}\mathbf{i} - 3\mathbf{j} - \dfrac{1}{2}\mathbf{k}\right) = 0$

$a_{\mathbf{N}} = \sqrt{\|\mathbf{a}\|^2 - a_{\mathbf{T}}^2} = \|\mathbf{a}\| = 1$

55. $r(t) = 2t\mathbf{i} - 3t\mathbf{j}, [0, 5]$

$r'(t) = 2\mathbf{i} - 3\mathbf{j}$

$s = \int_a^b \|r'(t)\| \, dt = \int_0^5 \sqrt{4 + 9} \, dt = \left[\sqrt{13}\, t\right]_0^5 = 5\sqrt{13}$

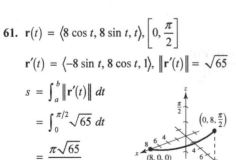

57. $r(t) = 2 \sin t\mathbf{i} + \mathbf{j}, \left[\dfrac{\pi}{2}, \pi\right]$

$r'(t) = 2 \cos t\mathbf{i}$

$\|r'(t)\| = \sqrt{4 \cos^2 t} = 2 \cos t$

$s = \int_{\pi/2}^\pi 2 \cos t \, dt = [2 \sin t]_{\pi/2}^\pi = [0 - (-2)] = 2$

59. $r(t) = -3t\mathbf{i} + 2t\mathbf{j} + 4t\mathbf{k}, [0, 3]$

$r'(t) = -3\mathbf{i} + 2\mathbf{j} + 4\mathbf{k}$

$s = \int_a^b \|r'(t)\| \, dt$

$= \int_0^3 \sqrt{9 + 4 + 16} \, dt$

$= \int_0^3 \sqrt{29} \, dt = 3\sqrt{29}$

61. $r(t) = \langle 8 \cos t, 8 \sin t, t \rangle, \left[0, \dfrac{\pi}{2}\right]$

$r'(t) = \langle -8 \sin t, 8 \cos t, 1 \rangle, \|r'(t)\| = \sqrt{65}$

$s = \int_a^b \|r'(t)\| \, dt$

$= \int_0^{\pi/2} \sqrt{65} \, dt$

$= \dfrac{\pi\sqrt{65}}{2}$

63. $r(t) = 3t\mathbf{i} + 2t\mathbf{j}$

Line

$K = 0$

65. $r(t) = 2t\mathbf{i} + \dfrac{1}{2}t^2\mathbf{j} + t^2\mathbf{k}$

$r'(t) = 2\mathbf{i} + t\mathbf{j} + 2t\mathbf{k}, \|r'\| = \sqrt{5t^2 + 4}$

$r''(t) = \mathbf{j} + 2\mathbf{k}$

$r' \times r'' = \begin{vmatrix} \mathbf{i} & \mathbf{j} & \mathbf{k} \\ 2 & t & 2t \\ 0 & 1 & 2 \end{vmatrix} = -4\mathbf{j} + 2\mathbf{k}, \|r' \times r''\| = \sqrt{20}$

$K = \dfrac{\|r' \times r''\|}{\|r'\|^3} = \dfrac{\sqrt{20}}{(5t^2 + 4)^{3/2}} = \dfrac{2\sqrt{5}}{(4 + 5t^2)^{3/2}}$

67. $r(t) = \dfrac{1}{2}t^2\mathbf{i} + t\mathbf{j} + \dfrac{1}{3}t^3\mathbf{k}, P\left(\dfrac{1}{2}, 1, \dfrac{1}{3}\right) \Rightarrow t = 1$

$r'(t) = t\mathbf{i} + \mathbf{j} + t^2\mathbf{k}, r'(1) = \mathbf{i} + \mathbf{j} + \mathbf{k}$

$r''(t) = \mathbf{i} + 2t\mathbf{k}, r''(1) = \mathbf{i} + 2\mathbf{k}$

$r' \times r'' = \begin{vmatrix} \mathbf{i} & \mathbf{j} & \mathbf{k} \\ 1 & 1 & 1 \\ 1 & 0 & 2 \end{vmatrix} = 2\mathbf{i} - \mathbf{j} - \mathbf{k}$

$K = \dfrac{\|r' \times r''\|}{\|r'\|^3} = \dfrac{\sqrt{4 + 1 + 1}}{(\sqrt{3})^3} = \dfrac{\sqrt{6}}{3\sqrt{3}} = \dfrac{\sqrt{2}}{3}$

69. $y = \dfrac{1}{2}x^2 + x, x = 4$

$y' = x + 1$

$y'' = 1$

$K = \dfrac{|y''|}{\left[1 + (y')^2\right]^{3/2}} = \dfrac{1}{(1 + x^2)^{3/2}}$

At $x = 4$, $K = \dfrac{1}{26^{3/2}}$ and $r = 26^{3/2} = 26\sqrt{26}$.

71. $y = \ln x, x = 1$

$y' = \dfrac{1}{x}$

$y'' = -\dfrac{1}{x^2}$

$K = \dfrac{|y''|}{\left[1 + (y')^2\right]^{3/2}} = \dfrac{1/x^2}{\left[1 + (1/x)^2\right]^{3/2}}$

At $x = 1$, $K = \dfrac{1}{2^{3/2}} = \dfrac{1}{2\sqrt{2}} = \dfrac{\sqrt{2}}{4}$ and $r = 2\sqrt{2}$.

73. $\mathbf{F} = ma_N = mk\left(\dfrac{ds}{dt}\right)^2 = \left(\dfrac{7200 \text{ lb}}{32 \text{ ft/sec}^2}\right)\left(\dfrac{1}{150 \text{ ft}}\right)\left(\dfrac{25(5280)\text{ft}}{3600 \text{ sec}}\right)^2 \approx 2016.67 \text{ lb}$

Problem Solving for Chapter 12

1. $x(t) = \int_0^t \cos\left(\frac{\pi u^2}{2}\right) du, \; y(t) = \int_0^t \sin\left(\frac{\pi u^2}{2}\right) du$

$x'(t) = \cos\left(\frac{\pi t^2}{2}\right), \; y'(t) = \sin\left(\frac{\pi t^2}{2}\right)$

(a) $s = \int_0^a \sqrt{x'(t)^2 + y'(t)^2}\, dt = \int_0^a dt = a$

(b) $x''(t) = -\pi t \sin\left(\frac{\pi t^2}{2}\right), \; y''(t) = \pi t \cos\left(\frac{\pi t^2}{2}\right)$

$K = \dfrac{\left|\pi t \cos^2\left(\frac{\pi t^2}{2}\right) + \pi t \sin^2\left(\frac{\pi t^2}{2}\right)\right|}{1} = \pi t$

At $t = a, \; K = \pi a$.

(c) $K = \pi a = \pi$ (length)

3. Bomb: $\mathbf{r}_1(t) = \langle 5000 - 400t, 3200 - 16t^2 \rangle$

Projectile: $\mathbf{r}_2(t) = \langle (v_0 \cos \theta)t, (v_0 \sin \theta)t - 16t^2 \rangle$

At 1600 ft: Bomb:

$3200 - 16t^2 = 1600 \Rightarrow t = 10$ seconds.

Projectile will travel 5 seconds:

$5(v_0 \sin \theta) - 16(25) = 1600$

$\qquad v_0 \sin \theta = 400$.

Horizontal position:

At $t = 10$, bomb is at $5000 - 400(10) = 1000$.

At $t = 5$, projectile is at $5v_0 \cos \theta$.

So, $v_0 \cos \theta = 200$.

Combining,

$\dfrac{v_0 \sin \theta}{v_0 \cos \theta} = \dfrac{400}{200} \Rightarrow \tan \theta = 2 \Rightarrow \theta \approx 63.43°$.

$v_0 = \dfrac{200}{\cos \theta} \approx 447.2$ ft/sec

5. $x'(\theta) = 1 - \cos \theta, \; y'(\theta) = \sin \theta, \; 0 \le \theta \le 2\pi$

$\sqrt{x'(\theta)^2\, y'(\theta)^2} = \sqrt{(1 - \cos \theta)^2 + \sin^2\theta} = \sqrt{2 - 2\cos \theta} = \sqrt{4 \sin^2\frac{\theta}{2}}$

$s(t) = \int_\pi^t 2 \sin \frac{\theta}{2}\, d\theta = \left[-4 \cos \frac{\theta}{2}\right]_\pi^t = -4 \cos \frac{t}{2}$

$x''(\theta) = \sin \theta, \; y''(\theta) = \cos \theta$

$K = \dfrac{\left|(1 - \cos \theta)\cos \theta - \sin \theta \sin \theta\right|}{\left(2 \sin \frac{\theta}{2}\right)^3} = \dfrac{\left|\cos \theta - 1\right|}{8 \sin^3 \frac{\theta}{2}} = \dfrac{1}{4 \sin \frac{\theta}{2}}$

So, $\rho = \dfrac{1}{K} = 4 \sin \frac{t}{2}$ and $s^2 + \rho^2 = 16 \cos^2\left(\frac{t}{2}\right) + 16 \sin^2\left(\frac{t}{2}\right) = 16$.

7. $\|\mathbf{r}(t)\|^2 = \mathbf{r}(t) \cdot \mathbf{r}(t)$

$\dfrac{d}{dt}\left(\|\mathbf{r}(t)\|\right)^2 = 2\|\mathbf{r}(t)\|\dfrac{d}{dt}\|\mathbf{r}(t)\| = \mathbf{r}(t) \cdot \mathbf{r}'(t) + \mathbf{r}'(t) \cdot \mathbf{r}(t) \Rightarrow \dfrac{d}{dt}\|\mathbf{r}(t)\| = \dfrac{\mathbf{r}(t) \cdot \mathbf{r}'(t)}{\|\mathbf{r}(t)\|}$

9. $\mathbf{r}(t) = 4\cos t\,\mathbf{i} + 4\sin t\,\mathbf{j} + 3t\,\mathbf{k},\ t = \dfrac{\pi}{2}$

$\mathbf{r}'(t) = -4\sin t\,\mathbf{i} + 4\cos t\,\mathbf{j} + 3\mathbf{k},\ \|\mathbf{r}'(t)\| = 5$

$\mathbf{r}''(t) = -4\cos t\,\mathbf{i} - 4\sin t\,\mathbf{j}$

$\mathbf{T} = -\dfrac{4}{5}\sin t\,\mathbf{i} + \dfrac{4}{5}\cos t\,\mathbf{j} + \dfrac{3}{5}\mathbf{k}$

$\mathbf{T}' = -\dfrac{4}{5}\cos t\,\mathbf{i} - \dfrac{4}{5}\sin t\,\mathbf{j}$

$\mathbf{N} = -\cos t\,\mathbf{i} - \sin t\,\mathbf{j}$

$\mathbf{B} = \mathbf{T} \times \mathbf{N} = \dfrac{3}{5}\sin t\,\mathbf{i} - \dfrac{3}{5}\cos t\,\mathbf{j} + \dfrac{4}{5}\mathbf{k}$

At $t = \dfrac{\pi}{2}$, $\mathbf{T}\left(\dfrac{\pi}{2}\right) = -\dfrac{4}{5}\mathbf{i} + \dfrac{3}{5}\mathbf{k}$

$\mathbf{N}\left(\dfrac{\pi}{2}\right) = -\mathbf{j}$

$\mathbf{B}\left(\dfrac{\pi}{2}\right) = \dfrac{3}{5}\mathbf{i} + \dfrac{4}{5}\mathbf{k}$

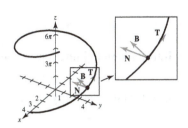

11. (a) $\|\mathbf{B}\| = \|\mathbf{T} \times \mathbf{N}\| = 1$ constant length $\Rightarrow \dfrac{d\mathbf{B}}{ds} \perp \mathbf{B}$

$\dfrac{d\mathbf{B}}{ds} = \dfrac{d}{ds}(\mathbf{T} \times \mathbf{N}) = (\mathbf{T} \times \mathbf{N}') + (\mathbf{T}' \times \mathbf{N})$

$\mathbf{T} \cdot \dfrac{d\mathbf{B}}{ds} = \mathbf{T} \cdot (\mathbf{T} \times \mathbf{N}') + \mathbf{T} \cdot (\mathbf{T}' \times \mathbf{N})$

$= (\mathbf{T} \times \mathbf{T}) \cdot \mathbf{N}' + \mathbf{T} \cdot \left(\mathbf{T}' \times \dfrac{\mathbf{T}'}{\|\mathbf{T}'\|}\right) = 0$

So, $\dfrac{d\mathbf{B}}{ds} \perp \mathbf{B}$ and $\dfrac{d\mathbf{B}}{ds} \perp \mathbf{T} \Rightarrow \dfrac{d\mathbf{B}}{ds} = \tau\mathbf{N}$

for some scalar τ.

(b) $\mathbf{B} = \mathbf{T} \times \mathbf{N}$. Using Section 11.4, exercise 58,

$\mathbf{B} \times \mathbf{N} = (\mathbf{T} \times \mathbf{N}) \times \mathbf{N} = -\mathbf{N} \times (\mathbf{T} \times \mathbf{N})$

$= -\big[(\mathbf{N} \cdot \mathbf{N})\mathbf{T} - (\mathbf{N} \cdot \mathbf{T})\mathbf{N}\big]$

$= -\mathbf{T}$

$\mathbf{B} \times \mathbf{T} = (\mathbf{T} \times \mathbf{N}) \times \mathbf{T} = -\mathbf{T} \times (\mathbf{T} \times \mathbf{N})$

$= -\big[(\mathbf{T} \cdot \mathbf{N})\mathbf{T} - (\mathbf{T} \cdot \mathbf{T})\mathbf{N}\big]$

$= \mathbf{N}.$

Now, $K\mathbf{N} = \left\|\dfrac{d\mathbf{T}}{ds}\right\| \dfrac{\mathbf{T}'(s)}{\|\mathbf{T}'(s)\|} = \mathbf{T}'(s) = \dfrac{d\mathbf{T}}{ds}$

Finally,

$\mathbf{N}'(s) = \dfrac{d}{ds}(\mathbf{B} \times \mathbf{T}) = (\mathbf{B} \times \mathbf{T}') + (\mathbf{B}' \times \mathbf{T})$

$= (\mathbf{B} \times K\mathbf{N}) + (-\tau\mathbf{N} \times \mathbf{T})$

$= -K\mathbf{T} + \tau\mathbf{B}.$

13. $\mathbf{r}(t) = \langle t\cos \pi t,\ t\sin \pi t \rangle,\ 0 \le t \le 2$

(a)

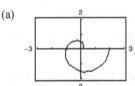

(b) Length $= \displaystyle\int_0^2 \|\mathbf{r}'(t)\|\,dt$

$= \displaystyle\int_0^2 \sqrt{\pi^2 t^2 + 1}\,dt$

≈ 6.766 (graphing utility)

(c) $K = \dfrac{\pi(\pi^2 t^2 + 2)}{\left[\pi^2 t^2 + 1\right]^{3/2}}$

$K(0) = 2\pi$

$K(1) = \dfrac{\pi(\pi^2 + 2)}{(\pi^2 + 1)^{3/2}} \approx 1.04$

$K(2) \approx 0.51$

(d)

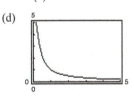

(e) $\displaystyle\lim_{t \to \infty} K = 0$

(f) As $t \to \infty$, the graph spirals outward and the curvature decreases

C H A P T E R   1 3
Functions of Several Variables

CHAPTER 13
Functions of Several Variables

Section 13.1   Introduction to Functions of Several Variables

1. There is not a unique value of z for each ordered pair (x, y). For example, if $x = y = 1$, then

$$z^2 = x + 3y = 4 \text{ and } z = \pm 2.$$

3. z is a function of x and y.

5. $x^2 z + 3y^2 - xy = 10$

$$x^2 z = 10 + xy - 3y^2$$

$$z = \frac{10 + xy - 3y^2}{x^2}$$

Yes, z is a function of x and y.

7. $\dfrac{x^2}{4} + \dfrac{y^2}{9} + z^2 = 1$

No, z is not a function of x and y. For example, $(x, y) = (0, 0)$ corresponds to both $z = \pm 1$.

9. $f(x, y) = 2x - y + 3$

(a) $f(0, 2) = 2(0) - 2 + 3 = 1$

(b) $f(-1, 0) = 2(-1) - 0 + 3 = 1$

(c) $f(5, 30) = 2(5) - 30 + 3 = -17$

(d) $f(3, y) = 2(3) - y + 3 = 9 - y$

(e) $f(x, 4) = 2x - 4 + 3 = 2x - 1$

(f) $f(5, t) = 2(5) - t + 3 = 13 - t$

11. $f(x, y) = xe^y$

(a) $f(-1, 0) = (-1)e^0 = -1$

(b) $f(0, 2) = 0e^2 = 0$

(c) $f(x, 3) = xe^3$

(d) $f(t, -y) = te^{-y}$

13. $h(x, y, z) = \dfrac{xy}{z}$

(a) $h(-1, 3, -1) = \dfrac{(-1)(3)}{-1} = 3$

(b) $h(2, 2, 2) = \dfrac{(2)(2)}{2} = 2$

(c) $h(4, 4t, t^2) = \dfrac{4(4t)}{t^2} = \dfrac{16}{t}$

(d) $h(-3, 2, 5) = \dfrac{(-3)(2)}{5} = -\dfrac{6}{5}$

15. $f(x, y) = x \sin y$

(a) $f\left(2, \dfrac{\pi}{4}\right) = 2 \sin \dfrac{\pi}{4} = 2\left(\dfrac{\sqrt{2}}{2}\right) = \sqrt{2}$

(b) $f(3, 1) = 3 \sin 1$

(c) $f(-3, 0) = -3 \sin 0 = 0$

(d) $f\left(4, \dfrac{\pi}{2}\right) = 4 \sin \dfrac{\pi}{2} = 4(1) = 4$

17. $g(x, y) = \int_x^y (2t - 3)\, dt$

$$= \left[t^2 - 3t\right]_x^y = y^2 - 3y - x^2 + 3x$$

(a) $g(4, 0) = 0 - 16 + 12 = -4$

(b) $g(4, 1) = (1 - 3) - 16 + 12 = -6$

(c) $g\left(4, \tfrac{3}{2}\right) = \left(\tfrac{9}{4} - \tfrac{9}{2}\right) - 16 + 12 = -\tfrac{25}{4}$

(d) $g\left(\tfrac{3}{2}, 0\right) = 0 - \tfrac{9}{4} + \tfrac{9}{2} = \tfrac{9}{4}$

19. $f(x, y) = 2x + y^2$

(a) $\dfrac{f(x + \Delta x, y) - f(x, y)}{\Delta x} = \dfrac{2(x + \Delta x) + y^2 - (2x + y^2)}{\Delta x} = \dfrac{2\Delta x}{\Delta x} = 2, \Delta x \neq 0$

(b) $\dfrac{f(x, y + \Delta y) - f(x, y)}{\Delta y} = \dfrac{2x + (y + \Delta y)^2 - 2x - y^2}{\Delta y} = \dfrac{2y\Delta y + (\Delta y)^2}{\Delta y} = 2y + \Delta y, \Delta y \neq 0$

21. $f(x, y) = 3x^2 - y$

Domain: $\{(x, y): x, y$ are real numbers.$\}$

Range: all real numbers

23. $g(x, y) = x\sqrt{y}$

Domain: $\{(x, y): y \geq 0\}$

Range: all real numbers

25. $z = \dfrac{x + y}{xy}$

Domain: $\{(x, y): x \neq 0$ and $y \neq 0\}$

Range: all real numbers

27. $f(x, y) = \sqrt{4 - x^2 - y^2}$

Domain: $4 - x^2 - y^2 \geq 0$

$x^2 + y^2 \leq 4$

$\{(x, y): x^2 + y^2 \leq 4\}$

Range: $0 \leq z \leq 2$

29. $f(x, y) = \arccos(x + y)$

Domain: $\{(x, y): -1 \leq x + y \leq 1\}$

Range: $0 \leq z \leq \pi$

31. $f(x, y) = \ln(5 - x - y)$

Domain: $5 - x - y > 0$

$x + y < 5$

$\{(x, y): y < -x + 5\}$

Range: all real numbers

33. $f(x, y) = \dfrac{-4x}{x^2 + y^2 + 1}$

(a) View from the positive x-axis: $(20, 0, 0)$

(b) View where x is negative, y and z are positive: $(-15, 10, 20)$

(c) View from the first octant: $(20, 15, 25)$

(d) View from the line $y = x$ in the xy-plane: $(20, 20, 0)$

35. $f(x, y) = 4$

Plane

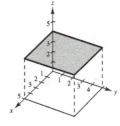

37. $f(x, y) = y^2$

Because the variable x is missing, the surface is a cylinder with rulings parallel to the x-axis. The generating curve is $z = y^2$. The domain is the entire xy-plane and the range is $z \geq 0$.

39. $z = -x^2 - y^2$

Paraboloid

Domain: entire xy-plane

Range: $z \leq 0$

41. $f(x, y) = e^{-x}$

Because the variable y is missing, the surface is a cylinder with rulings parallel to the y-axis. The generating curve is $z = e^{-x}$.

The domain is the entire xy-plane and the range is $z > 0$.

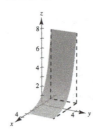

43. $z = y^2 - x^2 + 1$

Hyperbolic paraboloid

Domain: entire xy-plane

Range: $-\infty < z < \infty$

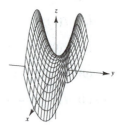

45. $f(x, y) = x^2 e^{(-xy/2)}$

47. $z = e^{1-x^2-y^2}$

Level curves:

$$c = e^{1-x^2-y^2}$$

$$\ln c = 1 - x^2 - y^2$$

$$x^2 + y^2 = 1 - \ln c$$

Circles centered at $(0, 0)$

Matches (c)

48. $z = e^{1-x^2+y^2}$

Level curves:

$$c = e^{1-x^2+y^2}$$

$$\ln c = 1 - x^2 + y^2$$

$$x^2 - y^2 = 1 - \ln c$$

Hyperbolas centered at $(0, 0)$

Matches (d)

49. $z = \ln\left|y - x^2\right|$

Level curves:

$$c = \ln\left|y - x^2\right|$$

$$\pm e^c = y - x^2$$

$$y = x^2 \pm e^c$$

Parabolas

Matches (b)

50. $z = \cos\left(\dfrac{x + 2y^2}{4}\right)$

Level curves:

$$c = \cos\left(\frac{x^2 + 2y^2}{4}\right)$$

$$\cos^{-1} c = \frac{x^2 + 2y^2}{4}$$

$$x^2 + 2y^2 = 4\cos^{-1} c$$

Ellipses

Matches (a)

51. $z = x + y$

Level curves are parallel lines of the form $x + y = c$.

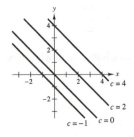

53. $z = x^2 + 4y^2$

The level curves are ellipses of the form $x^2 + 4y^2 = c$ $\left(\text{except } x^2 + 4y^2 = 0 \text{ is the point } (0, 0)\right)$.

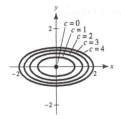

55. $f(x, y) = xy$

The level curves are hyperbolas of the form $xy = c$.

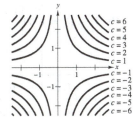

57. $f(x, y) = \dfrac{x}{x^2 + y^2}$

The level curves are of the form

$$c = \frac{x}{x^2 + y^2}$$

$$x^2 - \frac{x}{c} + y^2 = 0$$

$$\left(x - \frac{1}{2c}\right)^2 + y^2 = \left(\frac{1}{2c}\right)^2.$$

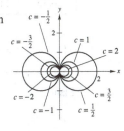

So, the level curves are circles passing through the origin and centered at $(\pm 1/2c, 0)$.

59. $f(x, y) = x^2 - y^2 + 2$

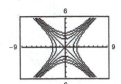

61. $g(x, y) = \dfrac{8}{1 + x^2 + y^2}$

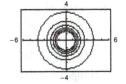

63. Yes. The definition of a function of two variables $z = f(x, y)$ requires that z be unique for each ordered pair (x, y) in the domain.

65. $f(x, y) = \dfrac{x}{y}$

The level curves are the lines $c = \dfrac{x}{y}$ or $y = \dfrac{1}{c}x$.

These lines all pass through the origin.

67. The surface is sloped like a saddle. The graph is not unique. Any vertical translation would have the same level curves.

One possible function is

$f(x, y) = |xy|$.

69. $V(I, R) = 1000\left[\dfrac{1 + 0.06(1 - R)}{1 + I}\right]^{10}$

		Inflation Rate	
Tax Rate	0	0.03	0.05
0	\$1790.85	\$1332.56	\$1099.43
0.28	\$1526.43	\$1135.80	\$937.09
0.35	\$1466.07	\$1090.90	\$900.04

71. $f(x, y, z) = x - y + z,\ c = 1$

$1 = x - y + z$

Plane

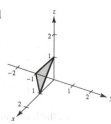

73. $f(x, y, z) = x^2 + y^2 + z^2$

$c = 9$

$9 = x^2 + y^2 + z^2$

Sphere

75. $f(x, y, z) = 4x^2 + 4y^2 - z^2$

$c = 0$

$0 = 4x^2 + 4y^2 - z^2$

Elliptic cone

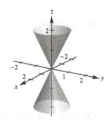

77. $N(d, L) = \left(\dfrac{d - 4}{4}\right)^2 L$

(a) $N(22, 12) = \left(\dfrac{22 - 4}{4}\right)^2 (12) = 243$ board-feet

(b) $N(30, 12) = \left(\dfrac{30 - 4}{4}\right)^2 (12) = 507$ board-feet

79. $T = 600 - 0.75x^2 - 0.75y^2$

The level curves are of the form

$$c = 600 - 0.75x^2 - 0.75y^2$$
$$x^2 + y^2 = \dfrac{600 - c}{0.75}.$$

The level curves are circles centered at the origin.

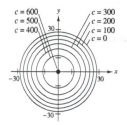

81. $f(x, y) = 80x^{0.5}y^{0.5}$

$$f(600, 350) = 80(600)^{0.5}(350)^{0.5}$$
$$= 8000\sqrt{21}$$
$$\approx 36{,}661 \text{ units}$$

83. Let $f(x, y) = Cx^a y^{1-a}$

$$f(2x, 2y) = C(2x)^a(2y)^{1-a}$$
$$= C(2)^a x^a (2)^{1-a}(y)^{1-a}$$
$$= 2Cx^a y^{1-a}$$
$$= 2f(x, y)$$

85. $PV = kT$

(a) $26(2000) = k(300) \Rightarrow k = \dfrac{520}{3}$

(b) $P = \dfrac{kT}{V} = \dfrac{520}{3}\left(\dfrac{T}{V}\right)$

The level curves are of the form

$$c = \dfrac{520}{3}\left(\dfrac{T}{V}\right), \text{ or } V = \dfrac{520}{3c}T.$$

These are lines through the origin with slope $\dfrac{520}{3c}$.

87. (a) Highest pressure at C

(b) Lowest pressure at A

(c) Highest wind velocity at B

89. $C = 4.50xy + 2.50(2)(yz) + 2.50(2)(xz)$

$\quad = 4.50xy + 5.00(yz + xz)$

91. False. Let

$\quad f(x, y) = 2xy$

$\quad f(1, 2) = f(2, 1)$, but $1 \neq 2$.

93. False. The equation of a sphere is not a function.

95. We claim that $g(x) = f(x, 0)$. First note that $x = y = z = 0$ implies $3f(0, 0) = 0 \Rightarrow f(0, 0) = 0$.

Letting $y = z = 0$ implies $f(x, 0) + f(0, 0) + f(0, x) = 0 \Rightarrow -f(0, x) = f(x, 0)$.

Letting $z = 0$ implies $f(x, y) + f(y, 0) + f(0, x) = 0 \Rightarrow f(x, y) = -f(y, 0) - f(0, x) = f(x, 0) - f(y, 0)$.

Hence, $f(x, y) = g(x) - g(y)$, as desired.

Section 13.2   Limits and Continuity

1. As x approaches -1 and y approaches 3, z approaches 1.

3. $\displaystyle\lim_{(x, y) \to (1, 0)} x = 1$

$f(x, y) = x, L = 1$

Show that for all $\varepsilon > 0$, there exists a δ-neighborhood about $(1, 0)$ such that

$\left| f(x, y) - L \right| = \left| x - 1 \right| < \varepsilon.$

whenever $(x, y) \neq (1, 0)$ lies in the neighborhood.

From $0 < \sqrt{(x - 1)^2 + (y - 0)^2} < \delta$, it follows that

$\left| x - 1 \right| = \sqrt{(x - 1)^2} \leq \sqrt{(x - 1)^2 + (y - 0)^2} < \delta.$

So, choose $\delta = \varepsilon$ and the limit is verified.

5. $\displaystyle\lim_{(x, y) \to (1, -3)} y = -3.$ $f(x, y) = y, L = -3$

Show that for all $\varepsilon > 0$, there exists a δ-neighborhood about $(1, -3)$ such that

$\left| f(x, y) - L \right| = \left| y + 3 \right| < \varepsilon$

whenever $(x, y) \neq (1, -3)$ lies in the neighborhood.

From $0 < \sqrt{(x - 1)^2 + (y + 3)^2} < \delta$ it follows that

$\left| y + 3 \right| = \sqrt{(y + 3)^2} \leq \sqrt{(x - 1)^2 + (y + 3)^2} < \delta.$

So, choose $\delta = \varepsilon$ and the limit is verified.

7. $\displaystyle\lim_{(x, y) \to (a, b)} \left[f(x, y) - g(x, y) \right] = 4 - (-5) = 9$

9. $\displaystyle\lim_{(x, y) \to (a, b)} \left[f(x, y)g(x, y) \right] = 4(-5) = -20$

11. $\displaystyle\lim_{(x, y) \to (3, 1)} \left(x^2 - 2y \right) = 3^2 - 2(1) = 9 - 2 = 7$

Continuous everywhere

13. $\displaystyle\lim_{(x, y) \to (1, 2)} e^{xy} = e^{1(2)} = e^2$

Continuous everywhere

15. $\displaystyle\lim_{(x, y) \to (0, 2)} \frac{x}{y} = \frac{0}{2} = 0$

Continuous for all $y \neq 0$

17. $\displaystyle\lim_{(x, y) \to (1, 1)} \frac{xy}{x^2 + y^2} = \frac{1}{2}$

Continuous except at $(0, 0)$

19. $\displaystyle\lim_{(x, y) \to (\pi/3, 2)} y \cos xy = 2 \cos\left(\frac{\pi}{3} \cdot 2 \right) = 2\left(-\frac{1}{2} \right) = -1$

Continuous everywhere

21. $\displaystyle\lim_{(x,y)\to(0,1)} \frac{\arcsin xy}{1-xy} = \frac{\arcsin 0}{1} = 0$

The domain of $f(x,y) = \dfrac{\arcsin xy}{1-xy}$ is $xy \neq 1$ and

$|xy| \leq 1 \Rightarrow -1 \leq xy < 1$. It is continuous on the
interior of its domain, which is $-1 < xy < 1$.

23. $\displaystyle\lim_{(x,y,z)\to(1,3,4)} \sqrt{x+y+z} = \sqrt{1+3+4} = 2\sqrt{2}$

Continuous for $x+y+z \geq 0$

25. $\displaystyle\lim_{(x,y)\to(1,1)} \frac{xy-1}{1+xy} = \frac{1-1}{1+1} = 0$

27. $\displaystyle\lim_{(x,y)\to(0,0)} \frac{1}{x+y}$ does not exist

Because the denominator $x+y$ approaches 0 as
$(x,y) \to (0,0)$.

29. $\displaystyle\lim_{(x,y)\to(0,0)} \frac{x-y}{\sqrt{x}-\sqrt{y}}$

does not exist because you cannot approach $(0,0)$ from
negative values of x and y.

31. The limit does not exist because along the line $y = 0$
you have

$$\lim_{(x,y)\to(0,0)} \frac{x+y}{x^2+y} = \lim_{(x,0)\to(0,0)} \frac{x}{x^2} = \lim_{(x,0)\to(0,0)} \frac{1}{x}$$

which does not exist.

33. $\displaystyle\lim_{(x,y)\to(0,0)} \frac{x^2}{(x^2+1)(y^2+1)} = \frac{0}{(1)(1)} = 0$

35. The limit does not exist because along the path $x = 0$,
$y = 0$, you have

$$\lim_{(x,y,z)\to(0,0,0)} \frac{xy+yz+xz}{x^2+y^2+z^2} = \lim_{(0,0,z)\to(0,0,0)} \frac{0}{z^2} = 0$$

whereas along the path $x = y = z$, you have

$$\lim_{(x,y,z)\to(0,0,0)} \frac{xy+yz+xz}{x^2+y^2+z^2} = \lim_{(x,x,x)\to(0,0,0)} \frac{x^2+x^2+x^2}{x^2+x^2+x^2}$$
$$= 1$$

37. No. The existence of $f(2,3)$ has no bearing on the
existence of the limit as $(x,y) \to (2,3)$.

39. Yes. $\displaystyle\lim_{(x,0)\to(0,0)} f(x,0) = 0$ if $f(x,0)$ exists.

41. $\displaystyle\lim_{(x,y)\to(0,0)} e^{xy} = 1$

Continuous everywhere

43. $f(x,y) = \dfrac{xy}{x^2+y^2}$

Continuous except at $(0,0)$

Path: $y = 0$

(x,y)	$(1,0)$	$(0.5,0)$	$(0.1,0)$	$(0.01,0)$	$(0.001,0)$
$f(x,y)$	0	0	0	0	0

Path: $y = x$

(x,y)	$(1,1)$	$(0.5,0.5)$	$(0.1,0.1)$	$(0.01,0.01)$	$(0.001,0.001)$
$f(x,y)$	$\frac{1}{2}$	$\frac{1}{2}$	$\frac{1}{2}$	$\frac{1}{2}$	$\frac{1}{2}$

The limit does not exist because along the path $y = 0$ the function equals 0, whereas along the path $y = x$
the function equals $\frac{1}{2}$.

45. $f(x, y) = \dfrac{y}{x^2 + y^2}$

Continuous except at $(0, 0)$

Path: $y = 0$

(x, y)	$(1, 0)$	$(0.5, 0)$	$(0.1, 0)$	$(0.01, 0)$	$(0.001, 0)$
$f(x, y)$	0	0	0	0	0

Path: $y = x$

(x, y)	$(1, 1)$	$(0.5, 0.5)$	$(0.1, 0.1)$	$(0.01, 0.01)$	$(0.001, 0.001)$
$f(x, y)$	$\frac{1}{2}$	1	5	50	500

The limit does not exist because along the path $y = 0$ the function equals 0, whereas along the path $y = x$ the function tends to infinity.

47. $\displaystyle\lim_{(x, y) \to (0, 0)} \frac{x^2 + y^2}{xy}$

(a) Along $y = ax$:

$$\lim_{(x, ax) \to (0, 0)} \frac{x^2 + (ax)^2}{x(ax)} = \lim_{x \to 0} \frac{x^2(1 + a^2)}{ax^2} = \frac{1 + a^2}{a}, a \neq 0$$

If $a = 0$, then $y = 0$ and the limit does not exist.

(b) Along $y = x^2$: $\displaystyle\lim_{(x, x^2) \to (0, 0)} \frac{x^2 + (x^2)^2}{x(x^2)} = \lim_{x \to 0} \frac{1 + x^2}{x}$

Limit does not exist.

(c) No, the limit does not exist. Different paths result in different limits.

49. $\displaystyle\lim_{(x, y) \to (0, 0)} \frac{x^4 - y^4}{x^2 + y^2} = \lim_{(x, y) \to (0, 0)} \frac{(x^2 + y^2)(x^2 - y^2)}{x^2 + y^2}$

$$= \lim_{(x, y) \to (0, 0)} (x^2 - y^2)$$

$$= 0$$

So, f is continuous everywhere, whereas g is continuous everywhere except at $(0, 0)$. g has a removable discontinuity at $(0, 0)$.

51. $\displaystyle\lim_{(x, y) \to (0, 0)} \frac{xy^2}{x^2 + y^2} = \lim_{r \to 0} \frac{(r \cos \theta)(r^2 \sin^2 \theta)}{r^2}$

$$= \lim_{r \to 0} (r \cos \theta \sin^2 \theta) = 0$$

53. $\displaystyle\lim_{(x, y) \to (0, 0)} \frac{x^2 y^2}{x^2 + y^2} = \lim_{r \to 0} \frac{r^4 \cos^2 \theta \sin^2 \theta}{r^2}$

$$= \lim_{r \to 0} r^2 \cos^2 \theta \sin^2 \theta = 0$$

55. $\displaystyle\lim_{(x, y) \to (0, 0)} \cos(x^2 + y^2) = \lim_{r \to 0} \cos(r^2) = \cos(0) = 1$

57. $\sqrt{x^2 + y^2} = r$

$$\lim_{(x, y) \to (0, 0)} \frac{\sin \sqrt{x^2 + y^2}}{\sqrt{x^2 + y^2}} = \lim_{r \to 0^+} \frac{\sin(r)}{r} = 1$$

59. $x^2 + y^2 = r^2$

$$\lim_{(x, y) \to (0, 0)} \frac{1 - \cos(x^2 + y^2)}{x^2 + y^2} = \lim_{x \to 0} \frac{1 - \cos(r^2)}{r^2} = 0$$

61. $f(x, y, z) = \dfrac{1}{\sqrt{x^2 + y^2 + z^2}}$

Continuous except at $(0, 0, 0)$

63. $f(x, y, z) = \dfrac{\sin z}{e^x + e^y}$

Continuous everywhere

65. For $xy \neq 0$, the function is clearly continuous.

For $xy \neq 0$, let $z = xy$. Then

$$\lim_{z \to 0} \frac{\sin z}{z} = 1$$

implies that f is continuous for all x, y.

67. $f(t) = t^2$, $g(x, y) = 2x - 3y$

$$f(g(x, y)) = f(2x - 3y) = (2x - 3y)^2$$

Continuous everywhere

69. $f(t) = \dfrac{1}{t}$, $g(x, y) = 2x - 3y$

$$f(g(x, y)) = f(2x - 3y) = \frac{1}{2x - 3y}$$

Continuous for all $y \neq \dfrac{2}{3}x$

71. $f(x, y) = x^2 - 4y$

(a) $\displaystyle\lim_{\Delta x \to 0} \frac{f(x + \Delta x, y) - f(x, y)}{\Delta x} = \lim_{\Delta x \to 0} \frac{\left[(x + \Delta x)^2 - 4y\right] - (x^2 - 4y)}{\Delta x} = \lim_{\Delta x \to 0} \frac{2x\Delta x + (\Delta x)^2}{\Delta x} = \lim_{\Delta x \to 0}(2x + \Delta x) = 2x$

(b) $\displaystyle\lim_{\Delta y \to 0} \frac{f(x, y + \Delta y) - f(x, y)}{\Delta y} = \lim_{\Delta y \to 0} \frac{\left[x^2 - 4(y + \Delta y)\right] - (x^2 - 4y)}{\Delta y} = \lim_{\Delta y \to 0} \frac{-4\Delta y}{\Delta y} = \lim_{\Delta y \to 0}(-4) = -4$

73. $f(x, y) = \dfrac{x}{y}$

(a) $\displaystyle\lim_{\Delta x \to 0} \frac{f(x + \Delta x, y) - f(x, y)}{\Delta x} = \lim_{\Delta x \to 0} \frac{\dfrac{x + \Delta x}{y} - \dfrac{x}{y}}{\Delta x} = \lim_{\Delta x \to 0} \frac{\dfrac{\Delta x}{y}}{\Delta x} = \lim_{\Delta x \to 0} \frac{1}{y} = \frac{1}{y}$

(b) $\displaystyle\lim_{\Delta y \to 0} \frac{f(x, y + \Delta y) - f(x, y)}{\Delta y} = \lim_{\Delta y \to 0} \frac{\dfrac{x}{y + \Delta y} - \dfrac{x}{y}}{\Delta y} = \lim_{\Delta y \to 0} \frac{xy - (xy + x\Delta y)}{(y + \Delta y)y\Delta y} = \lim_{\Delta y \to 0} \frac{-x\Delta y}{(y + \Delta y)y\Delta y} = \lim_{\Delta y \to 0} \frac{-x}{(y + \Delta y)y} = \frac{-x}{y^2}$

75. $f(x, y) = 3x + xy - 2y$

(a) $\displaystyle\lim_{\Delta x \to 0} \frac{f(x + \Delta x, y) - f(x, y)}{\Delta x} = \lim_{\Delta x \to 0} \frac{3(x + \Delta x) + (x + \Delta x)y - 2y - (3x + xy - 2y)}{\Delta x}$

$$= \lim_{\Delta x \to 0} \frac{3\Delta x + y\Delta x}{\Delta x} = \lim_{\Delta x \to 0}(3 + y) = 3 + y$$

(b) $\displaystyle\lim_{\Delta y \to 0} \frac{f(x, y + \Delta y) - f(x, y)}{\Delta y} = \lim_{\Delta y \to 0} \frac{3x + x(y + \Delta y) - 2(y + \Delta y) - (3x + xy - 2y)}{\Delta y}$

$$= \lim_{\Delta y \to 0} \frac{x\Delta y - 2\Delta y}{\Delta y} = \lim_{\Delta y \to 0}(x - 2) = x - 2$$

77. $\displaystyle\lim_{(x, y, z) \to (0, 0, 0)} \frac{xyz}{x^2 + y^2 + z^2} = \lim_{\rho \to 0^+} \frac{(\rho \sin \phi \cos \theta)(\rho \sin \phi \sin \theta)(\rho \cos \phi)}{\rho^2}$

$$= \lim_{\rho \to 0^+} \rho\left[\sin^2 \phi \cos \theta \sin \theta \cos \phi\right] = 0$$

79. True

81. False. Let $f(x, y) = \begin{cases} \ln(x^2 + y^2), & (x, y) \neq (0, 0) \\ 0, & x = 0, y = 0 \end{cases}$.

83. As $(x, y) \to (0, 1)$, $x^2 + 1 \to 1$ and $x^2 + (y - 1)^2 \to 0$.

So, $\displaystyle\lim_{(x, y) \to (0, 1)} \tan^{-1}\left[\frac{x^2 + 1}{x^2 + (y - 1)^2}\right] = \frac{\pi}{2}$.

85. Because $\lim\limits_{(x,y)\to(a,b)} f(x, y) = L_1$, then for $\varepsilon/2 > 0$, there corresponds $\delta_1 > 0$ such that $\left| f(x, y) - L_1 \right| < \varepsilon/2$ whenever

$$0 < \sqrt{(x - a)^2 + (y - b)^2} < \delta_1.$$

Because $\lim\limits_{(x,y)\to(a,b)} g(x, y) = L_2$, then for $\varepsilon/2 > 0$, there corresponds $\delta_2 > 0$ such that $\left| g(x, y) - L_2 \right| < \varepsilon/2$ whenever

$$0 < \sqrt{(x - a)^2 + (y - b)^2} < \delta_2.$$

Let δ be the smaller of δ_1 and δ_2. By the triangle inequality, whenever $\sqrt{(x - a)^2 + (y - b)^2} < \delta$, we have

$$\left| f(x, y) + g(x, y) - (L_1 + L_2) \right| = \left| (f(x, y) - L_1) + (g(x, y) - L_2) \right| \le \left| f(x, y) - L_1 \right| + \left| g(x, y) - L_2 \right| < \frac{\varepsilon}{2} + \frac{\varepsilon}{2} = \varepsilon.$$

So, $\lim\limits_{(x,y)\to(a,b)} \left[f(x, y) + g(x, y) \right] = L_1 + L_2$.

Section 13.3   Partial Derivatives

1. $z_x,\; f_x(x, y),\; \dfrac{\partial z}{\partial x}$

3. (a) Differentiate first with respect to y, then with respect to x, and last with respect to z.

   (b) Differentiate first with respect to z and then with respect to x.

5. No, x only occurs in the numerator.

7. No, y only occurs in the numerator.

9. Yes, x occurs in both the numerator and denominator.

11. $f(x, y) = 2x - 5y + 3$

$f_x(x, y) = 2$

$f_y(x, y) = -5$

13. $z = 6x - x^2 y + 8y^2$

$\dfrac{\partial z}{\partial x} = 6 - 2xy$

$\dfrac{\partial z}{\partial y} = -x^2 + 16y$

15. $z = x\sqrt{y}$

$\dfrac{\partial z}{\partial x} = \sqrt{y}$

$\dfrac{\partial z}{\partial y} = \dfrac{x}{2\sqrt{y}}$

17. $z = e^{xy}$

$\dfrac{\partial z}{\partial x} = ye^{xy}$

$\dfrac{\partial z}{\partial y} = xe^{xy}$

19. $z = x^2 e^{2y}$

$\dfrac{\partial z}{\partial x} = 2xe^{2y}$

$\dfrac{\partial z}{\partial y} = 2x^2 e^{2y}$

21. $z = \ln \dfrac{x}{y} = \ln x - \ln y$

$\dfrac{\partial z}{\partial x} = \dfrac{1}{x}$

$\dfrac{\partial z}{\partial y} = -\dfrac{1}{y}$

23. $z = \ln(x^2 + y^2)$

$\dfrac{\partial z}{\partial x} = \dfrac{2x}{x^2 + y^2}$

$\dfrac{\partial z}{\partial y} = \dfrac{2y}{x^2 + y^2}$

25. $z = \dfrac{x^2}{2y} + \dfrac{3y^2}{x}$

$\dfrac{\partial z}{\partial x} = \dfrac{2x}{2y} - \dfrac{3y^2}{x^2} = \dfrac{x^3 - 3y^3}{x^2 y}$

$\dfrac{\partial z}{\partial y} = \dfrac{-x^2}{2y^2} + \dfrac{6y}{x} = \dfrac{12y^3 - x^3}{2xy^2}$

27. $h(x, y) = e^{-(x^2 + y^2)}$

$h_x(x, y) = -2xe^{-(x^2 + y^2)}$

$h_y(x, y) = -2ye^{-(x^2 + y^2)}$

29. $f(x, y) = \sqrt{x^2 + y^2}$

$$f_x(x, y) = \frac{1}{2}(x^2 + y^2)^{-1/2}(2x) = \frac{x}{\sqrt{x^2 + y^2}}$$

$$f_y(x, y) = \frac{1}{2}(x^2 + y^2)^{-1/2}(2y) = \frac{y}{\sqrt{x^2 + y^2}}$$

31. $z = \cos xy$

$$\frac{\partial z}{\partial x} = -y \sin xy$$

$$\frac{\partial z}{\partial y} = -x \sin xy$$

33. $z = \tan(2x - y)$

$$\frac{\partial z}{\partial x} = 2 \sec^2(2x - y)$$

$$\frac{\partial z}{\partial y} = -\sec^2(2x - y)$$

35. $z = e^y \sin 8xy$

$$\frac{\partial z}{\partial x} = e^y \cos(8xy)(8y) = 8ye^y \cos 8xy$$

$$\frac{\partial z}{\partial y} = e^y \sin 8xy + e^y \cos(8xy)(8x)$$

$$= e^y(\sin 8xy + 8x \cos 8xy)$$

37. $z = \sinh(2x + 3y)$

$$\frac{\partial z}{\partial x} = 2 \cosh(2x + 3y)$$

$$\frac{\partial z}{\partial y} = 3 \cosh(2x + 3y)$$

39. $f(x, y) = \int_x^y (t^2 - 1) \, dt$

$$= \left[\frac{t^3}{3} - t\right]_x^y = \left(\frac{y^3}{3} - y\right) - \left(\frac{x^3}{3} - x\right)$$

$$f_x(x, y) = -x^2 + 1 = 1 - x^2$$

$$f_y(x, y) = y^2 - 1$$

[You could also use the Second Fundamental Theorem of Calculus.]

41. $f(x, y) = 3x + 2y$

$$\frac{\partial f}{\partial x} = \lim_{\Delta x \to 0} \frac{f(x + \Delta x, y) - f(x, y)}{\Delta x}$$

$$= \lim_{\Delta x \to 0} \frac{3(x + \Delta x) + 2y - (3x + 2y)}{\Delta x}$$

$$= \lim_{\Delta x \to 0} \frac{3\Delta x}{\Delta x} = 3$$

$$\frac{\partial f}{\partial y} = \lim_{\Delta y \to 0} \frac{f(x, y + \Delta y) - f(x, y)}{\Delta y}$$

$$= \lim_{\Delta y \to 0} \frac{3x + 2(y + \Delta y) - (3x + 2y)}{\Delta y}$$

$$= \lim_{\Delta y \to 0} \frac{2\Delta y}{\Delta y} = 2$$

43. $f(x, y) = \sqrt{x + y}$

$$\frac{\partial f}{\partial x} = \lim_{\Delta x \to 0} \frac{f(x + \Delta x, y) - f(x, y)}{\Delta x}$$

$$= \lim_{\Delta x \to 0} \frac{\sqrt{x + \Delta x + y} - \sqrt{x + y}}{\Delta x}$$

$$= \lim_{\Delta x \to 0} \frac{\left(\sqrt{x + \Delta x + y} - \sqrt{x + y}\right)\left(\sqrt{x + \Delta x + y} + \sqrt{x + y}\right)}{\Delta x\left(\sqrt{x + \Delta x + y} + \sqrt{x + y}\right)} = \lim_{\Delta x \to 0} \frac{1}{\sqrt{x + \Delta x + y} + \sqrt{x + y}} = \frac{1}{2\sqrt{x + y}}$$

$$\frac{\partial f}{\partial y} = \lim_{\Delta y \to 0} \frac{f(x, y + \Delta y) - f(x, y)}{\Delta y} = \lim_{\Delta y \to 0} \frac{\sqrt{x + y + \Delta y} - \sqrt{x + y}}{\Delta y}$$

$$= \lim_{\Delta y \to 0} \frac{\left(\sqrt{x + y + \Delta y} - \sqrt{x + y}\right)\left(\sqrt{x + y + \Delta y} + \sqrt{x + y}\right)}{\Delta y\left(\sqrt{x + y + \Delta y} + \sqrt{x + y}\right)}$$

$$= \lim_{\Delta y \to 0} \frac{1}{\sqrt{x + y + \Delta y} + \sqrt{x + y}} = \frac{1}{2\sqrt{x + y}}$$

45. $f(x, y) = e^x y^2$

$f_x(x, y) = e^x y^2$

At $(\ln 3, 2)$, $f_x(\ln 3, 2) = e^{\ln 3}(2)^2 = 3 \cdot 4 = 12$.

$f_y(x, y) = 2e^x y$

At $(\ln 3, 2)$, $f_y(\ln 3, 2) = 2e^{\ln 3}(2) = 6 \cdot 2 = 12$.

47. $f(x, y) = \cos(2x - y)$

$f_x(x, y) = -2 \sin(2x - y)$

$\text{At}\left(\frac{\pi}{4}, \frac{\pi}{3}\right)$, $f_x\left(\frac{\pi}{4}, \frac{\pi}{3}\right) = -2 \sin\left(\frac{\pi}{2} - \frac{\pi}{3}\right) = -1$.

$f_y(x, y) = \sin(2x - y)$

$\text{At}\left(\frac{\pi}{4}, \frac{\pi}{3}\right)$, $f_y\left(\frac{\pi}{4}, \frac{\pi}{3}\right) = \sin\left(\frac{\pi}{2} - \frac{\pi}{3}\right) = \frac{1}{2}$.

49. $f(x, y) = \arctan \frac{y}{x}$

$f_x(x, y) = \frac{1}{1 + (y^2/x^2)}\left(-\frac{y}{x^2}\right) = \frac{-y}{x^2 + y^2}$

At $(2, -2)$: $f_x(2, -2) = \frac{1}{4}$

$f_y(x, y) = \frac{1}{1 + (y^2/x^2)}\left(\frac{1}{x}\right) = \frac{x}{x^2 + y^2}$

At $(2, -2)$: $f_y(2, -2) = \frac{1}{4}$

51. $f(x, y) = \frac{xy}{x - y}$

$f_x(x, y) = \frac{y(x - y) - xy}{(x - y)^2} = \frac{-y^2}{(x - y)^2}$

At $(2, -2)$: $f_x(2, -2) = -\frac{1}{4}$

$f_y(x, y) = \frac{x(x - y) + xy}{(x - y)^2} = \frac{x^2}{(x - y)^2}$

At $(2, -2)$: $f_y(2, -2) = \frac{1}{4}$

53. $z = xy$

$\frac{\partial z}{\partial x} = y$

At $(1, 2, 2)$: $\frac{\partial z}{\partial x}(1, 2, 2) = 2$

$\frac{\partial z}{\partial y} = x$

At $(1, 2, 2)$: $\frac{\partial z}{\partial y}(1, 2, 2) = 1$

55. $g(x, y) = 4 - x^2 - y^2$

$g_x(x, y) = -2x$

At $(1, 1)$: $g_x(1, 1) = -2$

$g_y(x, y) = -2y$

At $(1, 1)$: $g_y(1, 1) = -2$

57. $H(x, y, z) = \sin(x + 2y + 3z)$

$H_x(x, y, z) = \cos(x + 2y + 3z)$

$H_y(x, y, z) = 2 \cos(x + 2y + 3z)$

$H_z(x, y, z) = 3 \cos(x + 2y + 3z)$

59. $w = \sqrt{x^2 + y^2 + z^2}$

$\frac{\partial w}{\partial x} = \frac{x}{\sqrt{x^2 + y^2 + z^2}}$

$\frac{\partial w}{\partial y} = \frac{y}{\sqrt{x^2 + y^2 + z^2}}$

$\frac{\partial w}{\partial z} = \frac{z}{\sqrt{x^2 + y^2 + z^2}}$

61. $F(x, y, z) = \ln\sqrt{x^2 + y^2 + z^2} = \frac{1}{2} \ln(x^2 + y^2 + z^2)$

$F_x(x, y, z) = \frac{x}{x^2 + y^2 + z^2}$

$F_y(x, y, z) = \frac{y}{x^2 + y^2 + z^2}$

$F_z(x, y, z) = \frac{z}{x^2 + y^2 + z^2}$

63. $f(x, y, z) = x^3 y z^2$

$f_x(x, y, z) = 3x^2 y z^2$

$f_x(1, 1, 1) = 3$

$f_y(x, y, z) = x^3 z^2$

$f_y(1, 1, 1) = 1$

$f_z(x, y, z) = 2x^3 y z$

$f_z(1, 1, 1) = 2$

65. $f(x, y, z) = \dfrac{\ln x}{yz}$

$f_x = \dfrac{1}{xyz}$

$f_x(1, -1, -1) = \dfrac{1}{1(-1)(-1)} = 1$

$f_y = \dfrac{-\ln x}{y^2 z}$

$f_y(1, -1, -1) = 0$

$f_z = \dfrac{-\ln x}{z^2 y}$

$f_z(1, -1, -1) = 0$

67. $f(x, y, z) = z \sin(x + 6y)$

$f_x = z \cos(x + 6y)$

$f_x\left(0, \dfrac{\pi}{2}, -4\right) = -4 \cos 3\pi = 4$

$f_y = 6z \cos(x + 6y)$

$f_y\left(0, \dfrac{\pi}{2}, -4\right) = 6(-4) \cos 3\pi = 24$

$f_z = \sin(x + 6y)$

$f_z\left(0, \dfrac{\pi}{2}, -4\right) = \sin 3\pi = 0$

69. $f_x(x, y) = 2x + y - 2 = 0$

$f_y(x, y) = x + 2y + 2 = 0$

$2x + y - 2 = 0 \Rightarrow y = 2 - 2x$

$x + 2(2 - 2x) + 2 = 0 \Rightarrow -3x + 6 = 0 \Rightarrow x = 2,$

$y = -2$

Point: $(2, -2)$

71. $f_x(x, y) = 2x + 4y - 4, f_y(x, y) = 4x + 2y + 16$

$f_x = f_y = 0: \quad 2x + 4y = 4$

$\qquad\qquad\qquad 4x + 2y = -16$

Solving for x and y, $x = -6$ and $y = 4$.

73. $f_x(x, y) = -\dfrac{1}{x^2} + y, f_y(x, y) = -\dfrac{1}{y^2} + x$

$f_x = f_y = 0: -\dfrac{1}{x^2} + y = 0 \text{ and } -\dfrac{1}{y^2} + x = 0$

$\qquad\qquad y = \dfrac{1}{x^2} \text{ and } x = \dfrac{1}{y^2}$

$y = y^4 \Rightarrow y = 1 = x$

Points: $(1, 1)$

75. $f_x(x, y) = (2x + y)e^{x^2 + xy + y^2} = 0$

$f_y(x, y) = (x + 2y)e^{x^2 + xy + y^2} = 0$

$2x + y = 0 \Rightarrow y = -2x$

$x + 2(-2x) = 0 \Rightarrow x = 0 \Rightarrow y = 0$

Point: $(0, 0)$

77. $z = 3xy^2$

$\dfrac{\partial z}{\partial x} = 3y^2, \dfrac{\partial^2 z}{\partial x^2} = 0, \dfrac{\partial^2 z}{\partial y \partial x} = 6y$

$\dfrac{\partial z}{\partial y} = 6xy, \dfrac{\partial^2 z}{\partial y^2} = 6x, \dfrac{\partial^2 z}{\partial x \partial y} = 6y$

79. $z = x^4 - 2xy + 3y^3$

$\dfrac{\partial z}{\partial x} = 4x^3 - 2y$

$\dfrac{\partial z}{\partial y} = -2x + 9y^2$

$\dfrac{\partial^2 z}{\partial x^2} = 12x^2$

$\dfrac{\partial^2 z}{\partial x \partial y} = -2$

$\dfrac{\partial^2 z}{\partial y \partial x} = -2$

$\dfrac{\partial^2 z}{\partial y^2} = 18y$

81. $z = \sqrt{x^2 + y^2}$

$\dfrac{\partial z}{\partial x} = \dfrac{x}{\sqrt{x^2 + y^2}}$

$\dfrac{\partial^2 z}{\partial x^2} = \dfrac{y^2}{\left(x^2 + y^2\right)^{3/2}}$

$\dfrac{\partial^2 z}{\partial y \partial x} = \dfrac{-xy}{\left(x^2 + y^2\right)^{3/2}}$

$\dfrac{\partial z}{\partial y} = \dfrac{y}{\sqrt{x^2 + y^2}}$

$\dfrac{\partial^2 z}{\partial y^2} = \dfrac{x^2}{\left(x^2 + y^2\right)^{3/2}}$

$\dfrac{\partial^2 z}{\partial x \partial y} = \dfrac{-xy}{\left(x^2 + y^2\right)^{3/2}}$

83. $z = e^x \tan y$

$$\frac{\partial z}{\partial x} = e^x \tan y$$

$$\frac{\partial^2 z}{\partial x^2} = e^x \tan y$$

$$\frac{\partial^2 z}{\partial y \partial x} = e^x \sec^2 y$$

$$\frac{\partial z}{\partial y} = e^x \sec^2 y$$

$$\frac{\partial^2 z}{\partial y^2} = 2e^x \sec^2 y \tan y$$

$$\frac{\partial^2 z}{\partial x \partial y} = e^x \sec^2 y$$

85. $z = \cos xy$

$$\frac{\partial z}{\partial x} = -y \sin xy, \frac{\partial^2 z}{\partial x^2} = -y^2 \cos xy$$

$$\frac{\partial^2 z}{\partial y \partial x} = -yx \cos xy - \sin xy$$

$$\frac{\partial z}{\partial y} = -x \sin xy, \frac{\partial^2 z}{\partial y^2} = -x^2 \cos xy$$

$$\frac{\partial^2 z}{\partial x \partial y} = -xy \cos xy - \sin xy$$

87. $z = x \sec y$

$$\frac{\partial z}{\partial x} = \sec y$$

$$\frac{\partial^2 z}{\partial x^2} = 0$$

$$\frac{\partial^2 z}{\partial y \partial x} = \sec y \tan y$$

$$\frac{\partial z}{\partial y} = x \sec y \tan y$$

$$\frac{\partial^2 z}{\partial y^2} = x \sec y \left(\sec^2 y + \tan^2 y\right)$$

$$\frac{\partial^2 z}{\partial x \partial y} = \sec y \tan y$$

So, $\dfrac{\partial^2 z}{\partial y \partial x} = \dfrac{\partial^2 z}{\partial x \partial y}$

There are no points for which $z_x = 0 = z_y$, because

$$\frac{\partial z}{\partial x} = \sec y \neq 0.$$

89. $z = \ln\left(\dfrac{x}{x^2 + y^2}\right) = \ln x - \ln\left(x^2 + y^2\right)$

$$\frac{\partial z}{\partial x} = \frac{1}{x} - \frac{2x}{x^2 + y^2} = \frac{y^2 - x^2}{x\left(x^2 + y^2\right)}$$

$$\frac{\partial^2 z}{\partial x^2} = \frac{x^4 - 4x^2 y^2 - y^4}{x^2\left(x^2 + y^2\right)^2}$$

$$\frac{\partial^2 z}{\partial y \partial x} = \frac{4xy}{\left(x^2 + y^2\right)^2}$$

$$\frac{\partial z}{\partial y} = -\frac{2y}{x^2 + y^2}$$

$$\frac{\partial^2 z}{\partial y^2} = \frac{2\left(y^2 - x^2\right)}{\left(x^2 + y^2\right)^2}$$

$$\frac{\partial^2 z}{\partial x \partial y} = \frac{4xy}{\left(x^2 + y^2\right)^2}$$

There are no points for which $z_x = z_y = 0$.

91. $f(x, y, z) = xyz$

$$f_x(x, y, z) = yz$$

$$f_y(x, y, z) = xz$$

$$f_{yy}(x, y, z) = 0$$

$$f_{xy}(x, y, z) = z$$

$$f_{yx}(x, y, z) = z$$

$$f_{yyx}(x, y, z) = 0$$

$$f_{xyy}(x, y, z) = 0$$

$$f_{yxy}(x, y, z) = 0$$

So, $f_{xyy} = f_{yxy} = f_{yyx} = 0$.

93. $f(x, y, z) = e^{-x} \sin yz$

$$f_x(x, y, z) = -e^{-x} \sin yz$$

$$f_y(x, y, z) = ze^{-x} \cos yz$$

$$f_{yy}(x, y, z) = -z^2 e^{-x} \sin yz$$

$$f_{xy}(x, y, z) = -ze^{-x} \cos yz$$

$$f_{yx}(x, y, z) = -ze^{-x} \cos yz$$

$$f_{yyx}(x, y, z) = z^2 e^{-x} \sin yz$$

$$f_{xyy}(x, y, z) = z^2 e^{-x} \sin yz$$

$$f_{yxy}(x, y, z) = z^2 e^{-x} \sin yz$$

So, $f_{xyy} = f_{yxy} = f_{yyz}$.

95. $z = 5xy$

$$\frac{\partial z}{\partial x} = 5y$$

$$\frac{\partial^2 z}{\partial x^2} = 0$$

$$\frac{\partial z}{\partial y} = 5x$$

$$\frac{\partial^2 z}{\partial y^2} = 0$$

So, $\dfrac{\partial^2 z}{\partial x^2} + \dfrac{\partial^2 z}{\partial y^2} = 0 + 0 = 0.$

97. $z = e^x \sin y$

$$\frac{\partial z}{\partial x} = e^x \sin y$$

$$\frac{\partial^2 z}{\partial x^2} = e^x \sin y$$

$$\frac{\partial z}{\partial y} = e^x \cos y$$

$$\frac{\partial^2 z}{\partial y^2} = -e^x \sin y$$

So, $\dfrac{\partial^2 z}{\partial x^2} + \dfrac{\partial^2 z}{\partial y^2} = e^x \sin y - e^x \sin y = 0.$

99. $z = \sin(x - ct)$

$$\frac{\partial z}{\partial t} = -c \cos(x - ct)$$

$$\frac{\partial^2 z}{\partial t^2} = -c^2 \sin(x - ct)$$

$$\frac{\partial z}{\partial x} = \cos(x - ct)$$

$$\frac{\partial^2 z}{\partial x^2} = -\sin(x - ct)$$

So, $\dfrac{\partial^2 z}{\partial t^2} = c^2\left(\dfrac{\partial^2 z}{\partial x^2}\right).$

101. $z = \ln(x + ct)$

$$\frac{\partial z}{\partial t} = \frac{c}{x + ct}$$

$$\frac{\partial^2 z}{\partial t^2} = \frac{-c^2}{(x + ct)^2}$$

$$\frac{\partial z}{\partial x} = \frac{1}{x + ct}$$

$$\frac{\partial^2 z}{\partial x^2} = \frac{-1}{(x + ct)^2}$$

$$\frac{\partial^2 z}{\partial t^2} = \frac{-c^2}{(x + ct)^2} = c^2\left(\frac{\partial^2 z}{\partial x^2}\right)$$

103. $z = e^{-t} \cos \dfrac{x}{c}$

$$\frac{\partial z}{\partial t} = -e^{-t} \cos \frac{x}{c}$$

$$\frac{\partial z}{\partial x} = -\frac{1}{c} e^{-t} \sin \frac{x}{c}$$

$$\frac{\partial^2 z}{\partial x^2} = -\frac{1}{c^2} e^{-t} \cos \frac{x}{c}$$

So, $\dfrac{\partial z}{\partial t} = c^2\left(\dfrac{\partial^2 z}{\partial x^2}\right).$

105. $u = x^2 - y^2,\ v = 2xy$

$$\frac{\partial u}{\partial x} = 2x = \frac{\partial v}{\partial y}$$

$$\frac{\partial u}{\partial y} = -2y = -\frac{\partial v}{\partial x}$$

107. Yes. The function $f(x, y) = \cos(3x - 2y)$ satisfies both equations.

109. No. For example, let $z = x + y + 1$.

$z_x = 1$ and $z_y = 1$, but $z \neq c(x + y)$.

111. The plane $z = -x + y = f(x, y)$ satisfies

$$\frac{\partial f}{\partial x} < 0 \text{ and } \frac{\partial f}{\partial y} > 0.$$

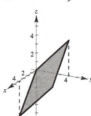

113. The units of $\dfrac{\partial P}{\partial A}$ are dollars per year. $\dfrac{\partial P}{\partial A}$ is negative, since the value of a car generally decreases with time.

115. $A = \dfrac{1}{2} ab \sin \theta$

$$\frac{\partial A}{\partial b} = \frac{1}{2} a \sin \theta$$

$$\frac{\partial A}{\partial \theta} = \frac{1}{2} ab \cos \theta$$

(a) $a = 4, b = 1, \theta = \dfrac{\pi}{4}:\ \dfrac{\partial A}{\partial b} = \dfrac{1}{2}(4) \sin \dfrac{\pi}{4} = \sqrt{2}$

(b) $a = 2, b = 5, \theta = \dfrac{\pi}{3}:\ \dfrac{\partial A}{\partial \theta} = \dfrac{1}{2}(2)(5) \cos \dfrac{\pi}{3} = \dfrac{5}{2}$

117. $R = 200x_1 + 200x_2 - 4x_1^2 - 8x_1x_2 - 4x_2^2$

(a) $\dfrac{\partial r}{\partial x_1} = 200 - 8x_1 - 8x_2$

At $(x_1, x_2) = (4, 12)$, $\dfrac{\partial R}{\partial x_1} = 200 - 32 - 96 = 72$.

(b) $\dfrac{\partial R}{\partial x_2} = 200 - 8x_1 - 8x_2$

At $(x_1, x_2) = (4, 12)$, $\dfrac{\partial R}{\partial x_2} = 72$.

119. $IQ(M, C) = 100\dfrac{M}{C}$

$IQ_M = \dfrac{100}{C}$, $IQ_M(12, 10) = 10$

$IQ_c = \dfrac{-100M}{C^2}$, $IQ_c(12, 10) = -12$

When the chronological age is constant, IQ increases at a rate of 10 points per mental age year.

When the mental age is constant, IQ decreases at a rate of 12 points per chronological age year.

121. An increase in either price will cause a decrease in demand.

123. $T = 500 - 0.6x^2 - 1.5y^2$

$\dfrac{\partial T}{\partial x} = -1.2x$, $\dfrac{\partial T}{\partial x}(2, 3) = -2.4°/\text{m}$

$\dfrac{\partial T}{\partial y} = -3y$, $\dfrac{\partial T}{\partial y}(2, 3) = -9°/\text{m}$

125.
$$PV = nRT$$

$T = \dfrac{PV}{nR} \Rightarrow \dfrac{\partial T}{\partial P} = \dfrac{V}{nR}$

$P = \dfrac{nRT}{V} \Rightarrow \dfrac{\partial P}{\partial V} = -\dfrac{nRT}{V^2}$

$V = \dfrac{nRT}{P} \Rightarrow \dfrac{\partial V}{\partial T} = \dfrac{nR}{P}$

$\dfrac{\partial T}{\partial P} \cdot \dfrac{\partial P}{\partial V} \cdot \dfrac{\partial V}{\partial T} = \left(\dfrac{V}{nR}\right)\left(-\dfrac{nRT}{V^2}\right)\left(\dfrac{nR}{P}\right)$

$= -\dfrac{nRT}{VP} = -\dfrac{nRT}{nRT} = -1$

127. $z = 0.23x + 0.14y + 6.85$

(a) $\dfrac{\partial z}{\partial x} = 0.23$, $\dfrac{\partial z}{\partial y} = 0.14$

(b) As the expenditures on amusement parks and campgrounds (x) increase, the expenditures on spectator sports (z) increase. As the expenditures on live entertainment (y) increase, the expenditures on spectator sports (z) also increase.

129. $f(x, y) = \begin{cases} \dfrac{xy(x^2 - y^2)}{x^2 + y^2}, & (x, y) \neq (0, 0) \\ 0, & (x, y) = (0, 0) \end{cases}$

(a) $f_x(x, y) = \dfrac{(x^2 + y^2)(3x^2y - y^3) - (x^3y - xy^3)(2x)}{(x^2 + y^2)^2} = \dfrac{y(x^4 + 4x^2y^2 - y^4)}{(x^2 + y^2)^2}$

$f_y(x, y) = \dfrac{(x^2 + y^2)(x^3 - 3xy^2) - (x^3y - xy^3)(2y)}{(x^2 + y^2)^2} = \dfrac{x(x^4 - 4x^2y^2 - y^4)}{(x^2 + y^2)^2}$

(b) $f_x(0, 0) = \lim\limits_{\Delta x \to 0} \dfrac{f(\Delta x, 0) - f(0, 0)}{\Delta x} = \lim\limits_{\Delta x \to 0} \dfrac{0/[(\Delta x)^2] - 0}{\Delta x} = 0$

$f_y(0, 0) = \lim\limits_{\Delta y \to 0} \dfrac{f(0, \Delta y) - f(0, 0)}{\Delta y} = \lim\limits_{\Delta y \to 0} \dfrac{0/[(\Delta y)^2] - 0}{\Delta y} = 0$

(c) $f_{xy}(0, 0) = \dfrac{\partial}{\partial y}\left(\dfrac{\partial f}{\partial x}\right)\bigg|_{(0, 0)} = \lim\limits_{\Delta y \to 0} \dfrac{f_x(0, \Delta y) - f_x(0, 0)}{\Delta y} = \lim\limits_{\Delta y \to 0} \dfrac{\Delta y(-(\Delta y)^4)}{((\Delta y)^2)^2(\Delta y)} = \lim\limits_{\Delta y \to 0}(-1) = -1$

$f_{yx}(0, 0) = \dfrac{\partial}{\partial x}\left(\dfrac{\partial f}{\partial y}\right)\bigg|_{(0, 0)} = \lim\limits_{\Delta x \to 0} \dfrac{f_y(\Delta x, 0) - f_y(0, 0)}{\Delta x} = \lim\limits_{\Delta x \to 0} \dfrac{\Delta x((\Delta x)^4)}{((\Delta x)^2)^2(\Delta x)} = \lim\limits_{\Delta x \to 0} 1 = 1$

(d) f_{yx} or f_{xy} or both are not continuous at $(0, 0)$.

131. $f(x, y) = (x^2 + y^2)^{2/3}$

For $(x, y) \neq (0, 0)$, $f_x(x, y) = \frac{2}{3}(x^2 + y^2)^{-1/3}(2x) = \frac{4x}{3(x^2 + y^2)^{1/3}}$.

For $(x, y) = (0, 0)$, use the definition of partial derivative.

$$f_x(0, 0) = \lim_{\Delta x \to 0} \frac{f(0 + \Delta x) - f(0, 0)}{\Delta x} = \lim_{\Delta x \to 0} \frac{(\Delta x)^{4/3}}{\Delta x} = \lim_{\Delta x \to 0} (\Delta x)^{1/3} = 0$$

Section 13.4   Differentials

1. In general, the accuracy worsens as Δx and Δy increase.

3. $z = 5x^3y^2$

$dz = 15x^2y^2\, dx + 10x^3y\, dy$

5. $z = \frac{-1}{x^2 + y^2}$

$$dz = \frac{2x}{(x^2 + y^2)^2}\, dx + \frac{2y}{(x^2 + y^2)^2}\, dy$$

$$= \frac{2}{(x^2 + y^2)^2}(x\, dx + y\, dy)$$

7. $w = x^2yz^2 + \sin yz$

$dw = 2xyz^2\, dx + (x^2z^2 + z \cos yz)\, dy + (2x^2yz + y \cos yz)\, dz$

9. $w = 2z^3y \sin x$

$dw = 2z^3y \cos x\, dx + 2z^3 \sin x\, dy + 6z^2y \sin x\, dz$

11. $w = x^2yz^2 + \sin yz$

$dw = 2xyz^2\, dx + (x^2z^2 + z \cos yz)dy$

$\quad + (2x^2yz + y \cos yz)dz$

13. $f(x, y) = 2x - 3y$

(a) $f(2, 1) = 1$

$f(2.1, 1.05) = 1.05$

$\Delta z = f(2.1, 1.05) - f(2, 1) = 0.05$

(b) $dz = 2\, dx - 3\, dy = 2(0.1) - 3(0.05) = 0.05$

15. $f(x, y) = 16 - x^2 - y^2$

(a) $f(2, 1) = 11$

$f(2.1, 1.05) = 10.4875$

$\Delta z = f(2.1, 1.05) - f(2, 1) = -0.5125$

(b) $dz = -2x\, dx - 2y\, dy$

$= -2(2)(0.1) - 2(1)(0.05) = -0.5$

17. $f(x, y) = ye^x$

(a) $f(2, 1) = e^2 \approx 7.3891$

$f(2.1, 1.05) = 1.05e^{2.1} \approx 8.5745$

$\Delta z = f(2.1, 1.05) - f(2, 1) = 1.1854$

(b) $dz = ye^x\, dx + e^x\, dy$

$= e^2(0.1) + e^2(0.05) \approx 1.1084$

19. Yes. Because f_x and f_y are continuous on R, you know that f is differentiable on R. Because f is differentiable on R, you know that f is continuous on R.

21. $A = lh$

$dA = l\, dh + h\, dl$

$\Delta A = (1 + dl)(h + dh) - lh = h\, dl + l\, dh + dl\, dh$

$\Delta A - dA = dl\, dh$

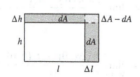

23. $V = xyz, dV = yz\,dx + xz\,dy + xy\,dz$

Propagated error $= dV = 5(12)(\pm 0.02) + 8(12)(\pm 0.02) + 8(5)(\pm 0.02)$

$$= (60 + 96 + 40)(\pm 0.02) = 196(\pm 0.02) = \pm 3.92 \text{ in.}^3$$

The measured volume is $V = 8(5)(12) = 480 \text{ in.}^3$

Relative error $= \dfrac{\Delta V}{V} \approx \dfrac{dV}{V} = \dfrac{3.92}{480} \approx 0.008167 \approx 0.82\%$

25. $V = \dfrac{\pi r^2 h}{3}, r = 4, h = 8$

$$dV = \frac{2\pi rh}{3}\,dr + \frac{\pi r^2}{3}\,dh = \frac{\pi r}{3}(2h\,dr + r\,dh) = \frac{4\pi}{3}(16\,dr + 4\,dh)$$

$$\Delta V = \frac{\pi}{3}\Big[(r + \Delta r)^2(h + \Delta h) - r^2 h\Big] = \frac{\pi}{3}\Big[(4 + \Delta r)^2(8 + \Delta h) - 128\Big]$$

Δr	Δh	dV	ΔV	$\Delta V - dV$
0.1	0.1	8.3776	8.5462	0.1686
0.1	−0.1	5.0265	5.0255	−0.0010
0.001	0.002	0.1005	0.1006	0.0001
−0.0001	0.0002	−0.0034	−0.0034	0.0000

27. $C = 35.74 + 0.6215T - 35.75v^{0.16} + 0.4275Tv^{0.16}$

$$\frac{\partial C}{\partial T} = 0.6215 + 0.4275v^{0.16}$$

$$\frac{\partial C}{\partial v} = -5.72v^{-0.84} + 0.0684Tv^{-0.84}$$

$$dC = \frac{\partial C}{\partial T}dT + \frac{\partial C}{\partial v}dv = \Big(0.6215 + 0.4275(23)^{0.16}\Big)(\pm 1) + \Big(-5.72(23)^{-0.84} + 0.0684(8)(23)^{-0.84}\Big)(\pm 3)$$

$$= \pm 1.3275 \pm 1.1143 = \pm 2.4418 \text{ Maximum propagated error}$$

$$\frac{dC}{C} = \frac{2.4418}{-12.6807} \approx 0.19 = 19\% \text{ Maximum relative error}$$

29. $P = \dfrac{E^2}{R}, \left|\dfrac{dE}{E}\right| = 3\% = 0.03, \left|\dfrac{dR}{R}\right| = 4\% = 0.04$

$$dP = \frac{2E}{R}\,dE - \frac{E^2}{R^2}\,dR$$

$$\frac{dP}{P} = \left[\frac{2E}{R}\,dE - \frac{E^2}{R^2}\,dR\right]\Big/ P = \left[\frac{2E}{R}\,dE - \frac{E^2}{R^2}\,dR\right]\Big/(E^2/R) = \frac{2}{E}\,dE - \frac{1}{R}\,dR$$

Using the worst case scenario, $\dfrac{dE}{E} = 0.03$ and $\dfrac{dR}{R} = -0.04$: $\dfrac{dP}{P} \le 2(0.03) - (-0.04) = 0.10 = 10\%$.

31. (a) $V = \dfrac{1}{2}bhl = \left(18\sin\dfrac{\theta}{2}\right)\left(18\cos\dfrac{\theta}{2}\right)(16)(12) = 31{,}104\sin\theta$ in.^{3} $= 18\sin\theta$ ft^3

V is maximum when $\sin\theta = 1$ or $\theta = \pi/2$.

(b) $V = \dfrac{s^2}{2}(\sin\theta)l$

$dV = s(\sin\theta)l\,ds + \dfrac{s^2}{2}l(\cos\theta)\,d\theta + \dfrac{s^2}{2}(\sin\theta)\,dl$

$= 18\left(\sin\dfrac{\pi}{2}\right)(16)(12)\left(\dfrac{1}{2}\right) + \dfrac{18^2}{2}(16)(12)\left(\cos\dfrac{\pi}{2}\right)\left(\dfrac{\pi}{90}\right) + \dfrac{18^2}{2}\left(\sin\dfrac{\pi}{2}\right)\left(\dfrac{1}{2}\right) = 1809$ in.^{3} ≈ 1.047 ft^3

33. $L = 0.00021\left(\ln\dfrac{2h}{r} - 0.75\right)$

$dL = 0.00021\left[\dfrac{dh}{h} - \dfrac{dr}{r}\right] = 0.00021\left[\dfrac{(\pm 1/100)}{100} - \dfrac{(\pm 1/16)}{2}\right] \approx (\pm 6.6)\times 10^{-6}$

$L = 0.00021(\ln 100 - 0.75) \pm dL \approx 8.096\times 10^{-4} \pm 6.6\times 10^{-6}$ micro henrys

35. $z = f(x, y) = x^2 - 2x + y$

$\Delta z = f(x + \Delta x, y + \Delta y) - f(x, y) = \left(x^2 + 2x(\Delta x) + (\Delta x)^2 - 2x - 2(\Delta x) + y + (\Delta y)\right) - \left(x^2 - 2x + y\right)$

$= 2x(\Delta x) + (\Delta x)^2 - 2(\Delta x) + (\Delta y) = (2x - 2)\,\Delta x + \Delta y + \Delta x(\Delta x) + 0(\Delta y)$

$= f_x(x, y)\,\Delta x + f_y(x, y)\,\Delta y + \varepsilon_1\Delta x + \varepsilon_2\Delta y$ where $\varepsilon_1 = \Delta x$ and $\varepsilon_2 = 0$.

As $(\Delta x, \Delta y) \to (0, 0)$, $\varepsilon_1 \to 0$ and $\varepsilon_2 \to 0$.

37. $z = f(x, y) = x^2 y$

$\Delta z = f(x + \Delta x, y + \Delta y) - f(x, y) = \left(x^2 + 2x(\Delta x) + (\Delta x)^2\right)(y + \Delta y) - x^2 y$

$= 2xy(\Delta x) + y(\Delta x)^2 + x^2\Delta y + 2x(\Delta x)(\Delta y) + (\Delta x)^2\Delta y = 2xy(\Delta x) + x^2\Delta y + (y\Delta x)\Delta x + \left[2x\Delta x + (\Delta x)^2\right]\Delta y$

$= f_x(x, y)\,\Delta x + f_y(x, y)\,\Delta y + \varepsilon_1\Delta x + \varepsilon_2\Delta y$ where $\varepsilon_1 = y(\Delta x)$ and $\varepsilon_2 = 2x\Delta x + (\Delta x)^2$.

As $(\Delta x, \Delta y) \to (0, 0)$, $\varepsilon_1 \to 0$ and $\varepsilon_2 \to 0$.

39. $f(x, y) = \begin{cases} \dfrac{3x^2 y}{x^4 + y^2}, & (x, y) \neq (0, 0) \\ 0, & (x, y) = (0, 0) \end{cases}$

$f_x(0, 0) = \lim\limits_{\Delta x \to 0}\dfrac{f(\Delta x, 0) - f(0, 0)}{\Delta x} = \lim\limits_{\Delta x \to 0}\dfrac{\dfrac{0}{(\Delta x)^4} - 0}{\Delta x} = 0$

$f_y(0, 0) = \lim\limits_{\Delta y \to 0}\dfrac{f(0, \Delta y) - f(0, 0)}{\Delta y} = \lim\limits_{\Delta y \to 0}\dfrac{\dfrac{0}{(\Delta y)^2} - 0}{\Delta y} = 0$

So, the partial derivatives exist at $(0, 0)$.

Along the line $y = x$: $\lim\limits_{(x, y)\to(0,0)} f(x, y) = \lim\limits_{x\to 0}\dfrac{3x^3}{x^4 + x^2} = \lim\limits_{x\to 0}\dfrac{3x}{x^2 + 1} = 0$

Along the curve $y = x^2$: $\lim\limits_{(x, y)\to(0,0)} f(x, y) = \dfrac{3x^4}{2x^4} = \dfrac{3}{2}$

f is not continuous at $(0, 0)$. So, f is not differentiable at $(0, 0)$. (See Theorem 12.5)

Section 13.5   Chain Rules for Functions of Several Variables

1. One way is to convert w into a function of s and t. The other way is to use the Chain Rule given in Theorem 13.7.

3. $w = x^2 + 5y$

   $x = 2t, \ y = t$

   $\dfrac{dw}{dt} = \dfrac{\partial w}{\partial x}\dfrac{dx}{dt} + \dfrac{\partial w}{\partial y}\dfrac{dy}{dt} = (2x)2 + 5(1) = 4x + 5 = 8t + 5$

   When $t = 2, \ \dfrac{dw}{dt} = 8(2) + 5 = 21.$

5. $w = x \sin y$

   $x = e^t, \ y = \pi - t$

   $\dfrac{dw}{dt} = \dfrac{\partial w}{\partial x}\dfrac{dx}{dt} + \dfrac{\partial w}{\partial y}\dfrac{dy}{dt} = \sin y(e^t) + x \cos y(-1) = \sin(\pi - t)e^t - e^t \cos(\pi - t) = e^t \sin t + e^t \cos t$

   When $t = 0, \ \dfrac{dw}{dt} = (1)(0) + (1)(1) = 0 + 1 = 1.$

7. $w = x - \dfrac{1}{y}, \ x = e^{2t}, \ y = t^3$

   (a) $\dfrac{dw}{dt} = \dfrac{\partial w}{\partial x}\dfrac{dx}{dt} + \dfrac{\partial w}{\partial y}\dfrac{dy}{dt} = 1(2e^{2t}) + \dfrac{1}{y^2}(3t^2) = 2e^{2t} + \dfrac{1}{t^6}(3t^2) = 2e^{2t} + \dfrac{3}{t^4}$

   (b) $w = x - \dfrac{1}{y} = e^{2t} - \dfrac{1}{t^3} = e^{2t} - t^{-3}$

   $\dfrac{dw}{dt} = 2e^{2t} + 3t^{-4} = 2e^{2t} + \dfrac{3}{t^4}$

9. $w = x^2 + y^2 + z^2, \ x = \cos t, \ y = \sin t, \ z = e^t$

   (a) $\dfrac{dw}{dt} = \dfrac{\partial w}{\partial x}\dfrac{dx}{dt} + \dfrac{\partial w}{\partial y}\dfrac{dy}{dt} + \dfrac{\partial w}{\partial z}\dfrac{dz}{dt} = 2x(-\sin t) + 2y(\cos t) + 2z(e^t) = -2\cos t \sin t + 2 \sin t \cos t + 2e^{2t} = 2e^{2t}$

   (b) $w = \cos^2 t + \sin^2 t + e^{2t} = 1 + e^{2t}$

   $\dfrac{dw}{dt} = 2e^{2t}$

11. $w = xy + xz + yz, \ x = t - 1, \ y = t^2 - 1, \ z = t$

   (a) $\dfrac{dw}{dt} = \dfrac{\partial w}{\partial x}\dfrac{dx}{dt} + \dfrac{\partial w}{\partial y}\dfrac{dy}{dt} + \dfrac{\partial w}{\partial z}\dfrac{dz}{dt} = (y + z) + (x + z)(2t) + (x + y)$

   $= (t^2 - 1 + t) + (t - 1 + t)(2t) + (t - 1 + t^2 - 1) = 3(2t^2 - 1)$

   (b) $w = (t - 1)(t^2 - 1) + (t - 1)t + (t^2 - 1)t$

   $\dfrac{dw}{dt} = 2t(t - 1) + (t^2 - 1) + 2t - 1 + 3t^2 - 1 = 3(2t^2 - 1)$

13. Distance $= f(t) = \sqrt{(x_1 - x_2)^2 + (y_1 - y_2)^2} = \sqrt{(10 \cos 2t - 7 \cos t)^2 + (6 \sin 2t - 4 \sin t)^2}$

$$f'(t) = \frac{1}{2}\left[(10 \cos 2t - 7 \cos t)^2 + (6 \sin 2t - 4 \sin t)^2\right]^{-1/2}$$

$$\left[\left[2(10 \cos 2t - 7 \cos t)(-20 \sin 2t + 7 \sin t)\right] + \left[2(6 \sin 2t - 4 \sin t)(12 \cos 2t - 4 \cos t)\right]\right]$$

$$f'\left(\frac{\pi}{2}\right) = \frac{1}{2}\left[(-10)^2 + 4^2\right]^{-1/2}\left[2(-10)(7)\right] + (2(-4)(-12)) = \frac{1}{2}(116)^{-1/2}(-44) = \frac{-22}{2\sqrt{29}} = \frac{-11\sqrt{29}}{29} \approx -2.04$$

15. $w = x^2 + y^2$

$x = s + t, y = s - t$

$\dfrac{\partial w}{\partial s} = 2x(1) + 2y(1) = 2(s + t) + 2(s - t) = 4s$

$\dfrac{\partial w}{\partial t} = 2x(1) + 2y(-1) = 2(s + t) - 2(s - t) = 4t$

When $s = 1$ and $t = 3$, $\dfrac{\partial w}{\partial s} = 4$ and $\dfrac{\partial w}{\partial t} = 12$.

17. $w = \sin(2x + 3y)$

$x = s + t$

$y = s - t$

$\dfrac{\partial w}{\partial s} = 2 \cos(2x + 3y) + 3 \cos(2x + 3y)$

$\quad = 5 \cos(2x + 3y) = 5 \cos(5s - t)$

$\dfrac{\partial w}{\partial t} = 2 \cos(2x + 3y) - 3 \cos(2x + 3y)$

$\quad = -\cos(2x + 3y) = -\cos(5s - t)$

When $s = 0$ and $t = \dfrac{\pi}{2}$, $\dfrac{\partial w}{\partial s} = 0$ and $\dfrac{\partial w}{\partial t} = 0$.

19. (a) $w = xyz, x = s + t, y = s - t, z = st^2$

$\dfrac{\partial w}{\partial s} = yz(1) + xz(1) + xy(t^2)$

$\quad = (s - t)st^2 + (s + t)st^2 + (s + t)(s - t)t^2 = 2s^2t^2 + s^2t^2 - t^4 = 3s^2t^2 - t^4 = t^2(3s^2 - t^2)$

$\dfrac{\partial w}{\partial t} = yz(1) + xz(-1) + xy(2st) = (s - t)st^2 - (s + t)st^2 + (s + t)(s - t)(2st) = -2st^3 + 2s^3t - 2st^3 = 2s^3t - 4st^3$

$\quad = 2st(s^2 - 2t^2)$

(b) $w = xyz = (s + t)(s - t)st^2 = (s^2 - t^2)st^2 = s^3t^2 - st^4$

$\dfrac{\partial w}{\partial s} = 3s^2t^2 - t^4 = t^2(3s^2 - t^2)$

$\dfrac{\partial w}{\partial t} = 2s^3t - 4st^3 = 2st(s^2 - 2t^2)$

21. (a) $w = ze^{xy}, x = s - t, y = s + t, z = st$

$\dfrac{\partial w}{\partial s} = yze^{xy}(1) + xze^{xy}(1) + e^{xy}(t) = e^{(s-t)(s+t)}\left[(s + t)st + (s - t)st + t\right] = e^{(s-t)(s+t)}\left[2s^2t + t\right] = te^{s^2-t^2}(2s^2 + 1)$

$\dfrac{\partial w}{\partial t} = yze^{xy}(-1) + xze^{xy}(1) + e^{xy}(s) = e^{(s-t)(s+t)}\left[-(s + t)(st) + (s - t)st + s\right] = e^{(s-t)(s+t)}\left[-2st^2 + s\right] = se^{s^2-t^2}(1 - 2t^2)$

(b) $w = ze^{xy} = ste^{(s-t)(s+t)} = ste^{s^2-t^2}$

$\dfrac{\partial w}{\partial s} = te^{s^2-t^2} + st(2s)e^{s^2-t^2} = te^{s^2-t^2}(1 + 2s^2)$

$\dfrac{\partial w}{\partial t} = se^{s^2-t^2} + st(-2t)e^{s^2-t^2} = se^{s^2-t^2}(1 - 2t^2)$

23. $x^2 - xy + y^2 - x + y = 0$

$$\frac{dy}{dx} = -\frac{F_x(x, y)}{F_y(x, y)} = -\frac{2x - y - 1}{-x + 2y + 1} = \frac{y - 2x + 1}{2y - x + 1}$$

25. $\ln\sqrt{x^2 + y^2} + x + y = 4$

$$\frac{1}{2}\ln(x^2 + y^2) + x + y - 4 = 0$$

$$\frac{dy}{dx} = -\frac{F_x(x, y)}{F_y(x, y)} = -\frac{\dfrac{x}{x^2 + y^2} + 1}{\dfrac{y}{x^2 + y^2} + 1} = -\frac{x + x^2 + y^2}{y + x^2 + y^2}$$

27. $F(x, y, z) = x^2 + y^2 + z^2 - 1$

$$F_x = 2x, \, F_y = 2y, \, F_z = 2z$$

$$\frac{\partial z}{\partial x} = -\frac{F_x}{F_z} = -\frac{x}{z}$$

$$\frac{\partial z}{\partial y} = -\frac{F_y}{F_z} = -\frac{y}{z}$$

29. $F(x, y, z) = x^2 + 2yz + z^2 - 1 = 0$

$$\frac{\partial z}{\partial x} = -\frac{F_x(x, y, z)}{F_z(x, y, z)} = \frac{-2x}{2y + 2z} = \frac{-x}{y + z}$$

$$\frac{\partial z}{\partial y} = -\frac{F_y(x, y, z)}{F_z(x, y, z)} = \frac{-2z}{2y + 2z} = \frac{-z}{y + z}$$

31. $F(x, y, z) = \tan(x + y) + \cos z - 2$

$$F_x = \sec^2(x + y)$$

$$F_y = \sec^2(x + y)$$

$$F_z = -\sin z$$

$$\frac{\partial z}{\partial x} = -\frac{F_x}{F_z} = \frac{-\sec^2(x + y)}{-\sin z} = \frac{\sec^2(x + y)}{\sin z}$$

$$\frac{\partial z}{\partial y} = -\frac{F_y}{F_z} = \frac{-\sec^2(x + y)}{-\sin z} = \frac{\sec^2(x + y)}{\sin z}$$

33. $F(x, y, z) = e^{xz} + xy = 0$

$$\frac{\partial z}{\partial x} = -\frac{F_x(x, y, z)}{F_z(x, y, z)} = -\frac{ze^{xz} + y}{xe^{xz}}$$

$$\frac{\partial z}{\partial y} = -\frac{F_y(x, y, z)}{F_z(x, y, z)} = \frac{-x}{xe^{xz}} = \frac{-1}{e^{xz}} = -e^{-xz}$$

35. $F(x, y, z, w) = 7xy + yz^2 - 4wz + w^2z + w^2x - 6$

$$F_x = 7y + w^2$$

$$F_y = 7x + z^2$$

$$F_z = 2yz - 4w + w^2$$

$$F_w = -4z + 2wz + 2wx$$

$$\frac{\partial w}{\partial x} = -\frac{F_x}{F_w} = \frac{-(7y + w^2)}{-4z + 2wz + 2wx} = \frac{7y + w^2}{4z - 2wz - 2wx}$$

$$\frac{\partial w}{\partial y} = \frac{-F_y}{F_w} = \frac{-(7x + z^2)}{-4z + 2wz + 2wx} = \frac{7x + z^2}{4z - 2wz - 2wx}$$

$$\frac{\partial w}{\partial z} = \frac{-F_z}{F_w} = \frac{-(2yz - 4w + w^2)}{-4z + 2wz + 2wx} = \frac{2yz - 4w + w^2}{4z - 2wz - 2wz}$$

37. $F(x, y, z, w) = \cos xy + \sin yz + wz - 20$

$$\frac{\partial w}{\partial x} = \frac{-F_x}{F_w} = \frac{y \sin xy}{z}$$

$$\frac{\partial w}{\partial y} = \frac{-F_y}{F_w} = \frac{x \sin xy - z \cos yz}{z}$$

$$\frac{\partial w}{\partial z} = \frac{-F_z}{F_w} = -\frac{y \cos zy + w}{z}$$

39. (a) $f(x, y) = 2x^2 - 5xy$

$$f(tx, ty) = 2(tx)^2 - 5(tx)(ty)$$
$$= t^2(2x^2 - 5xy) = t^2 f(x, y)$$

Degree: 2

(b) $x f_x(x, y) + y f_y(x, y) = x(4x - 5y) + y(-5x)$
$$= 4x^2 - 5xy - 5xy$$
$$= 2(2x^2 - 5xy)$$
$$= 2f(x, y)$$

41. (a) $f(x, y) = e^{x/y}$

$$f(tx, ty) = e^{tx/ty} = e^{x/y} = f(x, y)$$

Degree: 0

(b) $x f_x(x, y) + y f_y(x, y) = x\left(\frac{1}{y}e^{x/y}\right) + y\left(-\frac{x}{y^2}e^{x/y}\right)$
$$= 0$$

43. $\dfrac{dw}{dt} = \dfrac{\partial w}{\partial x}\dfrac{dx}{dt} + \dfrac{\partial w}{\partial y}\dfrac{dy}{dt} = \dfrac{\partial f}{\partial x}\dfrac{dg}{dt} + \dfrac{\partial f}{\partial y}\dfrac{dh}{dt}$

At $t = 2$, $x = 4$, $y = 3$, $f_x(4, 3) = -5$ and

$f_y(4, 3) = 7$.

So, $\dfrac{dw}{dt} = (-5)(-1) + (7)(6) = 47$.

45.
$$w = f(x, y)$$
$$x = u - v$$
$$y = v - u$$
$$\frac{\partial w}{\partial u} = \frac{\partial w}{\partial x}\frac{dx}{du} + \frac{\partial w}{\partial y}\frac{dy}{du} = \frac{\partial w}{\partial x} - \frac{\partial w}{\partial y}$$
$$\frac{\partial w}{\partial v} = \frac{\partial w}{\partial x}\frac{dx}{dv} + \frac{\partial w}{\partial y}\frac{dy}{dv} = -\frac{\partial w}{\partial x} + \frac{\partial w}{\partial y}$$
$$\frac{\partial w}{\partial u} + \frac{\partial w}{\partial v} = 0$$

47. (a) $f(x) = F(4x, 4),\ u = 4x,\ v = 4$
$$f'(x) = \frac{\partial F}{\partial x} = \frac{\partial F}{\partial u}\frac{\partial u}{\partial x} + \frac{\partial F}{\partial v}\frac{\partial v}{\partial x}$$
$$= \frac{\partial F}{\partial u}(4) + \frac{\partial F}{\partial v}(0) = 4\frac{\partial F}{\partial u}$$

(b) $f(x) = F(-2x, x^2)$
$$f'(x) = \frac{\partial F}{\partial u}\frac{\partial u}{\partial x} + \frac{\partial F}{\partial v}\frac{\partial v}{\partial x}$$
$$= \frac{\partial F}{\partial u}(-2) + \frac{\partial F}{\partial v}(2x)$$
$$= -2\frac{\partial F}{\partial u} + 2x\frac{\partial F}{\partial v}$$

49. $V = \pi r^2 h$
$$\frac{dV}{dt} = \pi\left(2rh\frac{dr}{dt} + r^2\frac{dh}{dt}\right) = \pi r\left(2h\frac{dr}{dt} + r\frac{dh}{dt}\right) = \pi(12)\left[2(36)(6) + 12(-4)\right] = 4608\pi \text{ in.}^3/\text{min}$$
$$S = 2\pi r(r + h)$$
$$\frac{dS}{dt} = 2\pi\left[(2r + h)\frac{dr}{dt} + r\frac{dh}{dt}\right] = 2\pi\left[(24 + 36)(6) + 12(-4)\right] = 624\pi \text{ in.}^2/\text{min}$$

51. $I = \frac{1}{2}m(r_1^2 + r_2^2)$
$$\frac{dI}{dt} = \frac{1}{2}m\left[2r_1\frac{dr_1}{dt} + 2r_2\frac{dr_2}{dt}\right] = m\left[(6)(2) + (8)(2)\right] = 28m \text{ cm}^2/\text{sec}$$

53. Given $\dfrac{\partial u}{\partial x} = \dfrac{\partial v}{\partial y}$ and $\dfrac{\partial u}{\partial y} = -\dfrac{\partial v}{\partial x}$, $x = r\cos\theta$ and $y = r\sin\theta$.
$$\frac{\partial u}{\partial r} = \frac{\partial u}{\partial x}\cos\theta + \frac{\partial u}{\partial y}\sin\theta = \frac{\partial v}{\partial y}\cos\theta - \frac{\partial v}{\partial x}\sin\theta$$
$$\frac{\partial v}{\partial\theta} = \frac{\partial v}{\partial x}(-r\sin\theta) + \frac{\partial v}{\partial y}(r\cos\theta) = r\left[\frac{\partial v}{\partial y}\cos\theta - \frac{\partial v}{\partial x}\sin\theta\right]$$
So, $\dfrac{\partial u}{\partial r} = \dfrac{1}{r}\dfrac{\partial v}{\partial\theta}$.
$$\frac{\partial v}{\partial r} = \frac{\partial v}{\partial x}\cos\theta + \frac{\partial v}{\partial y}\sin\theta = -\frac{\partial u}{\partial y}\cos\theta + \frac{\partial u}{\partial x}\sin\theta$$
$$\frac{\partial u}{\partial\theta} = \frac{\partial u}{\partial x}(-r\sin\theta) + \frac{\partial u}{\partial y}(r\cos\theta) = -r\left[-\frac{\partial u}{\partial y}\cos\theta + \frac{\partial u}{\partial x}\sin\theta\right]$$
So, $\dfrac{\partial v}{\partial r} = -\dfrac{1}{r}\dfrac{\partial u}{\partial\theta}$.

55. $g(t) = f(xt, yt) = t^n f(x, y)$
Let $u = xt,\ v = yt$, then
$$g'(t) = \frac{\partial f}{\partial u}\cdot\frac{du}{dt} + \frac{\partial f}{\partial v}\cdot\frac{dv}{dt} = \frac{\partial f}{\partial u}x + \frac{\partial f}{\partial v}y \text{ and } g'(t) = nt^{n-1}f(x, y).$$
Now, let $t = 1$ and we have $u = x,\ v = y$. Thus,
$$\frac{\partial f}{\partial x}x + \frac{\partial f}{\partial y}y = nf(x, y).$$

Section 13.6 Directional Derivatives and Gradients

1. The partial derivative with respect to x is the directional derivative in the direction of the positive x-axis.
   That is, the directional derivative for $\theta = 0$.

3. $f(x, y) = x^2 + y^2$, $P(1, -2)$, $\theta = \pi/4$

   $D_{\mathbf{u}} f(x, y) = f_x(x, y) \cos \theta + f_y(x, y) \sin \theta = 2x \cos \theta + 2y \sin \theta$

   At $\theta = \pi/4$, $x = 1$ and $y = -2$,

   $D_{\mathbf{u}} f(1, -2) = 2(1) \cos \pi/4 + 2(-2) \sin \pi/4 = \sqrt{2} - 2\sqrt{2} = -\sqrt{2}.$

5. $f(x, y) = \sin(2x + y)$, $P(0, \pi)$, $\theta = -\dfrac{5\pi}{6}$

   $D_{\mathbf{u}} f(x, y) = f_x(x, y) \cos \theta + f_y(x, y) \sin \theta = 2 \cos(2x + y) \cos \theta + \cos(2x + y) \sin \theta$

   At $\theta = -\dfrac{5\pi}{6}$, $x = 0$ and $y = \pi$.

   $D_{\mathbf{u}} f(0, \pi) = (-2)\left(-\dfrac{\sqrt{3}}{2}\right) + (-1)\left(-\dfrac{1}{2}\right) = \dfrac{1}{2} + \sqrt{3}$

7. $f(x, y) = 3x - 4xy + 9y$, $P(1, 2)$, $\mathbf{v} = \dfrac{3}{5}\mathbf{i} + \dfrac{4}{5}\mathbf{j}$

   $\mathbf{u} = \dfrac{\mathbf{v}}{\|\mathbf{v}\|} = \dfrac{3}{5}\mathbf{i} + \dfrac{4}{5}\mathbf{j} = \cos \theta \, \mathbf{i} + \sin \theta \, \mathbf{j}$

   $D_{\mathbf{u}} f(x, y) = (3 - 4y) \cos \theta + (-4x + 9) \sin \theta$

   $D_{\mathbf{u}}(1, 2) = (3 - 4(2))\dfrac{3}{5} + (-4(1) + 9)\dfrac{4}{5} = -3 + 4 = 1$

9. $g(x, y) = \sqrt{x^2 + y^2}$, $P(3, 4)$, $\mathbf{v} = 3\mathbf{i} - 4\mathbf{j}$

   $\mathbf{u} = \dfrac{\mathbf{v}}{\|\mathbf{v}\|} = \dfrac{3}{5}\mathbf{i} - \dfrac{4}{5}\mathbf{j}$

   $D_{\mathbf{u}} g(x, y) = \dfrac{x}{\sqrt{x^2 + y^2}}\left(\dfrac{3}{5}\right) + \dfrac{y}{\sqrt{x^2 + y^2}}\left(-\dfrac{4}{5}\right)$

   $D_{\mathbf{u}} g(3, 4) = \dfrac{3}{5}\left(\dfrac{3}{5}\right) + \dfrac{4}{5}\left(-\dfrac{4}{5}\right) = -\dfrac{7}{25}$

11. $f(x, y) = x^2 + 3y^2$, $P(1, 1)$, $Q(4, 5)$

    $\mathbf{v} = (4 - 1)\mathbf{i} + (5 - 1)\mathbf{j} = 3\mathbf{i} + 4\mathbf{j}$

    $\mathbf{u} = \dfrac{\mathbf{v}}{\|\mathbf{v}\|} = \dfrac{3}{5}\mathbf{i} + \dfrac{4}{5}\mathbf{j}$

    $D_{\mathbf{u}} f(x, y) = 2x\left(\dfrac{3}{5}\right) + 6y\left(\dfrac{4}{5}\right)$

    $D_{\mathbf{u}} f(1, 1) = 2\left(\dfrac{3}{5}\right) + 6\left(\dfrac{4}{5}\right) = 6$

13. $f(x, y) = e^y \sin x$, $P(0, 0)$, $Q(2, 1)$

    $\mathbf{v} = (2 - 0)\mathbf{i} + (1 - 0)\mathbf{j}$

    $\mathbf{v} = 2\mathbf{i} + \mathbf{j}$, $\mathbf{u} = \dfrac{\mathbf{v}}{\|\mathbf{v}\|} = \dfrac{2}{\sqrt{5}}\mathbf{i} + \dfrac{1}{\sqrt{5}}\mathbf{j}$

    $D_{\mathbf{u}} f(x, y) = e^y \cos x\left(\dfrac{2}{\sqrt{5}}\right) + e^y \sin x\left(\dfrac{1}{\sqrt{5}}\right)$

    $D_{\mathbf{u}} f(0, 0) = \dfrac{2}{\sqrt{5}} = \dfrac{2\sqrt{5}}{5}$

15. $f(x, y) = 3x + 5y^2 + 1$

    $\nabla f(x, y) = 3\mathbf{i} + 10y\mathbf{j}$

    $\nabla f(2, 1) = 3\mathbf{i} + 10\mathbf{j}$

17. $z = \dfrac{\ln(x^2 - y)}{x} - 4$

$$\nabla z(x, y) = \left(\dfrac{2}{x^2 - y} - \dfrac{\ln(x^2 - y)}{x^2} \right)\mathbf{i} + \left[\dfrac{1}{x(y - x^2)} \right]\mathbf{j}$$

$$\nabla z(2, 3) = 2\mathbf{i} - \dfrac{1}{2}\mathbf{j}$$

19. $w = 6xy - y^2 + 2xyz^3$

$$\nabla w(x, y, z) = (6y + 2yz^3)\mathbf{i} + (6x - 2y + 2xz^3)\mathbf{j} + (6xyz^2)\mathbf{k}$$

$$\nabla w(-1, 5, -1) = 20\mathbf{i} - 14\mathbf{j} - 30\mathbf{k}$$

21. $f(x, y) = xy$

$$\mathbf{v} = \dfrac{1}{2}(\mathbf{i} + \sqrt{3}\mathbf{j})$$

$$\nabla f(x, y) = y\mathbf{i} + x\mathbf{j}$$

$$\nabla f(0, -2) = -2\mathbf{i}$$

$$\mathbf{u} = \dfrac{\mathbf{v}}{\|\mathbf{v}\|} = \dfrac{1}{2}\mathbf{i} + \dfrac{\sqrt{3}}{2}\mathbf{j}$$

$$D_{\mathbf{u}}f(0, -2) = \nabla f(0, -2) \cdot \mathbf{u} = -1$$

23. $f(x, y, z) = x^2 + y^2 + z^2$

$$\mathbf{v} = \dfrac{\sqrt{3}}{3}(\mathbf{i} - \mathbf{j} + \mathbf{k})$$

$$\nabla f(x, y, z) = 2x\mathbf{i} + 2y\mathbf{j} + 2z\mathbf{k}$$

$$\nabla f(1, 1, 1) = 2\mathbf{i} + 2\mathbf{j} + 2\mathbf{k}$$

$$\mathbf{u} = \dfrac{\mathbf{v}}{\|\mathbf{v}\|} = \dfrac{\sqrt{3}}{3}\mathbf{i} - \dfrac{\sqrt{3}}{3}\mathbf{j} + \dfrac{\sqrt{3}}{3}\mathbf{k}$$

$$D_{\mathbf{u}}f(1, 1, 1) = \nabla f(1, 1, 1) \cdot \mathbf{u} = \dfrac{2}{3}\sqrt{3}$$

25. $\overrightarrow{PQ} = \mathbf{i} + \mathbf{j}, \; \mathbf{u} = \dfrac{\sqrt{2}}{2}\mathbf{i} + \dfrac{\sqrt{2}}{2}\mathbf{j}$

$$\nabla g(x, y) = 2x\mathbf{i} + 2y\mathbf{j}, \; \nabla g(1, 2) = 2\mathbf{i} + 4\mathbf{j}$$

$$D_{\mathbf{u}}g = \nabla g \cdot \mathbf{u} = \sqrt{2} + 2\sqrt{2} = 3\sqrt{2}$$

27. $g(x, y, z) = xye^z$

$$\mathbf{v} = -2\mathbf{i} - 4\mathbf{j}$$

$$\nabla g = ye^z\mathbf{i} + xe^z\mathbf{j} + xye^z\mathbf{k}$$

At $(2, 4, 0)$, $\nabla g = 4\mathbf{i} + 2\mathbf{j} + 8\mathbf{k}$.

$$\mathbf{u} = \dfrac{\mathbf{v}}{\|\mathbf{v}\|} = -\dfrac{1}{\sqrt{5}}\mathbf{i} - \dfrac{2}{\sqrt{5}}\mathbf{j}$$

$$D_{\mathbf{u}}g = \nabla g \cdot \mathbf{u} = -\dfrac{4}{\sqrt{5}} - \dfrac{4}{\sqrt{5}} = -\dfrac{8}{\sqrt{5}}$$

29. $f(x, y) = y^2 - x\sqrt{y}$

$$\nabla f(x, y) = -\sqrt{y}\mathbf{i} + \left(2y - \dfrac{x}{2\sqrt{y}} \right)\mathbf{j}$$

$$\nabla f(0, 3) = -\sqrt{3}\mathbf{i} + 6\mathbf{j}$$

$$\|\nabla f(0, 3)\| = \sqrt{3 + 36} = \sqrt{39}$$

31. $h(x, y) = x \tan y$

$$\nabla h(x, y) = \tan y\mathbf{i} + x \sec^2 y\mathbf{j}$$

$$\nabla h\left(2, \dfrac{\pi}{4} \right) = \mathbf{i} + 4\mathbf{j}$$

$$\left\| \nabla h\left(2, \dfrac{\pi}{4} \right) \right\| = \sqrt{17}$$

33. $f(x, y) = \sin x^2 y^3$

$$\nabla f(x, y) = 2x \cos(x^2 y^3)\mathbf{i} + 3y^2 \cos(x^2 y^3)\mathbf{j}$$

$$\nabla f\left(\dfrac{1}{\pi}, \pi \right) = \dfrac{2}{\pi}(-1)\mathbf{i} + 3\pi^2(-1)\mathbf{j}$$

$$\left\| \nabla f\left(\dfrac{1}{\pi}, \pi \right) \right\| = \sqrt{\dfrac{4}{\pi^2} + 9\pi^4} = \dfrac{1}{\pi}\sqrt{4 + 9\pi^6}$$

35. $f(x, y, z) = \sqrt{x^2 + y^2 + z^2}$

$$\nabla f(x, y, z) = \dfrac{1}{\sqrt{x^2 + y^2 + z^2}}(x\mathbf{i} + y\mathbf{j} + z\mathbf{k})$$

$$\nabla f(1, 4, 2) = \dfrac{1}{\sqrt{21}}(\mathbf{i} + 4\mathbf{j} + 2\mathbf{k})$$

$$\|\nabla f(1, 4, 2)\| = 1$$

37. $w = xy^2 z^2$

$$\nabla w = y^2 z^2\mathbf{i} + 2xyz^2\mathbf{j} + 2xy^2 z\mathbf{k}$$

$$\nabla w(2, 1, 1) = \mathbf{i} + 4\mathbf{j} + 4\mathbf{k}$$

$$\|\nabla w(2, 1, 1)\| = \sqrt{33}$$

39. $f(x, y) = 6 - 2x - 3y$

$c = 6, P = (0, 0)$

$\nabla f(x, y) = -2\mathbf{i} - 3\mathbf{j}$

$6 - 2x - 3y = 6$

$\qquad 0 = 2x + 3y$

$\nabla f(0, 0) = -2\mathbf{i} - 3\mathbf{j}$

41. $f(x, y) = xy$

$c = -3, P = (-1, 3)$

$\nabla f(x, y) = y\mathbf{i} + x\mathbf{j}$

$xy = -3$

$\nabla f(-1, 3) = 3\mathbf{i} - \mathbf{j}$

43. $f(x, y) = 4x^2 - y$

(a) $\nabla f(x, y) = 8x\mathbf{i} - \mathbf{j}$

$\quad \nabla f(2, 10) = 16\mathbf{i} - \mathbf{j}$

(b) $\|16\mathbf{i} - \mathbf{j}\| = \sqrt{257}$

$\dfrac{1}{\sqrt{257}}(16\mathbf{i} - \mathbf{j})$ is a unit vector normal to the level curve $4x^2 - y = 6$ at $(2, 10)$.

(c) The vector $\mathbf{i} + 16\mathbf{j}$ is tangent to the level curve. Slope $= \dfrac{16}{1} = 16$

$\quad y - 10 = 16(x - 2)$

$\qquad y = 16x - 22$ Tangent line

(d)

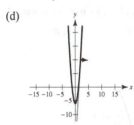

45. $f(x, y) = 3x^2 - 2y^2$

(a) $\nabla f = 6x\mathbf{i} - 4y\mathbf{j}$

$\quad \nabla f(1, 1) = 6\mathbf{i} - 4\mathbf{j}$

(b) $\|\nabla f(1, 1)\| = \sqrt{36 + 16} = 2\sqrt{13}$

$\dfrac{1}{\sqrt{13}}(3\mathbf{i} - 2\mathbf{j})$ is a unit vector normal to the level curve $3x^2 - 2y^2 = 1$ at $(1, 1)$.

(c) The vector $2\mathbf{i} + 3\mathbf{j}$ is tangent to the level curve. Slope $= \dfrac{3}{2}$.

$\quad y - 1 = \frac{3}{2}(x - 1)$

$\qquad y = \frac{3}{2}x - \frac{1}{2}$ tangent line

(d)

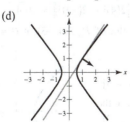

47. (a) $f(x, y) = 3 - \dfrac{x}{3} - \dfrac{y}{2}$

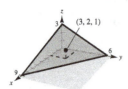

(b) $D_u f(x, y) = -\dfrac{1}{3}\cos\theta - \dfrac{1}{2}\sin\theta$

(i) $D_u f(3, 2) = -\left(\dfrac{1}{3}\right)\dfrac{\sqrt{2}}{2} - \left(\dfrac{1}{2}\right)\dfrac{\sqrt{2}}{2} = -\dfrac{5\sqrt{2}}{12}$

(ii) $D_u f(3, 2) = -\left(\dfrac{1}{3}\right)\left(-\dfrac{1}{2}\right) - \left(\dfrac{1}{2}\right)\dfrac{\sqrt{3}}{2} = \dfrac{2 - 3\sqrt{3}}{12}$

(iii) $D_u f(3, 2) = -\left(\dfrac{1}{3}\right)\left(-\dfrac{1}{2}\right) - \left(\dfrac{1}{2}\right)\left(-\dfrac{\sqrt{3}}{2}\right)$

$= \dfrac{2 + 3\sqrt{3}}{12}$

(iv) $D_u f(3, 2) = -\left(\dfrac{1}{3}\right)\left(\dfrac{\sqrt{3}}{2}\right) - \left(\dfrac{1}{2}\right)\left(-\dfrac{1}{2}\right)$

$= \dfrac{3 - 2\sqrt{3}}{12}$

(c) (i) $\mathbf{u} = \left(\dfrac{1}{\sqrt{2}}\right)(\mathbf{i} + \mathbf{j})$

$D_u f = \nabla f \cdot \mathbf{u}$

$= -\left(\dfrac{1}{3}\right)\dfrac{1}{\sqrt{2}} - \left(\dfrac{1}{2}\right)\dfrac{1}{\sqrt{2}} = -\dfrac{5\sqrt{2}}{12}$

(ii) $\mathbf{v} = -3\mathbf{i} - 4\mathbf{j}$

$\|\mathbf{v}\| = \sqrt{9 + 16} = 5$

$\mathbf{u} = -\dfrac{3}{5}\mathbf{i} - \dfrac{4}{5}\mathbf{j}$

$D_u f = \nabla f \cdot \mathbf{u} = \dfrac{1}{5} + \dfrac{2}{5} = \dfrac{3}{5}$

(iii) $\mathbf{v} - 3\mathbf{i} + 4\mathbf{j}$

$\|\mathbf{v}\| = \sqrt{9 + 16} = 5$

$\mathbf{u} = -\dfrac{3}{5}\mathbf{i} + \dfrac{4}{5}\mathbf{j}$

$D_u f = \nabla f \cdot \mathbf{u} = \dfrac{1}{5} - \dfrac{2}{5} = -\dfrac{1}{5}$

(iv) $\mathbf{v} = \mathbf{i} + 3\mathbf{j}$

$\|\mathbf{v}\| = \sqrt{10}$

$\mathbf{u} = \dfrac{1}{\sqrt{10}}\mathbf{i} + \dfrac{3}{\sqrt{10}}\mathbf{j}$

$D_u f = \nabla f \cdot \mathbf{u} = \dfrac{-11}{6\sqrt{10}} = -\dfrac{11\sqrt{10}}{60}$

(d) $\nabla f = -\left(\dfrac{1}{3}\right)\mathbf{i} - \left(\dfrac{1}{2}\right)\mathbf{j}$

(e) $\|\nabla f\| = \sqrt{\dfrac{1}{9} + \dfrac{1}{4}} = \dfrac{1}{6}\sqrt{13}$

(f) $\nabla f = -\dfrac{1}{3}\mathbf{i} - \dfrac{1}{2}\mathbf{j}$

$\dfrac{\nabla f}{\|\nabla f\|} = \dfrac{1}{\sqrt{13}}(-2\mathbf{i} - 3\mathbf{j})$

So, $\mathbf{u} = \left(1/\sqrt{13}\right)(3\mathbf{i} - 2\mathbf{j})$ and

$D_u f(3, 2) = \nabla f \cdot \mathbf{u} = 0.\, \nabla f$ is the direction of greatest rate of change of f. So, in a direction orthogonal to ∇f, the rate of change of f is 0.

49. $f(x, y) = x^2 - y^2,\ (4, -3, 7)$

(a)

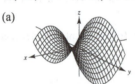

(b) $D_u f(x, y) = \nabla f(x, y) \cdot \mathbf{u} = 2x\cos\theta - 2y\sin\theta$

$D_u f(4, -3) = 8\cos\theta + 6\sin\theta$

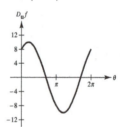

Generated by Mathematica

(c) Zeros: $\theta \approx 2.21, 5.36$

These are the angles θ for which $D_u f(4, 3)$ equals zero.

(d) $g(\theta) = D_u f(4, -3) = 8\cos\theta + 6\sin\theta$

$g'(\theta) = -8\sin\theta + 6\cos\theta$

Critical numbers: $\theta \approx 0.64, 3.79$

These are the angles for which $D_u f(4, -3)$ is a maximum (0.64) and minimum (3.79).

(e) $\|\nabla f(4, -3)\| = \|2(4)\mathbf{i} - 2(-3)\mathbf{j}\| = \sqrt{64 + 36} = 10$, the maximum value of $D_u f(4, -3)$, at $\theta \approx 0.64$.

(f) $f(x, y) = x^2 - y^2 = 7$

$\nabla f(4, -3) = 8\mathbf{i} + 6\mathbf{j}$

is perpendicular to the level curve at $(4, -3)$.

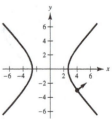

Generated by Mathematica

51. No. Answers will vary.

53. $h(x, y) = 5000 - 0.001x^2 - 0.004y^2$

$\nabla h = -0.002x\mathbf{i} - 0.008y\mathbf{j}$

$\nabla h(500, 300) = -\mathbf{i} - 2.4\mathbf{j}$ or

$5\nabla h = -(5\mathbf{i} + 12\mathbf{j})$

55. $T = \dfrac{x}{x^2 + y^2}$

$\nabla T = \dfrac{y^2 - x^2}{(x^2 + y^2)^2}\mathbf{i} - \dfrac{2xy}{(x^2 + y^2)^2}\mathbf{j}$

$\nabla T(3, 4) = \dfrac{7}{625}\mathbf{i} - \dfrac{24}{625}\mathbf{j} = \dfrac{1}{625}(7\mathbf{i} - 24\mathbf{j})$

57. $T(x, y) = 80 - 3x^2 - y^2,\ P(-1, 5)$

$\nabla T(x, y) = -6x\mathbf{i} - 2y\mathbf{j}$

Maximum increase in direction:

$\nabla T(-1, 5) = (-6)(-1)\mathbf{i} - 2(5)\mathbf{j} = 6\mathbf{i} - 10\mathbf{j}$

Maximum rate:

$\|\nabla T(-1, 5)\| = \sqrt{6^2 + (-10)^2} = 2\sqrt{34}$

$\approx 11.66°$ per centimeter

59. $T(x, y) = 400 - 2x^2 - y^2,\ P = (10, 10)$

$\dfrac{dx}{dt} = -4x \qquad\qquad \dfrac{dy}{dt} = -2y$

$x(t) = C_1 e^{-4t} \qquad\quad y(t) = C_2 e^{-2t}$

$10 = x(0) = C_1 \qquad 10 = y(0) = C_2$

$x(t) = 10e^{-4t} \qquad\quad y(t) = 10e^{-2t}$

$x = \dfrac{y^2}{10} \qquad\qquad\quad y^2(t) = 100e^{-4t}$

$y^2 = 10x$

61. True

63. True

65. Let $f(x, y, z) = e^x \cos y + \dfrac{z^2}{2} + C.$ Then

$\nabla f(x, y, z) = e^x \cos y\mathbf{i} - e^x \sin y\mathbf{j} + z\mathbf{k}.$

67. (a) $f(x, y) = \sqrt[3]{xy}$ is the composition of two
continuous functions, $h(x, y) = xy$ and
$g(z) = z^{1/3}$, and therefore continuous by
Theorem 13.2.

(b) $f_x(0, 0) = \lim\limits_{\Delta x \to 0} \dfrac{f(0 + \Delta x, 0) - f(0, 0)}{\Delta x}$

$= \lim\limits_{\Delta x \to 0} \dfrac{(0 \cdot \Delta x)^{1/3} - 0}{\Delta x} = 0$

$f_y(0, 0) = \lim\limits_{\Delta y \to 0} \dfrac{f(0, 0 + \Delta y) - f(0, 0)}{\Delta y}$

$= \lim\limits_{\Delta y \to 0} \dfrac{(0 \cdot \Delta y)^{1/3} - 0}{\Delta y} = 0$

Let $\mathbf{u} = \cos \theta \mathbf{i} + \sin \theta \mathbf{j},\ \theta \neq 0, \dfrac{\pi}{2}, \pi, \dfrac{3\pi}{2}.$ Then

$D_{\mathbf{u}} f(0, 0) = \lim\limits_{t \to 0} \dfrac{f(0 + t \cos \theta, 0 + t \sin \theta) - f(0, 0)}{t}$

$= \lim\limits_{t \to 0} \dfrac{\sqrt[3]{t^2 \cos \theta \sin \theta}}{t}$

$= \lim\limits_{t \to 0} \dfrac{\sqrt[3]{\cos \theta \sin \theta}}{t^{1/3}},$ does not exist.

(c)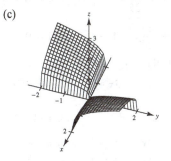

Section 13.7   Tangent Planes and Normal Lines

1. $\nabla F(x_0, y_0, z_0)$ is orthogonal to any tangent vector $\mathbf{v}$ at
the point (x_0, y_0, z_0). So, $\nabla F(x_0, y_0, z_0) \cdot \mathbf{v} = 0.$

3. $F(x, y, z) = 3x - 5y + 3z - 15 = 0$

$\qquad\qquad 3x - 5y + 3z = 15$ Plane

5. $F(x, y, z) = 4x^2 + 9y^2 - 4z^2 = 0$

$\qquad\qquad 4x^2 + 9y^2 = 4z^2$ Elliptic cone

7. $z = x^2 + y^2 + 3, (2, 1, 8)$

$F(x, y, z) = x^2 + y^2 + 3 - z$

$F_x(x, y, z) = 2x \quad F_y(x, y, z) = 2y \quad F_z(x, y, z) = -1$

$F_x(2, 1, 8) = 4 \qquad F_y(2, 1, 8) = 2 \qquad F_z(2, 1, 8) = -1$

$4(x - 2) + 2(y - 1) - 1(z - 8) = 0$

$4x + 2y - z = 2$

9. $z = \sqrt{x^2 + y^2}, (3, 4, 5)$

$F(x, y, z) = \sqrt{x^2 + y^2} - z$

$F_x(x, y, z) = \dfrac{x}{\sqrt{x^2 + y^2}}$ $F_y(x, y, z) = \dfrac{y}{\sqrt{x^2 + y^2}}$ $F_z(x, y, z) = -1$

$F_x(3, 4, 5) = \dfrac{3}{5}$ $F_y(3, 4, 5) = \dfrac{4}{5}$ $F_z(3, 4, 5) = -1$

$\dfrac{3}{5}(x - 3) + \dfrac{4}{5}(y - 4) - (z - 5) = 0$

$3(x - 3) + 4(y - 4) - 5(z - 5) = 0$

$\qquad\qquad 3x + 4y - 5z = 0$

11. $g(x, y) = x^2 + y^2, (1, -1, 2)$

$G(x, y, z) = x^2 + y^2 - z$

$G_x(x, y, z) = 2x$ $G_y(x, y, z) = 2y$ $G_z(x, y, z) = -1$

$G_x(1, -1, 2) = 2$ $G_y(1, -1, 2) = -2$ $G_z(1, -1, 2) = -1$

$2(x - 1) - 2(y + 1) - 1(z - 2) = 0$

$2x - 2y - z = 2$

13. $h(x, y) = \ln\sqrt{x^2 + y^2}, (3, 4, \ln 5)$

$H(x, y, z) = \ln\sqrt{x^2 + y^2} - z = \dfrac{1}{2}\ln(x^2 + y^2) - z$

$H_x(x, y, z) = \dfrac{x}{x^2 + y^2}$ $H_y(x, y, z) = \dfrac{y}{x^2 + y^2}$ $H_z(x, y, z) = -1$

$H_x(3, 4, \ln 5) = \dfrac{3}{25}$ $H_y(3, 4, \ln 5) = \dfrac{4}{25}$ $H_z(3, 4, \ln 5) = -1$

$\dfrac{3}{25}(x - 3) + \dfrac{4}{25}(y - 4) - (z - \ln 5) = 0$

$3(x - 3) + 4(y - 4) - 25(z - \ln 5) = 0$

$\qquad\qquad 3x + 4y - 25z = 25(1 - \ln 5)$

15. $F(x, y, z) = x^2 + y^2 - 5z^2 - 15, (-4, -2, 1)$

$F_x(x, y, z) = 2x$ $F_x(-4, -2, 1) = -8$

$F_y(x, y, z) = 2y$ $F_y(-4, -2, 1) = -4$

$F_z(x, y, z) = -10z$ $F_z(-4, -2, 1) = -10$

$-8(x + 4) - 4(y + 2) - 10(z - 1) = 0$

$\qquad 8x + 4y + 10z = -32 - 8 + 10$

$\qquad\qquad 4x + 2y + 5z = -15$

17. $x + y + z = 9, (3, 3, 3)$

$F(x, y, z) = x + y + z - 9$

$F_x(x, y, z) = 1$ $F_y(x, y, z) = 1$ $F_z(x, y, z) = 1$

$F_x(3, 3, 3) = 1$ $F_y(3, 3, 3) = 1$ $F_z(3, 3, 3) = 1$

Direction numbers: 1, 1, 1

(a) $(x - 3) + (y - 3) + (z - 3) = 0$

$\qquad\qquad x + y + z = 9$ (same plane!)

(b) Line: $x - 3 = y - 3 = z - 3$

19. $x^2 + 2y^2 + z^2 = 7, \ (1, -1, 2)$

$F(x, y, z) = x^2 + 2y^2 + z^2 - 7$

$F_x(x, y, z) = 2x \quad F_x(1, -1, 2) = 2$
$F_y(x, y, z) = 4y \quad F_y(1, -1, 2) = -4$
$F_z(x, y, z) = 2z \quad F_z(1, -1, 2) = 4$

Direction numbers: $1, -2, 2$

(a) $(x - 1) - 2(y + 1) + 2(z - 2) = 0$
$$x - 2y + 2z = 7$$

(b) $\dfrac{x - 1}{1} = \dfrac{y + 1}{-2} = \dfrac{z - 2}{2}$

21. $z = x^2 - y^2, (3, 2, 5)$

$F(x, y, z) = x^2 - y^2 - z$

$F_x(x, y, z) = 2x \quad F_y(x, y, z) = -2y \quad F_z(x, y, z) = -1$
$F_x(3, 2, 5) = 6 \quad F_y(3, 2, 5) = -4 \quad F_z(3, 2, 5) = -1$

Direction numbers: $6, -4, -1$

(a) $6(x - 3) - 4(y - 2) - (z - 5) = 0$
$$6x - 4y - z = 5$$

(b) $\dfrac{x - 3}{6} = \dfrac{y - 2}{-4} = \dfrac{z - 5}{-1}$

23. $xyz = 10, (1, 2, 5)$

$F(x, y, z) = xyz - 10$

$F_x(x, y, z) = yz \quad F_y(x, y, z) = xz \quad F_z(x, y, z) = xy$
$F_x(1, 2, 5) = 10 \quad F_y(1, 2, 5) = 5 \quad F_z(1, 2, 5) = 2$

Direction numbers: $10, 5, 2$

(a) $10(x - 1) + 5(y - 2) + 2(z - 5) = 0, 10x + 5y + 2z = 30$

(b) $\dfrac{x - 1}{10} = \dfrac{y - 2}{5} = \dfrac{z - 5}{2}$

25. $z = ye^{2xy}, (0, 2, 2)$

$F(x, y, z) = ye^{2xy} - z$

$F_x(x, y, z) = 2y^2e^{2xy} \quad F_y(x, y, z) = (1 + 2xy)e^{2xy} \quad F_z(x, y, z) = -1$
$F_x(0, 2, 2) = 8 \quad F_y(0, 2, 2) = 1 \quad F_z(0, 2, 2) = -1$

Direction numbers: $8, 1, -1$

(a) $8(x - 0) + (y - 2) - (z - 2) = 0$
$$8x + y - z = 0$$

(b) $\dfrac{x}{8} = \dfrac{y - 2}{1} = \dfrac{z - 2}{-1}$

27. $F(x, y, z) = x^2 + y^2 - 2 \quad G(x, y, z) = x - z$

$\nabla F(x, y, z) = 2x\mathbf{i} + 2y\mathbf{j} \quad \nabla G(x, y, z) = \mathbf{i} - \mathbf{k}$

$\nabla F(1, 1, 1) = 2\mathbf{i} + 2\mathbf{j} \quad \nabla G(1, 1, 1) = \mathbf{i} - \mathbf{k}$

$\nabla F \times \nabla G = \begin{vmatrix} \mathbf{i} & \mathbf{j} & \mathbf{k} \\ 2 & 2 & 0 \\ 1 & 0 & -1 \end{vmatrix} = -2\mathbf{i} + 2\mathbf{j} - 2\mathbf{k} = -2(\mathbf{i} - \mathbf{j} + \mathbf{k})$

Direction numbers: $1, -1, 1$

Line: $x = 1 + t, \ y = 1 - t, \ z = 1 + t$

29. $F(x, y, z) = x^2 + z^2 - 25 \quad G(x, y, z) = y^2 + z^2 - 25$

$\nabla F = 2x\mathbf{i} + 2z\mathbf{k} \qquad\qquad \nabla G = 2y\mathbf{j} + 2z\mathbf{k}$

$\nabla F(3, 3, 4) = 6\mathbf{i} + 8\mathbf{k} \qquad \nabla G(3, 3, 4) = 6\mathbf{j} + 8\mathbf{k}$

$\nabla F \times \nabla G = \begin{vmatrix} \mathbf{i} & \mathbf{j} & \mathbf{k} \\ 6 & 0 & 8 \\ 0 & 6 & 8 \end{vmatrix} = -48\mathbf{i} - 48\mathbf{j} + 36\mathbf{k} = -12(4\mathbf{i} + 4\mathbf{j} - 3\mathbf{k})$

Direction numbers: $4, 4, -3$

$x = 3 + 4t, \; y = 3 + 4t, \; z = 4 - 3t$

31. $F(x, y, z) = x^2 + y^2 + z^2 - 14 \quad G(x, y, z) = x - y - z$

$\nabla F(x, y, z) = 2x\mathbf{i} + 2y\mathbf{j} + 2z\mathbf{k} \quad \nabla G(x, y, z) = \mathbf{i} - \mathbf{j} - \mathbf{k}$

$\nabla F(3, 1, 2) = 6\mathbf{i} + 2\mathbf{j} + 4\mathbf{k} \qquad \nabla G(3, 1, 2) = \mathbf{i} - \mathbf{j} - \mathbf{k}$

$\nabla F \times \nabla G = \begin{vmatrix} \mathbf{i} & \mathbf{j} & \mathbf{k} \\ 6 & 2 & 4 \\ 1 & -1 & -1 \end{vmatrix} = 2\mathbf{i} + 10\mathbf{j} - 8\mathbf{k} = 2[\mathbf{i} + 5\mathbf{j} - 4\mathbf{k}]$

Direction numbers: $1, 5, -4$

$x = 3 + t, \; y = 1 + 5t, \; z = 2 - 4t$

33. $F(x, y, z) = 3x^2 + 2y^2 - z - 15, (2, 2, 5)$

$\nabla F(x, y, z) = 6x\mathbf{i} + 4y\mathbf{j} - \mathbf{k}$

$\nabla F(2, 2, 5) = 12\mathbf{i} + 8\mathbf{j} - \mathbf{k}$

$\cos \theta = \dfrac{|\nabla F(2, 2, 5) \cdot \mathbf{k}|}{\|\nabla F(2, 2, 5)\|} = \dfrac{1}{\sqrt{209}}$

$\theta = \arccos\left(\dfrac{1}{\sqrt{209}}\right) = 86.03°$

35. $F(x, y, z) = x^2 - y^2 + z, (1, 2, 3)$

$\nabla F(x, y, z) = 2x\mathbf{i} - 2y\mathbf{j} + \mathbf{k}$

$\nabla F(1, 2, 3) = 2\mathbf{i} - 4\mathbf{j} + \mathbf{k}$

$\cos \theta = \dfrac{|\nabla F(1, 2, 3) \cdot \mathbf{k}|}{\|\nabla F(1, 2, 3)\|} = \dfrac{1}{\sqrt{21}}$

$\theta = \arccos \dfrac{1}{\sqrt{21}} \approx 77.40°$

37. $F(x, y, z) = 3 - x^2 - y^2 + 6y - z$

$\nabla F(x, y, z) = -2x\mathbf{i} + (-2y + 6)\mathbf{j} - \mathbf{k}$

$-2x = 0, \; x = 0$

$-2y + 6 = 0, \; y = 3$

$z = 3 - 0^2 - 3^2 + 6(3) = 12$

$(0, 3, 12)$ (vertex of paraboloid)

39. $F(x, y, z) = x^2 - xy + y^2 - 2x - 2y - z$

$\nabla F(x, y, z) = (2x - y - 2)\mathbf{i} + (-x + 2y - 2)\mathbf{j} - \mathbf{k}$

$2x - y - 2 = 0$

$-x + 2y - 2 = 0$

$y = 2x - 2 \Rightarrow -x + 2(2x - 2) - 2$

$\qquad\quad = 3x - 6 = 0 \Rightarrow x = 2$

$y = 2, z = -4$

Point: $(2, 2, -4)$

41. $F(x, y, z) = 5xy - z$

$\nabla F(x, y, z) = 5y\mathbf{i} + 5x\mathbf{j} - \mathbf{k}$

$5y = 0$

$5x = 0$

$x = y = z = 0$

Point: $(0, 0, 0)$

43. $F(x, y, z) = x^2 + 2y^2 + 3z^2 - 3, (-1, 1, 0)$

$F_x(x, y, z) = 2x \quad F_y(x, y, z) = 4y \quad F_z(x, y, z) = 6z$

$F_x(-1, 1, 0) = -2 \quad F_y(-1, 1, 0) = 4 \quad F_z(-1, 1, 0) = 0$

$-2(x + 1) + 4(y - 1) + 0(z - 0) = 0$

$$-2x + 4y = 6$$

$$-x + 2y = 3$$

$G(x, y, z) = x^2 + y^2 + z^2 + 6x - 10y + 14, (-1, 1, 0)$

$G_x(x, y, z) = 2x + 6 \quad G_y(x, y, z) = 2y - 10 \quad G_z(x, y, z) = 2z$

$G_x(-1, 1, 0) = 4 \quad G_y(-1, 1, 0) = -8 \quad G_z(-1, 1, 0) = 0$

$4(x + 1) - 8(y - 1) + 0(z - 0) = 0$

$$4x - 8y + 12 = 0$$

$$-x + 2y = 3$$

The tangent planes are the same.

45. (a) $F(x, y, z) = 2xy^2 - z, F(1, 1, 2) = 2 - 2 = 0$

$G(x, y, z) = 8x^2 - 5y^2 - 8z + 13, G(1, 1, 2) = 8 - 5 - 16 + 13 = 0$

So, $(1, 1, 2)$ lies on both surfaces.

(b) $\nabla F = 2y^2\mathbf{i} + 4xy\mathbf{j} - \mathbf{k}, \nabla F(1, 1, 2) = 2\mathbf{i} + 4\mathbf{j} - \mathbf{k}$

$\nabla G = 16x\mathbf{i} - 10y\mathbf{j} - 8\mathbf{k}, \nabla G(1, 1, 2) = 16\mathbf{i} - 10\mathbf{j} - 8\mathbf{k}$

$\nabla F \cdot \nabla G = 2(16) + 4(-10) + (-1)(-8) = 0$

The tangent planes are perpendicular at $(1, 1, 2)$.

47. Not necessarily. The gradients only need to be parallel.

49. $F(x, y, z) = 3x^2 + y^2 + 3z^2 - 1$

$\nabla F = 6x\mathbf{i} + 2y\mathbf{j} + 6z\mathbf{k}$

The normal vector to the plane is $-6\mathbf{i} + \mathbf{j} + 3\mathbf{k}$.

This normal vector is parallel to ∇F.

$6x = -6t \Rightarrow x = -t$

$2y = t \Rightarrow y = \dfrac{t}{2}$

$6z = 3t \Rightarrow z = \dfrac{t}{2}$

$3x^2 + y^2 + 3z^2 = 3t^2 + \dfrac{t^2}{4} + \dfrac{3t^2}{4} = 1 \Rightarrow 16t^2 = 4$

$$\Rightarrow t = \pm\dfrac{1}{2}$$

There are two points: $\left(-\dfrac{1}{2}, \dfrac{1}{4}, \dfrac{1}{4}\right)$ and $\left(\dfrac{1}{2}, -\dfrac{1}{4}, -\dfrac{1}{4}\right)$.

51. $F(x, y, z) = x^2 + 4y^2 + z^2 - 9$

$\nabla F = 2x\mathbf{i} + 8y\mathbf{j} + 2z\mathbf{k}$

This normal vector is parallel to the line with direction number $-4, 8, -2$.

So, $2x = -4t \Rightarrow x = -2t$

$8y = 8t \Rightarrow y = t$

$2z = -2t \Rightarrow z = -t$

$x^2 + 4y^2 + z^2 - 9 = 4t^2 + 4t^2 + t^2 - 9$

$$= 0 \Rightarrow t$$

$$= \pm 1$$

There are two points on the ellipse where the tangent plane is perpendicular to the line:

$(-2, 1, -1) \ (t = 1)$

$(2, -1, 1) \ (t = -1)$

53. $z = f(x, y) = \dfrac{4xy}{(x^2 + 1)(y^2 + 1)}, \; -2 \le x \le 2, 0 \le y \le 3$

(a) Let $F(x, y, z) = \dfrac{4xy}{(x^2 + 1)(y^2 + 1)} - z$

$\nabla F(x, y, z) = \dfrac{4y}{y^2 + 1}\left(\dfrac{x^2 + 1 - 2x^2}{(x^2 + 1)^2}\right)\mathbf{i} + \dfrac{4x}{x^2 + 1}\left(\dfrac{y^2 + 1 - 2y^2}{(y^2 + 1)^2}\right)\mathbf{j} - \mathbf{k} = \dfrac{4y(1 - x^2)}{(y^2 + 1)(x^2 + 1)^2}\mathbf{i} + \dfrac{4x(1 - y^2)}{(x^2 + 1)(y^2 + 1)^2}\mathbf{j} - \mathbf{k}$

$\nabla F(1, 1, 1) = -\mathbf{k}$

Direction numbers: $0, 0, -1$

Line: $x = 1, y = 1, z = 1 - t$

Tangent plane: $0(x - 1) + 0(y - 1) - 1(z - 1) = 0 \Rightarrow z = 1$

(b) $\nabla F\left(-1, 2, -\dfrac{4}{5}\right) = 0\mathbf{i} + \dfrac{-4(-3)}{(2)(5)^2}\mathbf{j} - \mathbf{k} = \dfrac{6}{25}\mathbf{j} - \mathbf{k}$

Line: $x = -1, y = 2 + \dfrac{6}{25}t, z = -\dfrac{4}{5} - t$

Plane: $0(x + 1) + \dfrac{6}{25}(y - 2) - 1\left(z + \dfrac{4}{5}\right) = 0$

$6y - 12 - 25z - 20 = 0$

$6y - 25z - 32 = 0$

(c)

55. $f(x, y) = 6 - x^2 - \dfrac{y^2}{4}, g(x, y) = 2x + y$

(a) $F(x, y, z) = z + x^2 + \dfrac{y^2}{4} - 6 \quad G(x, y, z) = z - 2x - y$

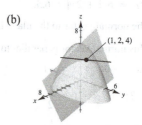

$\nabla F(x, y, z) = 2x\mathbf{i} + \dfrac{1}{2}y\mathbf{j} + \mathbf{k} \quad \nabla G(x, y, z) = -2\mathbf{i} - \mathbf{j} + \mathbf{k}$

$\nabla F(1, 2, 4) = 2\mathbf{i} + \mathbf{j} + \mathbf{k} \quad\quad \nabla G(1, 2, 4) = -2\mathbf{i} - \mathbf{j} + \mathbf{k}$

The cross product of these gradients is parallel to the curve of intersection.

$\nabla F(1, 2, 4) \times \nabla G(1, 2, 4) = \begin{vmatrix} \mathbf{i} & \mathbf{j} & \mathbf{k} \\ 2 & 1 & 1 \\ -2 & -1 & 1 \end{vmatrix} = 2\mathbf{i} - 4\mathbf{j}$

Using direction numbers $1, -2, 0$, you get $x = 1 + t, y = 2 - 2t, z = 4$.

$\cos \theta = \dfrac{\nabla F \cdot \nabla G}{\|\nabla F\|\|\nabla G\|} = \dfrac{-4 - 1 + 1}{\sqrt{6}\sqrt{6}} = \dfrac{-4}{6} \Rightarrow \theta \approx 48.2°$

57. $F(x, y, z) = \dfrac{x^2}{a^2} + \dfrac{y^2}{b^2} + \dfrac{z^2}{c^2} - 1$

$F_x(x, y, z) = \dfrac{2x}{a^2}$

$F_y(x, y, z) = \dfrac{2y}{b^2}$

$F_z(x, y, z) = \dfrac{2z}{c^2}$

Plane: $\dfrac{2x_0}{a^2}(x - x_0) + \dfrac{2y_0}{b^2}(y - y_0) + \dfrac{2z_0}{c^2}(z - z_0) = 0$

$\dfrac{x_0 x}{a^2} + \dfrac{y_0 y}{b^2} + \dfrac{z_0 z}{c^2} = \dfrac{x_0^2}{a^2} + \dfrac{y_0^2}{b^2} + \dfrac{z_0^2}{c^2} = 1$

59. $F(x, y, z) = a^2 x^2 + b^2 y^2 - z^2$

$F_x(x, y, z) = 2a^2 x$

$F_y(x, y, z) = 2b^2 y$

$F_z(x, y, z) = -2z$

Plane:

$2a^2 x_0(x - x_0) + 2b^2 y_0(y - y_0) - 2z_0(z - z_0) = 0$

$a^2 x_0 x + b^2 y_0 y - z_0 z = a^2 x_0^2 + b^2 y_0^2 - z_0^2 = 0$

So, the plane passes through the origin.

61. $f(x, y) = e^{x-y}$

$f_x(x, y) = e^{x-y}, \quad f_y(x, y) = -e^{x-y}$

$f_{xx}(x, y) = e^{x-y}, \quad f_{yy}(x, y) = e^{x-y}, \quad f_{xy}(x, y) = -e^{x-y}$

(a) $P_1(x, y) \approx f(0, 0) + f_x(0, 0)x + f_y(0, 0)y = 1 + x - y$

(b) $P_2(x, y) \approx f(0, 0) + f_x(0, 0)x + f_y(0, 0)y + \tfrac{1}{2}f_{xx}(0, 0)x^2 + f_{xy}(0, 0)xy + \tfrac{1}{2}f_{yy}(0, 0)y^2 = 1 + x - y + \tfrac{1}{2}x^2 - xy + \tfrac{1}{2}y^2$

(c) If $x = 0$, $P_2(0, y) = 1 - y + \tfrac{1}{2}y^2$. This is the second-degree Taylor polynomial for e^{-y}.

If $y = 0$, $P_2(x, 0) = 1 + x + \tfrac{1}{2}x^2$. This is the second-degree Taylor polynomial for e^x.

(d)

x	y	$f(x, y)$	$P_1(x, y)$	$P_2(x, y)$
0	0	1	1	1
0	0.1	0.9048	0.9000	0.9050
0.2	0.1	1.1052	1.1000	1.1050
0.2	0.5	0.7408	0.7000	0.7450
1	0.5	1.6487	1.5000	1.6250

63. Given $z = f(x, y)$, then:

$F(x, y, z) = f(x, y) - z = 0$

$\nabla F(x_0, y_0, z_0) = f_x(x_0, y_0)\mathbf{i} + f_y(x_0, y_0)\mathbf{j} - \mathbf{k}$

$\cos\theta = \dfrac{\left| \nabla F(x_0, y_0, z_0) \cdot \mathbf{k} \right|}{\left\| \nabla F(x_0, y_0, z_0) \right\| \left\| \mathbf{k} \right\|} = \dfrac{\left| -1 \right|}{\sqrt{\left[f_x(x_0, y_0) \right]^2 + \left[f_y(x_0, y_0) \right]^2 + (-1)^2}} = \dfrac{1}{\sqrt{\left[f_x(x_0, y_0) \right]^2 + \left[f_y(x_0, y_0) \right]^2 + 1}}$

Section 13.8   Extrema of Functions of Two Variables

1. (a) The function f has a relative minimum at (x_0, y_0) if $f(x, y) \geq f(x_0, y_0)$ for all (x, y) in an open disk containing (x_0, y_0).

(b) The function f has a relative maximum at (x_0, y_0) if $f(x, y) \leq f(x_0, y_0)$ for all (x, y) in an open disk containing (x_0, y_0).

(c) The point (x_0, y_0) is a critical point if either

(1) $f_x(x_0, y_0) = 0$ and $f_y(x_0, y_0) = 0$, or

(2) $f_x(x_0, y_0)$ or $f_y(x_0, y_0)$ does not exist.

(d) A critical point is a saddle point if it is neither a relative minimum nor a relative maximum.

3. $g(x, y) = (x - 1)^2 + (y - 3)^2 \geq 0$

Relative minimum: $(1, 3, 0)$

Check: $g_x = 2(x - 1) = 0 \Rightarrow x = 1$

$g_y = 2(y - 3) = 0 \Rightarrow y = 3$

$g_{xx} = 2, g_{yy} = 2, g_{xy} = 0, d = (2)(2) - 0 = 4 > 0$

At the critical point $(1, 3)$, $d > 0$ and $g_{xx} > 0 \Rightarrow$ relative minimum at $(1, 3, 0)$.

5. $f(x, y) = \sqrt{x^2 + y^2 + 1} \geq 1$

Relative minimum: $(0, 0, 1)$

Check: $f_x = \dfrac{x}{\sqrt{x^2 + y^2 + 1}} = 0 \Rightarrow x = 0$

$f_y = \dfrac{y}{\sqrt{x^2 + y^2 + 1}} = 0 \Rightarrow y = 0$

$f_{xx} = \dfrac{y^2 + 1}{\left(x^2 + y^2 + 1\right)^{3/2}}$

$f_{yy} = \dfrac{x^2 + 1}{\left(x^2 + y^2 + 1\right)^{3/2}}$

$f_{xy} = \dfrac{-xy}{\left(x^2 + y^2 + 1\right)^{3/2}}$

At the critical point $(0, 0)$, $f_{xx} > 0$ and

$f_{xx}f_{yy} - \left(f_{xy}\right)^2 > 0$.

So, $(0, 0, 1)$ is a relative minimum.

7. $f(x, y) = x^2 + y^2 + 2x - 6y + 6$

$= (x + 1)^2 + (y - 3)^2 - 4 \geq -4$

Relative minimum: $(-1, 3, -4)$

Check: $f_x = 2x + 2 = 0 \Rightarrow x = -1$

$f_y = 2y - 6 = 0 \Rightarrow y = 3$

$f_{xx} = 2, f_{yy} = 2, f_{xy} = 0$

At the critical point $(-1, 3)$, $f_{xx} > 0$ and

$f_{xx}f_{yy} - \left(f_{xy}\right)^2 > 0$. So, $(-1, 3, -4)$ is a relative minimum.

9. $f(x, y) = x^2 + y^2 + 8x - 12y - 3$

$\left.\begin{array}{l} f_x = 2x + 8 = 0 \\ f_y = 2y - 12 = 0 \end{array}\right\} x = -4, y = 6$

$f_{xx} = 2, f_{yy} = 2, f_{xy} = 0$

$d = (2)(2) - 0 = 4 > 0$

At the critical point $(-4, 6)$, $d > 0$ and

$f_{xx} > 0 \Rightarrow (-4, 6, -55)$ is a relative minimum.

11. $f(x, y) = -2x^4y^4$

$\left.\begin{array}{l} f_x = -8x^3y^4 = 0 \\ f_y = -8x^4y^3 = 0 \end{array}\right\} x = 0 \text{ or } y = 0$

Every point along the x-axis or y-axis is a critical point.

$f_{xx} = -24x^2y^4$

$f_{yy} = -24x^4y^2$

$f_{xy} = -32x^3y^3$

$d = 0$ at every critical number.

$f = 0$ for all critical points, which are relative maxima.

For all other points $xy \neq 0$, $f(x, y) < 0$.

13. $f(x, y) = -3x^2 - 2y^2 + 3x - 4y + 5$

$f_x = -6x + 3 = 0$ when $x = \frac{1}{2}$.

$f_y = -4y - 4 = 0$ when $y = -1$.

$f_{xx} = -6, f_{yy} = -4, f_{xy} = 0$

At the critical point $\left(\frac{1}{2}, -1\right)$, $f_{xx} < 0$

and $f_{xx}f_{yy} - \left(f_{xy}\right)^2 > 0$.

So, $\left(\frac{1}{2}, -1, \frac{31}{4}\right)$ is a relative maximum.

15. $f(x, y) = 7x^2 + 2y^2 - 7x + 16y - 13$

$\left.\begin{array}{l} f_x = 14x - 7 = 0 \\ f_y = 4y + 16 = 0 \end{array}\right\} x = \frac{1}{2}, y = -4$

$f_{xx} = 14, f_{yy} = 4, f_{xy} = 0$

$d = (14)(4) - 0 > 0$ and $f_x\left(\dfrac{1}{2}, -4\right) > 0$.

So, $\left(\dfrac{1}{2}, -4, \dfrac{-187}{4}\right)$ is a relative minimum.

17. $f(x, y) = z = x^2 + xy + \frac{1}{2}y^2 - 2x + y$

$\left.\begin{array}{l} f_x = 2x + y - 2 = 0 \\ f_y = x + y + 1 = 0 \end{array}\right\} \begin{array}{l} \text{Solving simultaneously} \\ \text{yields } x = 3, y = -4 \end{array}$

$f_{xx} = 2, f_{yy} = 1, f_{xy} = 1, d = 2(1) - 1 = 1 > 0$.

At the critical point $(3, -4)$, $d > 0$

and $f_{xx} > 0 \Rightarrow (3, -4, -5)$ is a relative minimum.

19. $f(x, y) = -4(x^2 + y^2 + 81)^{1/4}$

For $x = y = 0$, $f(0, 0) = -4(81)^{1/4} = -12$.

For all other (x, y) values, $f(x, y) < -12$.

So, $(0, 0, -12)$ is a relative maximum and $(0, 0)$ is the only critical point.

21. $f(x, y) = x^2 - xy - y^2 - 3x - y$

$f_x = 2x - y - 3 = 0$

$f_y = -x - 2y - 1 = 0$

Solving simultaneously yields $x = 1$, $y = -1$.

$f_{xx} = 2, f_{yy} = -2, f_{xy} = -1$

$d = (2)(-2) - (-1)^2 = -5 < 0$

At the critical point $(1, -1)$, $d < 0 \Rightarrow (1, -1, -1)$ is a saddle point.

23. $f(x, y) = e^{-x} \sin y$

$\left. \begin{array}{l} f_x = -e^{-x} \sin y = 0 \\ f_y = e^{-x} \cos y = 0 \end{array} \right\}$ Because $e^{-x} > 0$ for all x and $\sin y$ and $\cos y$ are never both zero for a given value of y, there are no critical points.

25. $z = \dfrac{-4x}{x^2 + y^2 + 1}$

Relative minimum: $(1, 0, -2)$

Relative maximum: $(-1, 0, 2)$

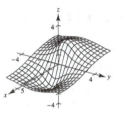

27. $z = (x^2 + 4y^2)e^{1-x^2-y^2}$

Relative minimum: $(0, 0, 0)$

Relative maxima: $(0, \pm 1, 4)$

Saddle points: $(\pm 1, 0, 1)$

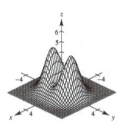

29. $z = \dfrac{(x - y)^4}{x^2 + y^2} \geq 0$. $z = 0$ if $x = y \neq 0$.

Relative minimum at all points $(x, x), x \neq 0$.

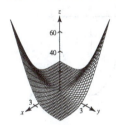

31. $f_{xx}f_{yy} - (f_{xy})^2 = (9)(4) - 6^2 = 0$

Insufficient information.

33. $f_{xx}f_{yy} - (f_{xy})^2 = (-9)(6) - 10^2 < 0$

f has a saddle point at (x_0, y_0).

35. $f(x, y) = x^3 + y^3$

(a) $\left. \begin{array}{l} f_x = 3x^2 = 0 \\ f_y = 3y^2 = 0 \end{array} \right\} x = y = 0$

Critical point: $(0, 0)$

(b) $f_{xx} = 6x, f_{yy} = 6y, f_{xy} = 0$

At $(0, 0)$, $f_{xx}f_{yy} - (f_{xy})^2 = 0$.

$(0, 0, 0)$ is a saddle point.

(c) Test fails at $(0, 0)$.

37. $f(x, y) = (x - 1)^2(y + 4)^2 \geq 0$

(a) $\left. \begin{array}{l} f_x = 2(x - 1)(y + 4)^2 = 0 \\ f_y = 2(x - 1)^2(y + 4) = 0 \end{array} \right\}$ critical points: $(1, a)$ and $(b, -4)$

(b) $f_{xx} = 2(y + 4)^2$

$f_{yy} = 2(x - 1)^2$

$f_{xy} = 4(x - 1)(y + 4)$

At both $(1, a)$ and $(b, -4)$, $f_{xx}f_{yy} - (f_{xy})^2 = 0$.

Because $f(x, y) \geq 0$, there are absolute minima at $(1, a, 0)$ and $(b, -4, 0)$.

(c) Test fails at $(1, a)$ and $(b, -4)$.

(d)

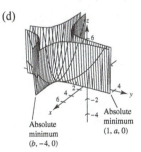

Absolute minimum $(b, -4, 0)$

Absolute minimum $(1, a, 0)$

39. $f(x, y) = x^2 - 4xy + 5, R = \{(x, y): 1 \leq x \leq 4, 0 \leq y \leq 2\}$

$\left.\begin{array}{l} f_x = 2x - 4y = 0 \\ f_y = -4x = 0 \end{array}\right\} x = y = 0 \quad \text{(not in region R)}$

Along $y = 0, 1 \leq x \leq 4$: $f = x^2 + 5, f(1, 0) = 6, f(4, 0) = 21$.

Along $y = 2, 1 \leq x \leq 4$: $f = x^2 - 8x + 5, f' = 2x - 8 = 0$

$\qquad f(1, 2) = -2, f(4, 2) = -11$.

Along $x = 1, 0 \leq y \leq 2$: $f = -4y + 6, f(1, 0) = 6, f(1, 2) = -2$.

Along $x = 4, 0 \leq y \leq 2$: $f = 21 - 16y, f(4, 0) = 21, f(4, 2) = -11$.

So, the maximum is $(4, 0, 21)$ and the minimum is $(4, 2, -11)$.

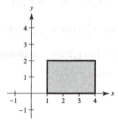

41. $f(x, y) = 12 - 3x - 2y$ has no critical points. On the line $y = x + 1, 0 \leq x \leq 1$,

$f(x, y) = f(x) = 12 - 3x - 2(x + 1) = -5x + 10$

and the maximum is 10, the minimum is 5. On the line $y = -2x + 4, 1 \leq x \leq 2$,

$f(x, y) = f(x) = 12 - 3x - 2(-2x + 4) = x + 4$

and the maximum is 6, the minimum is 5. On the line $y = -\frac{1}{2}x + 1, 0 \leq x \leq 2$,

$f(x, y) = f(x) = 12 - 3x - 2(-\frac{1}{2}x + 1) = -2x + 10$

and the maximum is 10, the minimum is 6.

Absolute maximum: 10 at $(0, 1)$

Absolute minimum: 5 at $(1, 2)$

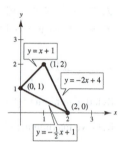

43. $f(x, y) = 3x^2 + 2y^2 - 4y$

$\left.\begin{array}{l} f_x = 6x = 0 \Rightarrow x = 0 \\ f_y = 4y - 4 = 0 \Rightarrow y = 1 \end{array}\right\} f(0, 1) = -2$

On the line $y = 4, -2 \leq x \leq 2$,

$f(x, y) = f(x) = 3x^2 + 32 - 16 = 3x^2 + 16$

and the maximum is 28, the minimum is 16. On the curve $y = x^2, -2 \leq x \leq 2$,

$f(x, y) = f(x) = 3x^2 + 2(x^2)^2 - 4x^2 = 2x^4 - x^2 = x^2(2x^2 - 1)$

and the maximum is 28, the minimum is $-\frac{1}{8}$.

Absolute maximum: 28 at $(\pm 2, 4)$

Absolute minimum: -2 at $(0, 1)$

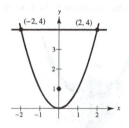

45. $f(x, y) = x^2 + 2xy + y^2$, $R = \{(x, y): |x| \le 2, |y| \le 1\}$

$\left.\begin{array}{l} f_x = 2x + 2y = 0 \\ f_y = 2x + 2y = 0 \end{array}\right\} y = -x$

$f(x, -x) = x^2 - 2x^2 + x^2 = 0$

Along $y = 1, -2 \le x \le 2$,

$f = x^2 + 2x + 1$, $f' = 2x + 2 = 0 \Rightarrow x = -1$, $f(-2, 1) = 1$, $f(-1, 1) = 0$, $f(2, 1) = 9$.

Along $y = -1, -2 \le x \le 2$,

$f = x^2 - 2x + 1$, $f' = 2x - 2 = 0 \Rightarrow x = 1$, $f(-2, -1) = 9$, $f(1, -1) = 0$, $f(2, -1) = 1$.

Along $x = 2, -1 \le y \le 1$, $f = 4 + 4y + y^2$, $f' = 2y + 4 \ne 0$.

Along $x = -2, -1 \le y \le 1$, $f = 4 - 4y + y^2$, $f' = 2y - 4 \ne 0$.

So, the maxima are $f(-2, -1) = 9$ and $f(2, 1) = 9$, and the minima are $f(x, -x) = 0, -1 \le x \le 1$.

47. $f(x, y, z) = x^2 + (y - 3)^2 + (z + 1)^2 \ge 0$

$\left.\begin{array}{l} f_x = 2x = 0 \\ f_y = 2(y - 3) = 0 \\ f_z = 2(z + 1) = 0 \end{array}\right\}$ Solving yields the critical point $(0, 3, -1)$.

Absolute minimum: 0 at $(0, 3, -1)$

49. $d = f_{xx}f_{yy} - f_{xy}{}^2 = (2)(8) - f_{xy}{}^2 = 16 - f_{xy}{}^2 > 0 \Rightarrow f_{xy}{}^2 < 16 \Rightarrow -4 < f_{xy} < 4$

51.

Extrema at all (x, y)

53. $f(x, y) = x^2 - y^2$, $g(x, y) = x^2 + y^2$

(a) $f_x = 2x = 0, f_y = -2y = 0 \Rightarrow (0, 0)$ is a critical point.

$g_x = 2x = 0, g_y = 2y = 0 \Rightarrow (0, 0)$ is a critical point.

(b) $f_{xx} = 2, f_{yy} = -2, f_{xy} = 0$

$d = 2(-2) - 0 < 0 \Rightarrow (0, 0)$ is a saddle point.

$g_{xx} = 2, g_{yy} = 2, g_{xy} = 0$

$d = 2(2) - 0 > 0 \Rightarrow (0, 0)$ is a relative minimum.

55. False.

Let $f(x, y) = 1 - |x| - |y|$.

$(0, 0, 1)$ is a relative maximum, but $f_x(0, 0)$ and

$f_y(0, 0)$ do not exist.

57. False. Let $f(x, y) = x^2y^2$ (See Example 4 on page 944).

Section 13.9   Applications of Extrema

1. Write the equation to be maximized or minimized as a function of two variables. Set the partial derivatives equal to zero (or undefined) to obtain the critical points. Use the Second Partials Test to test for relative extrema using the critical points. Check the boundary points, too.

3. A point on the plane is given by $(x, y, z) = (x, y, 3 - x + y)$. The square of the distance from $(1, -3, 2)$ to this point is

$$S = (x - 1)^2 + (y + 3)^2 + (3 - x + y - 2)^2$$
$$= (x - 1)^2 + (y + 3)^2 + (1 - x + y)^2.$$

$$S_x = 2(x - 1) - 2(1 - x + y) = 4x - 2y - 4$$
$$S_y = 2(y + 3) + 2(1 - x + y) = -2x + 4y + 8$$

From the equations $S_x = 0$ and $S_y = 0$, you obtain $x = 0$ and $y = -2$. So, $z = 1$ and the distance is

$$\sqrt{(0 - 1)^2 + (-2 + 3)^2 + (1 - 2)^2} = \sqrt{3}.$$

5. A point on the surface is given by $(x, y, z) = \left(x, y, \sqrt{1 - 2x - 2y}\right)$. The square of the distance from $(-2, -2, 0)$ to a point on the surface is given by

$$S = (x + 2)^2 + (y + 2)^2 + \left(\sqrt{1 - 2x - 2y} - 0\right)^2$$
$$= (x + 2)^2 + (y + 2)^2 + 1 - 2x - 2y.$$

$$S_x = 2(x + 2) - 2$$
$$S_y = 2(y + 2) - 2$$

From the equations $S_x = 0$ and $S_y = 0$, we obtain

$$\left.\begin{array}{r} 2x + 2 = 0 \\ 2y + 2 = 0 \end{array}\right\} \Rightarrow x = y = -1, z = \sqrt{5}.$$

So, the distance is

$$\sqrt{(-1 + 2)^2 + (-1 + 2)^2 + \left(\sqrt{5}\right)^2} = \sqrt{7}.$$

7. Let x, y, and z be the numbers. Because $xyz = 27$,

$$z = \frac{27}{xy}.$$

$$S = x + y + z = x + y + \frac{27}{xy}.$$

$$S_x = 1 - \frac{27}{x^2 y} = 0, \quad S_y = 1 - \frac{27}{xy^2} = 0.$$

$$\left.\begin{array}{r} x^2 y = 27 \\ xy^2 = 27 \end{array}\right\} x = y = 3$$

So, $x = y = z = 3$.

9. Let x, y, and z be the numbers and let $S = x^2 + y^2 + z^2$. Because $x + y + z = 30$, we have

$$S = x^2 + y^2 + (30 - x - y)^2$$

$$\left.\begin{array}{l} S_x = 2x + 2(30 - x - y)(-1) = 0 \\ S_y = 2y + 2(30 - x - y)(-1) = 0 \end{array}\right\} \begin{array}{l} 2x + y = 30 \\ x + 2y = 30. \end{array}$$

Solving simultaneously yields $x = 10$, $y = 10$, and $z = 10$.

11. The volume is $668.25 = xyz \Rightarrow z = \dfrac{668.25}{xy}$.

$$C = 0.06(2yz + 2xz) + 0.11(xy)$$
$$= 0.12\left(\frac{668.25}{x} + \frac{668.25}{y}\right) + 0.11(xy)$$

$$C = \frac{80.19}{x} + \frac{80.19}{y} + 0.11(xy)$$

$$C_x = \frac{-80.19}{x^2} + 0.11y = 0$$

$$C_y = \frac{-8.19}{y^2} + 0.11x = 0$$

Solving simultaneously, $x = y = 9$ and $z = 8.25$.

Minimum cost: $\dfrac{80.19}{9} + \dfrac{80.19}{9} + 0.11(xy) = \$26.73$

13. Let x, y, and z be the length, width, and height, respectively and let V_0 be the given volume. Then $V_0 = xyz$ and $z = V_0/xy$. The surface area is

$$S = 2xy + 2yz + 2xz = 2\left(xy + \frac{V_0}{x} + \frac{V_0}{y}\right)$$

$$\left.S_x = 2\left(y - \frac{V_0}{x^2}\right) = 0\right| x^2 y - V_0 = 0$$

$$\left.S_y = 2\left(x - \frac{V_0}{y^2}\right) = 0\right| xy^2 - V_0 = 0.$$

Solving simultaneously yields $x = \sqrt[3]{V_0}$, $y = \sqrt[3]{V_0}$, and $z = \sqrt[3]{V_0}$.

15. $R(x_1, x_2) = -5x_1^2 - 8x_2^2 - 2x_1 x_2 + 42x_1 + 102x_2$

$R_{x_1} = -10x_1 - 2x_2 + 42 = 0, 5x_1 + x_2 = 21$

$R_{x_2} = -16x_2 - 2x_1 + 102 = 0, x_1 + 8x_2 = 51$

Solving this system yields $x_1 = 3$ and $x_2 = 6$.

$R_{x_1 x_1} = -10$

$R_{x_1 x_2} = -2$

$R_{x_2 x_2} = -16$

$R_{x_1 x_1} < 0$ and $R_{x_1 x_1} R_{x_2 x_2} - \left(R_{x_1 x_2}\right)^2 > 0$

So, revenue is maximized when $x_1 = 3$ and $x_2 = 6$.

17. $P(p, q, r) = 2pq + 2pr + 2qr.$

$p + q + r = 1$ implies that $r = 1 - p - q.$

$P(p, q) = 2pq + 2p(1 - p - q) + 2q(1 - p - q)$

$\quad = 2pq + 2p - 2p^2 - 2pq + 2q - 2pq - 2q^2 = -2pq + 2p + 2q - 2p^2 - 2q^2$

$\dfrac{\partial P}{\partial p} = -2q + 2 - 4p; \dfrac{\partial P}{\partial q} = -2p + 2 - 4q$

Solving $\dfrac{\partial P}{\partial p} = \dfrac{\partial P}{\partial q} = 0$ gives $q + 2p = 1$

$\qquad\qquad\qquad\qquad\qquad p + 2q = 1$

and so $p = q = \dfrac{1}{3}$ and $P\left(\dfrac{1}{3}, \dfrac{1}{3}\right) = -2\left(\dfrac{1}{9}\right) + 2\left(\dfrac{1}{3}\right) + 2\left(\dfrac{1}{3}\right) - 2\left(\dfrac{1}{9}\right) - 2\left(\dfrac{1}{9}\right) = \dfrac{6}{9} = \dfrac{2}{3}.$

19. The distance from P to Q is $\sqrt{x^2 + 4}$. The distance from

Q to R is $\sqrt{(y - x)^2 + 1}$. The distance from R to S is

$10 - y.$

$C = 3k\sqrt{x^2 + 4} + 2k\sqrt{(y - x)^2 + 1} + k(10 - y)$

$C_x = 3k\left(\dfrac{x}{\sqrt{x^2 + 4}}\right) + 2k\left(\dfrac{-(y - x)}{\sqrt{(y - x)^2 + 1}}\right) = 0$

$C_y = 2k\left(\dfrac{y - x}{\sqrt{(y - x)^2 + 1}}\right) - k = 0$

$\Rightarrow \dfrac{y - x}{\sqrt{(y - x)^2 + 1}} = \dfrac{1}{2}$

$3k\left(\dfrac{x}{\sqrt{x^2 + 4}}\right) + 2k\left(-\dfrac{1}{2}\right) = 0$

$\qquad\qquad \dfrac{x}{\sqrt{x^2 + 4}} = \dfrac{1}{3}$

$\qquad\qquad 3x = \sqrt{x^2 + 4}$

$\qquad\qquad 9x^2 = x^2 + 4$

$\qquad\qquad x^2 = \dfrac{1}{2}$

$\qquad\qquad x = \dfrac{\sqrt{2}}{2}$

$2(y - x) = \sqrt{(y - x)^2 + 1}$

$4(y - x)^2 = (y - x)^2 + 1$

$(y - x)^2 = \dfrac{1}{3}$

$\qquad y = \dfrac{1}{\sqrt{3}} + \dfrac{1}{\sqrt{2}} = \dfrac{2\sqrt{3} + 3\sqrt{2}}{6}$

So, $x = \dfrac{\sqrt{2}}{2} \approx 0.707$ km and

$y = \dfrac{2\sqrt{3} + 3\sqrt{2}}{6} \approx 1.284$ km.

21. (a)

x	y	xy	x^2
-2	0	0	4
0	1	0	0
2	3	6	4
$\sum x_i = 0$	$\sum y_i = 4$	$\sum x_i y_i = 6$	$\sum x_i^2 = 8$

$a = \dfrac{3(6) - 0(4)}{3(8) - 0^2} = \dfrac{3}{4}, \; b = \dfrac{1}{3}\left[4 - \dfrac{3}{4}(0)\right] = \dfrac{4}{3},$

$y = \dfrac{3}{4}x + \dfrac{4}{3}$

(b) $S = \left(-\dfrac{3}{2} + \dfrac{4}{3} - 0\right)^2 + \left(\dfrac{4}{3} - 1\right)^2 + \left(\dfrac{3}{2} + \dfrac{4}{3} - 3\right)^2$

$\quad = \dfrac{1}{6}$

23. (a)

x	y	xy	x^2
-2	0	0	4
0	1	0	0
2	3	6	4
$\sum x_i = 0$	$\sum y_i = 4$	$\sum x_i y_i = 6$	$\sum x_i^2 = 8$

$a = \dfrac{4(4) - 4(8)}{4(6) - 4^2} = -2, \; b = \dfrac{1}{4}\left[8 + 2(4)\right] = 4,$

$y = -2x + 4$

(b) $S = (4 - 4)^2 + (2 - 3)^2 + (2 - 1)^2 + (0 - 0)^2$

$\quad = 2$

25.

x	y	xy	x^2
0	0	0	0
1	1	1	1
3	6	18	9
4	8	32	16
5	9	45	25
$\sum x_i = 13$	$\sum y_i = 24$	$\sum x_i y_i = 96$	$\sum x_i^2 = 51$

$$a = \frac{5(96) - (13)(24)}{5(51) - (13)^2} = \frac{84}{43}$$

$$b = \frac{1}{5}\left[24 - \frac{84}{43}(13)\right] = -\frac{12}{43}$$

$$y = \frac{84}{43}x - \frac{12}{43}$$

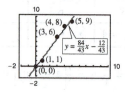

27.

x	y	xy	x^2
0	6	0	0
4	3	12	16
5	0	0	25
8	−4	−32	64
10	−5	−50	100
$\sum x_i = 27$	$\sum y_i = 0$	$\sum x_i y_i = 70$	$\sum x_i^2 = 205$

$$a = \frac{5(-70) - (27)(0)}{5(205) - (27)^2} = \frac{-350}{296} = -\frac{175}{148}$$

$$b = \frac{1}{5}\left[0 - \left(-\frac{175}{148}\right)(27)\right] = \frac{945}{148}$$

$$y = -\frac{175}{148}x + \frac{945}{148}$$

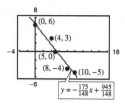

29. (a) Using a graphing utility, you obtain
$$y = 0.23x + 2.38.$$

(b) When $x = 1300$, $y \approx \$301.4$ billion.

(c) The new model is $y = 0.23x + 5.09$, so the constant increases.

31. $S(a, b, c) = \sum_{i=1}^{n}(y_i - ax_i^2 - bx_i - c)^2$

$$\frac{\partial S}{\partial a} = \sum_{i=1}^{n} -2x_i^2(y_i - ax_i^2 - bx_i - c) = 0$$

$$\frac{\partial S}{\partial b} = \sum_{i=1}^{n} -2x_i(y_i - ax_i^2 - bx_i - c) = 0$$

$$\frac{\partial S}{\partial c} = -2\sum_{i=1}^{n}(y_i - ax_i^2 - bx_i - c) = 0$$

$$a\sum_{i=1}^{n}x_i^4 + b\sum_{i=1}^{n}x_i^3 + c\sum_{i=1}^{n}x_i^2 = \sum_{i=1}^{n}x_i^2 y_i$$

$$a\sum_{i=1}^{n}x_i^3 + b\sum_{i=1}^{n}x_i^2 + c\sum_{i=1}^{n}x_i = \sum_{i=1}^{n}x_i y_i$$

$$a\sum_{i=1}^{n}x_i^2 + b\sum_{i=1}^{n}x_i + cn = \sum_{i=1}^{n}y_i$$

33. $(-2, 0), (-1, 0), (0, 1), (1, 2), (2, 5)$

$\sum x_i = 0$

$\sum y_i = 8$

$\sum x_i^2 = 10$

$\sum x_i^3 = 0$

$\sum x_i^4 = 34$

$\sum x_i y_i = 12$

$\sum x_i^2 y_i = 22$

$34a + 10c = 22, 10b = 12, 10a + 5c = 8$

$a = \frac{3}{7}, b = \frac{6}{5}, c = \frac{26}{35}, y = \frac{3}{7}x^2 + \frac{6}{5}x + \frac{26}{35}$

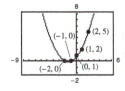

35. $(0, 0), (2, 2), (3, 6), (4, 12)$

$\sum x_i = 9$

$\sum y_i = 20$

$\sum x_i^2 = 29$

$\sum x_i^3 = 99$

$\sum x_i^4 = 353$

$\sum x_i y_i = 70$

$\sum x_i^2 y_i = 254$

$353a + 99b + 29c = 254$

$99a + 29b + 9c = 70$

$29a + 9b + 4c = 20$

$a = 1, b = -1, c = 0, y = x^2 - x$

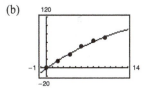

37. (a) $(0, 0), (2, 15), (4, 30), (6, 50), (8, 65), (10, 70)$

$\sum x_i = 30$

$\sum y_i = 230$

$\sum x_i^2 = 220$

$\sum x_i^3 = 1800$

$\sum x_i^4 = 15,664$

$\sum x_i y_i = 1670$

$\sum x_i^2 y_i = 13,500$

$15,664a + 1800b + 220c = 13,500$

$1800a + 220b + 30c = 1670$

$220a + 30b + 6c = 230$

$y = -\frac{25}{112}x^2 + \frac{541}{56}x - \frac{25}{14} \approx -0.22x^2 + 9.66x - 1.79$

(b)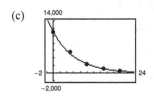

39. (a) $\ln P = -0.1499h + 9.3018$

(b) $\ln P = -0.1499h + 9.3018$

$P = e^{-0.1499h + 9.3018} = 10,957.7e^{-0.1499h}$

(c)

41. $S(a, b) = \sum\limits_{i=1}^{n} (ax_i + b - y_i)^2$

$S_a(a, b) = 2a\sum\limits_{i=1}^{n} x_i^2 + 2b\sum\limits_{i=1}^{n} x_i - 2\sum\limits_{i=1}^{n} x_i y_i$

$S_b(a, b) = 2a\sum\limits_{i=1}^{n} x_i + 2nb - 2\sum\limits_{i=1}^{n} y_i$

$S_{aa}(a, b) = 2\sum\limits_{i=1}^{n} x_i^2$

$S_{bb}(a, b) = 2n$

$S_{ab}(a, b) = 2\sum\limits_{i=1}^{n} x_i$

$S_{aa}(a, b) > 0$ as long as $x_i \neq 0$ for all i. (**Note:** If $x_i = 0$ for all i, then $x = 0$ is the least squares regression line.)

$$d = S_{aa}S_{bb} - S_{ab}{}^2 = 4n\sum\limits_{i=1}^{n} x_i^2 - 4\left(\sum\limits_{i=1}^{n} x_i\right)^2 = 4\left[n\sum\limits_{i=1}^{n} x_i^2 - \left(\sum\limits_{i=1}^{n} x_i\right)^2\right] \geq 0 \text{ since } n\sum\limits_{i=1}^{n} x_i^2 \geq \left(\sum\limits_{i=1}^{n} x_i\right)^2.$$

As long as $d \neq 0$, the given values for a and b yield a minimum.

Section 13.10   Lagrange Multipliers

1. Optimization problems that have restrictions or constraints on the values that can be used to produce the optimal solution are called constrained optimization problems.

3. Maximize $f(x, y) = xy$

   Constraint: $x + y = 10$

   $\nabla f = \lambda \nabla g$

   $y\mathbf{i} + x\mathbf{j} = \lambda(\mathbf{i} + \mathbf{j})$

   $\left.\begin{array}{l} y = \lambda \\ x = \lambda \\ x + y = 10 \end{array}\right\} \quad x = y = 5$

   $f(5, 5) = 25$

5. Minimize $f(x, y) = x^2 + y^2$.

   Constraint: $x + 2y - 5 = 0$

   $\nabla f = \lambda \nabla g$

   $2x\mathbf{i} + 2y\mathbf{j} = \lambda(\mathbf{i} + 2\mathbf{j})$

   $\left.\begin{array}{l} 2x = \lambda \\ 2y = 2\lambda \end{array}\right\} \begin{array}{l} x = \lambda/2 \\ y = \lambda \end{array}$

   $x + 2y - 5 = 0$

   $\dfrac{\lambda}{2} + 2\lambda = 5 \Rightarrow \lambda = 2, x = 1, y = 2$

   $f(1, 2) = 5$

7. Maximize $f(x, y) = 2x + 2xy + y$.

   Constraint: $2x + y = 100$

   $\nabla f = \lambda \nabla g$

   $(2 + 2y)\mathbf{i} + (2x + 1)\mathbf{j} = 2\lambda\mathbf{i} + \lambda\mathbf{j}$

   $\left.\begin{array}{l} 2 + 2y = 2\lambda \Rightarrow y = \lambda - 1 \\ 2x + 1 = \lambda \Rightarrow x = \dfrac{\lambda - 1}{2} \end{array}\right\} y = 2x$

   $2x + y = 100 \Rightarrow 4x = 100$

   $\hspace{3cm} x = 25, y = 50$

   $f(25, 50) = 2600$

9. **Note:** $f(x, y) = \sqrt{6 - x^2 - y^2}$ is maximum when $g(x, y)$ is maximum.

   Maximize $g(x, y) = 6 - x^2 - y^2$.

   Constraint: $x + y - 2 = 0$

   $\left.\begin{array}{l} -2x = \lambda \\ -2y = \lambda \end{array}\right\} x = y$

   $x + y = 2 \Rightarrow x = y = 1$

   $f(1, 1) = \sqrt{g(1, 1)} = 2$

11. Minimize $f(x, y, z) = x^2 + y^2 + z^2$.

Constraint: $x + y + z - 9 = 0$

$$\left.\begin{array}{l} 2x = \lambda \\ 2y = \lambda \\ 2z = \lambda \end{array}\right\} x = y = z$$

$x + y + z = 9 \Rightarrow x = y = z = 3$

$f(3, 3, 3) = 27$

13. Minimize $f(x, y, z) = x^2 + y^2 + z^2$.

Constraint: $x + y + z = 1$

$$\left.\begin{array}{l} 2x = \lambda \\ 2y = \lambda \\ 2z = \lambda \end{array}\right\} x = y = z$$

$x + y + z = 1 \Rightarrow x = y = z = \frac{1}{3}$

$f\left(\frac{1}{3}, \frac{1}{3}, \frac{1}{3}\right) = \frac{1}{3}$

15. Maximize or minimize $f(x, y) = x^2 + 3xy + y^2$.

Constraint: $x^2 + y^2 \le 1$

Case 1: On the circle $x^2 + y^2 = 1$

$$\left.\begin{array}{l} 2x + 3y = 2x\lambda \\ 3x + 2y = 2y\lambda \end{array}\right\} x^2 = y^2$$

$x^2 + y^2 = 1 \Rightarrow x = \pm\dfrac{\sqrt{2}}{2}, y = \pm\dfrac{\sqrt{2}}{2}$

Maxima: $f\left(\pm\dfrac{\sqrt{2}}{2}, \pm\dfrac{\sqrt{2}}{2}\right) = \dfrac{5}{2}$

Minima: $f\left(\pm\dfrac{\sqrt{2}}{2}, \mp\dfrac{\sqrt{2}}{2}\right) = -\dfrac{1}{2}$

Case 2: Inside the circle

$$\left.\begin{array}{l} f_x = 2x + 3y = 0 \\ f_y = 3x + 2y = 0 \end{array}\right\} x = y = 0$$

$f_{xx} = 2, f_{yy} = 2, f_{xy} = 3, f_{xx}f_{yy} - \left(f_{xy}\right)^2 \le 0$

Saddle point: $f(0, 0) = 0$

By combining these two cases, we have a maximum

of $\dfrac{5}{2}$ at $\left(\pm\dfrac{\sqrt{2}}{2}, \pm\dfrac{\sqrt{2}}{2}\right)$ and a minimum of

$-\dfrac{1}{2}$ at $\left(\pm\dfrac{\sqrt{2}}{2}, \mp\dfrac{\sqrt{2}}{2}\right)$.

17. Maximize $f(x, y, z) = xyz$.

Constraints: $x + y + z = 32$

$\qquad\qquad x - y + z = 0$

$\nabla f = \lambda \nabla g + \mu \nabla h$

$yz\mathbf{i} + xz\mathbf{j} + xy\mathbf{k} = \lambda(\mathbf{i} + \mathbf{j} + \mathbf{k}) + \mu(\mathbf{i} - \mathbf{j} + \mathbf{k})$

$$\left.\begin{array}{l} yz = \lambda + \mu \\ xz = \lambda - \mu \\ xy = \lambda + \mu \end{array}\right\} yz = xy \Rightarrow x = z$$

$$\left.\begin{array}{l} x + y + z = 32 \\ x - y + z = 0 \end{array}\right\} 2x + 2z = 32 \Rightarrow x = z = 8$$

$$y = 16$$

$f(8, 16, 8) = 1024$

19. Minimize the square of the distance

$f(x, y) = (x - 0)^2 + (y - 0)^2 = x^2 + y^2$ subject

to the constraint $x + y = 1$.

$$\left.\begin{array}{l} 2x = \lambda \\ 2y = \lambda \end{array}\right\} \begin{array}{l} x = \lambda/2 \\ y = \lambda/2 \end{array} \Rightarrow x = y$$

$x + y = 1$

$\qquad x = y = \dfrac{1}{2}$

The minimum distance is $d = \sqrt{\left(\dfrac{1}{2}\right)^2 + \left(\dfrac{1}{2}\right)^2} = \dfrac{\sqrt{2}}{2}$.

21. Minimize the square of the distance

$f(x, y) = x^2 + (y - 2)^2$ subject to the constraint

$x - y = 4$.

$$\left.\begin{array}{l} 2x = \lambda \\ 2(y - 2) = -\lambda \end{array}\right\} \begin{array}{l} x = \lambda/2 \\ y = \dfrac{4 - \lambda}{2} \end{array}$$

$x - y = 4$

$\dfrac{\lambda}{2} - \left(\dfrac{4 - \lambda}{2}\right) = 4$

$\qquad\qquad \lambda = 6$

$x = 3, y = -1$

The minimum distance is

$d = \sqrt{3^2 + (-1 - 2)^2} = 3\sqrt{2}$.

23. Minimize the square of the distance

$f(x, y) = x^2 + (y - 3)^2$ subject to the constraint

$y - x^2 = 0$

$2x = -2x\lambda$

$2(y - 3) = \lambda$

$y = x^2$

If $x = 0, y = 0,$ and $f(0, 0) = 9 \Rightarrow$ distance $= 3$.

If $x \neq 0, \lambda = -1, y = 5/2, x = \pm\sqrt{5/2}$

$f\left(\pm\sqrt{5/2}, 5/2\right) = 5/2 + \left(\dfrac{1}{2}\right)^2 = \dfrac{11}{4} < 3$

The minimum distance is. $d = \dfrac{\sqrt{11}}{2}$

25. Minimize the square of the distance

$f(x, y) = (x - 4)^2 + (y - 4)^2$ subject to the constraint

$x^2 + (y - 1)^2 = 9$.

$2(x - 4) = 2x\lambda$

$2(y - 4) = 2(y - 1)\lambda$

$x^2 + (y - 1)^2 = 9$

Solving these equations, you obtain

$x = 12/5, y = 14/5$ and $\lambda = -2/3$.

The minimum distance is

$d = \sqrt{\left(\dfrac{12}{5} - 4\right)^2 + \left(\dfrac{14}{5} - 4\right)^2} = \sqrt{\dfrac{64}{25} + \dfrac{36}{25}} = 2.$

27. Minimize the square of the distance

$f(x, y, z) = (x - 2)^2 + (y - 1)^2 + (z - 1)^2$

subject to the constraint $x + y + z = 1$.

$\left.\begin{array}{l} 2(x - 2) = \lambda \\ 2(y - 1) = \lambda \\ 2(z - 1) = \lambda \end{array}\right\} y = z$ and $y = x - 1$

$x + y + z = 1 \Rightarrow x + 2(x - 1) = 1$

$x = 1, y = z = 0$

The minimum distance is

$d = \sqrt{(1 - 2)^2 + (0 - 1)^2 + (0 - 1)^2} = \sqrt{3}.$

29. Maximize $f(x, y, z) = z$ subject to the constraints

$x^2 + y^2 - z^2 = 0$ and $x + 2z = 4$.

$0 = 2x\lambda + \mu$

$0 = 2y\lambda \Rightarrow y = 0$

$1 = -2z\lambda + 2\mu$

$x^2 + y^2 - z^2 = 0$

$x + 2z = 4 \Rightarrow x = 4 - 2z$

$(4 - 2z)^2 + 0^2 - z^2 = 0$

$3z^2 - 16z + 16 = 0$

$(3z - 4)(z - 4) = 0$

$z = \tfrac{4}{3}$ or $z = 4$

The maximum value of f occurs when $z = 4$ at the point of $(-4, 0, 4)$.

31. Minimize $f(x, y, z) = (x - 1)^2 + (y + 3)^2 + (z - 2)^2$.

Constraint: $x - y + z = 3$

$2(x - 1) = \lambda \Rightarrow x = 1 + \dfrac{\lambda}{2}$

$2(y + 3) = -\lambda \Rightarrow y = -3 - \dfrac{\lambda}{2}$

$2(z - 2) = \lambda \Rightarrow z = 2 + \dfrac{\lambda}{2}$

$x - y + z = 3 \Rightarrow \left(1 + \dfrac{\lambda}{2}\right) - \left(-3 - \dfrac{\lambda}{2}\right) + \left(2 + \dfrac{\lambda}{2}\right) = 3$

$\Rightarrow \dfrac{3}{2}\lambda = -3 \Rightarrow \lambda = -2$

$x = 0, y = -2, z = 1$

Minimum distance $= \sqrt{1^2 + 1^2 + (-1)^2} = \sqrt{3}$

33. Minimize $f(x, y, z) = x + y + z$.

Constraint: $g(x, y, z) = xyz = 27$

$\left.\begin{array}{l} 1 = \lambda yz \Rightarrow x = \lambda xyz \\ 1 = \lambda xz \Rightarrow y = \lambda xyz \\ 1 = \lambda xy \Rightarrow z = \lambda xyz \end{array}\right\} \Rightarrow x = y = z$

$xyz = 27$

$x^3 = 27 \Rightarrow x = y = z = 3$

35. Minimize $f(x, y, z) = 0.06(2yz + 2xz) + 0.11(xy)$.

Constraint: $g(x, y, z) = xyz = 668.25$

$0.12z + 0.11y = yz\lambda$

$0.12z + 0.11x = xz\lambda$

$0.12(y + x) = xy\lambda$

$\qquad xyz = 668.25$

$0.12xz + 0.11yx = xyz\lambda = 0.12yz + 0.11xy \Rightarrow x = y$

$0.12(2x) = x^2\lambda \Rightarrow \lambda = \dfrac{0.24}{x}$

$0.12z + 0.11x = xz\left(\dfrac{0.24}{x}\right) = 0.24z \Rightarrow z = \dfrac{0.11x}{0.12}$

$\qquad\qquad\qquad\qquad\qquad\qquad\qquad = \dfrac{11x}{12}$

$xyz = x^2\left(\dfrac{11}{12}x\right) = 668.25 \Rightarrow x = y = 9, z = \dfrac{33}{4}$

$f\left(9, 9, \dfrac{33}{4}\right) = \$26.73$

37. Maximize $P(p, q, r) = 2pq + 2pr + 2qr$.

Constraint: $g(p, q, r) = p + q + r = 1$

$\left.\begin{array}{l} 2q + 2r = \lambda \\ 2p + 2r = \lambda \\ 2p + 2q = \lambda \end{array}\right\} p = q = r$

$p + q + r = 3p = 1 \Rightarrow p = \frac{1}{3}$ and

$P\left(\frac{1}{3}, \frac{1}{3}, \frac{1}{3}\right) = 3\left(\frac{2}{9}\right) = \frac{2}{3}$.

39. Maximize $V(x, y, z) = (2x)(2y)(2z) = 8xyz$ subject to

the constraint $\dfrac{x^2}{a^2} + \dfrac{y^2}{b^2} + \dfrac{z^2}{c^2} = 1$.

$\left.\begin{array}{l} 8yz = \dfrac{2x}{a^2}\lambda \\[2mm] 8xz = \dfrac{2y}{b^2}\lambda \\[2mm] 8xy = \dfrac{2z}{c^2}\lambda \end{array}\right\} \dfrac{x^2}{a^2} = \dfrac{y^2}{b^2} = \dfrac{z^2}{c^2}$

$\dfrac{x^2}{a^2} + \dfrac{y^2}{b^2} + \dfrac{z^2}{c^2} = 1 \Rightarrow \dfrac{3x^2}{a^2} = 1, \dfrac{3y^2}{b^2} = 1, \dfrac{3z^2}{c^2} = 1$

$x = \dfrac{a}{\sqrt{3}}, y = \dfrac{b}{\sqrt{3}}, z = \dfrac{c}{\sqrt{3}}$

So, the dimensions of the box are

$\dfrac{2\sqrt{3}a}{3} \times \dfrac{2\sqrt{3}b}{3} \times \dfrac{2\sqrt{3}c}{3}$.

41. $f(x, y) = x$

Constraint: $y^2 + x^4 - x^3 = 0$

$1 = \lambda(4x^3 - 3x^2)$

$0 = 2y \qquad\qquad\quad \Rightarrow y = 0$

$y^2 + x^4 - x^3 = 0 \quad \Rightarrow x = 0$ or $x = 1$

$x = 1$ is clearly not a minimum and $x = 0$ does not

satisfy the equation $1 = \lambda(4x^3 - 3x^2)$.

Also, note that $\nabla g(0, 0) = 0$.

43. Minimize $C(x, y, z) = 5xy + 3(2xz + 2yz + xy)$

subject to the constraint $xyz = 480$.

$\left.\begin{array}{l} 8y + 6z = yz\lambda \\ 8x + 6z = xz\lambda \\ 6x + 6y = xy\lambda \end{array}\right\} x = y, 4y = 3z$

$xyz = 480 \Rightarrow \frac{4}{3}y^3 = 480$

$\qquad\qquad\qquad x = y = \sqrt[3]{360}, z = \frac{4}{3}\sqrt[3]{360}$

Dimensions: $\sqrt[3]{360} \times \sqrt[3]{360} \times \frac{4}{3}\sqrt[3]{360}$ ft

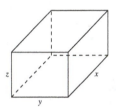

45. Minimize $A(\pi, r) = 2\pi rh + 2\pi r^2$ subject to the

constraint $\pi r^2 h = V_0$.

$\left.\begin{array}{l} 2\pi h + 4\pi r = 2\pi rh\lambda \\ 2\pi r = \pi r^2\lambda \end{array}\right\} h = 2r$

$\pi r^2 h = V_0 \Rightarrow 2\pi r^3 = V_0$

Dimensions: $r = \sqrt[3]{\dfrac{V_0}{2\pi}}$ and $h = 2\sqrt[3]{\dfrac{V_0}{2\pi}}$

47. Using the formula $\text{Time} = \dfrac{\text{Distance}}{\text{Rate}}$, minimize

$$T(x, y) = \frac{\sqrt{d_1^2 + x^2}}{v_1} + \frac{\sqrt{d_2^2 + y^2}}{v_2} \text{ subject to the}$$

constraint $x + y = a$.

$$\left.\begin{array}{r} \dfrac{x}{v_1\sqrt{d_1^2 + x^2}} = \lambda \\[3mm] \dfrac{y}{v_2\sqrt{d_2^2 + y^2}} = \lambda \end{array}\right\} \quad \dfrac{x}{v_1\sqrt{d_1^2 + x^2}} = \dfrac{y}{v_2\sqrt{d_2^2 + y^2}}$$

$$x + y = a$$

Because $\sin \theta_1 = \dfrac{x}{\sqrt{d_1^2 + x^2}}$

and $\sin \theta_2 = \dfrac{y}{\sqrt{d_2^2 + y^2}}$,

you have $\dfrac{x\big/\sqrt{d_1^2 + x^2}}{v_1} = \dfrac{y\big/\sqrt{d_2^2 + y^2}}{v_2}$ or

$$\frac{\sin \theta_1}{v_1} = \frac{\sin \theta_2}{v_2}.$$

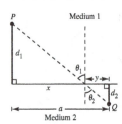

49. Maximize $P(x, y) = 100x^{0.25}y^{0.75}$ subject to the constraint $112x + 60y = 250,000$.

$$25x^{-0.75}y^{0.75} = 112\lambda \Rightarrow \left(\frac{y}{x}\right)^{0.75} = \frac{112}{25}\lambda$$

$$75x^{0.25}y^{-0.25} = 60\lambda \Rightarrow \left(\frac{x}{y}\right)^{0.25} = \frac{60}{75}\lambda$$

$$\left(\frac{y}{x}\right)^{0.75}\left(\frac{y}{x}\right)^{0.25} = \left(\frac{112}{25}\lambda\right)\left(\frac{75}{60\lambda}\right)$$

$$\frac{y}{x} = \frac{28}{5}$$

$$y = \frac{28}{5}x$$

$$112x + 60\left(\frac{28}{5}x\right) = 448x = 250,000 \Rightarrow x = \frac{15,625}{28}$$

and $y = 3125$

$$P\left(\frac{15,625}{28}, 3125\right) \approx 203,144$$

51. Minimize $C(x, y) = 72x + 80y$ subject to the constraint
$100x^{0.25}y^{0.75} = 50,000$.

$$72 = 25x^{-0.75}y^{0.75}\lambda \Rightarrow \left(\frac{y}{x}\right)^{0.75} = \frac{72}{25\lambda}$$

$$80 = 75x^{0.25}y^{-0.25}\lambda \Rightarrow \left(\frac{x}{y}\right)^{0.25} = \frac{80}{75\lambda}$$

$$\left(\frac{y}{x}\right)^{0.75}\left(\frac{y}{x}\right)^{0.25} = \left(\frac{72}{25\lambda}\right)\left(\frac{75\lambda}{80}\right)$$

$$\frac{y}{x} = \frac{27}{10}$$

$$y = \frac{27}{10}x = 2.7x$$

$$100x^{0.25}(2.7x)^{0.75} = 50,000$$

$$x = \frac{500}{2.7^{0.75}} \approx 237.38$$

$$y = 2.7x \approx 640.93$$

$C(237.38, 640.93) \approx \$68,366$ (Answers will vary.)

53. Let $r = $ radius of cylinder, and
$h = $ height of cylinder $= $ height of cone.

$$S = 2\pi rh + 2\pi r\sqrt{h^2 + r^2} = \text{constant surface area}$$

$$V = \pi r^2 h + \frac{2\pi r^2 h}{3} = \frac{5\pi r^2 h}{3} \text{ volume}$$

We maximize $f(r, h) = r^2 h$ subject to

$$g(r, h) = rh + r\sqrt{h^2 + r^2} = C.$$

$$(C - rh)^2 = r^2(h^2 + r^2)$$

$$C^2 - 2Crh = r^4$$

$$h = \frac{C^2 - r^4}{2Cr}$$

$$f(r, h) = F(r) = r^2\left[\frac{C^2 - r^4}{2Cr}\right] = \frac{Cr}{2} - \frac{r^5}{2C}$$

$$F'(r) = \frac{C}{2} - \frac{5r^4}{2C} = 0$$

$$C^2 = 5r^4$$

$$r^2 = \frac{C}{\sqrt{5}}$$

$$F''(r) = \frac{-10r^3}{C}$$

$$h = \frac{C^2 - r^4}{2Cr} = \frac{C^2 - C^2/5}{2C(C^2/5)^{1/4}} = \frac{(4/5)C}{2(C^2/5)^{1/4}}$$

$$= \frac{2C}{5r} = \frac{2}{5r}\left(\sqrt{5}r^2\right) = \frac{2\sqrt{5}}{5}r$$

So, $\dfrac{h}{r} = \dfrac{2\sqrt{5}}{5}$.

By the Second Derivative Test, this is a maximum.

Review Exercises for Chapter 13

1. $f(x, y) = x^2 y - 3$

   (a) $f(0, 4) = 0^2(4) - 3 = -3$

   (b) $f(2, -1) = 2^2(-1) - 3 = -7$

   (c) $f(-3, 2) = (-3)^2(2) - 3 = 15$

   (d) $f(x, 7) = x^2(7) - 3 = 7x^2 - 3$

3. $f(x, y) = \dfrac{\sqrt{x}}{y}$

The domain is $\{(x, y) : x \geq 0, y \neq 0\}$.

The range is all real numbers.

5. $f(x, y) = -2$

Plane parallel to xy-plane

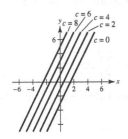

7. $z = 3 - 2x + y$

The level curves are parallel lines of the form
$y = 2x - 3 + c$.

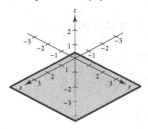

9. $f(x, y) = x^2 + y^2$

   (a)

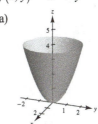

   (b) $g(x, y) = f(x, y) + 2$ is a vertical translation of f two units upward.

   (c) $g(x, y) = f(x, y - z)$ is a horizontal translation of f two units to the right. The vertex moves from $(0, 0, 0)$ to $(0, 2, 0)$.

   (d)

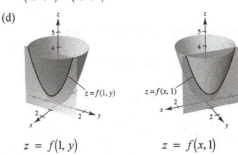

$$z = f(1, y) \qquad\qquad z = f(x, 1)$$

11. $f(x, y, z) = x^2 - y + z^2 = 2$

$$y = x^2 + z^2 - 2$$

Elliptic paraboloid

13. $\displaystyle\lim_{(x, y) \to (1, 1)} \frac{xy}{x^2 + y^2} = \frac{1}{2}$

Continuous except at $(0, 0)$.

15. $\displaystyle\lim_{(x, y) \to (0, 0)} \frac{y + xe^{-y^2}}{1 + x^2} = \frac{0 + 0}{1 + 0} = 0$

Continuous everywhere.

17. $\displaystyle\lim_{(x, y, z) \to (-3, 1, 2)} \frac{\ln z}{xy - z} = \frac{\ln 2}{(-3)(1) - 2} = \frac{\ln 2}{-5}$

Continuous for $x \neq \dfrac{z}{y}$

19. $f(x, y) = 5x^3 + 7y - 3$

$$\frac{\partial f}{\partial x} = 15x^2 \qquad \frac{\partial f}{\partial y} = 7$$

21. $f(x, y) = e^x \cos y$

$$f_x = e^x \cos y$$
$$f_y = -e^x \sin y$$

23. $f(x, y) = y^3 e^{y/x}$

$$f_x = y^3 e^{y/x} \left(\frac{-y}{x^2} \right) = \frac{-y^4 e^{y/x}}{x^2}$$

$$f_y = y^3 e^{y/x} \left(\frac{1}{x} \right) + 3y^2 e^{y/x}$$

$$= y^2 e^{y/x} \left(\frac{y}{x} + 3 \right)$$

25. $f(x, y, z) = 2xz^2 + 6xyz - 5xy^3$

$$\frac{\partial f}{\partial x} = 2z^2 + 6yz - 5y^3$$

$$\frac{\partial f}{\partial y} = 6xz - 15xy^2$$

$$\frac{\partial f}{\partial z} = 4xz + 6xy$$

27. $f(x, y) = x^2 - y, \ (0, 2)$

$$f_x = 2x \quad f_x(0, 2) = 0$$
$$f_y = -1 \quad f_y(0, 2) = -1$$

29. $f(x, y, z) = xy \cos xz, \ \left(2, 3, -\frac{\pi}{3} \right)$

$$f_x = y \cos xz - xyz \sin xz$$

$$f_x\left(2, 3, -\frac{\pi}{3} \right) = 3 \cos\left(-\frac{2\pi}{3} \right) - (2)(3)\left(-\frac{\pi}{3} \right) \sin\left(-\frac{2\pi}{3} \right)$$

$$= 3\left(-\frac{1}{2} \right) + 2\pi\left(\frac{-\sqrt{3}}{2} \right) = \frac{-3}{2} - \sqrt{3}\pi$$

$$f_y = x \cos xz$$

$$f_y\left(2, 3, -\frac{\pi}{3} \right) = 2 \cos\left(-\frac{2\pi}{3} \right) = -1$$

$$f_z = -x^2 y \sin xz$$

$$f_z\left(2, 3, -\frac{\pi}{3} \right) = -4(3) \sin\left(-\frac{2\pi}{3} \right) = 6\sqrt{3}$$

31. $f(x, y) = 3x^2 - xy + 2y^3$

$$f_x = 6x - y$$
$$f_y = -x + 6y^2$$
$$f_{xx} = 6$$
$$f_{yy} = 12y$$
$$f_{xy} = -1$$
$$f_{yx} = -1$$

33. $h(x, y) = x \sin y + y \cos x$

$$h_x = \sin y - y \sin x$$
$$h_y = x \cos y + \cos x$$
$$h_{xx} = -y \cos x$$
$$h_{yy} = -x \sin y$$
$$h_{xy} = \cos y - \sin x$$
$$h_{yx} = \cos y - \sin x$$

35. $z = x^2 \ln(y + 1)$

$$\frac{\partial z}{\partial x} = 2x \ln(y + 1). \ \text{At } (2, 0, 0), \frac{\partial z}{\partial x} = 0.$$

Slope in x-direction.

$$\frac{\partial z}{\partial y} = \frac{x^2}{1 + y}. \ \text{At } (2, 0, 0), \frac{\partial z}{\partial y} = 4.$$

Slope in y-direction.

37. $z = x \sin xy$

$$dz = \frac{\partial z}{\partial x} dx + \frac{\partial z}{\partial y} dy$$

$$= (xy \cos xy + \sin xy)dx + (x^2 \cos xy)dy$$

39. $w = 3xy^2 - 2x^3 yz^2$

$$dw = \frac{\partial w}{\partial x} dx + \frac{\partial w}{\partial y} dy + \frac{\partial w}{\partial z} dz$$

$$= (3y^2 - 6x^2 yz^2)dx + (6xy - 2x^3 z^2)dy - 4x^3 yz \ dz$$

41. $f(x, y) = 4x + 2y$

(a) $f(2, 1) = 4(2) + 2(1) = 10$

$\qquad f(2.1, 1.05) = 4(2.1) + 2(1.05) = 10.5$

$\qquad \Delta z = 10.5 - 10 = 0.5$

(b) $dz = 4dx + 2dy$

$\qquad = 4(0.1) + 2(0.05) = 0.5$

43. $V = \dfrac{1}{3}\pi r^2 h$

$$dV = \dfrac{2}{3}\pi r h\, dr + \dfrac{1}{3}\pi r^2\, dh$$

$$= \dfrac{2}{3}\pi(2)(5)\left(\pm\dfrac{1}{8}\right) + \dfrac{1}{3}\pi(2)^2\left(\pm\dfrac{1}{8}\right)$$

$$= \pm\dfrac{5}{6}\pi + \dfrac{1}{6}\pi = \pm\pi \text{ in.}^3 \qquad \text{Propogated error}$$

$$V = \dfrac{1}{3}\pi\,(2)^2 5 = \dfrac{20}{3}\pi \text{ in.}^3$$

$$\text{Relative error} = \dfrac{dV}{V} = \dfrac{\pm\pi}{\left(\dfrac{20}{3}\pi\right)} = \dfrac{3}{20} = 15\%$$

45. $z = f(x, y) = 6x - y^2$

$$\Delta z = f(x + \Delta x, y + \Delta y) - f(x, y)$$

$$= 6(x + \Delta x) - (y + \Delta y)^2 - (6x - y^2)$$

$$= 6x + 6\Delta x - y^2 - 2y\Delta y - (\Delta y)^2 - 6x + y^2$$

$$= 6(\Delta x) - 2y(\Delta y) - \Delta y(\Delta y)$$

$$= f_x(x, y)\Delta x - f_y(x, y)\Delta y + E_1\Delta x + E_2\Delta y,$$

where $E_1 = 0$ and $E_2 = -\Delta y$

as $(\Delta x, \Delta y) \to (0, 0)$, $E_1 = 0$ and $E_2 \to 0$.

51. $w = \dfrac{xy}{z},\ x = 2r + t,\ y = rt,\ z = 2r - t$

(a) Chain Rule:

$$\dfrac{\partial w}{\partial r} = \dfrac{\partial w}{\partial x}\dfrac{\partial x}{\partial r} + \dfrac{\partial w}{\partial y}\dfrac{\partial y}{\partial r} + \dfrac{\partial w}{\partial z}\dfrac{\partial z}{\partial r} = \dfrac{y}{z}(2) + \dfrac{x}{z}(t) - \dfrac{xy}{z^2}(2) = \dfrac{2rt}{2r - t} + \dfrac{(2r + t)t}{2r - t} - \dfrac{2(2r + t)(rt)}{(2r - t)^2} = \dfrac{4r^2 t - 4rt^2 - t^3}{(2r - t)^2}$$

$$\dfrac{\partial w}{\partial t} = \dfrac{\partial w}{\partial x}\dfrac{\partial x}{\partial t} + \dfrac{\partial w}{\partial y}\dfrac{\partial y}{\partial t} + \dfrac{\partial w}{\partial z}\dfrac{\partial z}{\partial t} = \dfrac{y}{z}(1) + \dfrac{x}{z}(r) = \dfrac{xy}{z^2}(-1) = \dfrac{4r^2 t - rt^2 + 4r^3}{(2r - t)^2}$$

(b) Substitution: $w = \dfrac{xy}{z} = \dfrac{(2r + t)(rt)}{2r - t} = \dfrac{2r^2 t + rt^2}{2r - t}$

$$\dfrac{\partial w}{\partial r} = \dfrac{4r^2 t - 4rt^2 - t^3}{(2r - t)^2}$$

$$\dfrac{\partial w}{\partial t} = \dfrac{4r^2 t - rt^2 + 4r^3}{(2r - t)^2}$$

53. $F(x, y) = x^3 - xy + 5y$

$$F_x(x, y) = 3x^2 - y$$

$$F_y(x, y) = -x + 5$$

$$\dfrac{dy}{dx} = \dfrac{-F_x(x, y)}{F_y(x, y)} = -\dfrac{3x^2 - y}{-x + 5} = \dfrac{y - 3x^2}{5 - x}$$

47. $w = \ln(x^2 + y),\ x = 2t,\ y = 4 - t$

(a) Chain Rule: $\dfrac{dw}{dt} = \dfrac{\partial w}{\partial x}\dfrac{dx}{dt} + \dfrac{\partial w}{\partial y}\dfrac{dy}{dt}$

$$= \dfrac{2x}{x^2 + y}(2) + \dfrac{1}{x^2 + y}(-1)$$

$$= \dfrac{8t - 1}{4t^2 + 4 - t}$$

(b) Substitution: $w = \ln(x^2 + y) = \ln(4t^2 + 4 - t)$

$$\dfrac{dw}{dt} = \dfrac{1}{4t^2 + 4 - t}(8t - 1)$$

49. $w = x^2 z + y + z,\ \lambda = e^t,\ y = t,\ z = t^2$

(a) Chain Rule: $\dfrac{dw}{dt} = \dfrac{\partial w}{\partial x}\dfrac{dx}{dt} + \dfrac{\partial w}{\partial y}\dfrac{dy}{dt} + \dfrac{\partial w}{\partial z}\dfrac{dz}{dt}$

$$= 2xze^t + 1(1) + (x^2 + 1)(2t)$$

$$= 2e^t t^2 e^t + 1 + (e^{2t} + 1)(2t)$$

$$= 2t^2 e^{2t} + 2te^{2t} + 2t + 1$$

(b) Substitution: $w = x^2 z + y + z = e^{2t}(t^2) + t + t^2$

$$\dfrac{dw}{dt} = 2e^{2t}(t^2) + 2te^{2t} + 1 + 2t$$

55. $x^2 + xy + y^2 + yz + z^2 = 0$

$$2x + y + y\dfrac{\partial z}{\partial x} + 2z\dfrac{\partial z}{\partial x} = 0$$

$$\dfrac{\partial z}{\partial x} = \dfrac{-2x - y}{y + 2z}$$

$$x + 2y + y\dfrac{\partial z}{\partial y} + z + 2z\dfrac{\partial z}{\partial y} = 0$$

$$\dfrac{\partial z}{\partial y} = \dfrac{-x - 2y - z}{y + 2z}$$

57. $f(x, y) = x^2y$, $P(-5, 5)$, $\mathbf{v} = 3\mathbf{i} - 4\mathbf{j}$

$$\mathbf{u} = \frac{\mathbf{v}}{\|\mathbf{v}\|} = \frac{3}{5}\mathbf{i} - \frac{4}{5}\mathbf{j}$$

$$D_{\mathbf{u}}f(x, y) = \frac{\partial f}{\partial x}\cos\theta + \frac{\partial f}{\partial y}\sin\theta$$

$$= 2xy\cos\theta + x^2\sin\theta$$

$$D_{\mathbf{u}}f(-5, 5) = 2(-5)(5)\left(\frac{3}{5}\right) + (-5)^2\left(-\frac{4}{5}\right)$$

$$= -30 - 20 = -50$$

59.
$$w = y^2 + xz$$
$$\nabla w = z\mathbf{i} + 2y\mathbf{j} + x\mathbf{k}$$
$$\nabla w(1, 2, 2) = 2\mathbf{i} + 4\mathbf{j} + \mathbf{k}$$
$$\mathbf{u} = \tfrac{1}{3}\mathbf{v} = \tfrac{2}{3}\mathbf{i} - \tfrac{1}{3}\mathbf{j} + \tfrac{2}{3}\mathbf{k}$$
$$D_{\mathbf{u}}w(1, 2, 2) = \nabla w(1, 2, 2) \cdot \mathbf{u} = \tfrac{4}{3} - \tfrac{4}{3} + \tfrac{2}{3} = \tfrac{2}{3}$$

61.
$$z = x^2y$$
$$\nabla z = 2xy\mathbf{i} + x^2\mathbf{j}$$
$$\nabla_z(2, 1) = 4\mathbf{i} + 4\mathbf{j}$$
$$\|\nabla z(2, 1)\| = 4\sqrt{2}$$

63.
$$z = \frac{y}{x^2 + y^2}$$
$$\nabla z = -\frac{2xy}{\left(x^2 + y^2\right)^2}\mathbf{i} + \frac{x^2 - y^2}{\left(x^2 + y^2\right)^2}\mathbf{j}$$
$$\nabla z(1, 1) = -\frac{1}{2}\mathbf{i} = \left\langle -\frac{1}{2}, 0 \right\rangle$$
$$\|\nabla z(1, 1)\| = \frac{1}{2}$$

65. $w = x^4y - y^2z^2$
$$\nabla w = 4x^3y\mathbf{i} + \left(x^4 - 2yz^2\right)\mathbf{j} - 2y^2z\mathbf{k}$$
$$\nabla w\left(-1, \tfrac{1}{2}, 2\right) = -2\mathbf{i} - 3\mathbf{j} - \mathbf{k} = \langle -2, -3, -1 \rangle$$
$$\left\|\nabla w\left(-1, \tfrac{1}{2}, 2\right)\right\| = \sqrt{4 + 9 + 1} = \sqrt{14}$$

67. $f(x, y) = 9x^2 - 4y^2$, $c = 65$, $P(3, 2)$

(a) $\nabla f(x, y) = 18x\mathbf{i} - 8y\mathbf{j}$

$\nabla f(3, 2) = 54\mathbf{i} - 16\mathbf{j}$

(b) Unit normal: $\dfrac{54\mathbf{i} - 16\mathbf{j}}{\|54\mathbf{i} - 16\mathbf{j}\|} = \dfrac{1}{\sqrt{793}}(27\mathbf{i} - 8\mathbf{j})$

(c) Slope $= \dfrac{27}{8}$.

$$y - z = \frac{27}{8}(x - 3)$$

$$y = \frac{27}{8}x - \frac{65}{8} \text{ Tangent line}$$

(d)

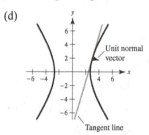

69. $F(x, y, z) = x^2 + y^2 + 2 - z = 0$, $(1, 3, 12)$

$$\nabla F = 2x\mathbf{i} + 2y\mathbf{j} - \mathbf{k}$$
$$\nabla F(1, 3, 12) = 2\mathbf{i} + 6\mathbf{j} - \mathbf{k}$$

Tangent Plane:

$$2(x - 1) + 6(y - 3) - (z - 12) = 0$$
$$2x + 6y - z = 8$$

71. $F(x, y, z) = x^2 + y^2 - 4x + 6y + z + 9 = 0$
$$\nabla F = (2x - 4)\mathbf{i} + (2y + 6)\mathbf{j} + \mathbf{k}$$
$$\nabla F(2, -3, 4) = \mathbf{k}$$

So, the equation of the tangent plane is

$$z - 4 = 0 \text{ or } z = 4.$$

73. $F(x, y, z) = x^2y - z = 0$
$$\nabla F = 2xy\mathbf{i} + x^2\mathbf{j} - \mathbf{k}$$
$$\nabla F(2, 1, 4) = 4\mathbf{i} + 4\mathbf{j} - \mathbf{k}$$

So, the equation of the tangent plane is

$$4(x - 2) + 4(y - 1) - (z - 4) = 0 \text{ or}$$

$$4x + 4y - z = 8,$$

and the equation of the normal line is

$$x = 4t + 2, y = 4t + 1, z = -t + 4.$$

Symmetric equations:

$$\frac{x - 2}{4} = \frac{y - 1}{4} = -\frac{z - 4}{1}$$

75. $f(x, y, z) = x^2 + y^2 + z^2 - 14$

$\nabla f(x, y, z) = 2x\mathbf{i} + 2y\mathbf{j} + 2z\mathbf{k}$

$\nabla f(2, 1, 3) = 4\mathbf{i} + 2\mathbf{j} + 6\mathbf{k},$ normal vector to plane

$\cos\theta = \dfrac{|\mathbf{n}\cdot\mathbf{k}|}{\|\mathbf{n}\|} = \dfrac{6}{\sqrt{56}} = \dfrac{3\sqrt{14}}{14}$

$\theta = 36.7°$

77. $F(x, y, z) = 9 - 2x^2 + y^3 - z$

$\nabla F(x, y, z) = -4x\mathbf{i} + 3y^2\mathbf{j} - \mathbf{k}$

$-4x = 0 \Rightarrow x = 0$

$3y^2 = 0 \Rightarrow y = 0$

$z = 9$

$(0, 0, 9)$

79. $f(x, y) = -x^2 - 4y^2 + 8x - 8y - 11$

$f_x = -2x + 8 = 0 \Rightarrow x = 4$

$f_y = -8y - 8 = 0 \Rightarrow y = -1$

$f_{xx} = -2,\ f_{yy} = -8,\ f_{xy} = 0$

$f_{xx}\,f_{yy} - (f_{xy})^2 = (-2)(-8) - 0 = 16 > 0$

So, $(4, -1, 9)$ is a relative maximum.

81. $f(x, y) = 2x^2 + 6xy + 9y^2 + 8x + 14$

$f_x = 4x + 6y + 8 = 0$

$f_y = 6x + 18y = 0,\ x = -3y$

$4(-3y) + 6y = -8 \Rightarrow y = \frac{4}{3}, x = -4$

$f_{xx} = 4$

$f_{yy} = 18$

$f_{xy} = 6$

$f_{xx}\,f_{yy} - (f_{xy})^2 = 4(18) - (6)^2 = 36 > 0.$

So, $\left(-4, \frac{4}{3}, -2\right)$ is a relative minimum.

83. $f(x, y) = xy + \dfrac{1}{x} + \dfrac{1}{y}$

$f_x = y - \dfrac{1}{x^2} = 0,\ x^2y = 1$

$f_y = x - \dfrac{1}{y^2} = 0,\ xy^2 = 1$

So, $x^2y = xy^2$ or $x = y$ and substitution yields the critical point $(1, 1)$.

$f_{xx} = \dfrac{2}{x^3}$

$f_{xy} = 1$

$f_{yy} = \dfrac{2}{y^3}$

At the critical point $(1, 1),\ f_{xx} = 2 > 0$ and

$f_{xx}\,f_{yy} - (f_{xy})^2 = 3 > 0.$

So, $(1, 1, 3)$ is a relative minimum.

85. A point on the plane is given by $(x, y, 4 - x - y)$

The square of the distance from $(2, 1, 4)$ to a point on the plane is

$S = (x - 2)^2 + (y - 1)^2 + (4 - x - y - 4)^2$

$\quad = (x - 2)^2 + (y - 1)^2 + (-x - y)^2.$

$S_x = 2(x - 2) - 2(-x - y) = 4x + 2y - 4$

$S_y = 2(y - 1) - 2(-x - y) = 2x + 4y - 2$

$S_x = S_y = 0 \Rightarrow \begin{cases} 4x + 2y = 4 \\ 2x + 4y = 2 \end{cases} \Rightarrow x = 1, y = 0, z = 3$

The distance is $\sqrt{(1-2)^2 + (0-1)^2 + (-1)^2} = \sqrt{3}.$

87. $R = -6x_1^2 - 10x_2^2 - 2x_1x_2 + 32x_1 + 84x_2$

$Rx_1 = -12x_1 - 2x_2 + 32 = 0 \Rightarrow 6x_1 + x_2 = 16$

$Rx_2 = -20x_2 - 2x_1 + 84 = 0 \Rightarrow x_1 + 10x_2 = 42$

Solving this system yields $x_1 = 2$ and $x_2 = 4$.

89. $(0, 4), (1, 5), (3, 6), (6, 8), (8, 10)$

$\sum x_i = 18 \qquad \sum y_i = 33$

$\sum x_iy_i = 151 \quad \sum x_i^2 = 110$

$a = \dfrac{5(151) - 18(33)}{5(110) - (18)^2} = \dfrac{161}{226} \approx 0.7124$

$b = \dfrac{1}{5}\left(33 - \dfrac{161}{226}(18)\right) = \dfrac{456}{113} \approx 4.0354$

$y = \dfrac{161}{226}x + \dfrac{456}{113}$

91. $(100, 35), (150, 44), (200, 50), (250, 56)$

    (a) Using a graphing utility, you obtain

        $y = 0.138x + 22.1.$

    (b) If $x = 175, y = 0.138(175) + 22.1 = 46.25$

        bushels per acre.

93. Minimize $f(x, y) = x^2 + y^2$

    Constraint: $x + y - 8 = 0$

    $\nabla f = \lambda \nabla g$

    $2x\mathbf{i} + 2y\mathbf{j} = \lambda(\mathbf{i} + \mathbf{j})$

    $\left.\begin{array}{l} 2x = \lambda \\ 2y = \lambda \end{array}\right\}$ $x = y$

    $x + y - 8 = 2x - 8 = 0 \Rightarrow x = y = 4$

    $f(4, 4) = 32$

95. Maximize $f(x, y) = 2x + 3xy + y$

    Constraint: $x + 2y = 29$

    $\nabla f = \lambda \nabla g$

    $\left.\begin{array}{l} 2 + 3y = \lambda \\ 3x + 1 = 2\lambda \end{array}\right\}$ $4 + 6y = 3x + 1 \Rightarrow x - 2y = 1$

    $\left.\begin{array}{l} x - 2y = 1 \\ x + 2y = 29 \end{array}\right\}$ $x = 15, y = 7$

    $f(15, 7) = 2(15) + 3(15)(7) + 7 = 352$

97. Maximize $f(x, y) = 2xy$

    Constraint: $2x + y = 12$

    $\nabla f = \lambda \nabla g$

    $\left.\begin{array}{l} 2y = 2\lambda \\ 2x = \lambda \end{array}\right\}$ $4x = 2y \Rightarrow y = 2x$

    $2x + y = 2x + 2x = 12 \Rightarrow x = 3, y = 6$

    $f(3, 6) = 2(3)(6) = 36$

99. $PQ = \sqrt{x^2 + 4},$

    $QR = \sqrt{y^2 + 1},$

    $RS = z; x + y + z = 10$

    $C = 3\sqrt{x^2 + 4} + 2\sqrt{y^2 + 1} + z$

    Constraint: $x + y + z = 10$

    $\nabla C = \lambda \nabla g$

    $\dfrac{3x}{\sqrt{x^2 + 4}}\mathbf{i} + \dfrac{2y}{\sqrt{y^2 + 1}}\mathbf{j} + \mathbf{k} = \lambda[\mathbf{i} + \mathbf{j} + \mathbf{k}]$

    $3x = \lambda\sqrt{x^2 + 4}$

    $2y = \lambda\sqrt{y^2 + 1}$

    $1 = \lambda$

    $9x^2 = x^2 + 4 \Rightarrow x^2 = \dfrac{1}{2}$

    $4y^2 = y^2 + 1 \Rightarrow y^2 = \dfrac{1}{3}$

    So, $x = \dfrac{\sqrt{2}}{2} \approx 0.707$ km,

    $y = \dfrac{\sqrt{3}}{3} \approx 0.577$ km,

    $z = 10 - \dfrac{\sqrt{2}}{2} - \dfrac{\sqrt{3}}{3} \approx 8.716$ km.

Problem Solving for Chapter 13

1. (a) The three sides have lengths 5, 6, and 5.

    Thus, $s = \frac{16}{2} = 8$ and $A = \sqrt{8(3)(2)(3)} = 12.$

    (b) Let $f(a, b, c) = (\text{area})^2 = s(s - a)(s - b)(s - c),$ subject to the constraint

      $a + b + c = $ constant (perimeter).

      Using Lagrange multipliers,

      $-s(s - b)(s - c) = \lambda$

      $-s(s - a)(s - c) = \lambda$

      $-s(s - a)(s - b) = \lambda.$

      From the first 2 equations

      $s - b = s - a \Rightarrow a = b.$

      Similarly, $b = c$ and hence $a = b = c$ which is an equilateral triangle.

(c) Let $f(a, b, c) = a + b + c$, subject to $(\text{Area})^2 = s(s - a)(s - b)(s - c)$ constant.

Using Lagrange multipliers,

$1 = -\lambda s(s - b)(s - c)$

$1 = -\lambda s(s - a)(s - c)$

$1 = -\lambda s(s - a)(s - b)$

So, $s - a = s - b \Rightarrow a = b$ and $a = b = c$.

3. (a) $F(x, y, z) = xyz - 1 = 0$

$F_x = yz, F_y = xz, F_z = xy$

Tangent plane: $y_0z_0(x - x_0) + x_0z_0(y - y_0) + x_0y_0(z - z_0) = 0$

$y_0z_0x + x_0z_0y + x_0y_0z = 3x_0y_0z_0 = 3$

(b) $V = \dfrac{1}{3}(\text{base})(\text{height})$

$= \dfrac{1}{3}\left(\dfrac{1}{2}\dfrac{3}{y_0z_0}\dfrac{3}{x_0z_0}\right)\left(\dfrac{3}{x_0y_0}\right)$

$= \dfrac{9}{2}$

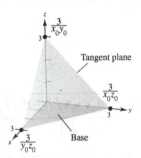

5. (a)

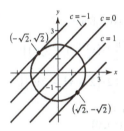

Maximum value of f is $f\left(\sqrt{2}, -\sqrt{2}\right) = 2\sqrt{2}$.

Maximize $f(x, y) = x - y$.

Constraint: $g(x, y) = x^2 + y^2 = 4$

$\nabla f = \lambda \nabla g$: $1 = 2\lambda x$

$-1 = 2\lambda y$

$x^2 + y^2 = 4$

$2\lambda x = -2\lambda y \Rightarrow x = -y$

$2x^2 = 4 \Rightarrow x = \pm\sqrt{2}, y = \mp\sqrt{2}$

$f\left(\sqrt{2}, -\sqrt{2}\right) = 2\sqrt{2}, f\left(-\sqrt{2}, \sqrt{2}\right) = -2\sqrt{2}$

(b) $f(x, y) = x - y$

Constraint: $x^2 + y^2 = 0 \Rightarrow (x, y) = (0, 0)$

Maximum and minimum values are 0.

Lagrange multipliers does not work:

$\left.\begin{array}{r} 1 = 2\lambda x \\ -1 = 2\lambda y \end{array}\right\} x = -y = 0$, a contradiction.

Note that $\nabla g(0, 0) = \mathbf{0}$.

7. $H = k(5xy + 6xz + 6yz)$

$z = \dfrac{1000}{xy} \Rightarrow H = k\left(5xy + \dfrac{6000}{y} + \dfrac{6000}{x}\right)$.

$H_x = 5y - \dfrac{6000}{x^2} = 0 \Rightarrow 5yx^2 = 6000$

By symmetry, $x = y \Rightarrow x^3 = y^3 = 1200$.

So, $x = y = 2\sqrt[3]{150}$ and $z = \dfrac{5}{3}\sqrt[3]{150}$.

9. (a) $\dfrac{\partial f}{\partial x} = Cax^{a-1}y^{1-a}, \dfrac{\partial f}{\partial y} = C(1 - a)x^a y^{-a}$

$x\dfrac{\partial f}{\partial x} + y\dfrac{\partial f}{\partial y} = Cax^a y^{1-a} + C(1 - a)x^a y^{1-a} = \left[Ca + C(1 - a)\right]x^a y^{1-a} = Cx^a y^{1-a} = f$

(b) $f(tx, ty) = C(tx)^a (ty)^{1-a} = Ct^a x^a t^{1-a} y^{1-a} = Cx^a y^{1-a}(t) = tf(x, y)$

11. (a) $x = 64(\cos 45°)t = 32\sqrt{2}\,t$

$y = 64(\sin 45°)t - 16t^2 = 32\sqrt{2}\,t - 16t^2$

(b) $\tan \alpha = \dfrac{y}{x + 50}$

$\alpha = \arctan\left(\dfrac{y}{x + 50}\right) = \arctan\left(\dfrac{32\sqrt{2}\,t - 16t^2}{32\sqrt{2}\,t + 50}\right)$

(c) $\dfrac{d\alpha}{dt} = \dfrac{1}{1 + \left(\dfrac{32\sqrt{2}\,t - 16t^2}{32\sqrt{2}\,t + 50}\right)^2} \cdot \dfrac{-64\left(8\sqrt{2}\,t^2 + 25t - 25\sqrt{2}\right)}{\left(32\sqrt{2}\,t + 50\right)^2} = \dfrac{-16\left(8\sqrt{2}\,t^2 + 25t - 25\sqrt{2}\right)}{64t^4 - 256\sqrt{2}\,t^3 + 1024t^2 + 800\sqrt{2}\,t + 625}$

(d)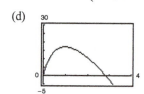

No. The rate of change of α is greatest when the projectile is closest to the camera.

(e) $\dfrac{d\alpha}{dt} = 0$ when $8\sqrt{2}\,t^2 + 25t - 25\sqrt{2} = 0$

$$t = \dfrac{-25 + \sqrt{25^2 - 4\left(8\sqrt{2}\right)\left(-25\sqrt{2}\right)}}{2\left(8\sqrt{2}\right)} \approx 0.98 \text{ second.}$$

No, the projectile is at its maximum height when $dy/dt = 32\sqrt{2} - 32t = 0$ or $t = \sqrt{2} \approx 1.41$ seconds.

13. (a) There is a minimum at $(0, 0, 0)$, maxima at $(0, \pm 1, 2/e)$ and saddle point at $(\pm 1, 0, 1/e)$:

$f_x = \left(x^2 + 2y^2\right)e^{-\left(x^2+y^2\right)}(-2x) + (2x)e^{-\left(x^2+y^2\right)}$

$\quad = e^{-\left(x^2+y^2\right)}\left[\left(x^2 + 2y^2\right)(-2x) + 2x\right]$

$\quad = e^{-\left(x^2+y^2\right)}\left[-2x^3 + 4xy^2 + 2x\right]$

$\quad = 0 \Rightarrow x^3 + 2xy^2 - x = 0$

$f_y = \left(x^2 + 2y^2\right)e^{-\left(x^2+y^2\right)}(-2y) + (4y)e^{-\left(x^2+y^2\right)}$

$\quad = e^{-\left(x^2+y^2\right)}\left[\left(x^2 + 2y^2\right)(-2y) + 4y\right]$

$\quad = e^{-\left(x^2+y^2\right)}\left[-4y^3 - 2x^2y + 4y\right]$

$\quad = 0 \Rightarrow 2y^3 + x^2y - 2y = 0$

Solving the two equations $x^3 + 2xy^2 - x = 0$ and $2y^3 + x^2y - 2y = 0$, you obtain the following critical points: $(0, \pm 1), (\pm 1, 0), (0, 0)$. Using the second derivative test, you obtain the results above.

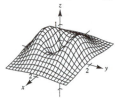

(b) As in part (a), you obtain

$f_x = e^{-\left(x^2+y^2\right)}\left[2x\left(x^2 - 1 - 2y^2\right)\right]$

$f_y = e^{-\left(x^2+y^2\right)}\left[2y\left(2 + x^2 - 2y^2\right)\right].$

The critical numbers are $(0, 0), (0, \pm 1), (\pm 1, 0)$.

These yield

$(\pm 1, 0, -1/e)$ minima

$(0, \pm 1, 2/e)$ maxima

$(0, 0, 0)$ saddle.

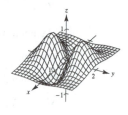

(c) In general, for $\alpha > 0$ you obtain

$(0, 0, 0)$ minimum

$(0, \pm 1, \beta/e)$ maxima

$(\pm 1, 0, \alpha/e)$ saddle.

For $\alpha < 0$, you obtain

$(\pm 1, 0, \alpha/e)$ minima

$(0, \pm 1, \beta/e)$ maxima

$(0, 0, 0)$ saddle.

15. (a)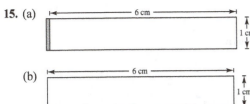

(b)

(c) The height has more effect since the shaded region in (b) is larger than the shaded region in (a).

(d) $A = hl \Rightarrow dA = l\,dh + h\,dl$

If $dl = 0.01$ and $dh = 0$, then $dA = 1(0.01) = 0.01$.

If $dh = 0.01$ and $dl = 0$, then $dA = 6(0.01) = 0.06$.

17. $\dfrac{\partial u}{\partial t} = \dfrac{1}{2}\Big[-\cos(x-t) + \cos(x+t)\Big]$

$\dfrac{\partial^2 u}{\partial t^2} = \dfrac{1}{2}\Big[-\sin(x-t) - \sin(x+t)\Big]$

$\dfrac{\partial u}{\partial x} = \dfrac{1}{2}\Big[\cos(x-t) + \cos(x+t)\Big]$

$\dfrac{\partial^2 u}{\partial x^2} = \dfrac{1}{2}\Big[-\sin(x-t) - \sin(x+t)\Big]$

Then, $\dfrac{\partial^2 u}{\partial t^2} = \dfrac{\partial^2 u}{\partial x^2}$.

19. $w = f(x, y),\ x = r\cos\theta,\ y = r\sin\theta$

$\dfrac{\partial w}{\partial r} = \dfrac{\partial w}{\partial x}\cos\theta + \dfrac{\partial w}{\partial y}\sin\theta$

$\dfrac{\partial w}{\partial \theta} = \dfrac{\partial w}{\partial x}(-r\sin\theta) + \dfrac{\partial w}{\partial y}(r\cos\theta)$

(a) $r\cos\theta\,\dfrac{\partial w}{\partial r} = \dfrac{\partial w}{\partial x}r\cos^2\theta + \dfrac{\partial w}{\partial y}r\sin\theta\cos\theta$

$-\sin\theta\,\dfrac{\partial w}{\partial \theta} = \dfrac{\partial w}{\partial x}\big(r\sin^2\theta\big) - \dfrac{\partial w}{\partial y}r\sin\theta\cos\theta$

$r\cos\theta\,\dfrac{\partial w}{\partial r} - \sin\theta\,\dfrac{\partial w}{\partial \theta} = \dfrac{\partial w}{\partial x}\big(r\cos^2\theta + r\sin^2\theta\big)$

$r\dfrac{\partial w}{\partial x} = \dfrac{\partial w}{\partial r}(r\cos\theta) - \dfrac{\partial w}{\partial \theta}\sin\theta$

$\dfrac{\partial w}{\partial x} = \dfrac{\partial w}{\partial r}\cos\theta - \dfrac{\partial w}{\partial \theta}\dfrac{\sin\theta}{r}$ (First Formula)

$r\sin\theta\,\dfrac{\partial w}{\partial r} = \dfrac{\partial w}{\partial x}r\sin\theta\cos\theta + \dfrac{\partial w}{\partial y}r\sin^2\theta$

$\cos\theta\,\dfrac{\partial w}{\partial \theta} = \dfrac{\partial w}{\partial x}(-r\sin\theta\cos\theta) + \dfrac{\partial w}{\partial y}\big(r\cos^2\theta\big)$

$r\sin\theta\dfrac{\partial w}{\partial r} + \cos\theta\dfrac{\partial w}{\partial \theta} = \dfrac{\partial w}{\partial y}\big(r\sin^2\theta + r\cos^2\theta\big)$

$r\dfrac{\partial w}{\partial y} = \dfrac{\partial w}{\partial r}r\sin\theta + \dfrac{\partial w}{\partial \theta}\cos\theta$

$\dfrac{\partial w}{\partial y} = \dfrac{\partial w}{\partial r}\sin\theta + \dfrac{\partial w}{\partial \theta}\dfrac{\cos\theta}{r}$ (Second Formula)

(b) $\left(\dfrac{\partial w}{\partial r}\right)^2 + \dfrac{1}{r^2}\left(\dfrac{\partial w}{\partial \theta}\right)^2 = \left(\dfrac{\partial w}{\partial x}\right)^2\cos^2\theta + 2\dfrac{\partial w}{\partial x}\dfrac{\partial w}{\partial y}\sin\theta\cos\theta + \left(\dfrac{\partial w}{\partial y}\right)^2\sin^2\theta + \left(\dfrac{\partial w}{\partial x}\right)^2\sin^2\theta$

$- 2\dfrac{\partial w}{\partial x}\dfrac{\partial w}{\partial y}\sin\theta\cos\theta + \left(\dfrac{\partial w}{\partial y}\right)^2\cos^2\theta = \left(\dfrac{\partial w}{\partial x}\right)^2 + \left(\dfrac{\partial w}{\partial y}\right)^2$

21. $x = r \cos \theta, y = r \sin \theta, z = z$

$$\frac{\partial u}{\partial \theta} = \frac{\partial u}{\partial x}\frac{\partial x}{\partial \theta} + \frac{\partial u}{\partial y}\frac{\partial y}{\partial \theta} + \frac{\partial u}{\partial z}\frac{\partial z}{\partial \theta} = \frac{\partial u}{\partial x}(-r \sin \theta) + \frac{\partial u}{\partial y}r \cos \theta \quad \text{Similarly,}$$

$$\frac{\partial u}{\partial r} = \frac{\partial u}{\partial x}\cos \theta + \frac{\partial u}{\partial y}\sin \theta.$$

$$\frac{\partial^2 u}{\partial \theta^2} = (-r \sin \theta)\left[\frac{\partial^2 u}{\partial x^2}\frac{\partial x}{\partial \theta} + \frac{\partial^2 u}{\partial x \partial y}\frac{\partial y}{\partial \theta} + \frac{\partial^2 u}{\partial x \partial z}\frac{\partial z}{\partial \theta}\right] - r\frac{\partial u}{\partial x}\cos \theta + (r \cos \theta)\left[\frac{\partial^2 u}{\partial y \partial x}\frac{\partial x}{\partial \theta} + \frac{\partial^2 u}{\partial y^2}\frac{\partial y}{\partial \theta} + \frac{\partial^2 u}{\partial y \partial z}\frac{\partial z}{\partial \theta}\right] - r\frac{\partial u}{\partial y}\sin \theta$$

$$= \frac{\partial^2 u}{\partial x^2}r^2 \sin^2 \theta + \frac{\partial^2 u}{\partial y^2}r^2 \cos^2 \theta - 2\frac{\partial^2 u}{\partial x \partial y}r^2 \sin \theta \cos \theta - \frac{\partial u}{\partial x}r \cos \theta - \frac{\partial u}{\partial y}r \sin \theta$$

Similarly, $\dfrac{\partial^2 u}{\partial r^2} = \dfrac{\partial^2 u}{\partial x^2}\cos^2 \theta + \dfrac{\partial^2 u}{\partial y^2}\sin^2 \theta + 2\dfrac{\partial^2 u}{\partial x \partial y}\cos \theta \sin \theta.$

Now observe that

$$\frac{\partial^2 u}{\partial r^2} + \frac{1}{r}\frac{\partial u}{\partial r} + \frac{1}{r^2}\frac{\partial^2 u}{\partial \theta^2} + \frac{\partial^2 u}{\partial z^2} = \left[\frac{\partial^2 u}{\partial x^2}\cos^2 \theta + \frac{\partial^2 u}{\partial y^2}\sin^2 \theta + 2\frac{\partial^2 u}{\partial x \partial y}\cos \theta \sin \theta\right] + \frac{1}{r}\left[\frac{\partial u}{\partial x}\cos \theta + \frac{\partial u}{\partial y}\sin \theta\right]$$

$$+ \left[\frac{\partial^2 u}{\partial x^2}\sin^2 \theta + \frac{\partial^2 u}{\partial y^2}\cos^2 \theta - 2\frac{\partial^2 u}{\partial x \partial y}\sin \theta \cos \theta - \frac{1}{r}\frac{\partial u}{\partial x}\cos \theta - \frac{1}{r}\frac{\partial u}{\partial y}\sin \theta\right] + \frac{\partial^2 u}{\partial z^2}$$

$$= \frac{\partial^2 u}{\partial x^2} + \frac{\partial^2 u}{\partial y^2} + \frac{\partial^2 u}{\partial z^2}.$$

So, Laplace's equation in cylindrical coordinates, is $\dfrac{\partial^2 u}{\partial r^2} + \dfrac{1}{r}\dfrac{\partial u}{\partial r} + \dfrac{1}{r^2}\dfrac{\partial^2 u}{\partial \theta^2} + \dfrac{\partial^2 u}{\partial z^2} = 0.$

CHAPTER 14
Multiple Integration

CHAPTER 14
Multiple Integration

Section 14.1   Iterated Integrals and Area in the Plane

1. An iterated integral is integration of a function of several variables. Integrate with respect to one variable while holding the other variables constant.

3. $\int_0^x (2x - y)\, dy = \left[2xy - \dfrac{y^2}{2} \right]_0^x = \dfrac{3x^2}{2}$

5. $\int_0^{\sqrt{4-x^2}} x^2 y\, dy = \left[\dfrac{1}{2}x^2 y^2 \right]_0^{\sqrt{4-x^2}} = \dfrac{4x^2 - x^4}{2}$

7. $\int_{e^y}^y \dfrac{y \ln x}{x}\, dx = \left[\dfrac{1}{2}y \ln^2 x \right]_{e^y}^y = \dfrac{1}{2}y\left[\ln^2 y - \ln^2 e^y \right] = \dfrac{y}{2}\left[(\ln y)^2 - y^2 \right], (y > 0)$

9. $\int_0^{x^3} ye^{-y/x}\, dy = \left[-xye^{-y/x} \right]_0^{x^3} + x\int_0^{x^3} e^{-y/x}\, dy = -x^4 e^{-x^2} - \left[x^2 e^{-y/x} \right]_0^{x^3} = x^2\left(1 - e^{-x^2} - x^2 e^{-x^2} \right)$

$u = y,\, du = dy,\, dv = e^{-y/x}\, dy,\, v = -xe^{-y/x}$

11. $\int_0^1 \int_0^2 (x + y)\, dy\, dx = \int_0^1 \left[xy + \tfrac{1}{2}y^2 \right]_0^2\, dx = \int_0^1 (2x + 2)\, dx = \left[x^2 + 2x \right]_0^1 = 3$

13. $\int_0^{\pi/4} \int_0^1 y \cos x\, dy\, dx = \int_0^{\pi/4} \left[\dfrac{y^2}{2}\cos x \right]_0^1\, dx = \int_0^{\pi/4} \dfrac{1}{2}\cos x\, dx = \left[\dfrac{1}{2}\sin x \right]_0^{\pi/4} = \dfrac{1}{2}\left(\dfrac{\sqrt{2}}{2} - 0 \right) = \dfrac{\sqrt{2}}{4}$

15. $\int_0^2 \int_0^{6x^2} x^3\, dy\, dx = \int_0^2 \left[x^3 y \right]_0^{6x^2}\, dx = \int_0^2 6x^5\, dx = \left[x^6 \right]_0^2 = 64$

17. $\int_0^{\pi/2} \int_0^{\cos x} (1 + \sin x)\, dy\, dx = \int_0^{\pi/2} \left[y + y \sin x \right]_0^{\cos x}\, dx = \int_0^{\pi/2} (\cos x + \cos x \sin x)\, dx = \left[\sin x + \dfrac{\sin^2 x}{2} \right]_0^{\pi/2} = 1 + \dfrac{1}{2} = \dfrac{3}{2}$

19. $\int_0^1 \int_0^x \sqrt{1 - x^2}\, dy\, dx = \int_0^1 \left[y\sqrt{1 - x^2} \right]_0^x\, dx = \int_0^1 x\sqrt{1 - x^2}\, dx = \left[-\tfrac{1}{2}\left(\tfrac{2}{3}\right)(1 - x^2)^{3/2} \right]_0^1 = \dfrac{1}{3}$

21. $\int_0^1 \int_0^{\sqrt{1-y^2}} (x + y)\, dx\, dy = \int_0^1 \left[\tfrac{1}{2}x^2 + xy \right]_0^{\sqrt{1-y^2}}\, dy = \int_0^1 \left[\tfrac{1}{2}(1 - y^2) + y\sqrt{1 - y^2} \right]\, dy = \left[\tfrac{1}{2}y - \tfrac{1}{6}y^3 - \tfrac{1}{2}\left(\tfrac{2}{3}\right)(1 - y^2)^{3/2} \right]_0^1 = \dfrac{2}{3}$

23. $\int_0^2 \int_0^{\sqrt{4-y^2}} \dfrac{2}{\sqrt{4 - y^2}}\, dx\, dy = \int_0^2 \left[\dfrac{2x}{\sqrt{4 - y^2}} \right]_0^{\sqrt{4-y^2}}\, dy = \int_0^2 2\, dy = \left[2y \right]_0^2 = 4$

25. $\int_0^{\pi/2} \int_0^{2\cos\theta} r\, dr\, d\theta = \int_0^{\pi/2} \left[\dfrac{r^2}{2} \right]_0^{2\cos\theta}\, d\theta = \int_0^{\pi/2} 2\cos^2\theta\, d\theta = \left[\theta - \dfrac{1}{2}\sin 2\theta \right]_0^{\pi/2} = \dfrac{\pi}{2}$

27. $\int_0^{\pi/2} \int_0^{\sin\theta} \theta r\, dr\, d\theta = \int_0^{\pi/2} \left[\theta \dfrac{r^2}{2} \right]_0^{\sin\theta}\, d\theta = \int_0^{\pi/2} \dfrac{1}{2}\theta \sin^2\theta\, d\theta$

$= \dfrac{1}{4}\int_0^{\pi/2} (\theta - \theta \cos 2\theta)\, d\theta = \dfrac{1}{4}\left[\dfrac{\theta^2}{2} - \left(\dfrac{1}{4}\cos 2\theta + \dfrac{\theta}{2}\sin 2\theta \right) \right]_0^{\pi/2} = \dfrac{\pi^2}{32} + \dfrac{1}{8}$

29. $\int_1^\infty \int_0^{1/x} y\, dy\, dx = \int_1^\infty \left[\dfrac{y^2}{2} \right]_0^{1/x}\, dx = \dfrac{1}{2}\int_1^\infty \dfrac{1}{x^2}\, dx = \left[-\dfrac{1}{2x} \right]_1^\infty = 0 + \dfrac{1}{2} = \dfrac{1}{2}$

31. $\int_1^\infty \int_1^\infty \frac{1}{xy}\, dx\, dy = \int_1^\infty \left[\frac{1}{y} \ln x\right]_1^\infty dy = \int_1^\infty \left[\frac{1}{y}(\infty) - \frac{1}{y}(0)\right] dy$

Diverges

33. $A = \int_0^4 \int_0^x dy\, dx = \int_0^4 [y]_0^x\, dx = \int_0^4 x\, dx = \left[\frac{1}{2}x^2\right]_0^4 = 8$

35. $A = \int_0^2 \int_0^{4-x^2} dy\, dx = \int_0^2 [y]_0^{4-x^2}\, dx = \int_0^2 (4 - x^2)\, dx = \left[4x - \frac{x^3}{3}\right]_0^2 = \frac{16}{3}$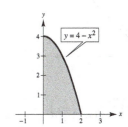

$A = \int_0^4 \int_0^{\sqrt{4-y}} dx\, dy = \int_0^4 [x]_0^{\sqrt{4-y}}\, dy = \int_0^4 \sqrt{4-y}\, dy$

$= -\int_0^4 (4-y)^{1/2}(-1)\, dy = \left[-\frac{2}{3}(4-y)^{3/2}\right]_0^4 = \frac{2}{3}(8) = \frac{16}{3}$

37. $A = \int_{-3}^3 \int_0^{9-x^2} dy\, dx$

$= \int_{-3}^3 (9 - x^2)\, dx$

$= \left[9x - \frac{x^3}{3}\right]_{-3}^3$

$= 36$

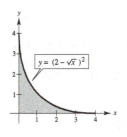

39. $A = \int_0^4 \int_0^{(2-\sqrt{x})^2} dy\, dx = \int_0^4 [y]_0^{(2-\sqrt{x})^2}\, dx = \int_0^4 (4 - 4\sqrt{x} + x)\, dx$

$= \left[4x - \frac{8}{3}x\sqrt{x} + \frac{x^2}{2}\right]_0^4 = \frac{8}{3}$

$A = \int_0^4 \int_0^{(2-\sqrt{y})^2} dx\, dy = \frac{8}{3}$

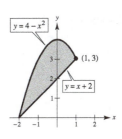

Integration steps are similar to those above.

41. $A = \int_{-2}^1 \int_{x+2}^{4-x^2} dy\, dx$

$= \int_{-2}^1 [y]_{x+2}^{4-x^2}\, dx$

$= \int_{-2}^1 (4 - x^2 - x - 2)\, dx$

$= \int_{-2}^1 (2 - x - x^2)\, dx$

$= \left[2x - \frac{1}{2}x^2 - \frac{1}{3}x^3\right]_{-2}^1 = \frac{9}{2}$

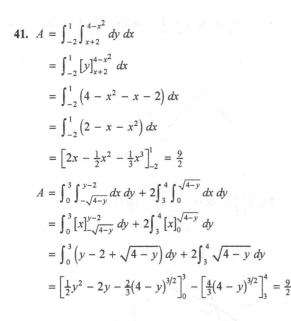

$A = \int_0^3 \int_{-\sqrt{4-y}}^{y-2} dx\, dy + 2\int_3^4 \int_0^{\sqrt{4-y}} dx\, dy$

$= \int_0^3 [x]_{-\sqrt{4-y}}^{y-2}\, dy + 2\int_3^4 [x]_0^{\sqrt{4-y}}\, dy$

$= \int_0^3 (y - 2 + \sqrt{4-y})\, dy + 2\int_3^4 \sqrt{4-y}\, dy$

$= \left[\frac{1}{2}y^2 - 2y - \frac{2}{3}(4-y)^{3/2}\right]_0^3 - \left[\frac{4}{3}(4-y)^{3/2}\right]_3^4 = \frac{9}{2}$

43. $\int_0^4 \int_0^y f(x, y)\, dx\, dy,\, 0 \le x \le y,\, 0 \le y \le 4$

$$= \int_0^4 \int_x^4 f(x, y)\, dy\, dx$$

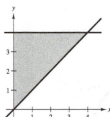

45. $\int_{-2}^2 \int_0^{\sqrt{4-x^2}} f(x, y)\, dy\, dx,\, 0 \le y \le \sqrt{4-x^2},\, -2 \le x \le 2$

$$= \int_0^2 \int_{-\sqrt{4-y^2}}^{\sqrt{4-y^2}} dx\, dy$$

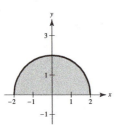

47. $\int_1^{10} \int_0^{\ln y} f(x, y)\, dx\, dy,\, 0 \le x \le \ln y,\, 1 \le y \le 10$

$$= \int_0^{\ln 10} \int_{e^x}^{10} f(x, y)\, dy\, dx$$

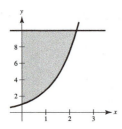

49. $\int_{-1}^1 \int_{x^2}^1 f(x, y)\, dy\, dx,\, x^2 \le y \le 1,\, -1 \le x \le 1$

$$= \int_0^1 \int_{-\sqrt{y}}^{\sqrt{y}} f(x, y)\, dx\, dy$$

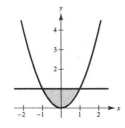

51. $\int_0^1 \int_0^2 dy\, dx = \int_0^2 \int_0^1 dx\, dy = 2$

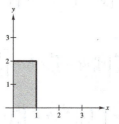

53. $\int_0^1 \int_{2y}^2 dx\, dy = \int_0^1 (2 - 2y)\, dy = \left[2y - y^2\right]_0^1 = 1$

$$\int_0^2 \int_0^{x/2} dy\, dx = \int_0^2 \frac{x}{2}\, dx = \left[\frac{x^2}{4}\right]_0^2 = 1$$

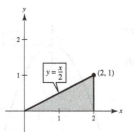

55. $\int_0^1 \int_{-\sqrt{1-y^2}}^{\sqrt{1-y^2}} dx\, dy = \int_{-1}^1 \int_0^{\sqrt{1-x^2}} dy\, dx = \frac{\pi}{2}$

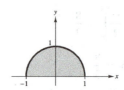

57. $\int_0^2 \int_0^x dy\, dx + \int_2^4 \int_0^{4-x} dy\, dx = \int_0^2 \int_y^{4-y} dx\, dy = 4$

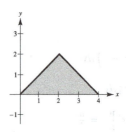

59. $\int_0^1 \int_{y^2}^{\sqrt[3]{y}} dx\, dy = \int_0^1 \int_{x^3}^{\sqrt{x}} dy\, dx = \frac{5}{12}$

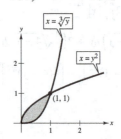

61. $\int_0^2 \int_x^2 x\sqrt{1+y^3}\,dy\,dx = \int_0^2 \int_0^y x\sqrt{1+y^3}\,dx\,dy$

$$= \int_0^2 \left[\sqrt{1+y^3} \cdot \frac{x^2}{2}\right]_0^y dy$$

$$= \frac{1}{2}\int_0^2 \sqrt{1+y^3}\, y^2\, dy$$

$$= \left[\frac{1}{2} \cdot \frac{1}{3} \cdot \frac{2}{3}(1+y^3)^{3/2}\right]_0^2$$

$$= \frac{1}{9}(27) - \frac{1}{9}(1) = \frac{26}{9}$$

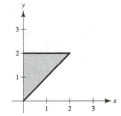

63. $\int_0^1 \int_{2x}^2 4e^{y^2}\,dy\,dx = \int_0^2 \int_0^{y/2} 4e^{y^2}\,dx\,dy$

$$= \int_0^2 \left[4xe^{y^2}\right]_0^{y/2} dy = \int_0^2 2ye^{y^2}\,dy$$

$$= \left[e^{y^2}\right]_0^2 = e^4 - 1$$

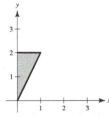

65. $\int_0^1 \int_y^1 \sin(x^2)\,dx\,dy = \int_0^1 \int_0^x \sin(x^2)\,dy\,dx$

$$= \int_0^1 \left[y\sin(x^2)\right]_0^x dx$$

$$= \int_0^1 x\sin(x^2)\,dx$$

$$= \left[-\frac{1}{2}\cos(x^2)\right]_0^1$$

$$= -\frac{1}{2}\cos 1 + \frac{1}{2}(1)$$

$$= \frac{1}{2}(1 - \cos 1) \approx 0.2298$$

67. $A = 2\int_{-5}^5 \int_0^{\sqrt{25-x^2}} dy\,dx = 4\int_0^5 \int_0^{\sqrt{25-x^2}} dy\,dx$

Evaluate the second integral using Theorem 8.2.

$$A = 4\int_0^5 \sqrt{25-x^2}\,dx$$

$$= 4\left[\frac{1}{2}\left(x\sqrt{25-x^2} + 25\arcsin\frac{x}{5}\right)\right]_0^5$$

$$= 2(0 + 25\arcsin 1)$$

$$= 50\left(\frac{\pi}{2}\right) = 25\pi \text{ square units}$$

69. (a) No. You cannot have the variable of integration in the limits of integration.

(b) Yes

(c) Yes

71. $\int_0^1 \int_y^{2y} \sin(x+y)\,dx\,dy = \dfrac{\sin 2}{2} - \dfrac{\sin 3}{3} \approx 0.4076$

73. $\int_0^4 \int_0^y \dfrac{2}{(x+1)(y+1)}\,dx\,dy = (\ln 5)^2 \approx 2.590$

75. $\int_0^{2\pi} \int_0^{1+\cos\theta} 6r^2\cos\theta\,dr\,d\theta = \dfrac{15\pi}{2}$

77. (a) $x = y^3 \Leftrightarrow y = x^{1/3}$

$$x = 4\sqrt{2y} \Leftrightarrow x^2 = 32y \Leftrightarrow y = \frac{x^2}{32}$$

(b) $\int_0^8 \int_{x^2/32}^{x^{1/3}} (x^2 y - xy^2)\,dy\,dx$

(c) Both integrals equal $\dfrac{67,520}{693} \approx 97.43.$

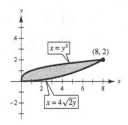

79. True

Section 14.2 Double Integrals and Volume

1. Use rectangular prisms to approximate the volume, where $f(x_i, y_i)$ is the height of prism i and ΔA_i is the area of the rectangular base of the prism. You can improve the approximation by using more rectangular prisms of smaller rectangular bases.

For Exercises 3–5, $\Delta x_i = \Delta y_i = 1$ and the midpoints of the squares are

$$\left(\tfrac{1}{2}, \tfrac{1}{2}\right), \left(\tfrac{3}{2}, \tfrac{1}{2}\right), \left(\tfrac{5}{2}, \tfrac{1}{2}\right), \left(\tfrac{7}{2}, \tfrac{1}{2}\right), \left(\tfrac{1}{2}, \tfrac{3}{2}\right), \left(\tfrac{3}{2}, \tfrac{3}{2}\right), \left(\tfrac{5}{2}, \tfrac{3}{2}\right), \left(\tfrac{7}{2}, \tfrac{3}{2}\right).$$

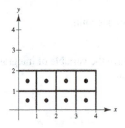

3. $f(x, y) = x + y$

$$\sum_{i=1}^{8} f(x_i, y_i) \, \Delta x_i \Delta y_i = 1 + 2 + 3 + 4 + 2 + 3 + 4 + 5 = 24$$

$$\int_0^4 \int_0^2 (x + y) \, dy \, dx = \int_0^4 \left[xy + \frac{y^2}{2} \right]_0^2 dx = \int_0^4 (2x + 2) \, dx = \left[x^2 + 2x \right]_0^4 = 24$$

5. $f(x, y) = x^2 + y^2$

$$\sum_{i=1}^{8} f(x_i, y_i) \, \Delta x_i \, \Delta y_i = \frac{2}{4} + \frac{10}{4} + \frac{26}{4} + \frac{50}{4} + \frac{10}{4} + \frac{18}{4} + \frac{34}{4} + \frac{58}{4} = 52$$

$$\int_0^4 \int_0^2 (x^2 + y^2) \, dy \, dx = \int_0^4 \left[x^2 y + \frac{y^3}{3} \right]_0^2 dx = \int_0^4 \left(2x^2 + \frac{8}{3} \right) dx = \left[\frac{2x^3}{3} + \frac{8x}{3} \right]_0^4 = \frac{160}{3}$$

7. $\int_0^2 \int_0^1 (1 - 4x + 8y) \, dy \, dx = \int_0^2 \left[y - 4xy + 4y^2 \right]_0^1 dx$

$$= \int_0^2 (1 - 4x + 4) \, dx$$

$$= \left[5x - 2x^2 \right]_0^2$$

$$= 10 - 8 = 2$$

9. $\int_0^6 \int_{y/2}^3 (x + y) \, dx \, dy = \int_0^6 \left[\frac{1}{2} x^2 + xy \right]_{y/2}^3 dy$

$$= \int_0^6 \left(\frac{9}{2} + 3y - \frac{5}{8} y^2 \right) dy$$

$$= \left[\frac{9}{2} y + \frac{3}{2} y^2 - \frac{5}{24} y^3 \right]_0^6$$

$$= 36$$

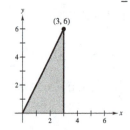

11. $\int_{-3}^{3} \int_{-\sqrt{9-x^2}}^{\sqrt{9-x^2}} (x + y)\, dy\, dx = \int_{-3}^{3} \left[xy + \dfrac{y^2}{2} \right]_{-\sqrt{9-x^2}}^{\sqrt{9-x^2}} dx$

$\qquad = \int_{-3}^{3} \left[\left(x\sqrt{9 - x^2} + \dfrac{9 - x^2}{2} \right) - \left(-x\sqrt{9 - x^2} + \dfrac{9 - x^2}{2} \right) \right] dx$

$\qquad = \int_{-3}^{3} 2x\sqrt{9 - x^2}\, dx \quad (u = 9 - x^2,\ du = -2x\, dx)$

$\qquad = \left[-\dfrac{2}{3}(9 - x^2)^{3/2} \right]_{-3}^{3}$

$\qquad = 0$

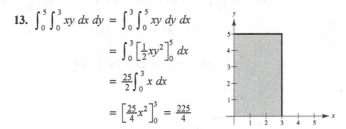

13. $\int_{0}^{5} \int_{0}^{3} xy\, dx\, dy = \int_{0}^{3} \int_{0}^{5} xy\, dy\, dx$

$\qquad = \int_{0}^{3} \left[\dfrac{1}{2}xy^2 \right]_{0}^{5} dx$

$\qquad = \dfrac{25}{2} \int_{0}^{3} x\, dx$

$\qquad = \left[\dfrac{25}{4}x^2 \right]_{0}^{3} = \dfrac{225}{4}$

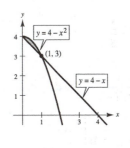

15. $\int_{1}^{2} \int_{1}^{y} \dfrac{y}{x^2 + y^2}\, dx\, dy + \int_{2}^{4} \int_{y/2}^{2} \dfrac{y}{x^2 + y^2}\, dx\, dy = \int_{1}^{2} \int_{x}^{2x} \dfrac{y}{x^2 + y^2}\, dy\, dx$

$\qquad = \dfrac{1}{2} \int_{1}^{2} \left[\ln(x^2 + y^2) \right]_{x}^{2x} dx$

$\qquad = \dfrac{1}{2} \int_{1}^{2} \left(\ln 5x^2 - \ln 2x^2 \right) dx$

$\qquad = \dfrac{1}{2} \ln \dfrac{5}{2} \int_{1}^{2} dx$

$\qquad = \left[\dfrac{1}{2}\left(\ln \dfrac{5}{2} \right)x \right]_{1}^{2} = \dfrac{1}{2} \ln \dfrac{5}{2}$

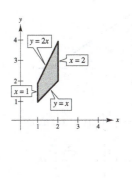

17. $\int_{3}^{4} \int_{4-y}^{\sqrt{4-y}} -2y\, dx\, dy = \int_{0}^{1} \int_{4-x}^{4-x^2} -2y\, dy\, dx$

$\qquad = \int_{0}^{1} \left[-y^2 \right]_{4-x}^{4-x^2} dx$

$\qquad = -\int_{0}^{1} \left[(4 - x^2)^2 - (4 - x)^2 \right] dx$

$\qquad = -\int_{0}^{1} \left[16 - 8x^2 + x^4 - (16 - 8x + x^2) \right] dx$

$\qquad = -\left[-3x^3 + \dfrac{x^5}{5} + 4x^2 \right]_{0}^{1} = -\dfrac{6}{5}$

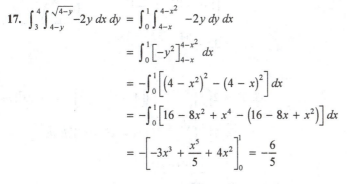

19. $\int_{0}^{4} \int_{0}^{3x/4} x\, dy\, dx + \int_{4}^{5} \int_{0}^{\sqrt{25-x^2}} x\, dy\, dx = \int_{0}^{3} \int_{4y/3}^{\sqrt{25-y^2}} x\, dx\, dy$

$\qquad = \int_{0}^{3} \left[\dfrac{1}{2}x^2 \right]_{4y/3}^{\sqrt{25-y^2}} dy$

$\qquad = \dfrac{25}{18} \int_{0}^{3} \left(9 - y^2 \right) dy$

$\qquad = \left[\dfrac{25}{18}\left(9y - \dfrac{1}{3}y^3 \right) \right]_{0}^{3} = 25$

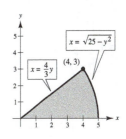

21. $V = \int_0^4 \int_0^2 \frac{y}{2} \, dy \, dx$

$= \int_0^4 \left[\frac{y^2}{4} \right]_0^2 dx$

$= \int_0^4 dx$

$= 4$

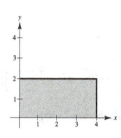

23. $V = \int_0^6 \int_0^{(-2/3)x+4} \left(\frac{12 - 2x - 3y}{4} \right) dy \, dx$

$= \int_0^6 \left[\frac{1}{4} \left(12y - 2xy - \frac{3}{2} y^2 \right) \right]_0^{(-2/3)x+4} dx$

$= \int_0^6 \left(\frac{1}{6} x^2 - 2x + 6 \right) dx$

$= \left[\frac{1}{18} x^3 - x^2 + 6x \right]_0^6 = 12$

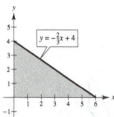

25. $V = \int_0^1 \int_0^y (1 - xy) \, dx \, dy$

$= \int_0^1 \left[x - \frac{x^2 y}{2} \right]_0^y dy$

$= \int_0^1 \left(y - \frac{y^3}{2} \right) dy$

$= \left[\frac{y^2}{2} - \frac{y^4}{8} \right]_0^1 = \frac{3}{8}$

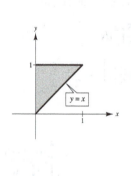

27. $V = \int_0^\infty \int_0^\infty \frac{1}{(x+1)^2(y+1)^2} \, dy \, dx$

$= \int_0^\infty \left[-\frac{1}{(x+1)^2(y+1)} \right]_0^\infty dx$

$= \int_0^\infty \frac{1}{(x+1)^2} \, dx$

$= \left[-\frac{1}{(x+1)} \right]_0^\infty = 1$

29. $V = \int_0^1 \int_0^{x^3} xy \, dy \, dx$

$= \int_0^1 \left[\frac{xy^2}{2} \right]_0^{x^3} dx$

$= \int_0^1 \frac{x^7}{2} \, dx$

$= \left[\frac{x^8}{16} \right]_0^1 = \frac{1}{16}$

31. $V = \int_0^2 \int_0^{\sqrt{4-x^2}} (x + y) \, dy \, dx = \int_0^2 \left[xy + \frac{1}{2} y^2 \right]_0^{\sqrt{4-x^2}} dx$

$= \int_0^2 \left(x\sqrt{4 - x^2} + 2 - \frac{1}{2} x^2 \right) dx$

$= \left[-\frac{1}{3} \left(4 - x^2 \right)^{3/2} + 2x - \frac{1}{6} x^3 \right]_0^2 = \frac{16}{3}$

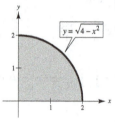

33. $V = \int_0^2 \int_0^{4-x^2} \left(4 - x^2 \right) dy \, dx$

$= \int_0^2 \left(4 - x^2 \right) \left(4 - x^2 \right) dx$

$= \int_0^2 \left(16 - 8x^2 + x^4 \right) dx$

$= \left[16x - \frac{8x^3}{3} + \frac{x^5}{5} \right]_0^2$

$= 32 - \frac{64}{3} + \frac{32}{5} = \frac{256}{15}$

35. $V = 2\int_0^2 \int_0^{\sqrt{1-(x-1)^2}} \left(\left[4 - x^2 - y^2 \right] - \left[4 - 2x \right] \right) dy \, dx$

$= 2\int_0^2 \int_0^{\sqrt{1-(x-1)^2}} \left(2x - x^2 - y^2 \right) dy \, dx$

37. $V = 4\int_0^2 \int_0^{\sqrt{4-x^2}} \left(x^2 + y^2\right) dy\, dx$

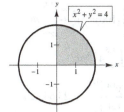

39. $V = \int_0^2 \int_{-\sqrt{2-2(y-1)^2}}^{\sqrt{2-2(y-1)^2}} \left[4y - \left(x^2 + 2y^2\right)\right] dx\, dy$

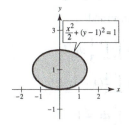

41. $V = 4\int_0^3 \int_0^{\sqrt{9-x^2}} \left(9 - x^2 - y^2\right) dy\, dx = \dfrac{81\pi}{2}$

43. $V = \int_0^2 \int_0^{-0.5x+1} \dfrac{2}{1 + x^2 + y^2}\, dy\, dx \approx 1.2315$

45. $\int_0^1 \int_{y/2}^{1/2} e^{-x^2}\, dx\, dy = \int_0^{1/2} \int_0^{2x} e^{-x^2}\, dy\, dx$

$= \int_0^{1/2} 2xe^{-x^2}\, dx$

$= \left[-e^{-x^2}\right]_0^{1/2}$

$= -e^{-1/4} + 1$

$= 1 - e^{-1/4} \approx 0.221$

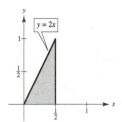

47. $\int_{-2}^2 \int_{-\sqrt{4-x^2}}^{\sqrt{4-x^2}} \sqrt{4 - y^2}\, dy\, dx = \int_{-2}^2 \int_{-\sqrt{4-y^2}}^{\sqrt{4-y^2}} \sqrt{4 - y^2}\, dx\, dy$

$= \int_{-2}^2 \left[x\sqrt{4 - y^2}\right]_{-\sqrt{4-y^2}}^{\sqrt{4-y^2}}\, dy$

$= \int_{-2}^2 2\left(4 - y^2\right) dy = \left[8y - \dfrac{2y^3}{3}\right]_{-2}^2$

$= \left(16 - \dfrac{16}{3}\right) - \left(-16 + \dfrac{16}{3}\right)$

$= \dfrac{64}{3}$

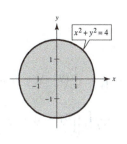

49. $\int_0^2 \int_{2x}^4 \sin y^2\, dy\, dx = \int_0^4 \int_0^{y/2} \sin y^2\, dx\, dy$

$= \int_0^4 \left[x \sin y^2\right]_0^{y/2}\, dy$

$= \int_0^4 \dfrac{y}{2} \sin y^2\, dy$

$= \left[-\dfrac{1}{4} \cos y^2\right]_0^4$

$= -\dfrac{1}{4} \cos 16 + \dfrac{1}{4} \approx 0.489$

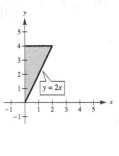

51. Average $= \dfrac{1}{8}\int_0^4 \int_0^2 x\, dy\, dx = \dfrac{1}{8}\int_0^4 2x\, dx = \left[\dfrac{x^2}{8}\right]_0^4 = 2$

53. Average $= \frac{1}{4}\int_0^2 \int_0^2 (x^2 + y^2) \, dx \, dy$

$= \frac{1}{4}\int_0^2 \left[\frac{x^3}{3} + xy^2\right]_0^2 dy$

$= \frac{1}{4}\int_0^2 \left(\frac{8}{3} + 2y^2\right) dy$

$= \left[\frac{1}{4}\left(\frac{8}{3}y + \frac{2}{3}y^3\right)\right]_0^2$

$= \frac{8}{3}$

55. Average $= \frac{1}{1/2}\int_0^1 \int_x^1 e^{x+y} \, dy \, dx$

$= 2\int_0^1 e^{x+1} - e^{2x} \, dx$

$= 2\left[e^{x+1} - \frac{1}{2}e^{2x}\right]_0^1$

$= 2\left[e^2 - \frac{1}{2}e^2 - e + \frac{1}{2}\right]$

$= e^2 - 2e + 1$

$= (e - 1)^2$

57. Average $= \frac{1}{1250}\int_{300}^{325} \int_{200}^{250} 100x^{0.6}y^{0.4} \, dx \, dy$

$= \frac{1}{1250}\int_{300}^{325} \left[(100y^{0.4})\frac{x^{1.6}}{1.6}\right]_{200}^{250} dy$

$= \frac{128,844.1}{1250}\int_{300}^{325} y^{0.4} \, dy$

$= 103.0753\left[\frac{y^{1.4}}{1.4}\right]_{300}^{325} \approx 25,645.24$

59. The value of $\int_R \int f(x, y) \, dA$ would be kB.

61. $f(x, y) \geq 0$ for all (x, y) and

$\int_{-\infty}^{\infty} \int_{-\infty}^{\infty} f(x, y) \, dA = \int_0^1 \int_1^4 \frac{1}{3} \, dy \, dx$

$= \int_0^1 \frac{1}{3}(4 - 1) \, dx$

$= \int_0^1 dx = 1$

$P(0 \leq x \leq 1, 1 \leq y \leq 3) = \int_0^1 \int_1^3 \frac{1}{3} \, dy \, dx$

$= \int_0^1 \frac{1}{3}(3 - 1) \, dx$

$= \int_0^1 \frac{2}{3} \, dx = \frac{2}{3}$

63. $f(x, y) \geq 0$ for all x and

$\int_{-\infty}^{\infty} \int_{-\infty}^{\infty} f(x, y) \, dA = \int_0^3 \int_3^6 \frac{1}{27}(9 - x - y) \, dy \, dx$

$= \frac{1}{27}\int_0^3 \left[\left(9y - xy - \frac{y^2}{2}\right)\right]_3^6 dx$

$= \frac{1}{27}\int_0^3 \left[(54 - 6x - 18) - \left(27 - 3x - \frac{9}{2}\right)\right] dx$

$= \frac{1}{27}\int_0^3 \left(\frac{27}{2} - 3x\right) dx$

$= \frac{1}{27}\left[\frac{27}{2}x - \frac{3x^2}{2}\right]_0^3$

$= \frac{1}{27}\left[\frac{81}{2} - \frac{27}{2}\right] = 1$

$P(0 \leq x < 1, 3 \leq y \leq 6) = \int_0^1 \int_3^6 \frac{1}{27}(9 - x - y) \, dy \, dx$

$= \frac{1}{27}\left[\frac{27}{2}x - \frac{3x^2}{2}\right]_0^1$

$= \frac{1}{27}\left(\frac{27}{2} - \frac{3}{2}\right) = \frac{4}{9}$

65. f is a continuous function such that $0 \leq f(x, y) \leq 1$ over a region R of area 1.

Let $f(m, n) =$ the minimum value of f over R and $f(M, N) =$ the maximum value of f over R.

Then $f(m, n) \int_R \int dA \leq \int_R \int f(x, y) \, dA \leq f(M, N) \int_R \int dA$.

Because $\int_R \int dA = 1$ and $0 \leq f(m, n) \leq f(M, N) \leq 1$, you have $0 \leq f(m, n)(1) \leq \int_R \int f(x, y) \, dA \leq f(M, N)(1) \leq 1$.

So, $0 \leq \int_R \int f(x, y) \, dA \leq 1$.

67. $\int_0^4 \int_0^4 f(x, y) \, dy \, dx \approx (32 + 31 + 28 + 23) + (31 + 30 + 27 + 22) + (28 + 27 + 24 + 19) + (23 + 22 + 19 + 14) = 400$

Using the corner of the ith square farthest from the origin, you obtain 272.

69. False

$$V = 8 \int_0^1 \int_0^{\sqrt{1-y^2}} \sqrt{1 - x^2 - y^2} \, dx \, dy$$

71. $z = 9 - x^2 - y^2$ is a paraboloid opening downward with vertex $(0, 0, 9)$. The double integral is maximized if $z \geq 0$. That is,

$R = \{(x, y) : x^2 + y^2 \leq 9\}$.

$$\left[\text{The maximum value is} \iint\limits_R (9 - x^2 - y^2) \, dA = \frac{81\pi}{2}. \right]$$

73. Average $= \int_0^1 f(x) \, dx$

$= \int_0^1 \int_1^x e^{t^2} \, dt \, dx$

$= -\int_0^1 \int_x^1 e^{t^2} \, dt \, dx$

$= -\int_0^1 \int_0^t e^{t^2} \, dx \, dt$

$= -\int_0^1 t e^{t^2} \, dt$

$= \left[-\frac{1}{2} e^{t^2} \right]_0^1$

$= -\frac{1}{2}(e - 1)$

$= \frac{1}{2}(1 - e)$

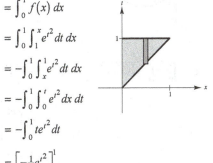

75. Let $I = \int_0^a \int_0^b e^{\max\{b^2 x^2, a^2 y^2\}} \, dy \, dx$.

Divide the rectangle into two parts by the diagonal line $ay = bx$. On lower triangle,

$b^2 x^2 \geq a^2 y^2$ because $y \leq \dfrac{b}{a} x$.

$I = \int_0^a \int_0^{bx/a} e^{b^2 x^2} \, dy \, dx + \int_0^b \int_0^{ay/b} e^{a^2 y^2} \, dx \, dy = \int_0^a \frac{bx}{a} e^{b^2 x^2} \, dx + \int_0^b \frac{ay}{b} e^{a^2 y^2} \, dy$

$= \frac{1}{2ab} \left[e^{b^2 x^2} \right]_0^a + \frac{1}{2ab} \left[e^{a^2 y^2} \right]_0^b = \frac{1}{2ab} \left[e^{b^2 a^2} - 1 + e^{a^2 b^2} - 1 \right] = \frac{e^{a^2 b^2} - 1}{ab}$

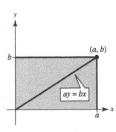

Section 14.3   Change of Variables: Polar Coordinates

1. Rectangular coordinates

3. r-simple regions have fixed bounds for θ.
θ-simple regions have fixed bounds for r.

5. $R = \{(r, \theta) : 0 \leq r \leq 8, 0 \leq \theta \leq \pi\}$

7. $R = \left\{ (r, \theta) : 4 \leq r \leq 8, 0 \leq \theta \leq \dfrac{\pi}{2} \right\}$

9. $\displaystyle\int_0^\pi \int_0^{2\cos\theta} r \, dr \, d\theta = \int_0^\pi \left[\frac{r^2}{2}\right]_0^{2\cos\theta} d\theta$

$\displaystyle = \int_0^\pi 2\cos^2\theta \, d\theta$

$\displaystyle = \int_0^\pi (1 + \cos 2\theta) \, d\theta$

$\displaystyle = \left[\theta + \frac{1}{2}\sin 2\theta\right]_0^\pi$

$= \pi$

11. $\displaystyle\int_0^{2\pi} \int_0^1 6r^2 \sin\theta \, dr \, d\theta = \int_0^{2\pi} \left[2r^3 \sin\theta\right]_0^1 d\theta$

$\displaystyle = \int_0^{2\pi} 2\sin\theta \, d\theta$

$\displaystyle = \left[-2\cos\theta\right]_0^{2\pi} = 0$

13. $\displaystyle\int_0^{\pi/2} \int_1^3 \sqrt{9 - r^2} \; r \, dr \, d\theta = \left(-\frac{1}{2}\right)\int_0^{\pi/2} \int_1^3 \left(9 - r^2\right)^{1/2}(-2r \, dr \, d\theta)$

$\displaystyle = -\frac{1}{2}\int_0^{\pi/2}\left[\frac{2\left(9 - r^2\right)^{3/2}}{3}\right]_1^3 d\theta$

$\displaystyle = -\frac{1}{3}\int_0^{\pi/2}\left(0 - 8^{3/2}\right)d\theta$

$\displaystyle = \frac{1}{3}\left(16\sqrt{2}\right)\frac{\pi}{2} = 8\sqrt{2}\frac{\pi}{3}$

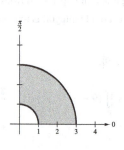

15. $\displaystyle\int_0^{\pi/2} \int_0^{1+\sin\theta} \theta r \, dr \, d\theta = \int_0^{\pi/2} \left[\frac{\theta r^2}{2}\right]_0^{1+\sin\theta} d\theta$

$\displaystyle = \int_0^{\pi/2} \frac{1}{2}\theta(1 + \sin\theta)^2 \, d\theta$

$\displaystyle = \left[\frac{1}{8}\theta^2 + \sin\theta - \theta\cos\theta + \frac{1}{2}\theta\left(-\frac{1}{2}\cos\theta\cdot\sin\theta + \frac{1}{2}\theta\right) + \frac{1}{8}\sin^2\theta\right]_0^{\pi/2}$

$\displaystyle = \frac{3}{32}\pi^2 + \frac{9}{8}$

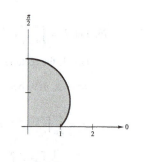

17. $\displaystyle\int_0^3 \int_0^{\sqrt{9-y^2}} y \, dx \, dy = \int_0^{\pi/2} \int_0^3 (r\sin\theta) \, r \, dr \, d\theta$

$\displaystyle = \int_0^{\pi/2} \left[\frac{r^3}{3}\sin\theta\right]_0^3 d\theta$

$\displaystyle = \int_0^{\pi/2} 9\sin\theta \, d\theta$

$\displaystyle = \left[-9\cos\theta\right]_0^{\pi/2} = 9$

19. $\displaystyle\int_{-2}^2 \int_0^{\sqrt{4-x^2}} \left(x^2 + y^2\right) dy \, dx = \int_0^\pi \int_0^2 r^2 \, r \, dr \, d\theta$

$\displaystyle = \int_0^\pi \left[\frac{r^4}{4}\right]_0^2 d\theta$

$\displaystyle = \int_0^\pi 4 \, d\theta$

$= 4\pi$

21. $\displaystyle\int_0^1 \int_0^{\sqrt{1-x^2}} \left(x^2 + y^2\right)^{3/2} dy \, dx = \int_0^{\pi/2} \int_0^1 r^3 \, r \, dr \, d\theta = \int_0^{\pi/2} \frac{1}{5} \, d\theta = \frac{\pi}{10}$

23. $\displaystyle\int_0^2 \int_0^{\sqrt{2x-x^2}} xy \, dy \, dx = \int_0^{\pi/2} \int_0^{2\cos\theta} r^3 \cos\theta \sin\theta \, dr \, d\theta = 4\int_0^{\pi/2} \cos^5\theta \sin\theta \, d\theta = \left[-\frac{4\cos^6\theta}{6}\right]_0^{\pi/2} = \frac{2}{3}$

25. $\int_{-1}^{1}\int_{0}^{\sqrt{1-x^2}}\cos(x^2+y^2)\,dy\,dx = \int_{0}^{\pi}\int_{0}^{1}\cos(r^2)r\,dr\,d\theta = \int_{0}^{\pi}\left[\frac{1}{2}\sin(r^2)\right]_{0}^{1}d\theta = \int_{0}^{\pi}\frac{1}{2}\sin(1)\,d\theta = \frac{\pi}{2}\sin(1) \approx 1.3218$

27. $\int_{0}^{2}\int_{0}^{x}\sqrt{x^2+y^2}\,dy\,dx + \int_{2}^{2\sqrt{2}}\int_{0}^{\sqrt{8-x^2}}\sqrt{x^2+y^2}\,dy\,dx = \int_{0}^{\pi/4}\int_{0}^{2\sqrt{2}}r^2\,dr\,d\theta$

$$= \int_{0}^{\pi/4}\frac{16\sqrt{2}}{3}\,d\theta$$

$$= \frac{4\sqrt{2}\pi}{3}$$

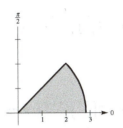

29. $\int_{0}^{2}\int_{0}^{\sqrt{6-x^2}}(x+y)\,dy\,dx = \int_{0}^{\pi/2}\int_{0}^{6}(r\cos\theta + r\sin\theta)r\,dr\,d\theta = \int_{0}^{\pi/2}\int_{0}^{6}(\cos\theta + \sin\theta)r^2\,dr\,d\theta$

$$= 72\int_{0}^{\pi/2}(\cos\theta + \sin\theta)\,d\theta = \left[72(\sin\theta - \cos\theta)\right]_{0}^{\pi/2} = 144$$

31. $\int_{0}^{1/\sqrt{2}}\int_{\sqrt{1-y^2}}^{\sqrt{4-y^2}}\arctan\frac{y}{x}\,dx\,dy + \int_{1/\sqrt{2}}^{\sqrt{2}}\int_{y}^{\sqrt{4-y^2}}\arctan\frac{y}{x}\,dx\,dy = \int_{0}^{\pi/4}\int_{1}^{2}\theta r\,dr\,d\theta$

$$= \int_{0}^{\pi/4}\frac{3}{2}\theta\,d\theta$$

$$= \left[\frac{3\theta^2}{4}\right]_{0}^{\pi/4}$$

$$= \frac{3\pi^2}{64}$$

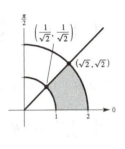

33. $V = \int_{0}^{\pi/2}\int_{0}^{1}(r\cos\theta)(r\sin\theta)r\,dr\,d\theta$

$$= \frac{1}{2}\int_{0}^{\pi/2}\int_{0}^{1}r^3\sin 2\theta\,dr\,d\theta = \frac{1}{8}\int_{0}^{\pi/2}\sin 2\theta\,d\theta = \left[-\frac{1}{16}\cos 2\theta\right]_{0}^{\pi/2} = \frac{1}{8}$$

35. $V = \int_{0}^{2\pi}\int_{0}^{5}r^2\,dr\,d\theta = \int_{0}^{2\pi}\frac{125}{3}\,d\theta = \frac{250\pi}{3}$

37. $V = 2\int_{0}^{\pi/2}\int_{0}^{4\cos\theta}\sqrt{16-r^2}\,r\,dr\,d\theta = 2\int_{0}^{\pi/2}\left[-\frac{1}{3}\left(\sqrt{16-r^2}\right)^3\right]_{0}^{4\cos\theta}d\theta = -\frac{2}{3}\int_{0}^{\pi/2}\left(64\sin^3\theta - 64\right)d\theta$

$$= \frac{128}{3}\int_{0}^{\pi/2}\left[1 - \sin\theta(1 - \cos^2\theta)\right]d\theta = \frac{128}{3}\left[\theta + \cos\theta - \frac{\cos^3\theta}{3}\right]_{0}^{\pi/2} = \frac{64}{9}(3\pi - 4)$$

39. $V = \int_{0}^{2\pi}\int_{a}^{4}\sqrt{16-r^2}\,r\,dr\,d\theta = \int_{0}^{2\pi}\left[-\frac{1}{3}\left(\sqrt{16-r^2}\right)^3\right]_{a}^{4}d\theta = \frac{1}{3}\left(\sqrt{16-a^2}\right)^3(2\pi)$

One-half the volume of the hemisphere is $(64\pi)/3$.

$$\frac{2\pi}{3}(16-a^2)^{3/2} = \frac{64\pi}{3}$$

$$(16-a^2)^{3/2} = 32$$

$$16-a^2 = 32^{2/3}$$

$$a^2 = 16 - 32^{2/3} = 16 - 8\sqrt[3]{2}$$

$$a = \sqrt{4(4 - 2\sqrt[3]{2})} = 2\sqrt{4 - 2\sqrt[3]{2}} \approx 2.4332$$

41. $A = \int_0^\pi \int_0^{6\cos\theta} r\,dr\,d\theta = \int_0^\pi 18\cos^2\theta\,d\theta = 9\int_0^\pi (1 + \cos 2\theta)\,d\theta = \left[9\left(\theta + \frac{1}{2}\sin 2\theta\right)\right]_0^\pi = 9\pi$

43. $A = \int_0^{2\pi} \int_0^{1+\cos\theta} r\,dr\,d\theta = \frac{1}{2}\int_0^{2\pi}\left(1 + 2\cos\theta + \cos^2\theta\right)d\theta$

$$= \frac{1}{2}\int_0^{2\pi}\left(1 + 2\cos\theta + \frac{1+\cos 2\theta}{2}\right)d\theta = \frac{1}{2}\left[\theta + 2\sin\theta + \frac{1}{2}\left(\theta + \frac{1}{2}\sin 2\theta\right)\right]_0^{2\pi} = \frac{3\pi}{2}$$

45. $A = 3\int_0^{\pi/3}\int_0^{2\sin 3\theta} r\,dr\,d\theta = \frac{3}{2}\int_0^{\pi/3} 4\sin^2 3\theta\,d\theta = 3\int_0^{\pi/3}(1 - \cos 6\theta)\,d\theta = 3\left[\theta - \frac{1}{6}\sin 6\theta\right]_0^{\pi/3} = \pi$

47. $r = 1 = 2\cos\theta \Rightarrow \theta = \pm\dfrac{\pi}{3}$

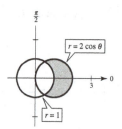

$A = 2\int_0^{\pi/3}\int_1^{2\cos\theta} r\,dr\,d\theta = 2\int_0^{\pi/3}\left[\frac{r^2}{2}\right]_1^{2\cos\theta} d\theta = 2\int_0^{\pi/3}\left(2\cos^2\theta - \frac{1}{2}\right)d\theta$

$= 2\int_0^{\pi/3}\left(1 + \cos 2\theta - \frac{1}{2}\right)d\theta = 2\left[\frac{1}{2}\theta + \frac{\sin 2\theta}{2}\right]_0^{\pi/3} = 2\left[\frac{\pi}{6} + \frac{\sqrt{3}}{4}\right] = \frac{\pi}{3} + \frac{\sqrt{3}}{2}$

49. $r = 3\cos\theta = 1 + \cos\theta \Rightarrow \cos\theta = \dfrac{1}{2} \Rightarrow \theta = \pm\dfrac{\pi}{3}$

$A = 2\int_0^{\pi/3}\int_{1+\cos\theta}^{3\cos\theta} r\,dr\,d\theta = 2\int_0^{\pi/3}\left[\frac{r^2}{2}\right]_{1+\cos\theta}^{3\cos\theta} d\theta = \int_0^{\pi/3}\left[9\cos^2\theta - (1+\cos\theta)^2\right]d\theta$

$= \int_0^{\pi/3}\left(8\cos^2\theta - 2\cos\theta - 1\right)d\theta = \int_0^{\pi/3}\left[4(1 + \cos 2\theta) - 2\cos\theta - 1\right]d\theta$

$= \left[3\theta + 2\sin 2\theta - 2\sin\theta\right]_0^{\pi/3} = 3\left(\frac{\pi}{3}\right) + \sqrt{3} - \sqrt{3} = \pi$

51. $r = 4\sin 3\theta = 2 \Rightarrow \sin 3\theta = \dfrac{1}{2} \Rightarrow 3\theta = \dfrac{\pi}{6}, \dfrac{5\pi}{6} \Rightarrow \theta = \dfrac{\pi}{18}, \dfrac{5\pi}{18}$

$A = 3\int_{\pi/18}^{5\pi/18}\int_2^{4\sin 3\theta} r\,dr\,d\theta = 3\int_{\pi/18}^{5\pi/18}\left[\frac{r^2}{2}\right]_2^{4\sin 3\theta} d\theta = \frac{3}{2}\int_{\pi/18}^{5\pi/18}\left[(4\sin 3\theta)^2 - 4\right]d\theta$

$= \frac{3}{2}\int_{\pi/18}^{5\pi/18}\left[8(1 - \cos 6\theta) - 4\right]d\theta = \frac{3}{2}\left[4\theta - \frac{4}{3}\sin 6\theta\right]_{\pi/18}^{5\pi/18}$

$= \frac{3}{2}\left[\left(\frac{10}{9}\pi - \frac{4}{3}\left(\frac{-\sqrt{3}}{2}\right)\right) - \left(\frac{2\pi}{9} - \frac{4}{3}\left(\frac{\sqrt{3}}{2}\right)\right)\right] = \frac{4}{3}\pi + 2\sqrt{3}$

53. Draw the line joining $(0,0)$ and $\left(\sqrt{3}, 1\right)$. The line makes an angle $\theta = \dfrac{\pi}{6}$ with the x-axis. Note that

$x = \sqrt{3} \Rightarrow r\cos\theta = \sqrt{3}$ or $r = \sqrt{3}\sec\theta$. Also, $y = 1 \Rightarrow r\sin\theta = 1$ or $r = \csc\theta$. So, the area is

$A = \int_0^{\pi/6}\int_1^{\sqrt{3}\sec\theta} r\,dr\,d\theta + \int_{\pi/6}^{\pi/2}\int_1^{\csc\theta} r\,dr\,d\theta$

$= \left(\frac{\sqrt{3}}{2} - \frac{\pi}{12}\right) + \left(\frac{\sqrt{3}}{2} - \frac{\pi}{6}\right) = \sqrt{3} - \frac{\pi}{4}.$

Using geometry, subtract the area of the quarter-circle from the area of the rectangle: $A = \sqrt{3} - \dfrac{\pi}{4}$.

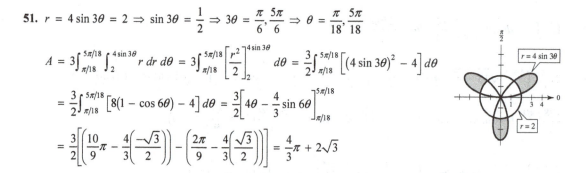

55. $\int_{-7}^{7} \int_{-\sqrt{49-x^2}}^{\sqrt{49-x^2}} 4000e^{-0.01\left(x^2+y^2\right)} \, dy \, dx = \int_{0}^{2\pi} \int_{0}^{7} 4000e^{-0.01r^2} \, r \, dr \, d\theta = \int_{0}^{2\pi} \left[-200{,}000e^{-0.01r^2}\right]_{0}^{7} d\theta$

$$= 2\pi(-200{,}000)\left(e^{-0.49}-1\right) = 400{,}000\pi\left(1-e^{-0.49}\right) \approx 486{,}788$$

57. Total volume $= V = \int_{0}^{2\pi} \int_{0}^{4} 25e^{-r^2/4} r \, dr \, d\theta = \int_{0}^{2\pi} \left[-50e^{-r^2/4}\right]_{0}^{4} d\theta = \int_{0}^{2\pi} -50\left(e^{-4}-1\right) d\theta = \left(1-e^{-4}\right)100\pi \approx 308.40524$

Let c be the radius of the hole that is removed.

$$\frac{1}{10}V = \int_{0}^{2\pi} \int_{0}^{c} 25e^{-r^2/4} r \, dr \, d\theta = \int_{0}^{2\pi} \left[-50e^{-r^2/4}\right]_{0}^{c} d\theta$$

$$= \int_{0}^{2\pi} -50\left(e^{-c^2/4}-1\right) d\theta \Rightarrow 30.84052 = 100\pi\left(1-e^{-c^2/4}\right)$$

$$\Rightarrow e^{-c^2/4} = 0.90183$$

$$-\frac{c^2}{4} = -0.10333$$

$$c^2 = 0.41331$$

$$c = 0.6429$$

$$\Rightarrow \text{diameter} = 2c = 1.2858$$

59. $\int_{\pi/4}^{\pi/2} \int_{0}^{5} r\sqrt{1+r^3} \, \sin\sqrt{\theta} \, dr \, d\theta \approx 56.051$

$\left[\textbf{Note:} \text{This integral equals } \left(\int_{\pi/4}^{\pi/2} \sin\sqrt{\theta} \, d\theta\right)\left(\int_{0}^{5} r\sqrt{1+r^3} \, dr\right).\right]$

61. False

Let $f(r, \theta) = r - 1$ where R is the circular sector $0 \le r \le 6$ and $0 \le \theta \le \pi$. Then,

$\int_{R} \int (r-1) \, dA > 0$ but $r - 1 \not> 0$ for all r.

63. (a) $I^2 = \int_{-\infty}^{\infty} \int_{-\infty}^{\infty} e^{-\left(x^2+y^2\right)/2} \, dA = 4\int_{0}^{\pi/2} \int_{0}^{\infty} e^{-r^2/2} r \, dr \, d\theta = 4\int_{0}^{\pi/2} \left[-e^{-r^2/2}\right]_{0}^{\infty} d\theta = 4\int_{0}^{\pi/2} d\theta = 2\pi$

(b) So, $I = \sqrt{2\pi}$.

65. (a) $\int_{2}^{4} \int_{y/\sqrt{3}}^{y} f \, dx \, dy$

(b) $\int_{2/\sqrt{3}}^{2} \int_{2}^{\sqrt{3}x} f \, dy \, dx$

$+ \int_{2}^{4/\sqrt{3}} \int_{x}^{\sqrt{3}x} f \, dy \, dx + \int_{4/\sqrt{3}}^{4} \int_{x}^{4} f \, dy \, dx$

(c) $\int_{\pi/4}^{\pi/3} \int_{2\csc\theta}^{4\csc\theta} fr \, dr \, d\theta$

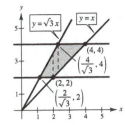

67. $\int_{0}^{\infty} \int_{0}^{\infty} ke^{-\left(x^2+y^2\right)} \, dy \, dx = \int_{0}^{\pi/2} \int_{0}^{\infty} ke^{-r^2} r \, dr \, d\theta$

$$= \int_{0}^{\pi/2} \left[-\frac{k}{2}e^{-r^2}\right]_{0}^{\infty} d\theta$$

$$= \int_{0}^{\pi/2} \frac{k}{2} \, d\theta = \frac{k\pi}{4}$$

For $f(x, y)$ to be a probability density function,

$$\frac{k\pi}{4} = 1$$

$$k = \frac{4}{\pi}.$$

Section 14.4   Center of Mass and Moments of Inertia

1. Use a double integral when the density of the lamina is not constant.

3. $m = \int_0^2 \int_0^2 xy \, dy \, dx$

$= \int_0^2 \left[\dfrac{xy^2}{2} \right]_0^2 dx$

$= \int_0^2 2x \, dx$

$= \left[x^2 \right]_0^2$

$= 4$

5. $m = \int_0^{\pi/2} \int_0^1 (r \cos \theta)(r \sin \theta) \, r \, dr \, d\theta$

$= \int_0^{\pi/2} \left[(\cos \theta \sin \theta) \dfrac{r^4}{4} \right]_0^1 d\theta$

$= \int_0^{\pi/2} \dfrac{1}{4} \cos \theta \sin \theta \, d\theta$

$= \left[\dfrac{1}{4} \cdot \dfrac{\sin^2 \theta}{2} \right]_0^{\pi/2}$

$= \dfrac{1}{8}$

7. (a) $m = \int_0^a \int_0^a k \, dy \, dx = ka^2$

$M_x = \int_0^a \int_0^a ky \, dy \, dx = \int_0^a \dfrac{ka^2}{2} \, dx = \dfrac{ka^3}{2}$

$M_y = \int_0^a \int_0^a kx \, dy \, dx = \dfrac{ka^3}{2}$

$\bar{x} = \dfrac{M_y}{m} = \dfrac{a}{2}, \bar{y} = \dfrac{M_x}{m} = \dfrac{a}{2}$

$(\bar{x}, \bar{y}) = \left(\dfrac{a}{2}, \dfrac{a}{2} \right)$ (center of square)

(b) $m = \int_0^a \int_0^a ky \, dy \, dx = \dfrac{1}{2} ka^3$

$M_x = \int_0^a \int_0^a ky^2 \, dy \, dx = \dfrac{1}{3} ka^4$

$M_y = \int_0^a \int_0^a kyx \, dy \, dx = \dfrac{1}{4} ka^4$

$\bar{x} = \dfrac{M_y}{m} = \dfrac{a}{2}, \qquad \bar{y} = \dfrac{M_x}{m} = \dfrac{2a}{3}$

$(\bar{x}, \bar{y}) = \left(\dfrac{a}{2}, \dfrac{2a}{3} \right)$

(c) $m = \int_0^a \int_0^a kx \, dy \, dx = \dfrac{1}{2} ka^3$

$M_x = \int_0^a \int_0^a kxy \, dy \, dx = \dfrac{1}{4} ka^4$

$M_y = \int_0^a \int_0^a kx^2 \, dy \, dx = \dfrac{1}{3} ka^3$

$\bar{x} = \dfrac{M_y}{m} = \dfrac{2a}{3}, \qquad \bar{y} = \dfrac{M_x}{m} = \dfrac{a}{2}$

$(\bar{x}, \bar{y}) = \left(\dfrac{2a}{3}, \dfrac{a}{2} \right)$

9. (a) $m = \int_0^a \int_0^y k \, dx \, dy = \dfrac{1}{2} ka^2$

$M_x = \int_0^a \int_0^y ky \, dx \, dy = \dfrac{1}{3} ka^3$

$M_y = \int_0^a \int_0^y kx \, dx \, dy = \dfrac{1}{6} ka^3$

$\bar{x} = \dfrac{M_y}{m} = \dfrac{a}{3} \qquad \bar{y} = \dfrac{M_x}{m} = \dfrac{2a}{3}$

$(\bar{x}, \bar{y}) = \left(\dfrac{a}{3}, \dfrac{2a}{3} \right)$

(b) $m = \int_0^a \int_0^y ky \, dx \, dy = \dfrac{1}{3} ka^3$

$M_x = \int_0^a \int_0^y ky^2 \, dx \, dy = \dfrac{1}{4} ka^4$

$M_y = \int_0^a \int_0^y kxy \, dx \, dy = \dfrac{1}{8} ka^4$

$x = \dfrac{M_y}{m} = \dfrac{3a}{8}, \qquad \bar{y} = \dfrac{M_x}{m} = \dfrac{3a}{4}$

$(\bar{x}, \bar{y}) = \left(\dfrac{3a}{8}, \dfrac{3a}{4} \right)$

(c) $m = \int_0^a \int_0^y kx \, dx \, dy = \dfrac{1}{6} ka^3$

$M_x = \int_0^a \int_0^y kxy \, dx \, dy = \dfrac{1}{8} ka^4$

$M_y = \int_0^a \int_0^y kx^2 \, dx \, dy = \dfrac{1}{12} ka^4$

$\bar{x} = \dfrac{M_y}{m} = \dfrac{a}{2} \qquad \bar{y} = \dfrac{M_x}{m} = \dfrac{3a}{4}$

$(\bar{x}, \bar{y}) = \left(\dfrac{a}{2}, \dfrac{3a}{4} \right)$

11. (a) The *x*-coordinate changes by 5: $(\bar{x}, \bar{y}) = \left(\dfrac{a}{2} + 5, \dfrac{a}{2}\right)$

(b) The *x*-coordinate changes by 5: $(\bar{x}, \bar{y}) = \left(\dfrac{a}{2} + 5, \dfrac{2a}{3}\right)$

(c) $\quad m = \displaystyle\int_5^{a+5}\int_0^a kx\, dy\, dx = \dfrac{1}{2}ka\left((a+5)^2 - 25\right)$

$\quad M_x = \displaystyle\int_5^{a+5}\int_0^a kxy\, dy\, dx = \dfrac{1}{4}ka^2\left((a+5)^2 - 25\right)$

$\quad M_y = \displaystyle\int_5^{a+5}\int_0^a kx^2\, dy\, dx = \dfrac{1}{3}ka\left((a+5)^3 - 125\right)$

$\quad \bar{x} = \dfrac{M_y}{m} = \dfrac{2\left[(a+5)^3 - 125\right]}{3\left[(a+5)^2 - 25\right]} = \dfrac{2(a^2 + 15a + 75)}{3(a + 10)}$

$\quad \bar{y} = \dfrac{M_x}{m} = \dfrac{a}{2}$

$\quad (\bar{x}, \bar{y}) = \left(\dfrac{2(a^2 + 15a + 75)}{3(a + 10)}, \dfrac{a}{2}\right)$

13. $\quad m = \displaystyle\int_0^1\int_0^{\sqrt{x}} ky\, dy\, dx = \dfrac{1}{4}k$

$\quad M_x = \displaystyle\int_0^1\int_0^{\sqrt{x}} ky^2\, dy\, dx = \dfrac{2}{15}k$

$\quad M_y = \displaystyle\int_0^1\int_0^{\sqrt{x}} kxy\, dy\, dx = \dfrac{1}{6}k$

$\quad \bar{x} = \dfrac{M_y}{m} = \dfrac{2}{3}$

$\quad \bar{y} = \dfrac{M_x}{m} = \dfrac{8}{15}$

$\quad (\bar{x}, \bar{y}) = \left(\dfrac{2}{3}, \dfrac{8}{15}\right)$

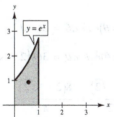

15. $\quad m = \displaystyle\int_1^4\int_0^{4/x} kx^2\, dy\, dx = 30k$

$\quad M_x = \displaystyle\int_1^4\int_0^{4/x} kx^2 y\, dy\, dx = 24k$

$\quad M_y = \displaystyle\int_1^4\int_0^{4/x} kx^3\, dy\, dx = 84k$

$\quad \bar{x} = \dfrac{M_y}{m} = \dfrac{84k}{30k} = \dfrac{14}{5}$

$\quad \bar{y} = \dfrac{M_y}{m} = \dfrac{24k}{30k} = \dfrac{4}{5}$

$\quad (\bar{x}, \bar{y}) = \left(\dfrac{14}{5}, \dfrac{4}{5}\right)$

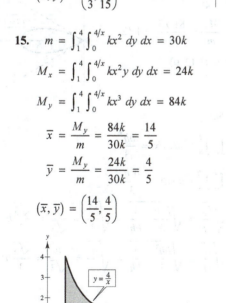

17. $\quad m = \displaystyle\int_0^1\int_0^{e^x} k\, dy\, dx = k(e - 1)$

$\quad M_x = \displaystyle\int_0^1\int_0^{e^x} ky\, dy\, dx = \dfrac{1}{4}k(e^2 - 1)$

$\quad M_y = \displaystyle\int_0^1\int_0^{e^x} kx\, dy\, dx = k$

$\quad \bar{x} = \dfrac{M_y}{m} = \dfrac{1}{e - 1},$

$\quad \bar{y} = \dfrac{M_x}{m} = \dfrac{e^2 - 1}{4(e - 1)} = \dfrac{e + 1}{4},$

$\quad (\bar{x}, \bar{y}) = \left(\dfrac{1}{e - 1}, \dfrac{e + 1}{4}\right)$

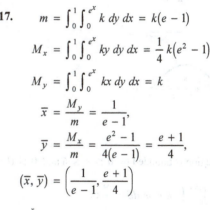

19. $\quad m = \displaystyle\int_{-2}^2\int_0^{4 - x^2} ky\, dy\, dx = \dfrac{256}{15}k$

$\quad M_x = \displaystyle\int_{-2}^2\int_0^{4 - x^2} ky^2\, dy\, dx = \dfrac{4096}{105}k$

$\quad \bar{x} = 0 \text{ (by symmetry)}$

$\quad \bar{y} = \dfrac{M_x}{m} = \dfrac{16}{7}$

$\quad (\bar{x}, \bar{y}) = \left(0, \dfrac{16}{7}\right)$

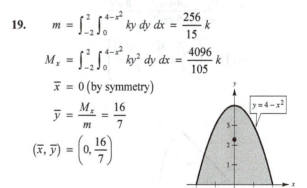

21. $m = \int_R \int k \, dA = \int_0^3 \int_0^{\sin(\pi x/3)} k \, dy \, dx$

$\qquad = \int_0^3 k \sin \dfrac{\pi x}{3} \, dx$

$\qquad = \left[-k \left(\dfrac{3}{\pi} \right) \cos \dfrac{\pi x}{3} \right]_0^3$

$\qquad = k \dfrac{3}{\pi} + k \dfrac{3}{\pi} = \dfrac{6k}{\pi}$

$M_x = \int_R \int ky \, dA = \int_0^3 \int_0^{\sin(\pi x/3)} ky \, dy \, dx = \dfrac{3k}{4}$

$M_y = \int_R \int kx \, dA = \dfrac{9k}{\pi}$

$\bar{x} = \dfrac{M_y}{m} = \dfrac{9k/\pi}{6k/\pi} = \dfrac{3}{2}$

$\bar{y} = \dfrac{M_x}{m} = \dfrac{3k/4}{6k/\pi} = \dfrac{\pi}{8}$

$(\bar{x}, \bar{y}) = \left(\dfrac{3}{2}, \dfrac{\pi}{8} \right)$

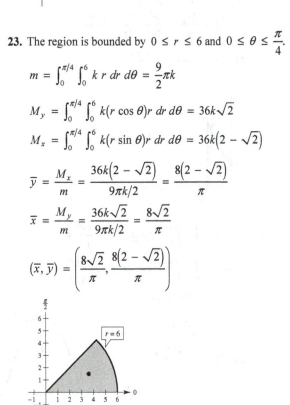

23. The region is bounded by $0 \le r \le 6$ and $0 \le \theta \le \dfrac{\pi}{4}$.

$m = \int_0^{\pi/4} \int_0^6 k \, r \, dr \, d\theta = \dfrac{9}{2}\pi k$

$M_y = \int_0^{\pi/4} \int_0^6 k(r \cos \theta) r \, dr \, d\theta = 36k\sqrt{2}$

$M_x = \int_0^{\pi/4} \int_0^6 k(r \sin \theta) r \, dr \, d\theta = 36k(2 - \sqrt{2})$

$\bar{y} = \dfrac{M_x}{m} = \dfrac{36k(2 - \sqrt{2})}{9\pi k/2} = \dfrac{8(2 - \sqrt{2})}{\pi}$

$\bar{x} = \dfrac{M_y}{m} = \dfrac{36k\sqrt{2}}{9\pi k/2} = \dfrac{8\sqrt{2}}{\pi}$

$(\bar{x}, \bar{y}) = \left(\dfrac{8\sqrt{2}}{\pi}, \dfrac{8(2 - \sqrt{2})}{\pi} \right)$

25. $m = \int_0^2 \int_0^{e^{-x}} kxy \, dy \, dx = \dfrac{1 - 5e^{-4}}{8} k$

$M_x = \int_0^2 \int_0^{e^{-x}} kxy^2 \, dy \, dx = \dfrac{1 - 7e^{-6}}{27} k$

$M_y = \int_0^2 \int_0^{e^{-x}} kx^2 y \, dy \, dx = \dfrac{1 - 13e^{-4}}{8} k$

$\bar{x} = \dfrac{M_y}{m} = \dfrac{e^4 - 13}{e^4 - 5}$

$\bar{y} = \dfrac{M_x}{m} = \dfrac{8(e^6 - 7)}{27(e^6 - 5e^2)}$

$(\bar{x}, \bar{y}) = \left(\dfrac{e^4 - 13}{e^4 - 5}, \dfrac{8(e^6 - 7)}{27(e^6 - 5e^2)} \right)$

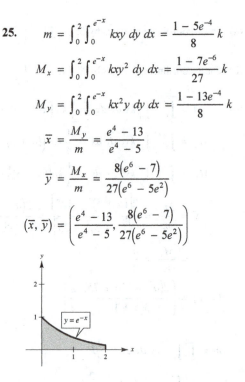

27. $\bar{y} = 0$ by symmetry

$m = \int_R \int k \, dA = \int_{-\pi/6}^{\pi/6} \int_0^{2\cos 3\theta} kr \, dr \, d\theta = \dfrac{k\pi}{3}$

$M_y = \int_R \int kx \, dA$

$\qquad = \int_{-\pi/6}^{\pi/6} \int_0^{2\cos 3\theta} kr^2 \cos \theta \, dr \, d\theta$

$\qquad = \dfrac{27\sqrt{3}}{40} k \approx 1.17k$

$\bar{x} = \dfrac{M_y}{m} = \dfrac{81\sqrt{3}}{40\pi} \approx 1.12$

$(\bar{x}, \bar{y}) \approx (1.12, 0)$

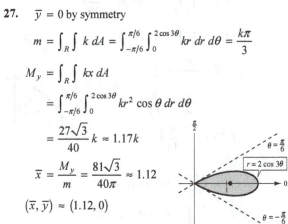

29. $m = bh$

$I_x = \int_0^b \int_0^h y^2 \, dy \, dx = \dfrac{bh^3}{3}$

$I_y = \int_0^b \int_0^h x^2 \, dy \, dx = \dfrac{b^3 h}{3}$

$\bar{\bar{x}} = \sqrt{\dfrac{I_y}{m}} = \sqrt{\dfrac{b^3 h}{3} \cdot \dfrac{1}{bh}} = \sqrt{\dfrac{b^2}{3}} = \dfrac{b}{\sqrt{3}} = \dfrac{\sqrt{3}}{3} b$

$\bar{\bar{y}} = \sqrt{\dfrac{I_x}{m}} = \sqrt{\dfrac{bh^3}{3} \cdot \dfrac{1}{bh}} = \sqrt{\dfrac{h^2}{3}} = \dfrac{h}{\sqrt{3}} = \dfrac{\sqrt{3}}{3} h$

31. $m = \pi a^2$

$$I_x = \int_R \int y^2 \, dA = \int_0^{2\pi} \int_0^a r^3 \sin^2 \theta \, dr \, d\theta = \frac{a^4 \pi}{4}$$

$$I_y = \int_R \int x^2 \, dA = \int_0^{2\pi} \int_0^a r^3 \cos^2 \theta \, dr \, d\theta = \frac{a^4 \pi}{4}$$

$$I_0 = I_x + I_y = \frac{a^4 \pi}{4} + \frac{a^4 \pi}{4} = \frac{a^4 \pi}{2}$$

$$\bar{\bar{x}} = \bar{\bar{y}} = \sqrt{\frac{I_x}{m}} = \sqrt{\frac{a^4 \pi}{4} \cdot \frac{1}{\pi a^2}} = \frac{a}{2}$$

33. $m = \dfrac{\pi a^2}{4}$

$$I_x = \int_R \int y^2 \, dA = \int_0^{\pi/2} \int_0^a r^3 \sin^2 \theta \, dr \, d\theta = \frac{\pi a^4}{16}$$

$$I_y = \int_R \int x^2 \, dA = \int_0^{\pi/2} \int_0^a r^3 \cos^2 \theta \, dr \, d\theta = \frac{\pi a^4}{16}$$

$$I_0 = I_x + I_y = \frac{\pi a^4}{16} + \frac{\pi a^4}{16} = \frac{\pi a^4}{8}$$

$$\bar{\bar{x}} = \bar{\bar{y}} = \sqrt{\frac{I_x}{m}} = \sqrt{\frac{\pi a^4}{16} \cdot \frac{4}{\pi a^2}} = \frac{a}{2}$$

35. $\rho = kx$

$$m = k\int_0^2 \int_0^{4-x^2} x \, dy \, dx = 4k$$

$$I_x = k\int_0^2 \int_0^{4-x^2} xy^2 \, dy \, dx = \frac{32k}{3}$$

$$I_y = k\int_0^2 \int_0^{4-x^2} x^3 \, dy \, dx = \frac{16k}{3}$$

$$I_0 = I_x + I_y = 16k$$

$$\bar{\bar{x}} = \sqrt{\frac{I_y}{m}} = \sqrt{\frac{16k/3}{4k}} = \sqrt{\frac{4}{3}} = \frac{2}{\sqrt{3}} = \frac{2\sqrt{3}}{3}$$

$$\bar{\bar{y}} = \sqrt{\frac{I_x}{m}} = \sqrt{\frac{32k/3}{4k}} = \sqrt{\frac{8}{3}} = \frac{4}{\sqrt{6}} = \frac{2\sqrt{6}}{3}$$

37. $\rho = kxy$

$$m = \int_0^4 \int_0^{\sqrt{x}} kxy \, dy \, dx = \frac{32k}{3}$$

$$I_x = \int_0^4 \int_0^{\sqrt{x}} kxy^3 \, dy \, dx = 16k$$

$$I_y = \int_0^4 \int_0^{\sqrt{x}} kx^3 y \, dy \, dx = \frac{512k}{5}$$

$$I_0 = I_x + I_y = \frac{592k}{5}$$

$$\bar{\bar{x}} = \sqrt{\frac{I_y}{m}} = \sqrt{\frac{512k}{5} \cdot \frac{3}{32k}} = \sqrt{\frac{48}{5}} = \frac{4\sqrt{15}}{5}$$

$$\bar{\bar{y}} = \sqrt{\frac{I_x}{m}} = \sqrt{\frac{16k}{1} \cdot \frac{3}{32k}} = \sqrt{\frac{3}{2}} = \frac{\sqrt{6}}{2}$$

39. $I = 2k\int_{-b}^{b} \int_0^{\sqrt{b^2 - x^2}} (x - a)^2 \, dy \, dx = 2k\int_{-b}^{b} (x - a)^2 \sqrt{b^2 - x^2} \, dx$

$$= 2k\left[\int_{-b}^{b} x^2 \sqrt{b^2 - x^2} \, dx - 2a\int_{-b}^{b} x\sqrt{b^2 - x^2} \, dx + a^2\int_{-b}^{b} \sqrt{b^2 - x^2} \, dx\right] = 2k\left[\frac{\pi b^4}{8} + 0 + \frac{\pi a^2 b^2}{2}\right] = \frac{k\pi b^2}{4}(b^2 + 4a^2)$$

41. $I = \int_{-a}^{a} \int_0^{\sqrt{a^2 - x^2}} ky(y - a)^2 \, dy \, dx$

$$= \int_{-a}^{a} k\left[\frac{y^4}{4} - \frac{2ay^3}{3} + \frac{a^2 y^2}{2}\right]_0^{\sqrt{a^2 - x^2}} dx$$

$$= \int_{-a}^{a} k\left[\frac{1}{4}(a^4 - 2a^2 x^2 + x^4) - \frac{2a}{3}\left(a^2\sqrt{a^2 - x^2} - x^2\sqrt{a^2 - x^2}\right) + \frac{a^2}{2}(a^2 - x^2)\right] dx$$

$$= k\left[\frac{1}{4}\left(a^4 x - \frac{2a^2 x^3}{3} + \frac{x^5}{5}\right) - \frac{2a}{3}\left[\frac{a^2}{2}\left(x\sqrt{a^2 - x^2} + a^2 \arcsin\frac{x}{a}\right)\right.\right.$$

$$\left.\left. - \frac{1}{8}\left(x(2x^2 - a^2)\sqrt{a^2 - x^2} + a^4 \arcsin\frac{x}{a}\right)\right] + \frac{a^2}{2}\left(a^2 x - \frac{x^3}{3}\right)\right]_{-a}^{a}$$

$$= 2k\left[\frac{1}{4}\left(a^5 - \frac{2}{3}a^5 + \frac{1}{5}a^5\right) - \frac{2a}{3}\left(\frac{a^4 \pi}{4} - \frac{a^4 \pi}{16}\right) + \frac{a^2}{2}\left(a^3 - \frac{a^3}{3}\right)\right] = 2k\left(\frac{7a^5}{15} - \frac{a^5 \pi}{8}\right) = ka^5\left(\frac{56 - 15\pi}{60}\right)$$

43. $\bar{y} = \dfrac{L}{2}, A = bL, h = \dfrac{L}{2}$

$$I_{\bar{y}} = \int_0^b \int_0^L \left(y - \dfrac{L}{2}\right)^2 dy \, dx$$

$$= \int_0^b \left[\dfrac{[y - (L/2)]^3}{3}\right]_0^L dx = \dfrac{L^3 b}{12}$$

$$y_a = \bar{y} - \dfrac{I_{\bar{y}}}{hA} = \dfrac{L}{2} - \dfrac{L^3 b/12}{(L/2)(bL)} = \dfrac{L}{3}$$

45. $\bar{y} = \dfrac{2L}{3}, A = \dfrac{bL}{2}, h = \dfrac{L}{3}$

$$I_{\bar{y}} = 2\int_0^{b/2} \int_{2Lx/b}^L \left(y - \dfrac{2L}{3}\right)^2 dy \, dx$$

$$= \dfrac{2}{3}\int_0^{b/2} \left[\left(y - \dfrac{2L}{3}\right)^3\right]_{2Lx/b}^L dx$$

$$= \dfrac{2}{3}\int_0^{b/2} \left[\dfrac{L}{27} - \left(\dfrac{2Lx}{b} - \dfrac{2L}{3}\right)^3\right] dx$$

$$= \dfrac{2}{3}\left[\dfrac{L^3 x}{27} - \dfrac{b}{8L}\left(\dfrac{2Lx}{b} - \dfrac{2L}{3}\right)^4\right]_0^{b/2} = \dfrac{L^3 b}{36}$$

$$y_a = \dfrac{2L}{3} - \dfrac{L^3 b/36}{L^2 b/6} = \dfrac{L}{2}$$

Section 14.5   Surface Area

1. If f and its first partial derivatives are continuous on the closed region R in the xy-plane, then the differential of the surface area given by $z = f(x, y)$ over R is

$$dS = \sqrt{1 + \left[f_x(x, y)\right]^2 + \left[f_y(x, y)\right]^2} \, dA.$$

3. $f(x, y) = 2x + 2y$

$f_x = f_y = 2$

$$\sqrt{1 + (f_x)^2 + (f_y)^2} = \sqrt{1 + 4 + 4} = 3$$

$$S = \int_0^4 \int_0^{4-x} 3 \, dy \, dx = 3\int_0^4 (4 - x) \, dx$$

$$= 3\left[4x - \dfrac{x^2}{2}\right]_0^4 = 24$$

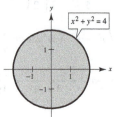

47. The object with a greater polar moment of inertia has more resistance, so more torque is required to twist the object.

49. Orient the xy-coordinate system so that L is along the y-axis and R is the first quadrant. Then the volume of the solid is

$$V = \int_R \int 2\pi x \, dA$$

$$= 2\pi \int_R \int x \, dA$$

$$= 2\pi \left(\dfrac{\int_R \int x \, dA}{\int_R \int dA}\right) \int_R \int dA$$

$$= 2\pi \bar{x} A.$$

By our positioning, $\bar{x} = r$. So, $V = 2\pi r A$.

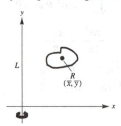

5. $f(x, y) = 4 + 5x + 6y$

$f_x = 5, f_y = 6$

$$\sqrt{1 + (f_x)^2 + (f_y)^2} = \sqrt{1 + 25 + 36} = \sqrt{62}$$

$$S = \int_0^{2\pi} \int_0^2 \sqrt{62} \, r \, dr \, d\theta$$

$$= \int_0^{2\pi} \left[\sqrt{62} \, \dfrac{r^2}{2}\right]_0^2 d\theta$$

$$= \int_0^{2\pi} 2\sqrt{62} \, d\theta$$

$$= 4\pi\sqrt{62}$$

7. $f(x, y) = 9 - x^2$

$f_x = -2x, f_y = 0$

$\sqrt{1 + (f_x)^2 + (f_y)^2} = \sqrt{1 + 4x^2}$

$S = \int_0^2 \int_0^2 \sqrt{1 + 4x^2} \, dy \, dx = 2\int_0^2 \sqrt{1 + 4x^2} \, dx$

$= 2\left[\frac{1}{4} \ln\left(\sqrt{1 + 4x^2} + 2x\right) + \frac{x}{2}\sqrt{1 + 4x^2}\right]_0^2$

$= 2\left[\frac{1}{4} \ln\left(\sqrt{17} + 4\right) + \sqrt{17}\right]$

$= 2\sqrt{17} + \frac{1}{2} \ln\left(4 + \sqrt{17}\right)$

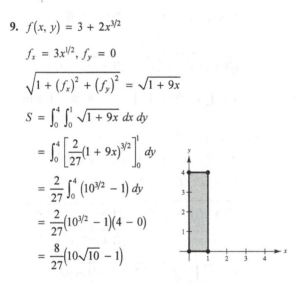

11. $f(x, y) = \ln|\sec x|$

$R = \left\{(x, y): 0 \le x \le \frac{\pi}{4}, 0 \le y \le \tan x\right\}$

$f_x = \tan x, f_y = 0$

$\sqrt{1 + (f_x)^2 + (f_y)^2} = \sqrt{1 + \tan^2 x} = \sec x$

$S = \int_0^{\pi/4} \int_0^{\tan x} \sec x \, dy \, dx$

$= \int_0^{\pi/4} \sec x \tan x \, dx$

$= [\sec x]_0^{\pi/4} = \sqrt{2} - 1$

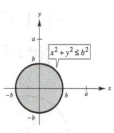

9. $f(x, y) = 3 + 2x^{3/2}$

$f_x = 3x^{1/2}, f_y = 0$

$\sqrt{1 + (f_x)^2 + (f_y)^2} = \sqrt{1 + 9x}$

$S = \int_0^4 \int_0^1 \sqrt{1 + 9x} \, dx \, dy$

$= \int_0^4 \left[\frac{2}{27}(1 + 9x)^{3/2}\right]_0^1 dy$

$= \frac{2}{27} \int_0^4 \left(10^{3/2} - 1\right) dy$

$= \frac{2}{27}\left(10^{3/2} - 1\right)(4 - 0)$

$= \frac{8}{27}\left(10\sqrt{10} - 1\right)$

13. $f(x, y) = \sqrt{x^2 + y^2}$

$R = \left\{(x, y): 0 \le f(x, y) \le 1\right\}$

$0 \le \sqrt{x^2 + y^2} \le 1, x^2 + y^2 \le 1$

$f_x = \frac{x}{\sqrt{x^2 + y^2}}, f_y = \frac{y}{\sqrt{x^2 + y^2}}$

$\sqrt{1 + (f_x)^2 + (f_y)^2} = \sqrt{1 + \frac{x^2}{x^2 + y^2} + \frac{y^2}{x^2 + y^2}} = \sqrt{2}$

$S = \int_{-1}^1 \int_{-\sqrt{1-x^2}}^{\sqrt{1-x^2}} \sqrt{2} \, dy \, dx = \int_0^{2\pi} \int_0^1 \sqrt{2}r \, dr \, d\theta = \sqrt{2}\pi$

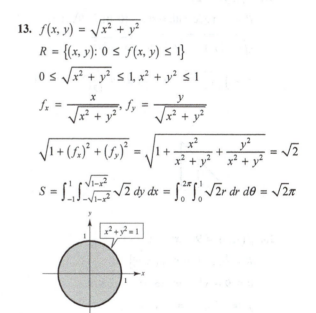

15. $f(x, y) = \sqrt{a^2 - x^2 - y^2}$

$R = \left\{(x, y): x^2 + y^2 \le b^2, 0 < b < a\right\}$

$f_x = \frac{-x}{\sqrt{a^2 - x^2 - y^2}}, f_y = \frac{-y}{\sqrt{a^2 - x^2 - y^2}}$

$\sqrt{1 + (f_x)^2 + (f_y)^2} = \sqrt{1 + \frac{x^2}{a^2 - x^2 - y^2} + \frac{y^2}{a^2 - x^2 - y^2}} = \frac{a}{\sqrt{a^2 - x^2 - y^2}}$

$S = \int_{-b}^b \int_{-\sqrt{b^2-x^2}}^{\sqrt{b^2-x^2}} \frac{a}{\sqrt{a^2 - x^2 - y^2}} \, dy \, dx = \int_0^{2\pi} \int_0^b \frac{a}{\sqrt{a^2 - r^2}} r \, dr \, d\theta = 2\pi a\left(a - \sqrt{a^2 - b^2}\right)$

17. $z = 12 - 3x - 2y$

$f_x = -3, f_y = -2$

$\sqrt{1 + (f_x)^2 + (f_y)^2} = \sqrt{1 + 9 + 4} = \sqrt{14}$

$S = \int_0^4 \int_0^{6-3/2x} \sqrt{14}\, dy\, dx = \int_0^4 \sqrt{14}\left(6 - \frac{3}{2}x\right) dx$

$\quad = \sqrt{14}\left[6x - \frac{3}{4}x^2\right]_0^4 = \sqrt{14}(24 - 12) = 12\sqrt{14}$

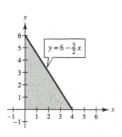

19. $z = \sqrt{25 - x^2 - y^2}$

$\sqrt{1 + (f_x)^2 + (f_y)^2} = \sqrt{1 + \dfrac{x^2}{25 - x^2 - y^2} + \dfrac{y^2}{25 - x^2 - y^2}} = \dfrac{5}{\sqrt{25 - x^2 - y^2}}$

$S = 2\int_{-3}^{3} \int_{-\sqrt{9-x^2}}^{\sqrt{9-x^2}} \dfrac{5}{\sqrt{25 - (x^2 + y^2)}}\, dy\, dx = 2\int_0^{2\pi} \int_0^3 \dfrac{5}{\sqrt{25 - r^2}} r\, dr\, d\theta = 20\pi$

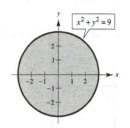

21. $f(x, y) = 2y + x^2$

$R = $ triangle with vertices $(0, 0), (1, 0), (1, 1)$

$\sqrt{1 + (f_x)^2 + (f_y)^2} = \sqrt{5 + 4x^2}$

$S = \int_0^1 \int_0^x \sqrt{5 + 4x^2}\, dy\, dx = \dfrac{1}{12}\left(27 - 5\sqrt{5}\right)$

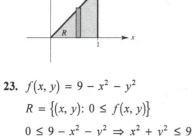

23. $f(x, y) = 9 - x^2 - y^2$

$R = \{(x, y): 0 \le f(x, y)\}$

$0 \le 9 - x^2 - y^2 \Rightarrow x^2 + y^2 \le 9$

$f_x = -2x, f_y = -2y$

$\sqrt{1 + (f_x)^2 + (f_y)^2} = \sqrt{1 + 4x^2 + 4y^2}$

$S = \int_{-3}^{3} \int_{-\sqrt{9-x^2}}^{\sqrt{9-x^2}} \sqrt{1 + 4x^2 + 4y^2}\, dy\, dx$

$\quad = \int_0^{2\pi} \int_0^3 \sqrt{1 + 4r^2}\, r\, dr\, d\theta$

$\quad = \dfrac{\pi}{6}\left(37\sqrt{37} - 1\right) \approx 117.3187$

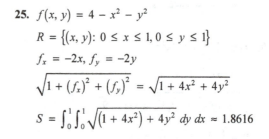

25. $f(x, y) = 4 - x^2 - y^2$

$R = \{(x, y): 0 \le x \le 1, 0 \le y \le 1\}$

$f_x = -2x, f_y = -2y$

$\sqrt{1 + (f_x)^2 + (f_y)^2} = \sqrt{1 + 4x^2 + 4y^2}$

$S = \int_0^1 \int_0^1 \sqrt{(1 + 4x^2) + 4y^2}\, dy\, dx \approx 1.8616$

27. $f(x, y) = e^{xy}$

$R = \{(x, y): 0 \le x \le 4, 0 \le y \le 10\}$

$f_x = ye^{xy}, f_y = xe^{xy}$

$\sqrt{1 + (f_x)^2 + (f_y)^2} = \sqrt{1 + y^2 e^{2xy} + x^2 e^{2xy}}$

$\quad = \sqrt{1 + e^{2xy}(x^2 + y^2)}$

$S = \int_0^4 \int_0^{10} \sqrt{1 + e^{2xy}(x^2 + y^2)}\, dy\, dx$

29. $f(x, y) = e^{-x} \sin y$

$f_x = -e^{-x} \sin y, f_y = e^{-x} \cos y$

$\sqrt{1 + (f_x^2) + (f_y^2)} = \sqrt{1 + e^{-2x} \sin^2 y + e^{-2x} \cos^2 y}$

$\quad = \sqrt{1 + e^{-2x}}$

$S = \int_{-2}^{2} \int_{-\sqrt{4-x^2}}^{\sqrt{4-x^2}} \sqrt{1 + e^{-2x}}\, dy\, dx$

31. No, the surface area is the same.

$z = f(x, y) \qquad$ and $\qquad z = f(x, y) + k$

have the same partial derivatives.

33. (a) Yes. For example, let R be the square given by $0 \le x \le 1, 0 \le y \le 1$,

and S the square parallel to R given by $0 \le x \le 1, 0 \le y \le 1, z = 1$.

(b) Yes. Let R be the region in part (a) and S the surface given by $f(x, y) = xy$.

(c) No.

35. (a) $V = \int_R \int f(x, y)$

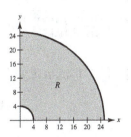

$= 8 \int_R \int \sqrt{625 - x^2 - y^2} \, dA$ where R is the region in the first quadrant

$= 8 \int_0^{\pi/2} \int_4^{25} \sqrt{625 - r^2} \, r \, dr \, d\theta = -4 \int_0^{\pi/2} \left[\frac{2}{3}(625 - r^2)^{3/2} \right]_4^{25} d\theta$

$= -\frac{8}{3} \left[0 - 609\sqrt{609} \right] \cdot \frac{\pi}{2} = 812\pi\sqrt{609} \ \text{cm}^3$

(b) $A = \int_R \int \sqrt{1 + (f_x)^2 + (f_y)^2} \, dA = 8 \int_R \int \sqrt{1 + \dfrac{x^2}{625 - x^2 - y^2} + \dfrac{y^2}{625 - x^2 - y^2}} \, dA$

$= 8 \int_R \int \dfrac{25}{\sqrt{625 - x^2 - y^2}} \, dA = 8 \int_0^{\pi/2} \int_4^{25} \dfrac{25}{\sqrt{625 - r^2}} \, r \, dr \, d\theta = \lim_{b \to 25^-} \left[-200\sqrt{625 - r^2} \right]_4^b \cdot \frac{\pi}{2} = 100\pi\sqrt{609} \ \text{cm}^2$

37. $f(x, y) = \sqrt{1 - x^2}; \ f_x = \dfrac{-x}{\sqrt{1^2 - x^2}}, \ f_y = 0$

$S = \int_R \int \sqrt{1 + (f_x)^2 + (f_y)^2} \, dA = 16 \int_0^1 \int_0^x \dfrac{1}{\sqrt{1 - x^2}} \, dy \, dx = 16 \int_0^1 \dfrac{x}{\sqrt{1 - x^2}} \, dx = \left[-16(1 - x^2)^{1/2} \right]_0^1 = 16$

Section 14.6   Triple Integrals and Applications

1. It represents the volume of the solid region Q.

3. $\int_0^3 \int_0^2 \int_0^1 (x + y + z) \, dx \, dz \, dy = \int_0^3 \int_0^2 \left[\dfrac{x^2}{2} + xy + xz \right]_0^1 dz \, dy = \int_0^3 \int_0^2 \left(\dfrac{1}{2} + y + z \right) dz \, dy = \int_0^3 \left[\dfrac{1}{2}z + yz + \dfrac{z^2}{2} \right]_0^2 dy$

$= \int_0^3 (1 + 2y + 2) \, dy = \left[3y + y^2 \right]_0^3 = 18$

5. $\int_0^1 \int_0^x \int_0^{\sqrt{x}\,y} x \, dz \, dy \, dx = \int_0^1 \int_0^x \left[xz \right]_0^{\sqrt{x}\,y} dy \, dx = \int_0^1 \int_0^x x^{3/2} y \, dy \, dx = \int_0^1 \left[x^{3/2} \dfrac{y^2}{2} \right]_0^x dx = \int_0^1 \dfrac{x^{7/2}}{2} \, dx = \left[\dfrac{x^{9/2}}{9} \right]_0^1 = \dfrac{1}{9}$

7. $\int_1^4 \int_0^1 \int_0^x 2ze^{-x^2} \, dy \, dx \, dz = \int_1^4 \int_0^1 \left[(2ze^{-x^2})y \right]_0^x dx \, dz = \int_1^4 \int_0^1 2zxe^{-x^2} \, dx \, dz$

$= \int_1^4 \left[-ze^{-x^2} \right]_0^1 dz = \int_1^4 z(1 - e^{-1}) \, dz = \left[(1 - e^{-1})\dfrac{z^2}{2} \right]_1^4 = \dfrac{15}{2}\left(1 - \dfrac{1}{e} \right)$

9. $\int_{-3}^4 \int_0^{\pi/2} \int_0^{1+3x} x \cos y \, dz \, dy \, dx = \int_{-3}^4 \int_0^{\pi/2} \left[z \, x \cos y \right]_0^{1+3x} dy \, dx = \int_{-3}^4 \int_0^{\pi/2} (1 + 3x)x \cos y \, dy \, dx = \int_{-3}^4 \left[(x + 3x^2) \sin y \right]_0^{\pi/2} dx$

$= \int_{-3}^4 (x + 3x^2) \, dx = \left[\dfrac{x^2}{2} + x^3 \right]_{-3}^4 = (8 + 64) - \left(\dfrac{9}{2} - 27 \right) = \dfrac{189}{2}$

11. $\int_0^3 \int_{-\sqrt{9-y^2}}^{\sqrt{9-y^2}} \int_0^{y^2} y \, dz \, dx \, dy = \dfrac{324}{5}$

13. $V = \int_0^7 \int_0^{(7-x)/2} \int_0^{7-x-2y} dz \, dy \, dx$

15. $V = \int_{-\sqrt{6}}^{\sqrt{6}} \int_{-\sqrt{6-x^2}}^{\sqrt{6-x^2}} \int_{0}^{6-x^2-y^2} dz\, dy\, dx = \int_{-\sqrt{6}}^{\sqrt{6}} \int_{-\sqrt{6-y^2}}^{\sqrt{6-y^2}} \int_{0}^{6-x^2-y^2} dz\, dx\, dy$

17. $z = \frac{1}{2}(x^2 + y^2) \Rightarrow 2z = x^2 + y^2$

$x^2 + y^2 + z^2 = 2z + z^2 = 80 \Rightarrow z^2 + 2z - 80 = 0 \Rightarrow (z - 8)(z + 10) = 0 \Rightarrow z = 8 \Rightarrow x^2 + y^2 = 2z = 16$

$V = \int_{-4}^{4} \int_{-\sqrt{16-x^2}}^{\sqrt{16-x^2}} \int_{1/2(x^2+y^2)}^{\sqrt{80-x^2-y^2}} dz\, dy\, dx$

19. $V = \int_{-2}^{2} \int_{0}^{4-y^2} \int_{0}^{x} dz\, dx\, dy = \int_{-2}^{2} \int_{0}^{4-y^2} x\, dx\, dy$

$= \frac{1}{2} \int_{-2}^{2} \left(4 - y^2\right)^2 dy = \int_{0}^{2} \left(16 - 8y^2 + y^4\right) dy = \left[16y - \frac{8}{3}y^3 + \frac{1}{5}y^5\right]_{0}^{2} = \frac{256}{15}$

21. $z = 6x^2,\ y = 3 - 3x;\ x, y, z \geq 0$

$V = \int_{0}^{1} \int_{0}^{3-3x} \int_{0}^{6x^2} dz\, dy\, dx$

$= \int_{0}^{1} \int_{0}^{3-3x} 6x^2\, dy\, dz$

$= \int_{0}^{1} 6x^2(3 - 3x)\, dx$

$= \int_{0}^{1} \left(18x^2 - 18x^3\right) dx$

$= 18\left[\frac{x^3}{3} - \frac{x^4}{4}\right]_{0}^{1}$

$= 18\left(\frac{1}{3} - \frac{1}{4}\right) = \frac{3}{2}$

23. $z = 2 - y,\ z = 4 - y^2,\ x = 0,\ x = 3,\ y = 0$

$V = \int_{0}^{3} \int_{0}^{2} \int_{2-y}^{4-y^2} dz\, dy\, dx$

$= \int_{0}^{3} \int_{0}^{2} \left[4 - y^2 - 2 + y\right] dy\, dx$

$= \int_{0}^{3} \left[2y - \frac{y^3}{3} + \frac{y^2}{2}\right]_{0}^{2} dx$

$= \int_{0}^{3} \left(4 - \frac{8}{3} + 2\right) dx$

$= \left[\frac{10}{3}x\right]_{0}^{3} = 10$

25.

$\int_{0}^{1} \int_{0}^{1} \int_{-1}^{-\sqrt{z}} dy\, dz\, dx$

27. Plane: $3x + 6y + 4z = 12$

$\int_{0}^{3} \int_{0}^{(12-4z)/3} \int_{0}^{(12-4z-3x)/6} dy\, dx\, dz$

29. Top cylinder: $y^2 + z^2 = 1$

Side plane: $x = y$

$\int_{0}^{1} \int_{0}^{x} \int_{0}^{\sqrt{1-y^2}} dz\, dy\, dx$

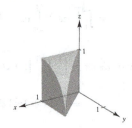

31. $\iint\int_Q xyz \, dV = \int_0^3 \int_0^1 \int_0^{5x} xyz \, dy \, dx \, dz$

$\qquad\qquad = \int_0^3 \int_0^5 \int_{y/5}^1 xyz \, dx \, dy \, dz$

$\qquad\qquad = \int_0^1 \int_0^{5x} \int_0^3 xyz \, dz \, dy \, dx$

$\qquad\qquad = \int_0^1 \int_0^3 \int_0^{5x} xyz \, dy \, dz \, dx$

$\qquad\qquad = \int_0^5 \int_0^3 \int_{y/5}^1 xyz \, dx \, dz \, dy$

$\qquad\qquad = \int_0^5 \int_{y/5}^1 \int_0^3 xyz \, dz \, dx \, dy$

Evaluating the first integral produces

$$\int_0^3 \int_0^1 \int_0^{5x} xyz \, dy \, dx \, dz = \int_0^3 \int_0^1 \left[\frac{1}{2}xy^2z\right]_0^{5x} dx \, dz = \int_0^3 \int_0^1 \frac{25}{2}x^3z \, dx \, dz = \int_0^3 \left[\frac{25}{8}x^4z\right]_0^1 dz = \int_0^3 \frac{25}{8}z \, dz = \left[\frac{25}{16}z^2\right]_0^3 = \frac{225}{16}.$$

33. $Q = \left\{(x, y, z): x^2 + y^2 \le 9, 0 \le z \le 4\right\}$

$\iint\int_Q xyz \, dV = \int_0^4 \int_{-3}^3 \int_{-\sqrt{9-x^2}}^{\sqrt{9-x^2}} xyz \, dy \, dx \, dz = \int_0^4 \int_{-3}^3 \int_{-\sqrt{9-y^2}}^{\sqrt{9-y^2}} xyz \, dx \, dy \, dz$

$\qquad\qquad = \int_{-3}^3 \int_0^4 \int_{-\sqrt{9-y^2}}^{\sqrt{9-y^2}} xyz \, dx \, dz \, dy = \int_{-3}^3 \int_{-\sqrt{9-y^2}}^{\sqrt{9-y^2}} \int_0^4 xyz \, dz \, dx \, dy$

$\qquad\qquad = \int_{-3}^3 \int_0^4 \int_{-\sqrt{9-x^2}}^{\sqrt{9-x^2}} xyz \, dy \, dz \, dx = \int_{-3}^3 \int_{-\sqrt{9-x^2}}^{\sqrt{9-x^2}} \int_0^4 xyz \, dz \, dy \, dx$

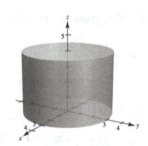

Evaluating the last integral produces

$$\int_{-3}^3 \int_{-\sqrt{9-x^2}}^{\sqrt{9-x^2}} \int_0^4 xyz \, dz \, dy \, dx = \int_{-3}^3 \int_{-\sqrt{9-x^2}}^{\sqrt{9-x^2}} \left[\frac{1}{2}xyz^2\right]_0^4 dy \, dx$$

$$= \int_{-3}^3 \int_{-\sqrt{9-x^2}}^{\sqrt{9-x^2}} 8xy \, dy \, dx$$

$$= \int_{-3}^3 \left[4xy^2\right]_{-\sqrt{9-x^2}}^{\sqrt{9-x^2}} dx$$

$$= \int_{-3}^3 0 \, dx$$

$$= 0.$$

35. $Q = \left\{(x, y, z): 0 \le y \le 1, 0 \le x \le 1 - y^2, 0 \le z \le 1 - y\right\}$

$\int_0^1 \int_0^{1-y^2} \int_0^{1-y} dz \, dx \, dy = \int_0^1 \int_0^{\sqrt{1-x}} \int_0^{1-y} dz \, dy \, dx$

$\qquad\qquad = \int_0^1 \int_0^{2z-z^2} \int_0^{1-z} dy \, dx \, dz + \int_0^1 \int_{2z-z^2}^1 \int_0^{\sqrt{1-x}} dy \, dx \, dz$

$\qquad\qquad = \int_0^1 \int_{1-\sqrt{1-x}}^1 \int_0^{1-z} dy \, dz \, dx + \int_0^1 \int_0^{1-\sqrt{1-x}} \int_0^{\sqrt{1-x}} dy \, dz \, dx$

$\qquad\qquad = \int_0^1 \int_0^{1-y} \int_0^{1-y^2} dx \, dz \, dy = \int_0^1 \int_0^{1-z} \int_0^{1-y^2} dx \, dy \, dz = \frac{5}{12}$

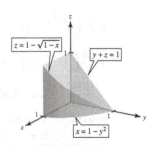

37. $m = k\int_0^6 \int_0^{4-(2x/3)} \int_0^{2-(y/2)-(x/3)} dz \, dy \, dx = 8k$

$\qquad M_{yz} = k\int_0^6 \int_0^{4-(2x/3)} \int_0^{2-(y/2)-(x/3)} x \, dz \, dy \, dx = 12k$

$\qquad \bar{x} = \dfrac{M_{yz}}{m} = \dfrac{12k}{8k} = \dfrac{3}{2}$

39. $m = k\int_0^4\int_0^4\int_0^{4-x} x\,dz\,dy\,dx = k\int_0^4\int_0^4 x(4-x)\,dy\,dx = \dfrac{kb^5}{4}$

$\qquad = 4k\int_0^4\left(4x - x^2\right)dx = \dfrac{128k}{3}$

$M_{xy} = k\int_0^4\int_0^4\int_0^{4-x} xz\,dz\,dy\,dx = k\int_0^4\int_0^4 x\dfrac{(4-x)^2}{2}\,dy\,dx$

$\qquad = 2k\int_0^4\left(16x - 8x^2 + x^3\right)dx = \dfrac{128k}{3}$

$\overline{z} = \dfrac{M_{xy}}{m} = 1$

41. $m = k\int_0^b\int_0^b\int_0^b xy\,dz\,dy\,dx = \dfrac{kb^5}{4}$

$M_{yz} = k\int_0^b\int_0^b\int_0^b x^2 y\,dz\,dy\,dx = \dfrac{kb^6}{6}$

$M_{xz} = k\int_0^b\int_0^b\int_0^b xy^2\,dz\,dy\,dx = \dfrac{kb^6}{6}$

$M_{xy} = k\int_0^b\int_0^b\int_0^b xyz\,dz\,dy\,dx = \dfrac{kb^6}{8}$

$\overline{x} = \dfrac{M_{yz}}{m} = \dfrac{kb^6/6}{kb^5/4} = \dfrac{2b}{3}$

$\overline{y} = \dfrac{M_{xz}}{m} = \dfrac{kb^6/6}{kb^5/4} = \dfrac{2b}{3}$

$\overline{z} = \dfrac{M_{xy}}{m} = \dfrac{kb^6/8}{kb^5/4} = \dfrac{b}{2}$

43. $\overline{x}$ will be greater than 2, whereas $\overline{y}$ and $\overline{z}$ will be unchanged.

45. $\overline{y}$ will be greater than 0, whereas $\overline{x}$ and $\overline{z}$ will be unchanged.

47. $m = \dfrac{1}{3}k\pi r^2 h$

$\overline{x} = \overline{y} = 0$ by symmetry

$M_{xy} = 4k\int_0^r\int_0^{\sqrt{r^2-x^2}}\int_{h\sqrt{x^2+y^2}/r}^{h} z\,dz\,dy\,dx = \dfrac{2kh^2}{r^2}\int_0^r\int_0^{\sqrt{r^2-x^2}}\left(r^2 - x^2 - y^2\right)dy\,dx = \dfrac{4kh^2}{3r^2}\int_0^r\left(r^2 - x^2\right)^{3/2}dx = \dfrac{k\pi r^2 h^2}{4}$

$\overline{z} = \dfrac{M_{xy}}{m} = \dfrac{k\pi r^2 h^2/4}{k\pi r^2 h/3} = \dfrac{3h}{4}$

$(\overline{x}, \overline{y}, \overline{z}) = \left(0, 0, \dfrac{3h}{4}\right)$

49. $m = \dfrac{128k\pi}{3}$

$\overline{x} = \overline{y} = 0$ by symmetry

$z = \sqrt{4^2 - x^2 - y^2}$

$M_{xy} = 4k\int_0^4\int_0^{\sqrt{4^2-x^2}}\int_0^{\sqrt{4^2-x^2-y^2}} z\,dz\,dy\,dx$

$\qquad = 2k\int_0^4\int_0^{\sqrt{4^2-x^2}}\left(4^2 - x^2 - y^2\right)dy\,dx = 2k\int_0^4\left[16y - x^2 y - \dfrac{1}{3}y^3\right]_0^{\sqrt{4^2-x^2}}dx$

$\qquad = \dfrac{4k}{3}\int_0^4\left(4^2 - x^2\right)^{3/2}dx$

$\qquad = \dfrac{1024k}{3}\int_0^{\pi/2}\cos^4\theta\,d\theta \quad (\text{let } x = 4\sin\theta)$

$\qquad = 64\pi k \quad \text{by Wallis's Formula}$

$\overline{z} = \dfrac{M_{xy}}{m} = \dfrac{64k\pi}{1}\cdot\dfrac{3}{128k\pi} = \dfrac{3}{2}$

$(\overline{x}, \overline{y}, \overline{z}) = \left(0, 0, \dfrac{3}{2}\right)$

51. $f(x, y) = \dfrac{5}{12}y$

$$m = k\int_0^{20}\int_0^{-(3/5)x+12}\int_0^{(5/12)y} dz\, dy\, dx = 200k$$

$$M_{yz} = k\int_0^{20}\int_0^{-(3/5)x+12}\int_0^{(5/12)y} x\, dz\, dy\, dx = 1000k$$

$$M_{xz} = k\int_0^{20}\int_0^{-(3/5)x+12}\int_0^{(5/12)y} y\, dz\, dy\, dx = 1200k$$

$$M_{xy} = k\int_0^{20}\int_0^{-(3/5)x+12}\int_0^{(5/12)y} z\, dz\, dy\, dx = 250k$$

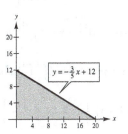

$$\overline{x} = \frac{M_{yz}}{m} = \frac{1000k}{200k} = 5$$

$$\overline{y} = \frac{M_{xz}}{m} = \frac{1200k}{200k} = 6$$

$$\overline{z} = \frac{M_{xy}}{m} = \frac{250k}{200k} = \frac{5}{4}$$

$$(\overline{x}, \overline{y}, \overline{z}) = \left(5, 6, \frac{5}{4}\right)$$

53. (a) $I_x = k\int_0^a\int_0^a\int_0^a (y^2 + z^2)\, dx\, dy\, dz = ka\int_0^a\int_0^a (y^2 + z^2)\, dy\, dz$

$$= ka\int_0^a \left[\frac{1}{3}y^3 + z^2 y\right]_0^a dz = ka\int_0^a \left(\frac{1}{3}a^3 + az^2\right) dz = \left[ka\left(\frac{1}{3}a^3 z + \frac{1}{3}az^3\right)\right]_0^a = \frac{2ka^5}{3}$$

$$I_x = I_y = I_z = \frac{2ka^5}{3} \text{ by symmetry}$$

(b) $I_x = k\int_0^a\int_0^a\int_0^a (y^2 + z^2)\, xyz\, dx\, dy\, dz = \dfrac{ka^2}{2}\int_0^a\int_0^a (y^3 z + yz^3)\, dy\, dz$

$$= \frac{ka^2}{2}\int_0^a \left[\frac{y^4 z}{4} + \frac{y^2 z^3}{2}\right]_0^a dz = \frac{ka^4}{8}\int_0^a (a^2 z + 2z^3)\, dz = \left[\frac{ka^4}{8}\left(\frac{a^2 z^2}{2} + \frac{2z^4}{4}\right)\right]_0^a = \frac{ka^8}{8}$$

$$I_x = I_y = I_z = \frac{ka^8}{8} \text{ by symmetry}$$

55. (a) $I_x = k\int_0^4\int_0^4\int_0^{4-x} (y^2 + z^2)\, dz\, dy\, dx = k\int_0^4\int_0^4 \left[y^2(4 - x) + \frac{1}{3}(4 - x)^3\right] dy\, dx$

$$= k\int_0^4 \left[\frac{y^3}{3}(4 - x) + \frac{y}{3}(4 - x)^3\right]_0^4 dx = k\int_0^4 \left[\frac{64}{3}(4 - x) + \frac{4}{3}(4 - x)^3\right] dx = k\left[-\frac{32}{3}(4 - x)^2 - \frac{1}{3}(4 - x)^4\right]_0^4 = 256k$$

$$I_y = k\int_0^4\int_0^4\int_0^{4-x} (x^2 + z^2)\, dz\, dy\, dx = k\int_0^4\int_0^4 \left[x^2(4 - x) + \frac{1}{3}(4 - x)^3\right] dy\, dx$$

$$= 4k\int_0^4 \left[4x^2 - x^3 + \frac{1}{3}(4 - x)^3\right] dx = 4k\left[\frac{4}{3}x^3 - \frac{1}{4}x^4 - \frac{1}{12}(4 - x)^4\right]_0^4 = \frac{512k}{3}$$

$$I_z = k\int_0^4\int_0^4\int_0^{4-x} (x^2 + y^2)\, dz\, dy\, dx = k\int_0^4\int_0^4 (x^2 + y^2)(4 - x)\, dy\, dx$$

$$= k\int_0^4 \left[\left(x^2 y + \frac{y^3}{3}\right)(4 - x)\right]_0^4 dx = k\int_0^4 \left(4x^2 + \frac{64}{3}\right)(4 - x)\, dx = 256k$$

(b) $I_x = k\int_0^4\int_0^4\int_0^{4-x} y(y^2 + z^2)\, dz\, dy\, dx = k\int_0^4\int_0^4 \left[y^3(4-x) + \frac{1}{3}y(4-x)^3\right] dy\, dx$

$\qquad = k\int_0^4 \left[\frac{y^4}{4}(4-x) + \frac{y^2}{6}(4-x)^3\right]_0^4 dx = k\int_0^4 \left[64(4-x) + \frac{8}{3}(4-x)^3\right] dx = k\left[-32(4-x)^2 - \frac{2}{3}(4-x)^4\right]_0^4 = \frac{2048k}{3}$

$I_y = k\int_0^4\int_0^4\int_0^{4-x} y(x^2 + z^2)\, dz\, dy\, dx = k\int_0^4\int_0^4 \left[x^2 y(4-x) + \frac{1}{3}y(4-x)^3\right] dy\, dx$

$\qquad = 8k\int_0^4 \left[4x^2 - x^3 + \frac{1}{3}(4-x)^3\right] dx = 8k\left[\frac{4}{3}x^3 - \frac{1}{4}x^4 - \frac{1}{12}(4-x)^4\right]_0^4 = \frac{1024k}{3}$

$I_z = k\int_0^4\int_0^4\int_0^{4-x} y(x^2 + y^2)\, dz\, dy\, dx = k\int_0^4\int_0^4 (x^2 y + y^3)(4-x)\, dx$

$\qquad = k\int_0^4 \left[\left(\frac{x^2 y^2}{2} + \frac{y^4}{4}\right)(4-x)\right]_0^4 dx = k\int_0^4 (8x^2 + 64)(4-x)\, dx$

$\qquad = 8k\int_0^4 (32 - 8x + 4x^2 - x^3)\, dx = \left[8k\left(32x - 4x^2 + \frac{4}{3}x^3 - \frac{1}{4}x^4\right)\right]_0^4 = \frac{2048k}{3}$

57. $I_{xy} = k\int_{-L/2}^{L/2}\int_{-a}^{a}\int_{-\sqrt{a^2-x^2}}^{\sqrt{a^2-x^2}} z^2\, dz\, dx\, dy = k\int_{-L/2}^{L/2}\int_{-a}^{a} \frac{2}{3}(a^2 - x^2)\sqrt{a^2 - x^2}\, dx\, dy$

$\qquad = \frac{2}{3}\int_{-L/2}^{L/2} k\left[\frac{a^2}{2}\left(x\sqrt{a^2 - x^2} + a^2 \arcsin\frac{x}{a}\right) - \frac{1}{8}\left(x(2x^2 - a^2)\sqrt{x^2 - a^2} + a^4 \arcsin\frac{x}{a}\right)\right]_{-a}^{a} dy$

$\qquad = \frac{2k}{3}\int_{-L/2}^{L/2} 2\left(\frac{a^4 \pi}{4} - \frac{a^4 \pi}{16}\right) dy = \frac{a^4 \pi L k}{4}$

Because $m = \pi a^2 Lk$, $I_{xy} = ma^2/4$.

$I_{xz} = k\int_{-L/2}^{L/2}\int_{-a}^{a}\int_{-\sqrt{a^2-x^2}}^{\sqrt{a^2-x^2}} y^2\, dz\, dx\, dy = 2k\int_{-L/2}^{L/2}\int_{-a}^{a} y^2\sqrt{a^2 - x^2}\, dx\, dy$

$\qquad = 2k\int_{-L/2}^{L/2} \left[\frac{y^2}{2}\left(x\sqrt{a^2 - x^2} + a^2 \arcsin\frac{x}{a}\right)\right]_{-a}^{a} dy = k\pi a^2\int_{-L/2}^{L/2} y^2\, dy = \frac{2k\pi a^2}{3}\left(\frac{L^3}{8}\right) = \frac{1}{12}mL^2$

$I_{yz} = k\int_{-L/2}^{L/2}\int_{-a}^{a}\int_{-\sqrt{a^2-x^2}}^{\sqrt{a^2-x^2}} x^2\, dz\, dx\, dy = 2k\int_{-L/2}^{L/2}\int_{-a}^{a} x^2\sqrt{a^2 - x^2}\, dx\, dy$

$\qquad = 2k\int_{-L/2}^{L/2} \frac{1}{8}\left[x(2x^2 - a^2)\sqrt{a^2 - x^2} + a^4 \arcsin\frac{x}{a}\right]_{-a}^{a} dy = \frac{ka^4 \pi}{4}\int_{-L/2}^{L/2} dy = \frac{ka^4 \pi L}{4} = \frac{ma^2}{4}$

$I_x = I_{xy} + I_{xz} = \frac{ma^2}{4} + \frac{mL^2}{12} = \frac{m}{12}(3a^2 + L^2)$

$I_y = I_{xy} + I_{yz} = \frac{ma^2}{4} + \frac{ma^2}{4} = \frac{ma^2}{2}$

$I_z = I_{xz} + I_{yz} = \frac{mL^2}{12} + \frac{ma^2}{4} = \frac{m}{12}(3a^2 + L^2)$

59. $\int_{-1}^{1}\int_{-1}^{1}\int_{0}^{1-x} (x^2 + y^2)\sqrt{x^2 + y^2 + z^2}\, dz\, dy\, dx$

61. $\rho = kz$

(a) $m = \int_{-2}^{2} \int_{-\sqrt{4-x^2}}^{\sqrt{4-x^2}} \int_{0}^{4-x^2-y^2} (kz)\, dz\, dy\, dx \left(= \dfrac{32k\pi}{3}\right)$

(b) $\bar{x} = \bar{y} = 0$ by symmetry

$\bar{z} = \dfrac{M_{xy}}{m} = \dfrac{1}{m}\int_{-2}^{2} \int_{-\sqrt{4-x^2}}^{\sqrt{4-x^2}} \int_{0}^{4-x^2-y^2} kz^2\, dz\, dy\, dx\ (= 2)$

(c) $I_z = \int_{-2}^{2} \int_{-\sqrt{4-x^2}}^{\sqrt{4-x^2}} \int_{0}^{4-x^2-y^2} (x^2 + y^2)kz\, dz\, dy\, dx \left(= \dfrac{32k\pi}{3}\right)$

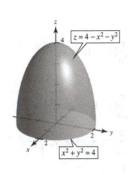

63. $V = 1$ (unit cube)

Average value $= \dfrac{1}{V}\iiint_Q f(x, y, z)\, dV = \int_0^1 \int_0^1 \int_0^1 (z^2 + 4)\, dx\, dy\, dz = \int_0^1 \int_0^1 (z^2 + 4)\, dy\, dz$

$= \int_0^1 (z^2 + 4)\, dz = \left[\dfrac{z^3}{3} + 4z\right]_0^1 = \dfrac{1}{3} + 4 = \dfrac{13}{3}$

65. $V = \dfrac{1}{3}$ base $\times$ height $= \dfrac{1}{3}\left(\dfrac{1}{2}(2)(2)\right)(2) = \dfrac{4}{3}$

$f(x, y, z) = x + y + z$

Plane: $x + y + z = 2$

Average value $= \dfrac{1}{V}\iiint_Q f(x, y, z)\, dV$

$= \dfrac{3}{4}\int_0^2 \int_0^{2-x} \int_0^{2-x-y} (x + y + z)\, dz\, dy\, dx$

$= \dfrac{3}{4}\int_0^2 \int_0^{2-x} \dfrac{1}{2}(2 - x - y)(x + y + 2)\, dy\, dx$

$= \dfrac{3}{4}\int_0^2 \dfrac{1}{6}(x + 4)(x - 2)^2\, dx = \dfrac{3}{4}(2) = \dfrac{3}{2}$

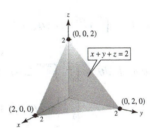

67. Because the density increases as you move away from the axis of symmetry, the moment of inertia will increase.

69. The region of integration is a cube:

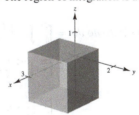

Answer: (b)

71. $1 - 2x^2 - y^2 - 3z^2 \geq 0$

$2x^2 + y^2 + 3z^2 \leq 1$

$Q = \{(x, y, z)\colon 2x^2 + y^2 + 3z^2 \leq 1\}$ ellipsoid

$\int_{-1/\sqrt{2}}^{1/\sqrt{2}} \int_{-\sqrt{1-2x^2}}^{\sqrt{1-2x^2}} \int_{-\sqrt{(1-2x^2-y^2)/3}}^{\sqrt{(1-2x^2-y^2)/3}} (1 - 2x^2 - y^2 - 3z^2)\, dz\, dy\, dx \approx 0.684$

Exact value: $\dfrac{4\sqrt{6}\pi}{45}$

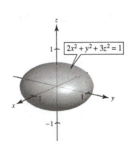

73. Let $y_k = 1 - x_k$.

$$\frac{\pi}{2n}(x_1 + \cdots + x_n) = \frac{\pi}{2n}(n - y_1 - y_2 - \cdots - y_n) = \frac{\pi}{2} - \frac{\pi}{2n}(y_1 + \cdots + y_n)$$

So,

$$I_1 = \int_0^1 \int_0^1 \cdots \int_0^1 \cos^2\left\{\frac{\pi}{2n}(x_1 + \cdots + x_n)\right\} dx_1 \, dx_2 \cdots dx_n$$

$$= \int_1^0 \int_1^0 \cdots \int_1^0 \sin^2\left\{\frac{\pi}{2n}(y_1 + \cdots + y_n)\right\}(-dy_1)(-dy_2)\cdots(-dy_n) = \int_0^1 \int_0^1 \cdots \int_0^1 \sin^2\left\{\frac{\pi}{2n}(x_1 + \cdots + x_n)\right\} dx_1 \, dx_2 \cdots dx_n = I_2$$

$$I_1 + I_2 = 1 \implies I_1 = \frac{1}{2}.$$

Finally, $\lim_{n \to \infty} I_1 = \frac{1}{2}$.

Section 14.7   Triple Integrals in Other Coordinates

1. Some solids are represented by equations involving x^2 and y^2. Often, converting these equations to cylindrical or spherical coordinates yields equations you can work with more easily.

3. $\int_{-1}^5 \int_0^{\pi/2} \int_0^3 r \cos \theta \, dr \, d\theta \, dz = \int_{-1}^5 \int_0^{\pi/2} \frac{9}{2} \cos \theta \, d\theta \, dz = \int_{-1}^5 \left[\frac{9}{2} \sin \theta\right]_0^{\pi/2} dz = \int_{-1}^5 \frac{9}{2} \, dz = \left[\frac{9}{2} z\right]_{-1}^5 = \frac{9}{2}(5 - (-1)) = 27$

5. $\int_0^{\pi/2} \int_0^{\cos \theta} \int_0^{3+r^2} 2r \sin \theta \, dz \, dr \, d\theta = \int_0^{\pi/2} \int_0^{\cos \theta} 2r \sin \theta\left(3 + r^2\right) dr \, d\theta$

$$= \int_0^{\pi/2} \left[\sin \theta\left(3r^2 + \frac{r^4}{2}\right)\right]_0^{\cos \theta} d\theta$$

$$= \int_0^{\pi/2} \sin \theta\left(3 \cos^2 \theta + \frac{1}{2} \cos^4 \theta\right) d\theta$$

$$= \left[-\cos^3 \theta - \frac{\cos^5 \theta}{10}\right]_0^{\pi/2}$$

$$= 1 + \frac{1}{10} = \frac{11}{10}$$

7. $\int_0^{2\pi} \int_0^{\pi/2} \int_0^{\sin \phi} \rho \cos \phi \, d\rho \, d\phi \, d\theta = \int_0^{2\pi} \int_0^{\pi/2} \left[\frac{1}{2}\rho^2 \cos \phi\right]_0^{\sin \phi} d\phi \, d\theta$

$$= \int_0^{2\pi} \int_0^{\pi/2} \frac{1}{2} \sin^2 \phi \cos \phi \, d\phi \, d\theta$$

$$= \int_0^{2\pi} \left[\frac{1}{6} \sin^3 \phi\right]_0^{\pi/2} d\theta$$

$$= \int_0^{2\pi} \frac{1}{6} \, d\theta = \frac{\pi}{3}$$

9. $\int_0^4 \int_0^z \int_0^{\pi/2} re^r \, d\theta \, dr \, dz = \pi\left(e^4 + 3\right)$

11. $\int_0^{\pi/2} \int_0^3 \int_0^{e^{-r^2}} r \, dz \, dr \, d\theta = \int_0^{\pi/2} \int_0^3 re^{-r^2} \, dr \, d\theta$

$$= \int_0^{\pi/2} \left[-\frac{1}{2} e^{-r^2} \right]_0^3 d\theta$$

$$= \int_0^{\pi/2} \frac{1}{2} (1 - e^{-9}) \, d\theta$$

$$= \frac{\pi}{4} (1 - e^{-9})$$

13. $\int_0^{2\pi} \int_{\pi/6}^{\pi/2} \int_0^4 \rho^2 \sin \phi \, d\rho \, d\phi \, d\theta = \frac{64}{3} \int_0^{2\pi} \int_{\pi/6}^{\pi/2} \sin \phi \, d\phi \, d\theta$

$$= \frac{64}{3} \int_0^{2\pi} [-\cos \phi]_{\pi/6}^{\pi/2} \, d\theta$$

$$= \frac{32\sqrt{3}}{3} \int_0^{2\pi} d\theta$$

$$= \frac{64\sqrt{3}\pi}{3}$$

15. Note that $(x - 3)^2 + y^2 = 9$ is equivalent to
$r = 6 \cos \theta, 0 \le \theta \le \pi$.

$V = 2\int_0^\pi \int_0^{6\cos\theta} \int_0^{\sqrt{36-r^2}} r \, dz \, dr \, d\theta$

$$= 2 \int_0^\pi \int_0^{6\cos\theta} r\sqrt{36 - r^2} \, dr \, d\theta$$

$$= 2 \int_0^\pi \left[-\frac{1}{3}(36 - r^2)^{3/2} \right]_0^{6\cos\theta} d\theta$$

$$= -\frac{2}{3} \int_0^\pi \left[(36 - 36\cos^2\theta)^{3/2} - 216 \right] d\theta$$

$$= -\frac{2}{3} \int_0^\pi (216 \sin^3\theta - 216) \, d\theta$$

$$= -144 \int_0^\pi \left[(1 - \cos^2\theta) \sin\theta - 1 \right] d\theta$$

$$= -144 \left[-\cos\theta + \frac{\cos^3\theta}{3} - \theta \right]_0^\pi$$

$$= -144 \left[\left(1 - \frac{1}{3} - \pi \right) - \left(-1 + \frac{1}{3} \right) \right]$$

$$= -144 \left(\frac{4}{3} - \pi \right)$$

$$= 48(3\pi - 4)$$

17. In the xy-plane, $2x = 2x^2 + 2y^2 \Rightarrow$

$0 = x^2 - x + y^2 \Rightarrow (x^2 - x + 1/4) + y^2 = 1/4$

$$\Rightarrow (x - 1/2)^2 + y^2 = (1/2)^2$$

In polar coordinates, use $r = \cos\theta$ for this circle.

$V = \int_0^\pi \int_0^{\cos\theta} \int_{2r^2}^{2r\cos\theta} r \, dz \, dr \, d\theta$

$$= \int_0^\pi \int_0^{\cos\theta} (2r^2 \cos\theta - 2r^3) \, dr \, d\theta$$

$$= \int_0^\pi \left[\frac{2r^3}{3} \cos\theta - \frac{r^4}{2} \right]_0^{\cos\theta} d\theta$$

$$= \int_0^\pi \left(\frac{2}{3} \cos^4\theta - \frac{\cos^4\theta}{2} \right) d\theta$$

$$= \frac{1}{6} \int_0^\pi \cos^4\theta \, d\theta = \frac{\pi}{16}$$

19. $V = 2 \int_0^\pi \int_0^{5\cos\theta} \int_0^{\sqrt{25-r^2}} r \, dz \, dr \, d\theta$

$$= 2 \int_0^\pi \int_0^{5\cos\theta} r\sqrt{25 - r^2} \, dr \, d\theta$$

$$= 2 \int_0^\pi \left[-\frac{1}{3}(25 - r^2)^{3/2} \right]_0^{5\cos\theta} d\theta$$

$$= -\frac{2}{3} \int_0^\pi \left[(25 - 25\cos^2\theta)^{3/2} - 125 \right] d\theta$$

$$= \frac{250}{3} \int_0^\pi (1 - \sin^3\theta) \, d\theta$$

$$= \frac{250}{3} \left[\theta + \cos\theta - \frac{\cos^3\theta}{3} \right]_0^\pi$$

$$= \frac{250}{3} \left[\left(\pi - 1 + \frac{1}{3} \right) - \left(1 - \frac{1}{3} \right) \right]$$

$$= \frac{250}{3} \left(\pi - \frac{4}{3} \right)$$

$$= \frac{250}{9}(3\pi - 4)$$

21. $m = \int_0^{2\pi} \int_0^2 \int_0^{9 - r\cos\theta - 2r\sin\theta} (kr)r \, dz \, dr \, d\theta$

$$= \int_0^{2\pi} \int_0^2 kr^2(9 - r\cos\theta - 2r\sin\theta) \, dr \, d\theta$$

$$= \int_0^{2\pi} k \left[3r^3 - \frac{r^4}{4} \cos\theta - \frac{r^4}{2} \sin\theta \right]_0^2 d\theta$$

$$= \int_0^{2\pi} k[24 - 4\cos\theta - 8\sin\theta] \, d\theta$$

$$= k[24\theta - 4\sin\theta + 8\cos\theta]_0^{2\pi}$$

$$= k[48\pi + 8 - 8] = 48k\pi$$

23. $z = h - \dfrac{h}{r_0}\sqrt{x^2 + y^2} = \dfrac{h}{r_0}(r_0 - r)$

$V = 4\displaystyle\int_0^{\pi/2}\int_0^{r_0}\int_0^{h(r_0-r)/r_0} r \, dz \, dr \, d\theta$

$\quad = \dfrac{4h}{r_0}\displaystyle\int_0^{\pi/2}\int_0^{r_0}\left(r_0 r - r^2\right) dr \, d\theta$

$\quad = \dfrac{4h}{r_0}\displaystyle\int_0^{\pi/2}\dfrac{r_0^3}{6}\, d\theta$

$\quad = \dfrac{4h}{r_0}\left(\dfrac{r_0^3}{6}\right)\left(\dfrac{\pi}{2}\right) = \dfrac{1}{3}\pi r_0^2 h$

25. $\rho = k\sqrt{x^2 + y^2} = kr$

$\overline{x} = \overline{y} = 0$ by symmetry

$m = 4k\displaystyle\int_0^{\pi/2}\int_0^{r_0}\int_0^{h(r_0-r)/r_0} r^2 \, dz \, dr \, d\theta = \dfrac{1}{6}k\pi r_0^3 h$

$M_{xy} = 4k\displaystyle\int_0^{\pi/2}\int_0^{r_0}\int_0^{h(r_0-r)/r_0} r^2 \, z \, dz \, dr \, d\theta$

$\quad = \dfrac{1}{30}k\pi r_0^3\, h^2$

$\overline{z} = \dfrac{M_{xy}}{m} = \dfrac{k\pi r_0^3 h^2/30}{k\pi r_0^3 h/6} = \dfrac{h}{5}$

$(\overline{x}, \overline{y}, \overline{z}) = \left(0, 0, \dfrac{h}{5}\right)$

27. $I_z = 4k\displaystyle\int_0^{\pi/2}\int_0^{r_0}\int_0^{h(r_0-r)/r_0} r^3 \, dz \, dr \, d\theta$

$\quad = \dfrac{4kh}{r_0}\displaystyle\int_0^{\pi/2}\int_0^{r_0}\left(r_0 r^3 - r^4\right) dr \, d\theta$

$\quad = \dfrac{4kh}{r_0}\left(\dfrac{r_0^5}{20}\right)\left(\dfrac{\pi}{2}\right) = \dfrac{1}{10}k\pi r_0^4 h$

Because the mass of the core is $m = kV = k\left(\dfrac{1}{3}\pi r_0^2 h\right)$

from Exercise 25, we have $k = 3m/\pi r_0^2 h$. So,

$I_z = \dfrac{1}{10}k\pi r_0^4 h = \dfrac{1}{10}\left(\dfrac{3m}{\pi r_0^2 h}\right)\pi r_0^4 h = \dfrac{3}{10}m r_0^2.$

29. $m = k\left(\pi b^2 h - \pi a^2 h\right) = k\pi h\left(b^2 - a^2\right)$

$I_z = 4k\displaystyle\int_0^{\pi/2}\int_a^b\int_0^h r^3 \, dz \, dr \, d\theta$

$\quad = 4kh\displaystyle\int_0^{\pi/2}\int_a^b r^3 \, dr \, d\theta = kh\int_0^{\pi/2}\left(b^4 - a^4\right) d\theta$

$\quad = \dfrac{k\pi\left(b^4 - a^4\right)h}{2} = \dfrac{k\pi\left(b^2 - a^2\right)\left(b^2 + a^2\right)h}{2}$

$\quad = \dfrac{1}{2}m\left(a^2 + b^2\right)$

31. $V = \displaystyle\int_0^{2\pi}\int_{\pi/4}^{\pi/2}\int_0^3 \rho^2 \sin\phi \, d\rho \, d\phi \, d\theta$

$\quad = \displaystyle\int_0^{2\pi}\int_{\pi/4}^{\pi/2} 9 \sin\phi \, d\phi \, d\theta$

$\quad = \displaystyle\int_0^{2\pi}\left[-9\cos\phi\right]_{\pi/4}^{\pi/2} d\theta$

$\quad = \displaystyle\int_0^{2\pi} 9\left(\dfrac{\sqrt{2}}{2}\right) d\theta = 18\pi\left(\dfrac{\sqrt{2}}{2}\right) = 9\pi\sqrt{2}$

33. $V = \displaystyle\int_0^{2\pi}\int_0^{\pi}\int_0^{4\sin\phi} \rho^2 \sin\phi \, d\rho \, d\phi \, d\theta = 16\pi^2$

35. $m = 8k\displaystyle\int_0^{\pi/2}\int_0^{\pi/2}\int_0^a \rho^3 \sin\phi \, d\rho \, d\theta \, d\phi$

$\quad = 2ka^4\displaystyle\int_0^{\pi/2}\int_0^{\pi/2} \sin\phi \, d\theta \, d\phi$

$\quad = k\pi a^4\displaystyle\int_0^{\pi/2}\sin\phi \, d\phi = \left[k\pi a^4(-\cos\phi)\right]_0^{\pi/2} = k\pi a^4$

37. $m = \dfrac{2}{3}k\pi r^3$

$\overline{x} = \overline{y} = 0$ by symmetry

$M_{xy} = 4k\displaystyle\int_0^{\pi/2}\int_0^{\pi/2}\int_0^r \rho^3 \cos\phi \sin\phi \, d\rho \, d\theta \, d\phi$

$\quad = \dfrac{1}{2}kr^4\displaystyle\int_0^{\pi/2}\int_0^{\pi/2}\sin 2\phi \, d\theta \, d\phi$

$\quad = \dfrac{kr^4\pi}{4}\displaystyle\int_0^{\pi/2}\sin 2\phi \, d\phi$

$\quad = \left[-\dfrac{1}{8}k\pi r^4 \cos 2\phi\right]_0^{\pi/2} = \dfrac{1}{4}k\pi r^4$

$\overline{z} = \dfrac{M_{xy}}{m} = \dfrac{k\pi r^4/4}{2k\pi r^3/3} = \dfrac{3r}{8}$

$(\overline{x}, \overline{y}, \overline{z}) = \left(0, 0, \dfrac{3r}{8}\right)$

39. $I_z = 4k\displaystyle\int_{\pi/4}^{\pi/2}\int_0^{\pi/2}\int_0^{\cos\phi} \rho^4 \sin^3\phi \, d\rho \, d\theta \, d\phi$

$\quad = \dfrac{4}{5}k\displaystyle\int_{\pi/4}^{\pi/2}\int_0^{\pi/2}\cos^5\phi \sin^3\phi \, d\theta \, d\phi$

$\quad = \dfrac{2}{5}k\pi\displaystyle\int_{\pi/4}^{\pi/2}\cos^5\phi\left(1 - \cos^2\phi\right)\sin\phi \, d\phi$

$\quad = \left[\dfrac{2}{5}k\pi\left(-\dfrac{1}{6}\cos^6\phi + \dfrac{1}{8}\cos^8\phi\right)\right]_{\pi/4}^{\pi/2} = \dfrac{k\pi}{192}$

41. Cylindrical: $\int_0^{2\pi} \int_0^2 \int_{r^2}^4 r^2 \cos \theta \, dz \, dr \, d\theta = 0$

Spherical: $\int_0^{2\pi} \int_0^{\arctan(1/2)} \int_0^{4 \sec \phi} \rho^3 \sin^2 \phi \cos \theta \, d\rho \, d\phi \, d\theta + \int_0^{2\pi} \int_{\arctan(1/2)}^{\pi/2} \int_0^{\cot \phi \csc \phi} \rho^3 \sin^2 \phi \cos \theta \, d\rho \, d\phi \, d\theta = 0$

43. Cylindrical: $1 \le z \le 1 + \sqrt{1 - x^2 - y^2} = 1 + \sqrt{1 - r^2}$

$0 \le r \le 1, 0 \le \theta \le 2\pi, x = r \cos \theta$

$\int_0^{2\pi} \int_0^1 \int_1^{1+\sqrt{1-r^2}} r^2 \cos \theta \, dz \, dr \, d\theta$

Spherical: $\sec \phi \le \rho \le 2 \cos \phi, 0 \le \phi \le \pi/4, x = \rho \sin \phi \cos \theta$

$\int_0^{\pi/4} \int_0^{2\pi} \int_{\sec \phi}^{2 \cos \phi} \rho^3 \sin^2 \phi \cos \theta \, d\rho \, d\theta \, d\phi = \int_0^{2\pi} \int_0^1 r^2 \cos \theta \left(\sqrt{1 - r^2}\right) dr \, d\theta = \int_0^{2\pi} \cos \theta \, d\theta \int_0^1 r^2 \sqrt{1 - r^2} \, dr = 0$

45. (a) $r = r_0$: right circular cylinder about z-axis

$\theta = \theta_0$: plane parallel to z-axis

$z = z_0$: plane parallel to xy-plane

(b) $\rho = \rho_0$: sphere of radius ρ_0

$\theta = \theta_0$: plane parallel to z-axis

$\phi = \phi_0$: cone

47. $\left(x^2 + y^2 + z^2 + 8\right)^2 \le 36\left(x^2 + y^2\right)$

In cylindrical coordinates,

$\left(r^2 + z^2 + 8\right)^2 \le 36r^2$

$r^2 + z^2 + 8 \le 6r$

$r^2 - 6r + 9 + z^2 - 1 \le 0$

$\left(r - 3\right)^2 + z^2 \le 1.$

This is a torus: rotate $(x - 3)^2 + z^2 = 1$ about the z-axis. By Pappus' Theorem, $V = 2\pi(3)\pi = 6\pi^2$.

Section 14.8   Change of Variables: Jacobians

1. The Jacobian is given by

$$\frac{\partial(x, y)}{\partial(u, v)} = \frac{\partial x}{\partial u} \frac{\partial y}{\partial v} - \frac{\partial y}{\partial u} \frac{\partial x}{\partial v} = \begin{vmatrix} \dfrac{\partial x}{\partial u} & \dfrac{\partial x}{\partial v} \\ \dfrac{\partial y}{\partial u} & \dfrac{\partial y}{\partial v} \end{vmatrix}.$$

3. $x = -\dfrac{1}{2}(u - v)$

$y = \dfrac{1}{2}(u + v)$

$\dfrac{\partial x}{\partial u} \dfrac{\partial y}{\partial v} - \dfrac{\partial y}{\partial u} \dfrac{\partial x}{\partial v} = \left(-\dfrac{1}{2}\right)\left(\dfrac{1}{2}\right) - \left(\dfrac{1}{2}\right)\left(\dfrac{1}{2}\right) = -\dfrac{1}{2}$

5. $x = u - v^2$

$y = u + v$

$\dfrac{\partial x}{\partial u} \dfrac{\partial y}{\partial v} - \dfrac{\partial y}{\partial u} \dfrac{\partial x}{\partial v} = (1)(1) - (1)(-2v) = 1 + 2v$

7. $x = u \cos \theta - v \sin \theta$

$y = u \sin \theta + v \cos \theta$

$\dfrac{\partial x}{\partial u} \dfrac{\partial y}{\partial v} - \dfrac{\partial y}{\partial u} \dfrac{\partial x}{\partial v} = \cos^2 \theta + \sin^2 \theta = 1$

9. $x = e^u \sin v$

$y = e^u \cos v$

$\dfrac{\partial x}{\partial u} \dfrac{\partial y}{\partial v} - \dfrac{\partial y}{\partial u} \dfrac{\partial x}{\partial v} = \left(e^u \sin v\right)\left(-e^u \sin v\right) - \left(e^u \cos v\right)\left(e^u \cos v\right) = -e^{2u}$

11. $x = 3u + 2v$

$y = 3v$

$v = \dfrac{y}{3}$

$u = \dfrac{x - 2v}{3} = \dfrac{x - 2(y/3)}{3} = \dfrac{x}{3} - \dfrac{2y}{9}$

(x, y)	(u, v)
$(0, 0)$	$(0, 0)$
$(3, 0)$	$(1, 0)$
$(2, 3)$	$(0, 1)$

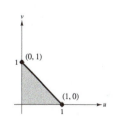

13. $x = \frac{1}{2}(u + v)$

$y = \frac{1}{2}(u - v)$

$u = x + y$

$v = x - y$

(x, y)	(u, v)
$\left(\frac{1}{2}, \frac{1}{2}\right)$	$(1, 0)$
$(0, 1)$	$(1, -1)$
$(1, 2)$	$(3, -1)$
$\left(\frac{3}{2}, \frac{3}{2}\right)$	$(3, 0)$

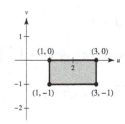

15. $\left.\begin{array}{r} x - 2y = 0 \\ x + y = 4 \end{array}\right\} \Rightarrow \begin{array}{l} 3y = 4 \\ y = \frac{4}{3}, \quad x = \frac{8}{3} \end{array}$

$\left.\begin{array}{r} x - 2y = -4 \\ x + y = 4 \end{array}\right\} \quad \begin{array}{l} 3y = 8 \\ y = \frac{8}{3}, \quad x = \frac{4}{3} \end{array}$

$\left.\begin{array}{r} x - 2y = -4 \\ x + y = 1 \end{array}\right\} \quad \begin{array}{l} 3y = 5 \\ y = \frac{5}{3}, \quad x = -\frac{2}{3} \end{array}$

$\left.\begin{array}{r} x - 2y = 0 \\ x + y = 1 \end{array}\right\} \quad \begin{array}{l} 3y = 1 \\ y = \frac{1}{2}, \quad x = \frac{2}{3} \end{array}$

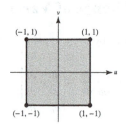

$\left.\begin{array}{l} u = x + y \\ v = x - 2y \end{array}\right\} \quad \begin{array}{l} u - v = 3y \Rightarrow y = \frac{1}{3}(u - v) \\ 2u + v = 3x \Rightarrow x = \frac{1}{3}(2u + v) \end{array}$

$\iint\limits_R 3\,xy\,dA = \int_{-2/3}^{2/3} \int_{1-x}^{(x+4)/2} 3\,xy\,dy\,dx + \int_{2/3}^{4/3} \int_{x/2}^{(x+4)/2} 3\,xy\,dy\,dx + \int_{4/3}^{8/3} \int_{x/2}^{4-x} 3\,xy\,dy\,dx = \frac{32}{27} + \frac{164}{27} + \frac{296}{27} = \frac{164}{9}$

17. $x = \frac{1}{2}(u + v)$

$y = \frac{1}{2}(u - v)$

$\frac{\partial x}{\partial u}\frac{\partial y}{\partial v} - \frac{\partial y}{\partial u}\frac{\partial x}{\partial v} = \left(\frac{1}{2}\right)\left(-\frac{1}{2}\right) - \left(\frac{1}{2}\right)\left(\frac{1}{2}\right) = -\frac{1}{2}$

$\iint\limits_R 4(x^2 + y^2)\,dA = \int_{-1}^{1} \int_{-1}^{1} 4\left[\frac{1}{4}(u + v)^2 + \frac{1}{4}(u - v)^2\right]\left(\frac{1}{2}\right)dv\,du$

$= \int_{-1}^{1} \int_{-1}^{1} (u^2 + v^2)\,dv\,du = \int_{-1}^{1} 2\left(u^2 + \frac{1}{3}\right)du = \left[2\left(\frac{u^3}{3} + \frac{u}{3}\right)\right]_{-1}^{1} = \frac{8}{3}$

19. $x = u + v$

$y = u$

$\frac{\partial x}{\partial u}\frac{\partial y}{\partial v} - \frac{\partial y}{\partial u}\frac{\partial x}{\partial v} = (1)(0) - (1)(1) = -1$

$\int_R\!\!\int y(x - y)\,dA = \int_0^3 \int_0^4 uv(1)\,dv\,du = \int_0^3 8u\,du = 36$

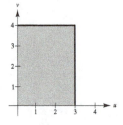

21. $\int_R \int e^{-xy/2} \, dA$

$R: y = \dfrac{x}{4}, \, y = 2x, \, y = \dfrac{1}{x}, \, y = \dfrac{4}{x}$

$x = \sqrt{v/u}, \, y = \sqrt{uv} \implies u = \dfrac{y}{x}, \, v = xy$

$\dfrac{\partial(x, y)}{\partial(u, v)} = \begin{vmatrix} \dfrac{\partial x}{\partial u} & \dfrac{\partial x}{\partial v} \\ \dfrac{\partial y}{\partial u} & \dfrac{\partial y}{\partial v} \end{vmatrix} = \begin{vmatrix} -\dfrac{1}{2}\dfrac{v^{1/2}}{u^{3/2}} & \dfrac{1}{2}\dfrac{1}{u^{1/2}v^{1/2}} \\ \dfrac{1}{2}\dfrac{v^{1/2}}{u^{1/2}} & \dfrac{1}{2}\dfrac{u^{1/2}}{v^{1/2}} \end{vmatrix} = -\dfrac{1}{4}\left(\dfrac{1}{u} + \dfrac{1}{u}\right) = -\dfrac{1}{2u}$

Transformed Region:

$y = \dfrac{1}{x} \implies yx = 1 \implies v = 1$

$y = \dfrac{4}{x} \implies ux = 4 \implies v = 4$

$y = 2x \implies \dfrac{y}{x} = 2 \implies u = 2$

$y = \dfrac{x}{4} \implies \dfrac{y}{x} = \dfrac{1}{4} \implies u = \dfrac{1}{4}$

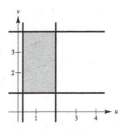

$\int_R \int e^{-xy/2} \, dA = \int_{1/4}^{2} \int_{1}^{4} e^{-v/2}\left(\dfrac{1}{2u}\right) dv \, du = -\int_{1/4}^{2} \left[\dfrac{e^{-v/2}}{u}\right]_{1}^{4} du = -\int_{1/4}^{2} \left(e^{-2} - e^{-1/2}\right)\dfrac{1}{u} \, du$

$\qquad = -\left[\left(e^{-2} - e^{-1/2}\right)\ln u\right]_{1/4}^{2} = -\left(e^{-2} - e^{-1/2}\right)\left(\ln 2 - \ln \dfrac{1}{4}\right) = \left(e^{-1/2} - e^{-2}\right)\ln 8 \approx 0.9798$

23. $x = \dfrac{1}{2}(u + v)$

$y = -\dfrac{1}{2}(u - v)$

$u = x - y$

$v = x + y$

(x, y)	(u, v)
$(0, 1)$	$(-1, 1)$
$(1, 0)$	$(1, 1)$
$(1, 2)$	$(-1, 3)$
$(2, 1)$	$(1, 3)$

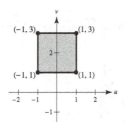

$\dfrac{\partial x}{\partial u}\dfrac{\partial y}{\partial v} - \dfrac{\partial y}{\partial u}\dfrac{\partial x}{\partial v} = \left(\dfrac{1}{2}\right)\left(\dfrac{1}{2}\right) - \left(-\dfrac{1}{2}\right)\left(\dfrac{1}{2}\right) = \dfrac{1}{2}$

$\int_R \int 9xy \, dA = \int_{-1}^{1} \int_{1}^{3} 9\left[\dfrac{1}{2}(u + v)\right]\left[-\dfrac{1}{2}(u - v)\right]\left(\dfrac{1}{2}\right) dv \, du$

$\qquad = \int_{-1}^{1} \int_{1}^{3} \dfrac{9}{8}(v^2 - u^2) \, dv \, du$

$\qquad = \dfrac{9}{8} \int_{-1}^{1} \left[\dfrac{v^3}{3} - u^2 v\right]_{1}^{3} du$

$\qquad = \dfrac{9}{8} \int_{-1}^{1} \left(\dfrac{26}{3} - 2u^2\right) du$

$\qquad = \dfrac{9}{8} \left[\dfrac{26}{3}u - \dfrac{2}{3}u^3\right]_{-1}^{1}$

$\qquad = \dfrac{9}{8}(8 + 8)$

$\qquad = 18$

25. $u = x + y = 4, \quad v = x - y = 0$

$u = x + y = 8, \quad v = x - y = 4$

$x = \dfrac{1}{2}(u + v) \qquad y = \dfrac{1}{2}(u - v)$

$\dfrac{\partial(x, y)}{\partial(u, v)} = -\dfrac{1}{2}$

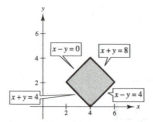

$\displaystyle\int_R\!\!\int (x + y)e^{x-y}\, dA = \int_4^8 \int_0^4 ue^v\left(\frac{1}{2}\right) dv\, du = \frac{1}{2}\int_4^8 u\left(e^4 - 1\right)du = \left[\frac{1}{4}u^2\left(e^4 - 1\right)\right]_4^8 = 12\left(e^4 - 1\right)$

27. $u = x + 4y = 0, \quad v = x - y = 0$

$u = x + 4y = 5, \quad v = x - y = 5$

$x = \dfrac{1}{5}(u + 4v), \qquad y = \dfrac{1}{5}(u - v)$

$\dfrac{\partial x}{\partial u}\dfrac{\partial y}{\partial v} - \dfrac{\partial y}{\partial u}\dfrac{\partial x}{\partial v} = \left(\frac{1}{5}\right)\!\left(-\frac{1}{5}\right) - \left(\frac{1}{5}\right)\!\left(\frac{4}{5}\right) = -\frac{1}{5}$

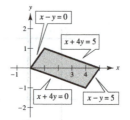

$\displaystyle\int_R\!\!\int \sqrt{(x - y)(x + 4y)}\, dA = \int_0^5 \int_0^5 \sqrt{uv}\left(\frac{1}{5}\right) du\, dv = \int_0^5 \left[\frac{1}{5}\left(\frac{2}{3}\right)u^{3/2}\sqrt{v}\right]_0^5 dv = \left[\frac{2\sqrt{5}}{3}\left(\frac{2}{3}\right)v^{3/2}\right]_0^5 = \frac{100}{9}$

29. $u = x + y,\ v = x - y,\ x = \dfrac{1}{2}(u + v),\ y = \dfrac{1}{2}(u - v)$

$\dfrac{\partial x}{\partial u}\dfrac{\partial y}{\partial v} - \dfrac{\partial y}{\partial u}\dfrac{\partial x}{\partial v} = -\dfrac{1}{2}$

$\displaystyle\int_R\!\!\int \sqrt{x + y}\, dA = \int_0^a \int_{-u}^u \sqrt{u}\left(\frac{1}{2}\right) dv\, du = \int_0^a u\sqrt{u}\, du = \left[\frac{2}{5}u^{5/2}\right]_0^a = \frac{2}{5}a^{5/2}$

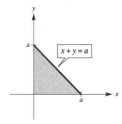
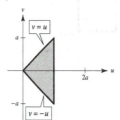

31. $u = 2x - y$

$v = x + y$

$3x = u + v \Rightarrow x = \frac{1}{3}(u + v)$

Then $y = v - x = v - \frac{1}{3}(u + v) = \frac{1}{3}(2v - u)$.

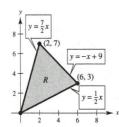

(x, y)	(u, v)
$(0, 0)$	$(0, 0)$
$(6, 3)$	$(9, 9)$
$(2, 7)$	$(-3, 9)$

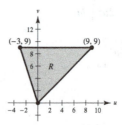

One side is parallel to the u-axis.

33. $\dfrac{x^2}{a^2} + \dfrac{y^2}{b^2} = 1$, $x = au$, $y = bv$

$$\dfrac{(au)^2}{a^2} + \dfrac{(bv)^2}{b^2} = 1$$

$$u^2 + v^2 = 1$$

(a) $\dfrac{x^2}{a^2} + \dfrac{y^2}{b^2} = 1$ $u^2 + v^2 = 1$

(b) $\dfrac{\partial(x,\,y)}{\partial(u,\,v)} = \dfrac{\partial x}{\partial u}\dfrac{\partial y}{\partial v} - \dfrac{\partial y}{\partial u}\dfrac{\partial x}{\partial v} = (a)(b) - (0)(0) = ab$

(c) $A = \displaystyle\int_S\!\!\int ab \; dS = ab\bigl(\pi(1)^2\bigr) = \pi ab$

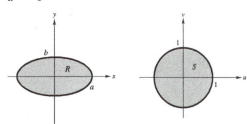

35. $x = u(1 - v)$, $y = uv(1 - w)$, $z = uvw$

$$\dfrac{\partial(x,\,y,\,z)}{\partial(u,\,v,\,w)} = \begin{vmatrix} 1 - v & -u & 0 \\ v(1 - w) & u(1 - w) & -uv \\ vw & uw & uv \end{vmatrix} = (1 - v)\bigl[u^2v(1 - w) + u^2vw\bigr] + u\bigl[uv^2(1 - w) + uv^2w\bigr] = (1 - v)(u^2v) + u(uv^2) = u^2v$$

37. $x = \dfrac{1}{2}(u + v)$, $y = \dfrac{1}{2}(u - v)$, $z = 2uvw$

$$\dfrac{\partial(x,\,y,\,z)}{\partial(u,\,v,\,w)} = \begin{vmatrix} 1/2 & 1/2 & 0 \\ 1/2 & -1/2 & 0 \\ 2vw & 2uw & 2uv \end{vmatrix} = 2uv[-1/4 - 1/4] = -uv$$

39. $x = \rho \sin\phi \cos\theta$, $y = \rho \sin\phi \sin\theta$, $z = \rho \cos\phi$

$$\dfrac{\partial(x,\,y,\,z)}{\partial(\rho,\,\theta,\,\phi)} = \begin{vmatrix} \sin\phi\cos\theta & -\rho\sin\phi\sin\theta & \rho\cos\phi\cos\theta \\ \sin\phi\sin\theta & \rho\sin\phi\cos\theta & \rho\cos\phi\sin\theta \\ \cos\phi & 0 & -\rho\sin\phi \end{vmatrix}$$

$$= \cos\phi\bigl[-\rho^2\sin\phi\cos\phi\sin^2\theta - \rho^2\sin\phi\cos\phi\cos^2\theta\bigr] - \rho\sin\phi\bigl[\rho\sin^2\phi\cos^2\theta + \rho\sin^2\phi\sin^2\theta\bigr]$$

$$= \cos\phi\bigl[-\rho^2\sin\phi\cos\phi(\sin^2\theta + \cos^2\theta)\bigr] - \rho\sin\phi\bigl[\rho\sin^2\phi(\cos^2\theta + \sin^2\theta)\bigr]$$

$$= -\rho^2\sin\phi\cos^2\phi - \rho^2\sin^3\phi = -\rho^2\sin\phi(\cos^2\phi + \sin^2\phi) = -\rho^2\sin\phi$$

41.   Let $u = \dfrac{x}{3}$, $v = y \Rightarrow \dfrac{\partial(x,y)}{\partial(u,v)} = \begin{vmatrix} 3 & 0 \\ 0 & 1 \end{vmatrix} = 3$, $y = \dfrac{x}{2} \Rightarrow v = \dfrac{3u}{2}$.

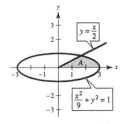   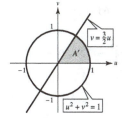   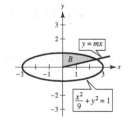   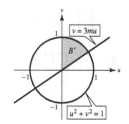

Region A is transformed to region A', and region B is transformed to region B'.

$$A' = B' \Rightarrow \dfrac{2}{3} = 3m \Rightarrow m = \dfrac{2}{9}$$

Note: You could also calculate the integrals directly.

Review Exercises for Chapter 14

1. $\int_0^{3x} \sin(xy)\, dy = \left[-\frac{1}{x} \cos(xy) \right]_0^{3x}$

$= -\frac{1}{x} \cos 3x^2 + \frac{1}{x}$

$= \frac{1 - \cos 3x^2}{x}$

3. $\int_0^1 \int_0^{1+x} (3x + 2y)\, dy\, dx = \int_0^1 \left[3xy + y^2 \right]_0^{1+x} dx$

$= \int_0^1 \left(4x^2 + 5x + 1 \right) dx$

$= \left[\frac{4}{3}x^3 + \frac{5}{2}x^2 + x \right]_0^1$

$= \frac{29}{6}$

5. $\int_0^1 \int_0^{\sqrt{1-x^4}} x^3\, dy\, dx = \int_0^1 x^3 \left(1 - x^4\right)^{1/2} dx = \left[-\frac{1}{6}\left(1 - x^4\right)^{3/2} \right]_0^1 = -\frac{1}{6}(0 - 1) = \frac{1}{6}$

7. $A = \int_0^1 \int_0^{3-3y} dx\, dy = \int_0^1 (3 - 3y)\, dy = \left[3y - \frac{3y^2}{2} \right]_0^1 = \frac{3}{2}$

9. $A = \int_0^4 \int_x^{2x+2} dy\, dx = \int_0^4 (x + 2)\, dx = \left[\frac{x^2}{2} + 2x \right]_0^4 = 16$

11.

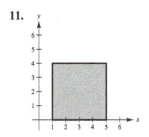

$\int_1^5 \int_0^4 dy\, dx = \int_1^5 [y]_0^4\, dx = \int_1^5 4\, dx = 4(5 - 1) = 16$

$\int_0^4 \int_1^5 dx\, dy = \int_0^4 (5 - 1)\, dy = [4y]_0^4 = 16$

13.

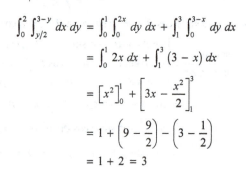

$\int_0^2 \int_{y/2}^{3-y} dx\, dy = \int_0^1 \int_0^{2x} dy\, dx + \int_1^3 \int_0^{3-x} dy\, dx$

$= \int_0^1 2x\, dx + \int_1^3 (3 - x)\, dx$

$= \left[x^2 \right]_0^1 + \left[3x - \frac{x^2}{2} \right]_1^3$

$= 1 + \left(9 - \frac{9}{2} \right) - \left(3 - \frac{1}{2} \right)$

$= 1 + 2 = 3$

15. $\iint_R 4xy\, dA = \int_0^4 \int_0^2 4xy\, dx\, dy = \int_0^2 \int_0^4 4xy\, dy\, dx$

$\int_0^4 \int_0^2 4xy\, dx\, dy = \int_0^4 \left[2x^2 y \right]_0^2 dy = \int_0^4 8y\, dy = \left[4y^2 \right]_0^4 = 64$

17. $V = \int_0^3 \int_0^2 (5 - x)\, dy\, dx = \int_0^3 (10 - 2x)\, dx = \left[10x - x^2 \right]_0^3 = 30 - 9 = 21$

19. $V = \int_{-1}^{1}\int_{-1}^{1}\left(4 - x^2 - y^2\right) dy\, dx$

$$= \int_{-1}^{1}\left[4y - x^2 y - \frac{y^3}{3}\right]_{-1}^{1} dx$$

$$= \int_{-1}^{1}\left[\left(4 - x^2 - \frac{1}{3}\right) - \left(-4 + x^2 - \frac{1}{3}\right)\right] dx$$

$$= \int_{-1}^{1}\left[\frac{22}{3} - 2x^2\right] dx$$

$$= \left[\frac{22}{3}x - \frac{2x^3}{3}\right]_{-1}^{1}$$

$$= \frac{40}{3}$$

Alternate Solution:

$$V = 4\int_{0}^{1}\int_{0}^{1}\left(4 - x^2 - y^2\right) dy\, dx$$

$$= 4\int_{0}^{1}\left(4 - x^2 - \frac{1}{3}\right) dx$$

$$= 4\int_{0}^{1}\left(\frac{11}{3} - x^2\right) dx$$

$$= 4\left[\frac{11}{3}x - \frac{1}{3}x^3\right]_{0}^{1}$$

$$= \frac{40}{3}$$

21. Area $R = 16$

Average Value $= \dfrac{1}{16}\displaystyle\int_{-2}^{2}\int_{-2}^{2}\left(16 - x^2 - y^2\right) dy\, dx$

$$= \frac{1}{16}\int_{-2}^{2}\left[16y - x^2 y - \frac{y^3}{3}\right]_{-2}^{2} dx$$

$$= \frac{1}{16}\int_{-2}^{2}\left[64 - 4x^2 - \frac{16}{3}\right] dx$$

$$= \frac{1}{16}\left[64x - \frac{4x^3}{3} - \frac{16}{3}x\right]_{-2}^{2}$$

$$= \frac{1}{16}\left[256 - \frac{64}{3} - \frac{64}{3}\right] = \frac{40}{3}$$

23. Area $R = 3(5) = 15$

Average temperature $= \dfrac{1}{15}\displaystyle\int_{0}^{3}\int_{0}^{5}\left(40 - 6x^2 - y^2\right) dy\, dx$

$$= \frac{1}{15}\int_{0}^{3}\left[40y - 6x^2 y - \frac{y^3}{3}\right]_{0}^{5} dx$$

$$= \frac{1}{15}\int_{0}^{3}\left[200 - 30x^2 - \frac{125}{3}\right] dx$$

$$= \frac{1}{15}\left[200x - 10x^3 - \frac{125x}{3}\right]_{0}^{3}$$

$$= \frac{1}{15}\left[600 - 270 - 125\right]$$

$$= 13\frac{2}{3}°C$$

25. The region of integration is the quarter-circle of radius $\sqrt{5}$ in the first quadrant.

$$\int_{0}^{\sqrt{5}}\int_{0}^{\sqrt{5-x^2}} \sqrt{x^2 + y^2}\, dy\, dx = \int_{0}^{\pi/2}\int_{0}^{\sqrt{5}} r^2\, dr\, d\theta = \int_{0}^{\pi/2}\left[\frac{r^3}{3}\right]_{0}^{\sqrt{5}} d\theta = \int_{0}^{\pi/2}\frac{5\sqrt{5}}{3}\, d\theta = 5\sqrt{5}\,\frac{\pi}{6}$$

27. $V = \displaystyle\int_{0}^{\pi/2}\int_{0}^{3} (r\cos\theta)(r\sin\theta)^2\, r\, dr\, d\theta$

$$= \int_{0}^{\pi/2}\int_{0}^{3} \cos\theta \sin^2\theta\, r^4\, dr\, d\theta$$

$$= \int_{0}^{\pi/2} \cos\theta \sin^2\theta \left[\frac{r^5}{5}\right]_{0}^{3} d\theta$$

$$= \frac{243}{5}\int_{0}^{\pi/2} \sin^2\theta \cos\theta\, d\theta$$

$$= \frac{243}{5}\left[\frac{\sin^3\theta}{3}\right]_{0}^{\pi/2} = \frac{81}{5}$$

29. One loop is traced out on $0 \le \theta \le \dfrac{2\pi}{3}$.

$$A = 3\int_{0}^{2\pi/3}\int_{0}^{1-\cos 3\theta} r\, dr\, d\theta$$

$$= 3\int_{0}^{2\pi/3} \frac{1}{2}\left(1 - \cos 3\theta^2\right) d\theta$$

$$= \frac{3}{2}\int_{0}^{2\pi/3} \left(1 - 2\cos 3\theta + \cos^2 3\theta\right) d\theta$$

$$= \frac{3}{2}\int_{0}^{2\pi/3} \left(1 - 2\cos 3\theta + \frac{1 + \cos 6\theta}{2}\right) d\theta$$

$$= \frac{3}{2}\left[\theta - \frac{2}{3}\sin 3\theta + \frac{1}{2}\theta + \frac{\sin 6\theta}{12}\right]_{0}^{2\pi/3}$$

$$= \frac{3}{2}\left[\frac{2\pi}{3} + \frac{\pi}{3}\right] = \frac{3}{2}\pi$$

31.

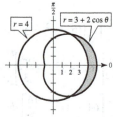

Intersection points: $\theta = -\dfrac{\pi}{3}, \dfrac{\pi}{3}$

$$A = \int_{-\pi/3}^{\pi/3} \int_{4}^{3+2\cos\theta} r \, dr \, d\theta$$

$$= 2\int_{0}^{\pi/3} \left[\frac{1}{2}(3 + 2\cos\theta)^2 - 8\right] d\theta$$

$$= 2\int_{0}^{\pi/3} \left[\frac{1}{2}(9 + 12\cos\theta + 4\cos^2\theta) - 8\right] d\theta$$

$$= 2\int_{0}^{\pi/3} \left(\frac{9}{2} + 6\cos\theta + 1 + \cos 2\theta - 8\right) d\theta$$

$$= 2\int_{0}^{\pi/3} \left(-\frac{5}{2} + 6\cos\theta + \cos 2\theta\right) d\theta$$

$$= 2\left[-\frac{5}{2}\theta + 6\sin\theta + \frac{1}{2}\sin 2\theta\right]_{0}^{\pi/3}$$

$$= 2\left[-\frac{5}{6}\pi + 6\left(\frac{\sqrt{3}}{2}\right) + \frac{1}{2}\frac{\sqrt{3}}{2}\right]$$

$$= \frac{13\sqrt{3}}{2} - \frac{5\pi}{3}$$

33. (a) $(x^2 + y^2)^2 = 9(x^2 - y^2)$

$$(r^2)^2 = 9(r^2\cos^2\theta - r^2\sin^2\theta)$$
$$r^2 = 9(\cos^2\theta - \sin^2\theta)$$
$$= 9\cos 2\theta$$
$$r = 3\sqrt{\cos 2\theta}$$

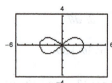

(b) $A = 4\int_{0}^{\pi/4}\int_{0}^{3\sqrt{\cos 2\theta}} r \, dr \, d\theta = 9$

(c) $V = 4\int_{0}^{\pi/4}\int_{0}^{3\sqrt{\cos 2\theta}} \sqrt{9 - r^2} \, r \, dr \, d\theta \approx 20.392$

35. $m = \int_{0}^{1}\int_{0}^{2} (x + 3y) \, dy \, dx$

$$= \int_{0}^{1} \left[xy + \frac{3}{2}y^2\right]_{0}^{2} dx$$

$$= \int_{0}^{1} (2x + 6) \, dx$$

$$= \left[x^2 + 6x\right]_{0}^{1} = 7$$

37. $m = \int_{0}^{2}\int_{0}^{x^3} kx \, dy \, dx = \dfrac{32k}{5}$

$$M_x = \int_{0}^{2}\int_{0}^{x^3} kxy \, dy \, dx = 16k$$

$$M_y = \int_{0}^{2}\int_{0}^{x^3} kx^2 \, dy \, dx = \frac{32k}{3}$$

$$\bar{x} = \frac{M_y}{m} = \frac{5}{3}$$

$$\bar{y} = \frac{M_x}{m} = \frac{5}{2}$$

$$(\bar{x}, \bar{y}) = \left(\frac{5}{3}, \frac{5}{2}\right)$$

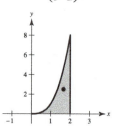

39. $m = k\int_{0}^{1}\int_{2x^3}^{2x} xy \, dy \, dx = \dfrac{k}{4}$

$$M_x = k\int_{0}^{1}\int_{2x^3}^{2x} xy^2 \, dy \, dx = \frac{16k}{55}$$

$$M_y = k\int_{0}^{1}\int_{2x^3}^{2x} x^2y \, dy \, dx = \frac{8k}{45}$$

$$\bar{x} = \frac{M_y}{m} = \frac{32}{45}$$

$$\bar{y} = \frac{M_x}{m} = \frac{64}{55}$$

$$(\bar{x}, \bar{y}) = \left(\frac{32}{45}, \frac{64}{55}\right)$$

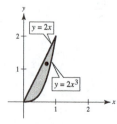

41. $I_x = \int_R \int y^2 \rho(x, y) \, dA = \int_0^3 \int_0^2 y^2 (kx) \, dy \, dx = \int_0^3 \frac{8}{3} kx \, dx = \left[\frac{8}{3} k \frac{x^2}{2}\right]_0^3 = 12k$

$I_y = \int_0^3 \int_0^2 x^2 (kx) \, dy \, dx = \int_0^3 2kx^3 \, dx = \left[\frac{x^4}{2} k\right]_0^3 = \frac{81}{2} k$

$I_0 = \int_0^3 \int_0^2 (x^2 + y^2) \, dy \, dx = 12k + \frac{81}{2} k = \frac{105k}{2}$

$m = \int_0^3 \int_0^2 kx \, dy \, dx = \int_0^3 2kx \, dx = 9k$

$\bar{\bar{x}} = \sqrt{\frac{I_y}{m}} = \sqrt{\frac{81k}{2} \Big/ 9k} = \frac{3\sqrt{2}}{2}$

$\bar{\bar{y}} = \sqrt{\frac{I_x}{m}} = \sqrt{\frac{12k}{9k}} = \frac{2\sqrt{3}}{3}$

43. $f(x, y) = 25 - x^2 - y^2$

$f_x = -2x, \ f_y = -2y$

$S = \int_R \int \sqrt{1 + (f_x)^2 + (f_y)^2} \, dA$

$= \int_R \int \sqrt{1 + 4x^2 + 4y^2} \, dA$

$= 4\int_0^{\pi/2} \int_0^5 \sqrt{1 + 4r^2} \, r \, dr \, d\theta$

$= \frac{1}{3} \int_0^{\pi/2} \left[(1 + 4r^2)^{3/2}\right]_0^5 d\theta$

$= \frac{1}{3} \int_0^{\pi/2} \left[(101)^{3/2} - 1\right] d\theta$

$= \frac{\pi}{6} \left[101\sqrt{101} - 1\right]$

45. $f(x, y) = 9 - y^2$

$f_x = 0, \ f_y = -2y$

$\sqrt{1 + (f_x)^2 + (f_y)^2} = \sqrt{1 + 4y^2}$

$S = \int_0^3 \int_{-y}^y \sqrt{1 + 4y^2} \, dx \, dy$

$= \int_0^3 \left[\sqrt{1 + 4y^2} \, x\right]_{-y}^y dy$

$= \int_0^3 2y\sqrt{4y^2 + 1} \, dy$

$= \left[\frac{1}{6}(4y^2 + 1)^{3/2}\right]_0^3$

$= \frac{37\sqrt{37} - 1}{6}$

47. (a) $V = \int_0^{50} \int_0^{\sqrt{50^2 - x^2}} \left(20 + \frac{xy}{100} - \frac{x + y}{5}\right) dy \, dx = \int_0^{50} \left[20\sqrt{50^2 - x^2} + \frac{x}{200}(50^2 - x^2) - \frac{x}{5}\sqrt{50^2 - x^2} - \frac{50^2 - x^2}{10}\right] dy$

$= \left[10\left(x\sqrt{50 - x^2} + 50^2 \arcsin\frac{x}{50}\right) + \frac{25}{4} x^2 - \frac{x^4}{800} + \frac{1}{15}(50^2 - x^2)^{3/2} - 250x + \frac{x^3}{30}\right]_0^{50} \approx 30{,}415.74 \text{ ft}^3$

(b) $z = 20 + \frac{xy}{100}$

$\sqrt{1 + (f_x)^2 + (f_y)^2} = \sqrt{1 + \frac{y^2}{100^2} + \frac{x^2}{100^2}} = \frac{\sqrt{100^2 + x^2 + y^2}}{100}$

$S = \frac{1}{100} \int_0^{50} \int_0^{\sqrt{50^2 - x^2}} \sqrt{100^2 + x^2 + y^2} \, dy \, dx = \frac{1}{100} \int_0^{\pi/2} \int_0^{50} \sqrt{100^2 + r^2} \, r \, dr \, d\theta \approx 2081.53 \text{ ft}^2$

49. $\int_0^4 \int_0^1 \int_0^2 (2x + y + 4z) \, dy \, dz \, dx = \int_0^4 \int_0^1 \left[2xy + \frac{y^2}{2} + 4zy\right]_0^2 dz \, dx$

$= \int_0^4 \int_0^1 (4x + 2 + 8z) \, dz \, dx = \int_0^4 \left[4xz + 2z + 4z^2\right]_0^1 dx$

$= \int_0^4 (4x + 2 + 4) \, dx = \left[2x^2 + 6x\right]_0^4 = 56$

51. $\int_0^2 \int_1^2 \int_0^1 \left(e^x + y^2 + z^2\right) dx\,dy\,dz = \int_0^2 \int_1^2 \left[e^x + xy^2 + xz^2\right]_0^1 dy\,dz$

$$= \int_0^2 \int_1^2 \left(e - 1 + y^2 + z^2\right) dy\,dz$$

$$= \int_0^2 \left[ey - y + \frac{1}{3}y^3 + yz^2\right]_1^2 dz$$

$$= \int_0^2 \left[\left(2e - 2 + \frac{8}{3} + 2z^2\right) - \left(e - 1 + \frac{1}{3} + z^2\right)\right] dz$$

$$= \int_0^2 \left(e + \frac{4}{3} + z^2\right) dz$$

$$= \left[ez + \frac{4}{3}z + \frac{z^3}{3}\right]_0^2$$

$$= 2e + \frac{8}{3} + \frac{8}{3}$$

$$= 2e + \frac{16}{3}$$

53. $\int_{-1}^1 \int_{-\sqrt{1-x^2}}^{\sqrt{1-x^2}} \int_{-\sqrt{1-x^2-y^2}}^{\sqrt{1-x^2-y^2}} \left(x^2 + y^2\right) dz\,dy\,dx = \int_0^{2\pi} \int_0^1 \int_{-\sqrt{1-r^2}}^{\sqrt{1-r^2}} r^3\,dz\,dr\,d\theta = \frac{8\pi}{15}$

55. $V = \int_0^3 \int_0^4 \int_0^{xy} dz\,dy\,dx$

$$= \int_0^3 \int_0^4 xy\,dy\,dx$$

$$= \int_0^3 \left[\frac{xy^2}{2}\right]_0^4 dx$$

$$= \int_0^3 8x\,dx$$

$$= \left[4x^2\right]_0^3 = 36$$

57. $\int_0^1 \int_0^y \int_0^{\sqrt{1-x^2}} dz\,dx\,dy = \int_0^1 \int_x^1 \int_0^{\sqrt{1-x^2}} dz\,dy\,dx$

59. $m = \int_0^{10} \int_0^{10-x} \int_0^{10-x-y} k\,dz\,dy\,dx = \frac{500}{3}k$

$Myz = \int_0^{10} \int_0^{10-x} \int_0^{10-x-y} kx\,dz\,dy\,dx = \frac{1250}{3}k$

$\bar{x} = \frac{Myz}{m} = \frac{5}{2}$

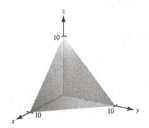

61. $\int_0^3 \int_{\pi/6}^{\pi/3} \int_0^4 r\cos\theta\,dr\,d\theta\,dz = \int_0^3 \int_{\pi/6}^{\pi/3} \left[\frac{r^2}{2}\cos\theta\right]_0^4 d\theta\,dz$

$$= \int_0^3 \int_{\pi/6}^{\pi/3} 8\cos\theta\,d\theta\,dz$$

$$= \int_0^3 \left[8\sin\theta\right]_{\pi/6}^{\pi/3} dz$$

$$= \int_0^3 8\left(\frac{\sqrt{3}}{2} - \frac{1}{2}\right) dz$$

$$= 24\left(\frac{\sqrt{3}}{2} - \frac{1}{2}\right)$$

$$= 12\sqrt{3} - 12$$

63. $\int_0^\pi \int_0^{\pi/2} \int_0^{\sin\theta} \rho^2 \sin\theta \cos\theta \, d\rho \, d\theta \, d\phi = \int_0^\pi \int_0^{\pi/2} \left[\dfrac{\rho^3}{3}\right]_0^{\sin\theta} d\theta \, d\phi = \int_0^\pi \int_0^{\pi/2} \dfrac{1}{3} \sin^4\theta \cos\theta \, d\theta \, d\phi$

$$= \int_0^\pi \left[\dfrac{1}{15} \sin^5\theta\right]_0^{\pi/2} d\phi = \int_0^\pi \dfrac{1}{15} \, d\phi = \dfrac{\pi}{15}$$

65. $\int_0^\pi \int_0^2 \int_0^3 \sqrt{z^2 + 4} \, dz \, dr \, d\theta \approx 48.995$

67. $z = 8 - x^2 - y^2 = x^2 + y^2$

$$8 = 2\left(x^2 + y^2\right)$$

$$x^2 + y^2 = 4$$

$V = \int_0^{2\pi} \int_0^2 \int_{r^2}^{8-r^2} r \, dz \, dr \, d\theta = \int_0^{2\pi} \int_0^2 r\left(8 - r^2 - r^2\right) dr \, d\theta = \int_0^{2\pi} \int_0^2 \left(8r - 2r^3\right) dr \, d\theta$

$$= \int_0^{2\pi} \left[4r^2 - \dfrac{r^4}{2}\right]_0^2 d\theta = \int_0^{2\pi} 8 \, d\theta = 16\pi$$

69. $z^2 = 4 - x^2 - y^2 = 3x^2 + 3y^2 \Rightarrow x^2 + y^2 = 1, z = \sqrt{3}$

$$\rho^2 = x^2 + y^2 + z^2 = 4 \Rightarrow \rho = 2$$

$$z = \rho \cos\phi \Rightarrow \cos\phi = \dfrac{\sqrt{3}}{2} \Rightarrow \phi = \dfrac{\pi}{6}$$

$V = \int_0^{2\pi} \int_0^{\pi/6} \int_0^2 \rho^2 \sin\phi \, d\rho \, d\phi \, d\theta = \int_0^{2\pi} \int_0^{\pi/6} \left[\dfrac{\rho^3}{3} \sin\phi\right]_0^2 d\phi \, d\theta = \int_0^{2\pi} \int_0^{\pi/6} \dfrac{8}{3} \sin\phi \, d\phi \, d\theta$

$$= \int_0^{2\pi} \left[-\dfrac{8}{3} \cos\phi\right]_0^{\pi/6} d\theta = \int_0^{2\pi} \dfrac{8}{3}\left(1 - \dfrac{\sqrt{3}}{2}\right) d\theta = 2\pi\left(\dfrac{8}{3}\right)\left(1 - \dfrac{\sqrt{3}}{2}\right) = \dfrac{8\pi}{3}\left(2 - \sqrt{3}\right)$$

71. $\dfrac{\partial(x, y)}{\partial(u, v)} = \dfrac{\partial x}{\partial u} \dfrac{\partial y}{\partial v} - \dfrac{\partial y}{\partial u} \dfrac{\partial x}{\partial v} = (3v)(-2) - (2)(3u) = -6(u + v)$

73. $\dfrac{\partial(x, y)}{\partial(u, v)} = \dfrac{\partial x}{\partial u} \dfrac{\partial y}{\partial v} - \dfrac{\partial y}{\partial u} \dfrac{\partial x}{\partial v} = (\sin\theta)(\sin\theta) - (\cos\theta)(\cos\theta) = \sin^2\theta - \cos^2\theta$

75. $\dfrac{\partial(x, y)}{\partial(u, v)} = \dfrac{\partial x}{\partial u} \dfrac{\partial y}{\partial v} - \dfrac{\partial x}{\partial v} \dfrac{\partial y}{\partial u} = \dfrac{1}{2}\left(-\dfrac{1}{2}\right) - \dfrac{1}{2}\left(\dfrac{1}{2}\right) = -\dfrac{1}{2}$

$x = \dfrac{1}{2}(u + v), y = \dfrac{1}{2}(u - v) \Rightarrow u = x + y, v = x - y$

Boundaries in xy-plane	Boundaries in uv-plane
$x + y = 3$	$u = 3$
$x + y = 5$	$u = 5$
$x - y = -1$	$v = -1$
$x - y = 1$	$v = 1$

$\int_R \int \ln(x + y) \, dA = \int_3^5 \int_{-1}^1 \ln\left(\dfrac{1}{2}(u + v) + \dfrac{1}{2}(u - v)\right)\left(\dfrac{1}{2}\right) dv \, du = \int_3^5 \int_{-1}^1 \dfrac{1}{2} \ln u \, dv \, du = \int_3^5 \ln u \, du = \left[u \ln u - u\right]_3^5$

$$= (5 \ln 5 - 5) - (3 \ln u - 3) = 5 \ln 5 - 3 \ln 3 - 2 \approx 2.751$$

77. $\dfrac{\partial(x, y)}{\partial(u, v)} = \dfrac{\partial x}{\partial u}\dfrac{\partial y}{\partial v} - \dfrac{\partial y}{\partial u}\dfrac{\partial x}{\partial v} = 1\left(-\dfrac{1}{3}\right) - \dfrac{1}{3}(0) = -\dfrac{1}{3}$

$x = u, y = \dfrac{1}{3}(u - v) \Rightarrow u = x, v = x - 3y$

Boundary in xy-plane	Boundary in uv-plane
$x = 1$	$u = 1$
$x = 4$	$u = 4$
$3y - x = 8$	$v = -8$
$3y - x = 2$	$v = -2$

$\displaystyle\int_R\!\int (xy + x^2)\, dA = \int_1^4 \int_{-2}^{-8}\left[u\frac{1}{3}(u - v) + u^2\right]\left(-\frac{1}{3}\right) dv\, du$

$= \left(-\dfrac{1}{3}\right)\int_1^4 \int_{-2}^{-8}\left(\dfrac{4}{3}u^2 - \dfrac{1}{3}uv\right) dv\, du = \left(-\dfrac{1}{3}\right)\int_1^4 \left[\dfrac{4}{3}u^2 v - \dfrac{1}{6}uv^2\right]_{-2}^{-8} du$

$= \left(-\dfrac{1}{3}\right)\int_1^4 \left(-8u^2 - 10u\right) du = 81$

Problem Solving for Chapter 14

1. $V = 16\displaystyle\int_R\!\int \sqrt{1 - x^2}\, dA$

$= 16\displaystyle\int_0^{\pi/4} \int_0^1 \sqrt{1 - r^2 \cos^2 \theta}\, r\, dr\, d\theta$

$= -\dfrac{16}{3}\displaystyle\int_0^{\pi/4} \dfrac{1}{\cos^2 \theta}\left[(1 - \cos^2 \theta)^{3/2} - 1\right] d\theta$

$= -\dfrac{16}{3}[\sec \theta + \cos \theta - \tan \theta]_0^{\pi/4}$

$= 8(2 - \sqrt{2}) \approx 4.6863$

3.

Boundary in xy-plane	Boundary in uv-plane
$y = \sqrt{x}$	$u = 1$
$y = \sqrt{2x}$	$u = 2$
$y = \dfrac{1}{3}x^2$	$v = 3$
$y = \dfrac{1}{4}x^2$	$v = 4$

$\dfrac{\partial(x, y)}{\partial(u, v)} = \begin{vmatrix} \dfrac{1}{3}\left(\dfrac{v}{u}\right)^{2/3} & \dfrac{2}{3}\left(\dfrac{u}{v}\right)^{1/3} \\[2mm] \dfrac{2}{3}\left(\dfrac{v}{u}\right)^{1/3} & \dfrac{1}{3}\left(\dfrac{u}{v}\right)^{2/3} \end{vmatrix} = -\dfrac{1}{3}$

$A = \displaystyle\int_R\!\int 1\, dA = \int_S\!\int 1\left|\dfrac{\partial(x, y)}{\partial(u, v)}\right| dA = \dfrac{1}{3}$

5. (a) $\displaystyle\int \dfrac{du}{a^2 + u^2} = \dfrac{1}{a}\arctan\dfrac{u}{a} + c.$ Let $a^2 = 2 - u^2, u = v.$

Then $\displaystyle\int \dfrac{1}{(2 - u^2) + v^2}\, dv = \dfrac{1}{\sqrt{2 - u^2}}\arctan\dfrac{v}{\sqrt{2 - u^2}} + C.$

(b) $I_1 = \displaystyle\int_0^{\sqrt{2}/2} \left[\dfrac{2}{\sqrt{2 - u^2}}\arctan\dfrac{v}{\sqrt{2 - u^2}}\right]_{-u}^{u} du$

$= \displaystyle\int_0^{\sqrt{2}/2} \dfrac{2}{\sqrt{2 - u^2}}\left(\arctan\dfrac{u}{\sqrt{2 - u^2}} - \arctan\dfrac{-u}{\sqrt{2 - u^2}}\right) du = \int_0^{\sqrt{2}/2} \dfrac{4}{\sqrt{2 - u^2}}\arctan\dfrac{u}{\sqrt{2 - u^2}}\, du$

Let $u = \sqrt{2}\sin\theta, du = \sqrt{2}\cos\theta\, d\theta, 2 - u^2 = 2 - 2\sin^2\theta = 2\cos^2\theta.$

$I_1 = 4\displaystyle\int_0^{\pi/6} \dfrac{1}{\sqrt{2}\cos\theta}\arctan\left(\dfrac{\sqrt{2}\sin\theta}{\sqrt{2}\cos\theta}\right)\cdot\sqrt{2}\cos\theta\, d\theta = 4\int_0^{\pi/6}\arctan(\tan\theta)d\theta = \dfrac{4\theta^2}{2}\Big]_0^{\pi/6} = 2\left(\dfrac{\pi}{6}\right)^2 = \dfrac{\pi^2}{18}$

(c) $I_2 = \int_{\sqrt{2}/2}^{\sqrt{2}} \left[\dfrac{2}{\sqrt{2 - u^2}} \arctan \dfrac{v}{\sqrt{2 - u^2}} \right]_{u - \sqrt{2}}^{-u + \sqrt{2}} du$

$= \int_{\sqrt{2}/2}^{\sqrt{2}} \dfrac{2}{\sqrt{2 - u^2}} \left[\arctan\left(\dfrac{-u + \sqrt{2}}{\sqrt{2 - u^2}} \right) - \arctan\left(\dfrac{u - \sqrt{2}}{\sqrt{2 - u^2}} \right) \right] du = \int_{\sqrt{2}/2}^{\sqrt{2}} \dfrac{4}{\sqrt{2 - u^2}} \arctan\left(\dfrac{\sqrt{2} - u}{\sqrt{2 - u^2}} \right) du$

Let $u = \sqrt{2} \sin \theta$.

$I_2 = 4\int_{\pi/6}^{\pi/2} \dfrac{1}{\sqrt{2} \cos \theta} \arctan\left(\dfrac{\sqrt{2} - \sqrt{2} \sin \theta}{\sqrt{2} \cos \theta} \right) \cdot \sqrt{2} \cos \theta \, d\theta = 4\int_{\pi/6}^{\pi/2} \arctan\left(\dfrac{1 - \sin \theta}{\cos \theta} \right) d\theta$

(d) $\tan\left(\dfrac{1}{2}\left(\dfrac{\pi}{2} - \theta \right) \right) = \sqrt{\dfrac{1 - \cos((\pi/2) - \theta)}{1 + \cos((\pi/2) - \theta)}} = \sqrt{\dfrac{1 - \sin \theta}{1 + \sin \theta}} = \sqrt{\dfrac{(1 - \sin \theta)^2}{(1 + \sin \theta)(1 - \sin \theta)}} = \sqrt{\dfrac{(1 - \sin \theta)^2}{\cos^2 \theta}} = \dfrac{1 - \sin \theta}{\cos \theta}$

(e) $I_2 = 4\int_{\pi/6}^{\pi/2} \arctan\left(\dfrac{1 - \sin \theta}{\cos \theta} \right) d\theta = 4\int_{\pi/6}^{\pi/2} \arctan\left(\tan\left(\dfrac{1}{2}\left(\dfrac{\pi}{2} - \theta \right) \right) \right) d\theta = 4\int_{\pi/6}^{\pi/2} \dfrac{1}{2}\left(\dfrac{\pi}{2} - \theta \right) d\theta = 2\int_{\pi/6}^{\pi/2} \left(\dfrac{\pi}{2} - \theta \right) d\theta$

$= 2\left[\dfrac{\pi}{2}\theta - \dfrac{\theta^2}{2} \right]_{\pi/6}^{\pi/2} = 2\left[\left(\dfrac{\pi^2}{4} - \dfrac{\pi^2}{8} \right) - \left(\dfrac{\pi^2}{12} - \dfrac{\pi^2}{72} \right) \right] = 2\left[\dfrac{18 - 9 - 6 + 1}{72}\pi^2 \right] = \dfrac{4}{36}\pi^2 = \dfrac{\pi^2}{9}$

(f) $\dfrac{1}{1 - xy} = 1 + (xy) + (xy)^2 + \cdots \qquad |xy| < 1$

$\int_0^1 \int_0^1 \dfrac{1}{1 - xy}\, dx\, dy = \int_0^1 \int_0^1 \left[1 + (xy) + (xy)^2 + \cdots \right] dx\, dy = \int_0^1 \int_0^1 \sum_{K=0}^{\infty}(xy)^K \, dx\, dy = \sum_{K=0}^{\infty}\int_0^1 \dfrac{x^{K+1}y^K}{K+1}\Big|_0^1 dy$

$= \sum_{K=0}^{\infty}\int_0^1 \dfrac{y^K}{K+1}\, dy = \sum_{K=0}^{\infty} \dfrac{y^{K+1}}{(K+1)^2}\Big|_0^1 = \sum_{K=0}^{\infty} \dfrac{1}{(K+1)^2} = \sum_{n=1}^{\infty} \dfrac{1}{n^2}$

(g) $u = \dfrac{x + y}{\sqrt{2}}, v = \dfrac{y - x}{\sqrt{2}}$

$u - v = \dfrac{2x}{\sqrt{2}} \Rightarrow x = \dfrac{u - v}{\sqrt{2}}$

$u + v = \dfrac{2y}{\sqrt{2}} \Rightarrow y = \dfrac{u + v}{\sqrt{2}}$

$\dfrac{\partial(x, y)}{\partial(u, v)} = \begin{vmatrix} 1/\sqrt{2} & -1/\sqrt{2} \\ 1/\sqrt{2} & 1/\sqrt{2} \end{vmatrix} = 1$

R	S
$(0, 0)$	$\leftrightarrow$ $(0, 0)$
$(1, 0)$	$\leftrightarrow$ $\left(\dfrac{1}{\sqrt{2}}, -\dfrac{1}{\sqrt{2}} \right)$
$(0, 1)$	$\leftrightarrow$ $\left(\dfrac{1}{\sqrt{2}}, \dfrac{1}{\sqrt{2}} \right)$
$(1, 1)$	$\leftrightarrow$ $\left(\sqrt{2}, 0 \right)$

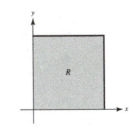
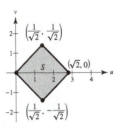

$\int_0^1 \int_0^1 \dfrac{1}{1 - xy}\, dx\, dy = \int_0^{\sqrt{2}/2} \int_{-u}^{u} \dfrac{1}{1 - \dfrac{u^2}{2} + \dfrac{v^2}{2}}\, dv\, du + \int_{\sqrt{2}/2}^{\sqrt{2}} \int_{u - \sqrt{2}}^{-u + \sqrt{2}} \dfrac{1}{1 - \dfrac{u^2}{2} + \dfrac{v^2}{2}}\, dv\, du = I_1 + I_2 = \dfrac{\pi^2}{18} + \dfrac{\pi^2}{9} = \dfrac{\pi^2}{6}$

7. $\int_0^1 \int_0^1 \frac{x-y}{(x+y)^3} \, dx \, dy = -\frac{1}{2}$

$\int_0^1 \int_0^1 \frac{(x-y)}{(x+y)^3} \, dy \, dx = \frac{1}{2}$

The results are not the same. Fubini's Theorem is not valid because f is not continuous on the region $0 \le x \le 1, 0 \le y \le 1$.

9. From Exercise 65, Section 14.3,

$\int_{-\infty}^{\infty} e^{-x^2/2} \, dx = \sqrt{2\pi}.$

So, $\int_0^{\infty} e^{-x^2/2} \, dx = \frac{\sqrt{2\pi}}{2}$ and $\int_0^{\infty} e^{-x^2} \, dx = \frac{\sqrt{\pi}}{2}$

$\int_0^{\infty} x^2 e^{-x^2} \, dx = \left[-\frac{1}{2} x e^{-x^2} \right]_0^{\infty} + \frac{1}{2} \int_0^{\infty} e^{-x^2} \, dx$

$= \frac{1}{2} \cdot \frac{\sqrt{\pi}}{2} = \frac{\sqrt{\pi}}{4}$

11. $f(x, y) = \begin{cases} ke^{-(x+y)/a} & x \ge 0, y \ge 0 \\ 0 & \text{elsewhere} \end{cases}$

$\int_{-\infty}^{\infty} \int_{-\infty}^{\infty} f(x, y) \, dA = \int_0^{\infty} \int_0^{\infty} ke^{-(x+y)/a} \, dx \, dy$

$= k \int_0^{\infty} e^{-x/a} \, dx \cdot \int_0^{\infty} e^{-y/a} \, dy$

These two integrals are equal to

$\int_0^{\infty} e^{-x/a} \, dx = \lim_{b \to \infty} \left[(-a)e^{-x/a} \right]_0^b = a.$

So, assuming $a, k > 0$, you obtain

$1 = ka^2 \text{ or } a = \frac{1}{\sqrt{k}}.$

13. Essay

15. The greater the angle between the given plane and the xy-plane, the greater the surface area. So:

$z_2 < z_1 < z_4 < z_3$

17. $V = \int_0^3 \int_0^{2x} \int_x^{6-x} dy \, dz \, dx = 18$

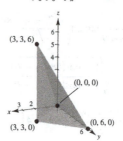

$V = \int_0^3 \int_0^2 \int_x^{6-x} dy \, dz \, dx$

$= \int_0^3 \int_0^2 [y]_x^{6-x} \, dz \, dx$

$= \int_0^3 \int_0^2 (6 - 2x) \, dz \, dx$

$= \int_0^3 [6z - 2zx]_0^2 \, dx$

$= \int_0^3 (12 - 4x) \, dx$

$= \left[12x - 2x^2 \right]_0^3 = 18$

C H A P T E R 1 5
Vector Analysis

CHAPTER 15
Vector Analysis

Section 15.1 Vector Fields

1. See "Definition of Vector Field" on Page 1044. Some physical examples of vector fields include velocity fields, gravitational fields, and electric force fields.

3. Reconstruct the potential function from its partial derivatives by integrating and comparing versions of the resulting function to determine constants. See Example 6.

5. All vectors are parallel to x-axis.
   Matches (d)

6. All vectors are parallel to y-axis.
   Matches (c)

7. Vectors are in rotational pattern.
   Matches (a)

8. All vectors point outward.
   Matches (b)

9. $\mathbf{F}(x, y) = \mathbf{i} + \mathbf{j}$

   $\|\mathbf{F}\| = \sqrt{2}$

   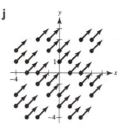

11. $\mathbf{F}(x, y) = -\mathbf{i} + 3y\mathbf{j}$

    $\|\mathbf{F}\| = \sqrt{1 + 9y^2}$

    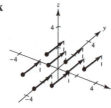

13. $\mathbf{F}(x, y, z) = \mathbf{i} + \mathbf{j} + \mathbf{k}$

    $\|\mathbf{F}\| = \sqrt{3}$

15. $\mathbf{F}(x, y) = \dfrac{1}{8}\left(2xy\mathbf{i} + y^2\mathbf{j}\right)$

17. $\mathbf{F}(x, y, z) = \dfrac{x\mathbf{i} + y\mathbf{j} + z\mathbf{k}}{\sqrt{x^2 + y^2 + z^2}}$

    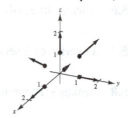

19. $f(x, y) = x^2 + 2y^2$

    $f_x(x, y) = 2x$

    $f_y(x, y) = 4y$

    $\mathbf{F}(x, y) = 2x\mathbf{i} + 4y\mathbf{j}$

    Note that $\nabla f = \mathbf{F}$.

21. $g(x, y) = 5x^2 + 3xy + y^2$

    $g_x(x, y) = 10x + 3y$

    $g_y(x, y) = 3x + 2y$

    $\mathbf{G}(x, y) = (10x + 3y)\mathbf{i} + (3x + 2y)\mathbf{j}$

23. $f(x, y, z) = 6xyz$

    $f_x(x, y, z) = 6yz$

    $f_y(x, y, z) = 6xz$

    $f_z(x, y, z) = 6xy$

    $\mathbf{F}(x, y, z) = 6yz\mathbf{i} + 6xz\mathbf{j} + 6xy\mathbf{k}$

25. $g(x, y, z) = z + ye^{x^2}$

$g_x(x, y, z) = 2xye^{x^2}$

$g_y(x, y, z) = e^{x^2}$

$g_z(x, y, z) = 1$

$\mathbf{G}(x, y, z) = 2xye^{x^2}\mathbf{i} + e^{x^2}\mathbf{j} + \mathbf{k}$

27. $h(x, y, z) = xy \ln(x + y)$

$h_x(x, y, z) = y \ln(x + y) + \dfrac{xy}{x + y}$

$h_y(x, y, z) = x \ln(x + y) + \dfrac{xy}{x + y}$

$h_z(x, y, z) = 0$

$\mathbf{H}(x, y, z) = \left[\dfrac{xy}{x + y} + y \ln(x + y)\right]\mathbf{i} + \left[\dfrac{xy}{x + y} + x \ln(x + y)\right]\mathbf{j}$

29. $\mathbf{F}(x, y) = xy^2\mathbf{i} + x^2y\mathbf{j}$

$M = xy^2$ and $N = x^2y$ have continuous first partial derivatives.

$\dfrac{\partial N}{\partial x} = 2xy = \dfrac{\partial M}{\partial y} \implies \mathbf{F}$ is conservative.

31. $\mathbf{F}(x, y) = \sin y\,\mathbf{i} + x \sin y\,\mathbf{j}$

$M = \sin y$ and $N = x \sin y$ have continuous first partial derivatives.

$\dfrac{\partial N}{\partial x} = \sin y \neq \dfrac{\partial M}{\partial y} = \cos y \implies \mathbf{F}$ is not conservative.

33. $\mathbf{F}(x, y) = \dfrac{1}{xy}(y\mathbf{i} - x\mathbf{j}) = \dfrac{1}{x}\mathbf{i} - \dfrac{1}{y}\mathbf{j}$

$M = 1/x$ and $N = -1/y$ have continuous first partial derivatives for all $x, y \neq 0$.

$\dfrac{\partial N}{\partial x} = 0 = \dfrac{\partial M}{\partial y} \implies \mathbf{F}$ is conservative.

35. $M = \dfrac{1}{\sqrt{x^2 + y^2}}, \ N = \dfrac{1}{\sqrt{x^2 + y^2}}$

$\dfrac{\partial N}{\partial x} = \dfrac{-x}{\left(x^2 + y^2\right)^{3/2}} \neq \dfrac{\partial M}{\partial y} = \dfrac{-y}{\left(x^2 + y^2\right)^{3/2}}$

$\implies$ Not conservative

37. $\mathbf{F}(x, y) = (3y - x^2)\mathbf{i} + (3x + y)\mathbf{j}$

$\dfrac{\partial N}{\partial x} = 3 = \dfrac{\partial M}{\partial y}$

Conservative

$f_x = 3y - x^2$

$f_y = 3x + y$

$f(x, y) = 3xy + \dfrac{y^2}{2} - \dfrac{x^3}{3} + K$

39. $\mathbf{F}(x, y) = xe^{x^2y}(2y\mathbf{i} + x\mathbf{j})$

$\dfrac{\partial}{\partial y}\left[2xye^{x^2y}\right] = 2xe^{x^2y} + 2x^3ye^{x^2y}$

$\dfrac{\partial}{\partial x}\left[x^2e^{x^2y}\right] = 2xe^{x^2y} + 2x^3ye^{x^2y}$

Conservative

$f_x(x, y) = 2xye^{x^2y}$

$f_y(x, y) = x^2e^{x^2y}$

$f(x, y) = e^{x^2y} + K$

41. $\mathbf{F}(x, y) = \dfrac{2y}{x}\mathbf{i} - \dfrac{x^2}{y^2}\mathbf{j}$

$\dfrac{\partial}{\partial y}\left[\dfrac{2y}{x}\right] = \dfrac{2}{x}$

$\dfrac{\partial}{\partial x}\left[-\dfrac{x^2}{y^2}\right] = -\dfrac{2x}{y^2}$

Not conservative

43. $\mathbf{F}(x, y) = \sin y \, \mathbf{i} + x \cos y \, \mathbf{j}$

$$\frac{\partial N}{\partial x} = \cos y = \frac{\partial M}{\partial y}$$

Conservative

$f_x = \sin y$

$f_y = x \cos y$

$f(x, y) = x \sin y + K$

45. $\mathbf{F}(x, y, z) = xyz \, \mathbf{i} + xyz \, \mathbf{j} + xyz \, \mathbf{k}, \ (2, 1, 3)$

$$\text{curl } \mathbf{F} = \begin{vmatrix} \mathbf{i} & \mathbf{j} & \mathbf{k} \\ \dfrac{\partial}{\partial x} & \dfrac{\partial}{\partial y} & \dfrac{\partial}{\partial z} \\ xyz & xyz & xyz \end{vmatrix}$$

$$= (xz - xy)\mathbf{i} - (yz - xy)\mathbf{j} + (yz - xz)\mathbf{k}$$

$$\text{curl } \mathbf{F} \,(2, 1, 3) = (6 - 2)\mathbf{i} - (3 - 2)\mathbf{j} + (3 - 6)\mathbf{k}$$

$$= 4\mathbf{i} - \mathbf{j} - 3\mathbf{k}$$

47. $\mathbf{F}(x, y, z) = e^x \sin y \, \mathbf{i} - e^x \cos y \, \mathbf{j}, \ (0, 0, 1)$

$$\text{curl } \mathbf{F} = \begin{vmatrix} \mathbf{i} & \mathbf{j} & \mathbf{k} \\ \dfrac{\partial}{\partial x} & \dfrac{\partial}{\partial y} & \dfrac{\partial}{\partial z} \\ e^x \sin y & -e^x \cos y & 0 \end{vmatrix}$$

$$= (-e^x \cos y - e^x \cos y)\mathbf{k}$$

$$= -2e^x \cos y \, \mathbf{k}$$

$$\text{curl } \mathbf{F} \,(0, 0, 1) = -2\mathbf{k}$$

49. $\mathbf{F}(x, y, z) = \arctan\left(\dfrac{x}{y}\right)\mathbf{i} + \ln\sqrt{x^2 + y^2}\,\mathbf{j} + \mathbf{k}$

$$\text{curl } \mathbf{F} = \begin{vmatrix} \mathbf{i} & \mathbf{j} & \mathbf{k} \\ \dfrac{\partial}{\partial x} & \dfrac{\partial}{\partial y} & \dfrac{\partial}{\partial z} \\ \arctan\left(\dfrac{x}{y}\right) & \dfrac{1}{2}\ln(x^2 + y^2) & 1 \end{vmatrix}$$

$$= \left[\frac{x}{x^2 + y^2} - \frac{(-x/y^2)}{1 + (x/y)^2}\right]\mathbf{k}$$

$$= \frac{2x}{x^2 + y^2}\mathbf{k}$$

51. $\mathbf{F}(x, y, z) = (3x^2 + yz)\mathbf{i} + (3y^2 + xz)\mathbf{j} + (3z^2 + xy)\mathbf{k}$

$$\text{curl } \mathbf{F} = \begin{vmatrix} \mathbf{i} & \mathbf{j} & \mathbf{k} \\ \dfrac{\partial}{\partial x} & \dfrac{\partial}{\partial y} & \dfrac{\partial}{\partial z} \\ 3x^2 + yz & 3y^2 + xz & 3z^2 + xy \end{vmatrix}$$

$$= (x - x)\mathbf{i} - (y - y)\mathbf{j} + (z - z)\mathbf{k} = \mathbf{0}$$

Conservative

$f_x(x, y, z) = 3x^2 + yz$

$f_y(x, y, z) = 3y^2 + xz$

$f_z(x, y, z) = 3z^2 + xy$

$f(x, y, z) = x^3 + y^3 + z^3 + xyz + K$

53. $\mathbf{F}(x, y, z) = \sin z \, \mathbf{i} + \sin x \, \mathbf{j} + \sin y \, \mathbf{k}$

$$\text{curl } \mathbf{F} = \begin{vmatrix} \mathbf{i} & \mathbf{j} & \mathbf{k} \\ \dfrac{\partial}{\partial x} & \dfrac{\partial}{\partial y} & \dfrac{\partial}{\partial z} \\ \sin z & \sin x & \sin y \end{vmatrix}$$

$$= \cos y \, \mathbf{i} + \cos z \, \mathbf{j} + \cos x \, \mathbf{k} \neq \mathbf{0}$$

Not conservative

55. $\mathbf{F}(x, y, z) = \dfrac{z}{y}\mathbf{i} - \dfrac{xz}{y^2}\mathbf{j} + \left(\dfrac{x}{y} - 1\right)\mathbf{k}$

$$\text{curl } \mathbf{F} = \begin{vmatrix} \mathbf{i} & \mathbf{j} & \mathbf{k} \\ \dfrac{\partial}{\partial x} & \dfrac{\partial}{\partial y} & \dfrac{\partial}{\partial z} \\ \dfrac{z}{y} & -\dfrac{xz}{y^2} & \dfrac{x}{y} - 1 \end{vmatrix}$$

$$= \left(-\frac{x}{y^2} + \frac{x}{y^2}\right)\mathbf{i} - \left(\frac{1}{y} - \frac{1}{y}\right)\mathbf{j} + \left(-\frac{z}{y^2} + \frac{z}{y^2}\right)\mathbf{k}$$

$$= \mathbf{0}$$

Conservative

$f_x(x, y, z) = \dfrac{z}{y}$

$f_y(x, y, z) = -\dfrac{xz}{y^2}$

$f_z(x, y, z) = \dfrac{x}{y} - 1$

$f(x, y, z) = \dfrac{xz}{y} - z + K$

57. $F(x, y) = x^2 \mathbf{i} + 2y^2 \mathbf{j}$

$\text{div } F(x, y) = \dfrac{\partial}{\partial x}(x^2) + \dfrac{\partial}{\partial y}(2y^2) = 2x + 4y$

59. $F(x, y, z) = \sin^2 x \mathbf{i} + z \cos z \mathbf{j} + z^3 \mathbf{k}$

$\text{div } F = 2 \sin x \cos x + 0 + 3z^2 = 2 \sin x \cos x + 3z^2$

61. $F(x, y, z) = xyz \mathbf{i} + xz^2 \mathbf{j} + 3yz^2 \mathbf{k}$

$\text{div } F = yz + 0 + 6yz$

$\text{div } F(2, 4, 1) = 4(1) + 6(4)(1) = 28$

63. $F(x, y, z) = e^x \sin y \mathbf{i} - e^x \cos y \mathbf{j} + z^2 \mathbf{k}$

$\text{div } F(x, y, z) = e^x \sin y + e^x \sin y + 2z$

$\text{div } F(3, 0, 0) = e^3 \sin 0 + e^3 \sin 0 + 2(0) = 0$

65. $\text{curl}(\nabla f)$ is a vector field. The curl of a vector field (∇f) is a vector field.

67. $\text{curl}(\text{div } F)$ is neither a vector field nor a scalar function. The expression $\text{curl}(\text{div } F)$ is meaningless because you cannot take the curl of a scalar function.

69. $F(x, y, z) = \mathbf{i} + 3x \mathbf{j} + 2y \mathbf{k}$

$G(x, y, z) = x \mathbf{i} - y \mathbf{j} + z \mathbf{k}$

$F \times G = \begin{vmatrix} \mathbf{i} & \mathbf{j} & \mathbf{k} \\ 1 & 3x & 2y \\ x & -y & z \end{vmatrix} = (3xz + 2y^2)\mathbf{i} - (z - 2xy)\mathbf{j} + (-y - 3x^2)\mathbf{k}$

$\text{curl}(F \times G) = \begin{vmatrix} \mathbf{i} & \mathbf{j} & \mathbf{k} \\ \dfrac{\partial}{\partial x} & \dfrac{\partial}{\partial y} & \dfrac{\partial}{\partial z} \\ 3xz + 2y^2 & -z + 2xy & -y - 3x^2 \end{vmatrix} = (-1 + 1)\mathbf{i} - (-6x - 3x)\mathbf{j} + (2y - 4y)\mathbf{k} = 9x\mathbf{j} - 2y\mathbf{k}$

71. $F(x, y, z) = xyz \mathbf{i} + y \mathbf{j} + z \mathbf{k}$

$\text{curl } F = \begin{vmatrix} \mathbf{i} & \mathbf{j} & \mathbf{k} \\ \dfrac{\partial}{\partial x} & \dfrac{\partial}{\partial y} & \dfrac{\partial}{\partial z} \\ xyz & y & z \end{vmatrix} = xy\mathbf{j} - xz\mathbf{k}$

$\text{curl}(\text{curl } F) = \begin{vmatrix} \mathbf{i} & \mathbf{j} & \mathbf{k} \\ \dfrac{\partial}{\partial x} & \dfrac{\partial}{\partial y} & \dfrac{\partial}{\partial z} \\ 0 & xy & -xz \end{vmatrix} = z\mathbf{j} + y\mathbf{k}$

75. $F(x, y, z) = xyz \mathbf{i} + y \mathbf{j} + z \mathbf{k}$

$\text{curl } F = \begin{vmatrix} \mathbf{i} & \mathbf{j} & \mathbf{k} \\ \dfrac{\partial}{\partial x} & \dfrac{\partial}{\partial y} & \dfrac{\partial}{\partial z} \\ xyz & y & z \end{vmatrix} = xy\mathbf{j} - xz\mathbf{k}$

$\text{div}(\text{curl } F) = x - x = 0$

73. $F(x, y, z) = \mathbf{i} + 3x \mathbf{j} + 2y \mathbf{k}$

$G(x, y, z) = x \mathbf{i} - y \mathbf{j} + z \mathbf{k}$

$F \times G = \begin{vmatrix} \mathbf{i} & \mathbf{j} & \mathbf{k} \\ 1 & 3x & 2y \\ x & -y & z \end{vmatrix}$

$\qquad = (3xz + 2y^2)\mathbf{i} - (z - 2xy)\mathbf{j} + (-y - 3x^2)\mathbf{k}$

$\text{div}(F \times G) = 3z + 2x$

77. (a) Let $\mathbf{F} = M\mathbf{i} + N\mathbf{j} + P\mathbf{k}$ and $\mathbf{G} = Q\mathbf{i} + R\mathbf{j} + S\mathbf{k}$ where $M, N, P, Q, R,$ and S have continuous partial derivatives.

$$\mathbf{F} + \mathbf{G} = (M + Q)\mathbf{i} + (N + R)\mathbf{j} + (P + S)\mathbf{k}$$

$$\operatorname{curl}(\mathbf{F} + \mathbf{G}) = \begin{vmatrix} \mathbf{i} & \mathbf{j} & \mathbf{k} \\ \dfrac{\partial}{\partial x} & \dfrac{\partial}{\partial y} & \dfrac{\partial}{\partial z} \\ M + Q & N + R & P + S \end{vmatrix}$$

$$= \left[\frac{\partial}{\partial y}(P + S) - \frac{\partial}{\partial z}(N + R)\right]\mathbf{i} - \left[\frac{\partial}{\partial x}(P + S) - \frac{\partial}{\partial z}(M + Q)\right]\mathbf{j} + \left[\frac{\partial}{\partial x}(N + R) - \frac{\partial}{\partial y}(M + Q)\right]\mathbf{k}$$

$$= \left(\frac{\partial P}{\partial y} - \frac{\partial N}{\partial z}\right)\mathbf{i} - \left(\frac{\partial P}{\partial x} - \frac{\partial M}{\partial z}\right)\mathbf{j} + \left(\frac{\partial N}{\partial x} - \frac{\partial M}{\partial y}\right)\mathbf{k} + \left(\frac{\partial S}{\partial y} - \frac{\partial R}{\partial z}\right)\mathbf{i} - \left(\frac{\partial S}{\partial x} - \frac{\partial Q}{\partial z}\right)\mathbf{j} + \left(\frac{\partial R}{\partial x} - \frac{\partial Q}{\partial y}\right)\mathbf{k}$$

$$= \operatorname{curl}\mathbf{F} + \operatorname{curl}\mathbf{G}$$

(b) Let $f(x, y, z)$ be a scalar function whose second partial derivatives are continuous.

$$\nabla f = \frac{\partial f}{\partial x}\mathbf{i} + \frac{\partial f}{\partial y}\mathbf{j} + \frac{\partial f}{\partial z}\mathbf{k}$$

$$\operatorname{curl}(\nabla f) = \begin{vmatrix} \mathbf{i} & \mathbf{j} & \mathbf{k} \\ \dfrac{\partial}{\partial x} & \dfrac{\partial}{\partial y} & \dfrac{\partial}{\partial z} \\ \dfrac{\partial f}{\partial x} & \dfrac{\partial f}{\partial y} & \dfrac{\partial f}{\partial z} \end{vmatrix} = \left(\frac{\partial^2 f}{\partial y\partial z} - \frac{\partial^2 f}{\partial z\partial y}\right)\mathbf{i} - \left(\frac{\partial^2 f}{\partial x\partial z} - \frac{\partial^2 f}{\partial z\partial x}\right)\mathbf{j} + \left(\frac{\partial^2 f}{\partial x\partial y} - \frac{\partial^2 f}{\partial y\partial x}\right)\mathbf{k} = 0$$

(c) Let $\mathbf{F} = M\mathbf{i} + N\mathbf{j} + P\mathbf{k}$ and $\mathbf{G} = R\mathbf{i} + S\mathbf{j} + T\mathbf{k}$.

$$\operatorname{div}(\mathbf{F} + \mathbf{G}) = \frac{\partial}{\partial x}(M + R) + \frac{\partial}{\partial y}(N + S) + \frac{\partial}{\partial z}(P + T) = \frac{\partial M}{\partial x} + \frac{\partial R}{\partial x} + \frac{\partial N}{\partial y} + \frac{\partial S}{\partial y} + \frac{\partial P}{\partial z} + \frac{\partial T}{\partial z}$$

$$= \left[\frac{\partial M}{\partial x} + \frac{\partial N}{\partial y} + \frac{\partial P}{\partial z}\right] + \left[\frac{\partial R}{\partial x} + \frac{\partial S}{\partial y} + \frac{\partial T}{\partial z}\right]$$

$$= \operatorname{div}\mathbf{F} + \operatorname{div}\mathbf{G}$$

(d) Let $\mathbf{F} = M\mathbf{i} + N\mathbf{j} + P\mathbf{k}$ and $\mathbf{G} = R\mathbf{i} + S\mathbf{j} + T\mathbf{k}$.

$$\mathbf{F} \times \mathbf{G} = \begin{vmatrix} \mathbf{i} & \mathbf{j} & \mathbf{k} \\ M & N & P \\ R & S & T \end{vmatrix} = (NT - PS)\mathbf{i} - (MT - PR)\mathbf{j} + (MS - NR)\mathbf{k}$$

$$\operatorname{div}(\mathbf{F} \times \mathbf{G}) = \frac{\partial}{\partial x}(NT - PS) + \frac{\partial}{\partial y}(PR - MT) + \frac{\partial}{\partial z}(MS - NR)$$

$$= N\frac{\partial T}{\partial x} + T\frac{\partial N}{\partial x} - P\frac{\partial S}{\partial x} - S\frac{\partial P}{\partial x} + P\frac{\partial R}{\partial y} + R\frac{\partial P}{\partial y} - M\frac{\partial T}{\partial y} - T\frac{\partial M}{\partial y} + M\frac{\partial S}{\partial z} + S\frac{\partial M}{\partial z} - N\frac{\partial R}{\partial z} - R\frac{\partial N}{\partial z}$$

$$= \left[\left(\frac{\partial P}{\partial y} - \frac{\partial N}{\partial z}\right)R + \left(\frac{\partial M}{\partial z} - \frac{\partial P}{\partial x}\right)S + \left(\frac{\partial N}{\partial x} - \frac{\partial M}{\partial y}\right)T\right] - \left[M\left(\frac{\partial T}{\partial y} - \frac{\partial S}{\partial z}\right) + N\left(\frac{\partial R}{\partial z} - \frac{\partial T}{\partial x}\right) + P\left(\frac{\partial S}{\partial x} - \frac{\partial R}{\partial y}\right)\right]$$

$$= (\operatorname{curl}\mathbf{F}) \cdot \mathbf{G} - \mathbf{F} \cdot (\operatorname{curl}\mathbf{G})$$

(e) $\mathbf{F} = M\mathbf{i} + N\mathbf{j} + P\mathbf{k}$

$$\nabla \times \left[\nabla f + (\nabla \times \mathbf{F})\right] = \operatorname{curl}(\nabla f + (\nabla \times \mathbf{F}))$$

$$= \operatorname{curl}(\nabla f) + \operatorname{curl}(\nabla \times \mathbf{F}) \qquad \text{(Part (a))}$$

$$= \operatorname{curl}(\nabla \times \mathbf{F}) \qquad \text{(Part (b))}$$

$$= \nabla \times (\nabla \times \mathbf{F})$$

(f) Let $\mathbf{F} = M\mathbf{i} + N\mathbf{j} + P\mathbf{k}$.

$$\nabla \times (f\mathbf{F}) = \begin{vmatrix} \mathbf{i} & \mathbf{j} & \mathbf{k} \\ \dfrac{\partial}{\partial x} & \dfrac{\partial}{\partial y} & \dfrac{\partial}{\partial z} \\ fM & fN & fP \end{vmatrix}$$

$$= \left(\frac{\partial f}{\partial y}P + f\frac{\partial P}{\partial y} - \frac{\partial f}{\partial z}N - f\frac{\partial N}{\partial z} \right)\mathbf{i} - \left(\frac{\partial f}{\partial x}P + f\frac{\partial P}{\partial x} - \frac{\partial f}{\partial z}M - f\frac{\partial M}{\partial z} \right)\mathbf{j} + \left(\frac{\partial f}{\partial x}N + f\frac{\partial N}{\partial x} - \frac{\partial f}{\partial y}M - f\frac{\partial M}{\partial y} \right)\mathbf{k}$$

$$= f\left[\left(\frac{\partial P}{\partial y} - \frac{\partial N}{\partial z} \right)\mathbf{i} - \left(\frac{\partial P}{\partial x} - \frac{\partial M}{\partial z} \right)\mathbf{j} + \left(\frac{\partial N}{\partial x} - \frac{\partial M}{\partial y} \right)\mathbf{k} \right] + \begin{vmatrix} \mathbf{i} & \mathbf{j} & \mathbf{k} \\ \dfrac{\partial f}{\partial x} & \dfrac{\partial f}{\partial y} & \dfrac{\partial f}{\partial z} \\ M & N & P \end{vmatrix} = f[\nabla \times \mathbf{F}] + (\nabla f) \times \mathbf{F}$$

(g) Let $\mathbf{F} = M\mathbf{i} + N\mathbf{j} + P\mathbf{k}$, then $f\mathbf{F} = fM\mathbf{i} + fN\mathbf{j} + fP\mathbf{k}$.

$$\text{div}(f\mathbf{F}) = \frac{\partial}{\partial x}(fM) + \frac{\partial}{\partial y}(fN) + \frac{\partial}{\partial z}(fP) = f\frac{\partial M}{\partial x} + M\frac{\partial f}{\partial x} + f\frac{\partial N}{\partial y} + N\frac{\partial f}{\partial y} + f\frac{\partial P}{\partial z} + P\frac{\partial f}{\partial z}$$

$$= f\left(\frac{\partial M}{\partial x} + \frac{\partial N}{\partial y} + \frac{\partial N}{\partial z} \right) + \left(\frac{\partial f}{\partial x}M + \frac{\partial f}{\partial y}N + \frac{\partial f}{\partial z}P \right) = f\,\text{div}\,\mathbf{F} + \nabla f \cdot \mathbf{F}$$

(h) Let $\mathbf{F} = M\mathbf{i} + N\mathbf{j} + P\mathbf{k}$.

$$\text{curl}\,\mathbf{F} = \left(\frac{\partial P}{\partial y} - \frac{\partial N}{\partial z} \right)\mathbf{i} - \left(\frac{\partial P}{\partial x} - \frac{\partial M}{\partial z} \right)\mathbf{j} + \left(\frac{\partial N}{\partial x} - \frac{\partial M}{\partial y} \right)\mathbf{k}$$

$$\text{div}(\text{curl}\,\mathbf{F}) = \frac{\partial}{\partial x}\left[\frac{\partial P}{\partial y} - \frac{\partial N}{\partial z} \right] - \frac{\partial}{\partial y}\left[\frac{\partial P}{\partial x} - \frac{\partial M}{\partial z} \right] + \frac{\partial}{\partial z}\left[\frac{\partial N}{\partial x} - \frac{\partial M}{\partial y} \right]$$

$$= \frac{\partial^2 P}{\partial x \partial y} - \frac{\partial^2 N}{\partial x \partial z} - \frac{\partial^2 P}{\partial y \partial x} + \frac{\partial^2 M}{\partial y \partial z} + \frac{\partial^2 N}{\partial z \partial x} - \frac{\partial^2 M}{\partial z \partial y} = 0 \quad \text{(because the mixed partials are equal)}$$

Section 15.2   Line Integrals

1. (a) $\int_C 1\,ds$ is arc length of the curve C.

(b) $\int_C f(x, y, z)\,ds$, where $f(x, y, z)$ is the density of a string of finite length is the mass of the string.

3. $\mathbf{r}(t) = \begin{cases} t\mathbf{i} + t\mathbf{j}, & 0 \le t \le 1 \\ (2 - t)\mathbf{i} + \sqrt{2 - t}\,\mathbf{j}, & 1 \le t \le 2 \end{cases}$

5. $\mathbf{r}(t) = \begin{cases} t\mathbf{i}, & 0 \le t \le 3 \\ 3\mathbf{i} + (t - 3)\mathbf{j}, & 3 \le t \le 6 \\ (9 - t)\mathbf{i} + 3\mathbf{j}, & 6 \le t \le 9 \\ (12 - t)\mathbf{j}, & 9 \le t \le 12 \end{cases}$

7.
$$x^2 + y^2 = 9$$
$$\frac{x^2}{9} + \frac{y^2}{9} = 1$$
$$\cos^2 t + \sin^2 t = 1$$
$$\cos^2 t = \frac{x^2}{9}$$
$$\sin^2 t = \frac{y^2}{9}$$
$$x = 3\cos t$$
$$y = 3\sin t$$
$$\mathbf{r}(t) = 3\cos t\mathbf{i} + 3\sin t\mathbf{j}$$
$$0 \le t \le 2\pi$$

9. (a) $\mathbf{r}(t) = t\mathbf{i} + t\mathbf{j}, \quad 0 \le t \le 1$

(b) $\mathbf{r}'(t) = \mathbf{i} + \mathbf{j}, \|\mathbf{r}'(t)\| = \sqrt{2}$

$$\int_C (x^2 + y^2)\,ds = \int_0^1 (t^2 + t^2)\sqrt{2}\,dt$$

$$= 2\sqrt{2}\left[\frac{t^3}{3} \right]_0^1 = \frac{2\sqrt{2}}{3}$$

11. (a) $\mathbf{r}(t) = \cos t\mathbf{i} + \sin t\mathbf{j}, \quad 0 \le t \le \dfrac{\pi}{2}$

(b) $\displaystyle\int_C \left(x^2 + y^2\right)ds = \int_0^{\pi/2}\left[\cos^2 t + \sin^2 t\right]\sqrt{\left(-\sin t\right)^2 + \left(\cos t\right)^2}\,dt = \int_0^{\pi/2} dt = \dfrac{\pi}{2}$

13. (a) $C: \mathbf{r}(t) = \begin{cases} t\mathbf{i}, & 0 \le t \le 1 \\ t\mathbf{i} + (4t - 4)\mathbf{j}, & 1 \le t \le 2 \end{cases}$

(b) $\displaystyle\int_{C_1} \left(2x + 3\sqrt{y}\right)ds = \int_0^1 2t\,dt = 1$

$\displaystyle\int_{C_2} \left(2x + 3\sqrt{y}\right)ds = \int_1^2 \left(2t + 3\sqrt{4t - 4}\right)\sqrt{1 + 4^2}\,dt = \sqrt{17}\left[t^2 + 4(t - 1)^{3/2}\right]_1^2 = \sqrt{17}\left[(4 + 4) - (1)\right] = 7\sqrt{17}$

$\displaystyle\int_C \left(2x + 3\sqrt{y}\right)ds = 1 + 7\sqrt{17}$

15. (a) $C: \mathbf{r}(t) = \begin{cases} t\mathbf{i}, 0 \le t \le 1 \\ (2 - t)\mathbf{i} + (t - 1)\mathbf{j}, 1 \le t \le 2 \\ (3 - t)\mathbf{j}, 2 \le t \le 3 \end{cases}$

(b) $\displaystyle\int_{C_1} \left(2x + 3\sqrt{y}\right)ds = \int_0^1 2t\,dt = 1$

$\displaystyle\int_{C_2} \left(2x + 3\sqrt{y}\right)ds = \int_1^2 \left[2(2 - t) + 3\sqrt{t - 1}\right]\sqrt{1 + 1}\,dt$

$= 3\sqrt{2}$

$\displaystyle\int_{C_3} \left(2x + 3\sqrt{y}\right)ds = \int_2^3 3\sqrt{3 - t}\,dt = 2$

$\displaystyle\int_C \left(2x + 3\sqrt{y}\right)ds = 1 + 3\sqrt{2} + 2 = 3 + 3\sqrt{2}$

17. (a) $C_1: (0, 0, 0)$ to $(1, 0, 0): \mathbf{r}(t) = t\mathbf{i}, 0 \le t \le 1, \mathbf{r}'(t) = \mathbf{i}, \left\|\mathbf{r}'(t)\right\| = 1$

$\displaystyle\int_{C_1} \left(2x + y^2 - z\right)ds = \int_0^1 2t\,dt = t^2\Big]_0^1 = 1$

$C_2: (1, 0, 0)$ to $(1, 0, 1): \mathbf{r}(t) = \mathbf{i} + t\mathbf{k}, 0 \le t \le 1, \mathbf{r}'(t) = \mathbf{k}, \left\|\mathbf{r}'(t)\right\| = 1$

$\displaystyle\int_{C_2} \left(2x + y^2 - z\right)ds = \int_0^1 (2 - t)\,dt = \left[2t - \dfrac{t^2}{2}\right]_0^1 = \dfrac{3}{2}$

$C_3: (1, 0, 1)$ to $(1, 1, 1): \mathbf{r}(t) = \mathbf{i} + t\mathbf{j} + \mathbf{k}, 0 \le t \le 1, \mathbf{r}'(t) = \mathbf{j}, \left\|\mathbf{r}'(t)\right\| = 1$

$\displaystyle\int_{C_3} \left(2x + y^2 - z\right)ds = \int_0^1 \left(2 + t^2 - 1\right)dt = \left[t + \dfrac{t^3}{3}\right]_0^1 = \dfrac{4}{3}$

(b) Combining, $\displaystyle\int_C \left(2x + y^2 - z\right)ds = 1 + \dfrac{3}{2} + \dfrac{4}{3} = \dfrac{23}{6}$.

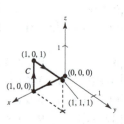

19. $\mathbf{r}(t) = 4t\mathbf{i} + 3t\mathbf{j}, \quad 0 \le t \le 1$

$\mathbf{r}'(t) = 4\mathbf{i} + 3\mathbf{j}$

$\displaystyle\int_C xy\,ds = \int_0^1 (4t)(3t)\sqrt{4^2 + 3^2}\,dt = \int_0^1 60t^2\,dt = \left[20t^3\right]_0^1 = 20$

21. $\mathbf{r}(t) = \sin t\mathbf{i} + \cos t\mathbf{j} + 2\mathbf{k}, \quad 0 \le t \le \dfrac{\pi}{2}$

$\mathbf{r}'(t) = \cos t\mathbf{i} - \sin t\mathbf{j}$

$\displaystyle\int_C \left(x^2 + y^2 + z^2\right) ds = \int_0^{\pi/2} \left(\sin^2 t + \cos^2 t + 4\right)\sqrt{\cos^2 t + \sin^2 t}\, dt = \int_0^{\pi/2} 5\, dt = \dfrac{5\pi}{2}$

23. $\rho(x, y, z) = \dfrac{1}{2}\left(x^2 + y^2 + z^2\right)$

$\mathbf{r}(t) = 2\cos t\mathbf{i} + 2\sin t\mathbf{j} + t\mathbf{k}, 0 \le t \le 4\pi$

$\mathbf{r}'(t) = -2\sin t\mathbf{i} + 2\cos t\mathbf{j} + \mathbf{k}$

$\|\mathbf{r}'(t)\| = \sqrt{4\sin^2 t + 4\cos^2 t + 1} = \sqrt{5}$

Mass $= \displaystyle\int_C \rho(x, y, z)\, ds$

$= \displaystyle\int_0^{4\pi} \dfrac{1}{2}\left(4\cos^2 t + 4\sin^2 t + t^2\right)\sqrt{5}\, dt$

$= \dfrac{\sqrt{5}}{2}\displaystyle\int_0^{4\pi} \left(4 + t^2\right) dt = \dfrac{\sqrt{5}}{2}\left[4t + \dfrac{t^3}{3}\right]_0^{4\pi}$

$= \dfrac{\sqrt{5}}{2}\left[16\pi + \dfrac{64\pi^3}{3}\right] = \dfrac{8\pi\sqrt{5}}{3}\left(4\pi^2 + 3\right)$

≈ 795.7

25. $\mathbf{r}(t) = \cos t\mathbf{i} + \sin t\mathbf{j}, \quad 0 \le t \le \pi$

$\mathbf{r}'(t) = -\sin t\mathbf{i} + \cos t\mathbf{j}, \quad \|\mathbf{r}'(t)\| = 1$

Mass $= \displaystyle\int_C \rho(x, y)\, ds = \int_C (x + y + 2)\, ds$

$= \displaystyle\int_0^{\pi} (\cos t + \sin t + 2)\, dt$

$= \left[\sin t - \cos + 2\,t\right]_0^{\pi}$

$= (1 + 2\pi) - (-1) = 2 + 2\pi$

27. $\mathbf{r}(t) = t^2\mathbf{i} + 2t\mathbf{j} + t\mathbf{k}, \quad 1 \le t \le 3$

$\mathbf{r}'(t) = 2t\mathbf{i} + 2\mathbf{j} + \mathbf{k}, \quad \|\mathbf{r}'(t)\| = \sqrt{4t^2 + 5}$

Mass $= \displaystyle\int_C \rho(x, y, z)\, ds = \int_C kz\, ds$

$= \displaystyle\int_1^3 kt\sqrt{4t^2 + 5}\, dt$

$= \dfrac{k\left(4t^2 + 5\right)^{3/2}}{12}\Bigg]_1^3$

$= \dfrac{k}{12}\left[41\sqrt{41} - 27\right]$

29. $\mathbf{F}(x, y) = x\mathbf{i} + y\mathbf{j}$

$C: \mathbf{r}(t) = (3t + 1)\mathbf{i} + t\mathbf{j}, 0 \le t \le 1$

$\mathbf{F}(t) = (3t + 1)\mathbf{i} + t\mathbf{j}$

$\mathbf{r}'(t) = 3\mathbf{i} + \mathbf{j}$

$\displaystyle\int_C \mathbf{F} \cdot d\mathbf{r} = \int_0^1 \left[(3t + 1)(3) + t\right] dt$

$= \displaystyle\int_0^1 (10t + 3)\, dt$

$= \left[5t^2 + 3t\right]_0^1 = 8$

31. $\mathbf{F}(x, y) = x^2\mathbf{i} + 4y\mathbf{j}$

$C: \mathbf{r}(t) = e^t\mathbf{i} + t^2\mathbf{j}, 0 \le t \le 2$

$\mathbf{F}(t) = e^{2t}\mathbf{i} + 4t^2\mathbf{j}$

$\mathbf{r}'(t) = e^t\mathbf{i} + 2t\mathbf{j}$

$\displaystyle\int_C \mathbf{F} \cdot d\mathbf{r} = \int_0^2 \left[e^{2t} \cdot e^t + 4t^2(2t)\right] dt$

$= \displaystyle\int_0^2 \left(e^{3t} + 8t^3\right) dt$

$= \left[\dfrac{1}{3}e^{3t} + 2t^4\right]_0^2$

$= \dfrac{1}{3}e^6 + 32 - \dfrac{1}{3}$

$= \dfrac{1}{3}e^6 + \dfrac{95}{3}$

33. $\mathbf{F}(x, y, z) = xy\mathbf{i} + xz\mathbf{j} + yz\mathbf{k}$

$C: \mathbf{r}(t) = t\mathbf{i} + t^2\mathbf{j} + 2t\mathbf{k}, \quad 0 \le t \le 1$

$\mathbf{F}(t) = t^3\mathbf{i} + 2t^2\mathbf{j} + 2t^3\mathbf{k}$

$\mathbf{r}'(t) = \mathbf{i} + 2t\mathbf{j} + 2\mathbf{k}$

$\displaystyle\int_C \mathbf{F} \cdot d\mathbf{r} = \int_0^1 \left(t^3 + 4t^3 + 4t^3\right) dt = \left[\dfrac{9t^4}{4}\right]_0^1 = \dfrac{9}{4}$

35. $\mathbf{F}(x, y, z) = x^2 z\mathbf{i} + 6y\mathbf{j} + yz^2\mathbf{k}$

$\mathbf{r}(t) = t\mathbf{i} + t^2\mathbf{j} + \ln t\mathbf{k}, \quad 1 \le t \le 3$

$\mathbf{F}(t) = t^2 \ln t\mathbf{i} + 6t^2\mathbf{j} + t^2 \ln^2 t\mathbf{k}$

$d\mathbf{r} = \left(\mathbf{i} + 2t\mathbf{j} + \dfrac{1}{t}\mathbf{k}\right) dt$

$\displaystyle\int_C \mathbf{F} \cdot d\mathbf{r} = \int_1^3 \left[t^2 \ln t + 12t^3 + t(\ln t)^2\right] dt \approx 249.49$

37. $\mathbf{F}(x, y) = x\mathbf{i} + 2y\mathbf{j}$

$C: \mathbf{r}(t) = t\mathbf{i} + t^3\mathbf{j}, \quad 0 \le t \le 2$

$\mathbf{r}'(t) = \mathbf{i} + 3t^2\mathbf{j}$

$\mathbf{F}(t) = t\mathbf{i} + 2t^3\mathbf{j}$

Work $= \int_C \mathbf{F} \cdot d\mathbf{r} = \int_0^2 \left(t + 6t^5\right) dt = \left[\dfrac{t^2}{2} + t^6\right]_0^2 = 66$

39. $\mathbf{F}(x, y) = x\mathbf{i} + y\mathbf{j}$

$C: \mathbf{r}(t) = \begin{cases} t\mathbf{i} & 0 \le t \le 1 \\ (2 - t)\mathbf{i} + (t - 1)\mathbf{j}, & 1 \le t \le 2 \\ (3 - t)\mathbf{j} & 2 \le t \le 3 \end{cases}$

On C_1, $\mathbf{F}(t) = t\mathbf{i}$, $\mathbf{r}'(t) = \mathbf{i}$

Work $= \int_{C_1} \mathbf{F} \cdot d\mathbf{r} = \int_0^1 t\, dt = \dfrac{1}{2}$

On C_2, $\mathbf{F}(t) = (2 - t)\mathbf{i} + (t - 1)\mathbf{j}$, $\mathbf{r}'(t) = -\mathbf{i} + \mathbf{j}$

Work $= \int_{C_2} \mathbf{F} \cdot d\mathbf{r} = \int_1^2 \left[(t - 2) + (t - 1)\right] dt$

$= \left[t^2 - 3t\right]_1^2$

$= (4 - 6) - (1 - 3) = 0$

On C_3, $\mathbf{F}(t) = (3 - t)\mathbf{j}$, $\mathbf{r}'(t) = -\mathbf{j}$

Work $= \int_{C_3} \mathbf{F} \cdot d\mathbf{r} = \int_2^3 (t - 3)\, dt = \left[\dfrac{t^2}{2} - 3t\right]_2^3$

$= \left(\dfrac{9}{2} - 9\right) - (2 - 6) = -\dfrac{1}{2}$

Total work $= \dfrac{1}{2} + 0 - \dfrac{1}{2} = 0$

41. $\mathbf{F}(x, y, z) = x\mathbf{i} + y\mathbf{j} - 5z\mathbf{k}$

$C: \mathbf{r}(t) = 2\cos t\mathbf{i} + 2\sin t\mathbf{j} + t\mathbf{k}, \quad 0 \le t \le 2\pi$

$\mathbf{r}'(t) = -2\sin t\mathbf{i} + 2\cos t\mathbf{j} + \mathbf{k}$

$\mathbf{F}(t) = 2\cos t\mathbf{i} + 2\sin t\mathbf{j} - 5t\mathbf{k}$

$\mathbf{F} \cdot \mathbf{r}' = -5t$

Work $= \int_C \mathbf{F} \cdot d\mathbf{r} = \int_0^{2\pi} -5t\, dt = -10\pi^2$

43. Because the vector field determined by $\mathbf{F}$ points in the general direction of the path C, $\mathbf{F} \cdot \mathbf{T} > 0$ and work will be positive.

45. Because the vector field determined by $\mathbf{F}$ is perpendicular to the path, work will be 0.

47. $\mathbf{F}(x, y) = x^2\mathbf{i} + xy\mathbf{j}$

(a) $\mathbf{r}_1(t) = 2t\mathbf{i} + (t - 1)\mathbf{j}, \quad 1 \le t \le 3$

$\mathbf{r}_1'(t) = 2\mathbf{i} + \mathbf{j}$

$\mathbf{F}(t) = 4t^2\mathbf{i} + 2t(t - 1)\mathbf{j}$

$\int_{C_1} \mathbf{F} \cdot d\mathbf{r} = \int_1^3 \left(8t^2 + 2t(t - 1)\right) dt = \dfrac{236}{3}$

Both paths join $(2, 0)$ and $(6, 2)$. The integrals are negatives of each other because the orientations are different.

(b) $\mathbf{r}_2(t) = 2(3 - t)\mathbf{i} + (2 - t)\mathbf{j}, \quad 0 \le t \le 2$

$\mathbf{r}_2'(t) = -2\mathbf{i} - \mathbf{j}$

$\mathbf{F}(t) = 4(3 - t)^2\mathbf{i} + 2(3 - t)(2 - t)\mathbf{j}$

$\int_{C_2} \mathbf{F} \cdot d\mathbf{r} = \int_0^2 \left[-8(3 - t)^2 - 2(3 - t)(2 - t)\right] dt$

$= -\dfrac{236}{3}$

49. $\mathbf{F}(x, y) = y\mathbf{i} - x\mathbf{j}$

$C: \mathbf{r}(t) = t\mathbf{i} - 2t\mathbf{j}$

$\mathbf{r}'(t) = \mathbf{i} - 2\mathbf{j}$

$\mathbf{F}(t) = -2t\mathbf{i} - t\mathbf{j}$

$\mathbf{F} \cdot \mathbf{r}' = -2t + 2t = 0$

So, $\int_C \mathbf{F} \cdot d\mathbf{r} = 0$.

51. $\mathbf{F}(x, y) = \left(x^3 - 2x^2\right)\mathbf{i} + \left(x - \dfrac{y}{2}\right)\mathbf{j}$

$C: \mathbf{r}(t) = t\mathbf{i} + t^2\mathbf{j}$

$\mathbf{r}'(t) = \mathbf{i} + 2t\mathbf{j}$

$\mathbf{F}(t) = \left(t^3 - 2t^2\right)\mathbf{i} + \left(t - \dfrac{t^2}{2}\right)\mathbf{j}$

$\mathbf{F} \cdot \mathbf{r}' = \left(t^3 - 2t^2\right) + 2t\left(t - \dfrac{t^2}{2}\right) = 0$

So, $\int_C \mathbf{F} \cdot d\mathbf{r} = 0$.

53. $x = 2t, y = 4t, 0 \le t \le 1 \Rightarrow y = 2x, x = \dfrac{y}{2}, 0 \le y \le 4$

$\int_C \left(x + 3y^2\right) dy = \int_0^4 \left(\dfrac{y}{2} + 3y^2\right) dy = \left[\dfrac{y^2}{4} + y^3\right]_0^4 = 4 + 64 = 68$

55. $x = 2t$, $y = 4t$, $0 \le t \le 1 \Rightarrow y = 2x$, $x = \dfrac{y}{2}$, $dx = \dfrac{dy}{2}$, $0 \le y \le 4$

$$\int_C xy\, dx + y\, dy = \int_0^4 \left[\frac{y}{2}(y)\left(\frac{dy}{2}\right) + y\, dy \right]$$

$$= \int_0^4 \left(\frac{y^2}{4} + y \right) dy$$

$$= \left[\frac{y^3}{12} + \frac{y^2}{2} \right]_0^4$$

$$= \frac{64}{12} + \frac{16}{2} = \frac{40}{3}$$

57. $\mathbf{r}(t) = t\mathbf{i}$, $0 \le t \le 5$

$x(t) = t$, $y(t) = 0$

$dx = dt$, $dy = 0$

$$\int_C (2x - y)\, dx + (x + 3y)\, dy = \int_0^5 2t\, dt = 25$$

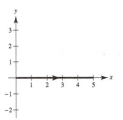

59. $\mathbf{r}(t) = \begin{cases} t\mathbf{i}, & 0 \le t \le 3 \\ 3\mathbf{i} + (t-3)\mathbf{j}, & 3 \le t \le 6 \end{cases}$

C_1: $x(t) = t$, $y(t) = 0$,

$\quad dx = dt$, $dy = 0$

$$\int_{C_1} (2x - y)\, dx + (x + 3y)\, dy = \int_0^3 2t\, dt = 9$$

C_2: $x(t) = 3$, $y(t) = t - 3$

$\quad dx = 0$, $dy = dt$

$$\int_{C_2} (2x - y)\, dx + (x + 3y)\, dy = \int_3^6 \left[3 + 3(t - 3) \right] dt = \left[\frac{3t^2}{2} - 6t \right]_3^6 = \frac{45}{2}$$

$$\int_C (2x - y)\, dx + (x + 3y)\, dy = 9 + \frac{45}{2} = \frac{63}{2}$$

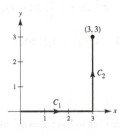

61. $x(t) = t$, $y(t) = 1 - t^2$, $0 \le t \le 1$, $dx = dt$, $dy = -2t\, dt$

$$\int_C (2x - y)\, dx + (x + 3y)\, dy = \int_0^1 \left[(2t - 1 + t^2) + (t + 3 - 3t^2)(-2t) \right] dt$$

$$= \int_0^1 \left(6t^3 - t^2 - 4t - 1 \right) dt = \left[\frac{3t^4}{2} - \frac{t^3}{3} - 2t^2 - t \right]_0^1 = -\frac{11}{6}$$

63. $x(t) = t$, $y(t) = 2t^2$, $0 \le t \le 2$

$dx = dt$, $dy = 4t\, dt$

$$\int_C (2x - y)\, dx + (x + 3y)\, dy = \int_0^2 (2t - 2t^2)\, dt + (t + 6t^2)4t\, dt$$

$$= \int_0^2 \left(24t^3 + 2t^2 + 2t \right) dt$$

$$= \left[6t^4 + \tfrac{2}{3}t^3 + t^2 \right]_0^2$$

$$= \frac{316}{3}$$

65. $f(x, y) = h$

   C: line from $(0, 0)$ to $(3, 4)$

   $$\mathbf{r} = 3t\mathbf{i} + 4t\mathbf{j}, \quad 0 \leq t \leq 1$$
   $$\mathbf{r}'(t) = 3\mathbf{i} + 4\mathbf{j}$$
   $$\|\mathbf{r}'(t)\| = 5$$

   Lateral surface area:

   $$\int_C f(x, y) \, ds = \int_0^1 5h \, dt = 5h$$

67. $f(x, y) = xy$

   $C: x^2 + y^2 = 1$ from $(1, 0)$ to $(0, 1)$

   $$\mathbf{r}(t) = \cos t\mathbf{i} + \sin t\mathbf{j}, \quad 0 \leq t \leq \frac{\pi}{2}$$
   $$\mathbf{r}'(t) = -\sin t\mathbf{i} + \cos t\mathbf{j}$$
   $$\|\mathbf{r}'(t)\| = 1$$

   Lateral surface area:

   $$\int_C f(x, y) \, ds = \int_0^{\pi/2} \cos t \sin t \, dt = \left[\frac{\sin^2 t}{2}\right]_0^{\pi/2} = \frac{1}{2}$$

69. $f(x, y) = h$

   $C: y = 1 - x^2$ from $(1, 0)$ to $(0, 1)$

   $$\mathbf{r}(t) = (1 - t)\mathbf{i} + \left[1 - (1 - t)^2\right]\mathbf{j}, \quad 0 \leq t \leq 1$$
   $$\mathbf{r}'(t) = -\mathbf{i} + 2(1 - t)\mathbf{j}$$
   $$\|\mathbf{r}'(t)\| = \sqrt{1 + 4(1 - t)^2}$$

   Lateral surface area:

   $$\int_C f(x, y) \, ds = \int_0^1 h\sqrt{1 + 4(1 - t)^2} \, dt = -\frac{h}{4}\left[2(1 - t)\sqrt{1 + 4(1 - t)^2} + \ln\left|2(1 - t) + \sqrt{1 + 4(1 - t)^2}\right|\right]_0^1$$
   $$= \frac{h}{4}\left[2\sqrt{5} + \ln\left(2 + \sqrt{5}\right)\right] \approx 1.4789h$$

71. $f(x, y) = xy$

   $C: y = 1 - x^2$ from $(1, 0)$ to $(0, 1)$

   You could parameterize the curve C as in Exercises 67 and 68. Alternatively, let $x = \cos t$, then:

   $$y = 1 - \cos^2 t = \sin^2 t$$

   $$\mathbf{r}(t) = \cos t\mathbf{i} + \sin^2 t\mathbf{j}, \quad 0 \leq t \leq \frac{\pi}{2}$$
   $$\mathbf{r}'(t) = -\sin t\mathbf{i} + 2\sin t \cos t\mathbf{j}$$
   $$\|\mathbf{r}'(t)\| = \sqrt{\sin^2 t + 4\sin^2 t \cos^2 t} = \sin t\sqrt{1 + 4\cos^2 t}$$

   Lateral surface area:

   $$\int_C f(x, y) \, ds = \int_0^{\pi/2} \cos t \sin^2 t\left(\sin t\sqrt{1 + 4\cos^2 t}\right) dt = \int_0^{\pi/2} \sin^2 t\left[\left(1 + 4\cos^2 t\right)^{1/2} \sin t \cos t\right] dt$$

   Let $u = \sin^2 t$ and $dv = \left(1 + 4\cos^2 t\right)^{1/2} \sin t \cos t$, then $du = 2\sin t \cos t \, dt$ and $v = -\frac{1}{12}\left(1 + 4\cos^2 t\right)^{3/2}$.

   $$\int_C f(x, y) \, ds = \left[-\frac{1}{12}\sin^2 t\left(1 + 4\cos^2 t\right)^{3/2}\right]_0^{\pi/2} + \frac{1}{6}\int_0^{\pi/2}\left(1 + 4\cos^2 t\right)^{3/2} \sin t \cos t \, dt$$
   $$= \left[-\frac{1}{12}\sin^2 t\left(1 + 4\cos^2 t\right)^{3/2} - \frac{1}{120}\left(1 + 4\cos^2 t\right)^{5/2}\right]_0^{\pi/2}$$
   $$= \left(\frac{1}{12} - \frac{1}{120}\right) + \frac{1}{120}(5)^{5/2}$$
   $$= \frac{1}{120}\left(25\sqrt{5} - 11\right)$$
   $$\approx 0.3742$$

73. (a) $f(x, y) = 1 + y^2$

$\mathbf{r}(t) = 2 \cos t\mathbf{i} + 2 \sin t\mathbf{j}, \ 0 \le t \le 2\pi$

$\mathbf{r}'(t) = -2 \sin t\mathbf{i} + 2 \cos t\mathbf{j}$

$\|\mathbf{r}'(t)\| = 2$

$S = \int_C f(x, y) \, ds = \int_0^{2\pi} (1 + 4 \sin^2 t)(2) \, dt = \left[2t + 4(t - \sin t \cos t)\right]_0^{2\pi} = 12\pi \approx 37.70 \text{ cm}^2$

(b) $0.2(12\pi) = \dfrac{12\pi}{5} \approx 7.54 \text{ cm}^3$

(c)

75. $\mathbf{r}(t) = a \cos t\mathbf{i} + a \sin t\mathbf{j}, \ 0 \le t \le 2\pi$

$\mathbf{r}'(t) = -a \sin t\mathbf{i} + a \cos t\mathbf{j}, \ \|\mathbf{r}'(t)\| = a$

$I_x = \int_C y^2 \rho(x, y) \, ds = \int_0^{2\pi} (a^2 \sin^2 t)(1)a \, dt = a^3 \int_0^{2\pi} \sin^2 t \, dt = a^3\pi$

$I_y = \int_C x^2 \rho(x, y) \, ds = \int_0^{2\pi} (a^2 \cos^2 t)(1)a \, dt = a^3 \int_0^{2\pi} \cos^2 t \, dt = a^3\pi$

77. (a) Graph of: $\mathbf{r}(t) = 3 \cos t\mathbf{i} + 3 \sin t\mathbf{j} + (1 + \sin^2 2t)\mathbf{k}, \ 0 \le t \le 2\pi$

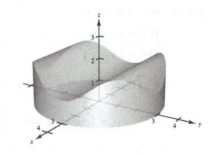

For $y = b$ constant, $3 \sin t = b \Rightarrow \sin t = \dfrac{b}{3}$ and

$1 + \sin^2 2t = 1 + (2 \sin t \cos t)^2$

$\qquad\qquad = 1 + 4 \sin^2 t \cos^2 t$

$\qquad\qquad = 1 + 4 \sin^2 t(1 - \sin^2 t) = 1 + \dfrac{4}{9}b^2\left(1 - \dfrac{b^2}{9}\right).$

(b) Consider the portion of the surface in the first quadrant. The curve $z = 1 + \sin^2 2t$ is over the curve

$\mathbf{r}_1(t) = 3 \cos t\mathbf{i} + 3 \sin t\mathbf{j}, 0 \le t \le \pi/2.$ So, the total lateral surface area is

$4\int_C f(x, y) \, ds = 4\int_0^{\pi/2} (1 + \sin^2 2t)3 \, dt = 12\left(\dfrac{3\pi}{4}\right) = 9\pi \text{ cm}^2.$

(c) The cross sections parallel to the xz-plane are rectangles of height $1 + 4(y/3)^2(1 - y^2/9)$ and base $2\sqrt{9 - y^2}.$ So,

$\text{Volume} = 2\int_0^3 2\sqrt{9 - y^2}\left[1 + \dfrac{4y^2}{9}\left(1 - \dfrac{y^2}{9}\right)\right] dy = \dfrac{27\pi}{2} \approx 42.412 \text{ cm}^3.$

79. $\mathbf{r}(t) = 3 \sin t\mathbf{i} + 3 \cos t\mathbf{j} + \dfrac{10}{2\pi} t\mathbf{k}, \quad 0 \le t \le 2\pi$

$\mathbf{F} = 175\mathbf{k}$

$d\mathbf{r} = \left(3 \cos t\mathbf{i} - 3 \sin t\mathbf{j} + \dfrac{10}{2\pi}\mathbf{k}\right) dt$

$\int_C \mathbf{F} \cdot d\mathbf{r} = \int_0^{2\pi} \dfrac{1750}{2\pi} \, dt = \left[\dfrac{1750}{2\pi}t\right]_0^{2\pi} = 1750 \text{ ft} \cdot \text{lb}$

81. No. $y = 2t = 2x,$ so $dy = 2 \, dx.$

83. The greater the height of the surface over the curve, the greater the lateral surface area.

So, $z_3 < z_1 < z_2 < z_4.$

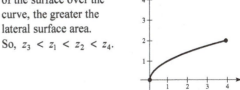

85. False

$$\int_C xy\, ds = \sqrt{2} \int_0^1 t^2\, dt$$

87. $\mathbf{F}(x, y) = (y - x)\mathbf{i} + xy\mathbf{j}$

$\mathbf{r}(t) = kt(1 - t)\mathbf{i} + t\mathbf{j}, \quad 0 \le t \le 1$

$\mathbf{r}'(t) = k(1 - 2t)\mathbf{i} + \mathbf{j}$

Work $= 1 = \int_C \mathbf{F} \cdot d\mathbf{r}$

$= \int_0^1 \left[(t - kt(1 - t))\mathbf{i} + kt^2(1 - t)\mathbf{j} \right] \cdot \left[k(1 - 2t)\mathbf{i} + \mathbf{j} \right] dt$

$= \int_0^1 \left[(t - kt(1 - t))k(1 - 2t) + kt^2(1 - t) \right] dt$

$= \int_0^1 \left(-2k^2t^3 - kt^3 - kt^2 + 3k^2t^2 - k^2t + kt \right) dt = \dfrac{-k}{12}$

$k = -12$

Section 15.3   Conservative Vector Fields and Independence of Path

1. First, verify that the vector field is conservative. Then find a potential function. Calculate the difference of the values of the potential function evaluated at the endpoints.

3. $\mathbf{F}(x, y) = x^2\mathbf{i} + y\mathbf{j}$

(a) $\dfrac{\partial N}{\partial x} = 0 = \dfrac{\partial M}{\partial y} \Rightarrow$ Conservative

(b) (i) $\mathbf{r}_1(t) = t\mathbf{i} + t^2\mathbf{j}, 0 \le t \le 1$

$\mathbf{r}_1'(t) = \mathbf{i} + 2t\mathbf{j}$

$\mathbf{F}(t) = t^2\mathbf{i} + t^2\mathbf{j}$

$\displaystyle\int_{C_1} \mathbf{F} \cdot d\mathbf{r} = \int_0^1 \left(t^2\mathbf{i} + t^2\mathbf{j} \right) \cdot (\mathbf{i} + 2t\mathbf{j})\, dt$

$\qquad = \int_0^1 \left(t^2 + 2t^3 \right) dt = \left[\dfrac{t^3}{3} + \dfrac{t^4}{2} \right]_0^1 = \dfrac{1}{3} + \dfrac{1}{2} = \dfrac{5}{6}$

(ii) $\mathbf{r}_2(\theta) = \sin\theta\,\mathbf{i} + \sin^2\theta\,\mathbf{j}, 0 \le \theta \le \dfrac{\pi}{2}$

$\mathbf{r}_2'(\theta) = \cos\theta\,\mathbf{i} + 2\sin\theta\cos\theta\,\mathbf{j}$

$\mathbf{F}(t) = \sin^2\theta\,\mathbf{i} + \sin^2\theta\,\mathbf{j}$

$\displaystyle\int_{C_2} \mathbf{F} \cdot d\mathbf{r} = \int_0^{\pi/2} \left(\sin^2\theta\,\mathbf{i} + \sin^2\theta\,\mathbf{j} \right) \cdot (\cos\theta\,\mathbf{i} + 2\sin\theta\cos\theta\,\mathbf{j})\, dt$

$\qquad = \int_0^{\pi/2} \left(\sin^2\theta\cos\theta + 2\sin^3\theta\cos\theta \right) d\theta$

$\qquad = \left[\dfrac{\sin^3\theta}{3} + \dfrac{\sin^4\theta}{2} \right]_0^{\pi/2} = \dfrac{1}{3} + \dfrac{1}{2} = \dfrac{5}{6}$

5. $\mathbf{F}(x, y) = 3y\mathbf{i} + 3x\mathbf{j}$

(a) $\dfrac{\partial N}{\partial x} = 3 = \dfrac{\partial M}{\partial y} \Rightarrow$ Conservative

(b) (i) $\mathbf{r}_1(\theta) = \sec\theta\mathbf{i} + \tan\theta\mathbf{j}, \ 0 \le \theta \le \dfrac{\pi}{3}$

$\mathbf{r}_1'(\theta) = \sec\theta\tan\theta\mathbf{i} + \sec^2\theta\mathbf{j}$

$\mathbf{F}(\theta) = 3\tan\theta\mathbf{i} + 3\sec\theta\mathbf{j}$

$\displaystyle\int_{C_1} \mathbf{F} \cdot d\mathbf{r} = \int_0^{\pi/3} \left(3\tan^2\theta\sec\theta + 3\sec^3\theta\right) d\theta$

$\displaystyle = \int_0^{\pi/3} \left[3(\sec^2\theta - 1)\sec\theta + 3\sec^2\theta\right] d\theta$

$\displaystyle = \int_0^{\pi/3} \left[6\sec^3\theta - 3\sec\theta\right] d\theta$

$= 6\sqrt{3} \approx 10.3923$

(ii) $\mathbf{r}_2(t) = \sqrt{t + 1}\,\mathbf{i} + \sqrt{t}\,\mathbf{j}, \ 0 \le t \le 3$

$\mathbf{r}_2'(t) = \dfrac{1}{2\sqrt{t+1}}\mathbf{i} + \dfrac{1}{2\sqrt{t}}\mathbf{j}$

$\mathbf{F}(t) = 3\sqrt{t}\,\mathbf{i} + 3\sqrt{t+1}\,\mathbf{j}$

$\displaystyle\int_{C_2} \mathbf{F} \cdot d\mathbf{r} = \int_0^3 \left[\dfrac{3\sqrt{t}}{2\sqrt{t+1}} + \dfrac{3\sqrt{t+1}}{2\sqrt{t}}\right] dt = 6\sqrt{3} \approx 10.3923$

7. $\mathbf{F}(x, y, z) = y^2z\mathbf{i} + 2xyz\mathbf{j} + xy^2\mathbf{k}$

(a) $\operatorname{curl} \mathbf{F} = \begin{vmatrix} \mathbf{i} & \mathbf{j} & \mathbf{k} \\ \dfrac{\partial}{\partial x} & \dfrac{\partial}{\partial y} & \dfrac{\partial}{\partial z} \\ y^2z & 2xyz & xy^2 \end{vmatrix} = \mathbf{0} \Rightarrow$ Conservative

(b) (i) $\mathbf{r}_1(t) = t\mathbf{i} + 2t\mathbf{j} + 4t\mathbf{k}, \ 0 \le t \le 1$

$\mathbf{r}_1'(t) = \mathbf{i} + 2\mathbf{j} + 4\mathbf{k}$

$\mathbf{F}(t) = 16t^3\mathbf{i} + 16t^3\mathbf{j} + 4t^3\mathbf{k}$

$\displaystyle\int_{C_1} \mathbf{F} \cdot d\mathbf{r} = \int_0^1 \left[16t^3 + 32t^3 + 16t^3\right] dt = \int_0^1 64t^3 \, dt = \left[16t^4\right]_0^1 = 16$

(ii) $\mathbf{r}_2(\theta) = \sin\theta\mathbf{i} + 2\sin\theta\mathbf{j} + 4\sin\theta\mathbf{k}, \ 0 \le \theta \le \dfrac{\pi}{2}$

$\mathbf{r}_2'(\theta) = \cos\theta\mathbf{i} + 2\cos\theta\mathbf{j} + 4\cos\theta\,\mathbf{k}$

$\mathbf{F}(\theta) = 16\sin^3\theta\mathbf{i} + 16\sin^3\theta\mathbf{j} + 4\sin^3\theta\mathbf{k}$

$\displaystyle\int_{C_2} \mathbf{F} \cdot d\mathbf{r} = \int_0^{\pi/2} 64\sin^3\theta\cos\theta \, d\theta = \left[16\sin^4\theta\right]_0^{\pi/2} = 16$

9. $\displaystyle\int_C (3y\mathbf{i} + 3x\mathbf{j}) \cdot d\mathbf{r} = \left[3xy\right]_{(0,0)}^{(3,8)} = 72$

11. $\displaystyle\int_C \cos x \sin y \, dx + \sin x \cos y \, dy = \left[\sin x \sin y\right]_{(0,-\pi)}^{(3\pi/2,\,\pi/2)} = -1$

13. $\displaystyle\int_C e^x \sin y \, dx + e^x \cos y \, dy = \left[e^x \sin y\right]_{(0,0)}^{(2\pi,0)} = 0$

15. $\int_C (z + 2y)\, dx + (2x - z)\, dy + (x - y)\, dz$

$\mathbf{F}(x, y, z)$ is conservative and the potential function is $f(x, y, z) = xz + 2xy - yz$

(a) $\left[xz + 2xy - yz \right]_{(0,0,0)}^{(1,1,1)} = 2 - 0 = 2$

(b) $\left[xz + 2xy - yz \right]_{(0,0,0)}^{(0,0,1)} + \left[xz + 2xy - yz \right]_{(0,0,1)}^{(1,1,1)} = 0 + 2 = 2$

(c) $\left[xz + 2xy - yz \right]_{(0,0,0)}^{(1,0,0)} + \left[xz + 2xy - yz \right]_{(1,0,0)}^{(1,1,0)} + \left[xz + 2xy - yz \right]_{(1,1,0)}^{(1,1,1)} = 0 + 2 + (2 - 2) = 2$

17. $\int_C -\sin x\, dx + z\, dy + y\, dz = \left[\cos x + yz \right]_{(0,0,0)}^{(\pi/2, 3, 4)} = 12 - 1 = 11$

19. $\mathbf{F}(x, y) = 9x^2 y^2 \mathbf{i} + \left(6x^3 y - 1\right)\mathbf{j}$

(a) $\dfrac{\partial N}{\partial x} = 18x^2 y = \dfrac{\partial M}{\partial y} \Rightarrow$ Conservative

(b) $f(x, y) = 3x^3 y^2 - y$

Work $= \left[3x^3 y^2 - y \right]_{(0,0)}^{(5,9)} = 30{,}366$

21. $\mathbf{F}(x, y, z) = 3\mathbf{i} + 4y\mathbf{j} - \sin z\,\mathbf{k}$

(a) curl $\mathbf{F} = \begin{vmatrix} \mathbf{i} & \mathbf{j} & \mathbf{k} \\ \dfrac{\partial}{\partial x} & \dfrac{\partial}{\partial y} & \dfrac{\partial}{\partial z} \\ 3 & 4y & -\sin z \end{vmatrix} = \mathbf{0} \Rightarrow$ Conservative

(b) $f(x, y, z) = 3x + 2y^2 + \cos z$

Work $= \left[3x + 2y^2 + \cos z \right]_{(0,1,\pi/2)}^{(1,4,\pi)}$

$= (3 + 32 - 1) - (0 + 2 + 0)$

$= 32$

23. $\mathbf{F}(x, y) = 2xy\mathbf{i} + x^2 \mathbf{j}$

(a) $\mathbf{r}_1(t) = t\mathbf{i} + t^2\mathbf{j}, \quad 0 \le t \le 1$

$\mathbf{r}_1'(t) = \mathbf{i} + 2t\mathbf{j}$

$\mathbf{F}(t) = 2t^3\mathbf{i} + t^2\mathbf{j}$

$\int_C \mathbf{F} \cdot d\mathbf{r} = \int_0^1 4t^3\, dt = 1$

(b) $\mathbf{r}_2(t) = t\mathbf{i} + t^3\mathbf{j}, \quad 0 \le t \le 1$

$\mathbf{r}_2'(t) = \mathbf{i} + 3t^2\mathbf{j}$

$\mathbf{F}(t) = 2t^4\mathbf{i} + t^2\mathbf{j}$

$\int_C \mathbf{F} \cdot d\mathbf{r} = \int_0^1 5t^4\, dt = 1$

25. $\int_C y^2\, dx + 2xy\, dy$

Because $\partial M / \partial y = \partial N / \partial x = 2y$, $\mathbf{F}(x, y) = y^2\mathbf{i} + 2xy\mathbf{j}$ is conservative. The potential function is $f(x, y) = xy^2 + k$. So, you can use the Fundamental Theorem of Line Integrals.

(a) $\int_C y^2\, dx + 2xy\, dy = \left[x^2 y \right]_{(0,0)}^{(4,4)} = 64$

(b) $\int_C y^2\, dx + 2xy\, dy = \left[x^2 y \right]_{(-1,0)}^{(1,0)} = 0$

(c) and (d) Because C is a closed curve,

$\int_C y^2\, dx + 2xy\, dy = 0.$

27. $\mathbf{F}(x, y, z) = yz\mathbf{i} + xz\mathbf{j} + xy\mathbf{k}$

Because curl $\mathbf{F} = \mathbf{0}$, $\mathbf{F}(x, y, z)$ is conservative. The potential function is $f(x, y, z) = xyz + k$.

(a) $\mathbf{r}_1(t) = t\mathbf{i} + 2\mathbf{j} + t\mathbf{k}, \quad 0 \le t \le 4$

$\int_C \mathbf{F} \cdot d\mathbf{r} = \left[xyz \right]_{(0,2,0)}^{(4,2,4)} = 32$

(b) $\mathbf{r}_2(t) = t^2\mathbf{i} + t\mathbf{j} + t^2\mathbf{k}, \quad 0 \le t \le 2$

$\int_C \mathbf{F} \cdot d\mathbf{r} = \left[xyz \right]_{(0,0,0)}^{(4,2,4)} = 32$

29. $\mathbf{F}(x, y, z) = (2y + x)\mathbf{i} + (x^2 - z)\mathbf{j} + (2y - 4z)\mathbf{k}$

$\mathbf{F}(x, y, z)$ is not conservative.

(a) $\mathbf{r}_1(t) = t\mathbf{i} + t^2\mathbf{j} + \mathbf{k}, \ \ 0 \le t \le 1$

$\mathbf{r}_1'(t) = \mathbf{i} + 2t\mathbf{j}$

$\mathbf{F}(t) = (2t^2 + t)\mathbf{i} + (t^2 - 1)\mathbf{j} + (2t^2 - 4)\mathbf{k}$

$\int_C \mathbf{F} \cdot d\mathbf{r} = \int_0^1 (2t^3 + 2t^2 - t) \, dt = \dfrac{2}{3}$

(b) $\mathbf{r}_2(t) = t\mathbf{i} + t\mathbf{j} + (2t - 1)^2\mathbf{k}, \ \ 0 \le t \le 1$

$\mathbf{r}_2'(t) = \mathbf{i} + \mathbf{j} + 4(2t - 1)\mathbf{k}$

$\mathbf{F}(t) = 3t\mathbf{i} + \left[t^2 - (2t - 1)^2 \right]\mathbf{j} + \left[2t - 4(2t - 1)^2 \right]\mathbf{k}$

$\int_C \mathbf{F} \cdot d\mathbf{r} = \int_0^1 \left[3t + t^2 - (2t - 1)^2 + 8t(2t - 1) - 16(2t - 1)^3 \right] dt$

$= \int_0^1 \left[17t^2 - 5t - (2t - 1)^2 - 16(2t - 1)^3 \right] dt = \left[\dfrac{17t^3}{3} - \dfrac{5t^2}{2} - \dfrac{(2t - 1)^3}{6} - 2(2t - 1)^4 \right]_0^1 = \dfrac{17}{6}$

31. $\mathbf{F}(x, y, z) = e^z(y\mathbf{i} + x\mathbf{j} + xy\mathbf{k})$

$\mathbf{F}(x, y, z)$ is conservative. The potential function is $f(x, y, z) = xye^z + k$.

(a) $\mathbf{r}_1(t) = 4 \cos t\mathbf{i} + 4 \sin t\mathbf{j} + 3\mathbf{k}, \ \ 0 \le t \le \pi$

$\int_C \mathbf{F} \cdot d\mathbf{r} = \left[xye^z \right]_{(4,0,3)}^{(-4,0,3)} = 0$

(b) $\mathbf{r}_2(t) = (4 - 8t)\mathbf{i} + 3\mathbf{k}, \ \ 0 \le t \le 1$

$\int_C \mathbf{F} \cdot d\mathbf{r} = \left[xye^z \right]_{(4,0,3)}^{(-4,0,3)} = 0$

33. $\mathbf{r}(t) = 2 \cos 2\pi t\mathbf{i} + 2 \sin 2\pi t\mathbf{j}$

$\mathbf{r}'(t) = -4\pi \sin 2\pi t\mathbf{i} + 4\pi \cos 2\pi t\mathbf{j}$

$\mathbf{a}(t) = -8\pi^2 \cos 2\pi t\mathbf{i} - 8\pi^2 \sin 2\pi t\mathbf{j}$

$\mathbf{F}(t) = m\mathbf{a}(t) = \dfrac{1}{32}\mathbf{a}(t) = -\dfrac{\pi^2}{4}(\cos 2\pi t\mathbf{i} + \sin 2\pi t\mathbf{j})$

$W = \int_C \mathbf{F} \cdot d\mathbf{r} = \int_C -\dfrac{\pi^2}{4}(\cos 2\pi t\mathbf{i} + \sin 2\pi t\mathbf{j}) \cdot 4\pi(-\sin 2\pi t\mathbf{i} + \cos 2\pi t\mathbf{j}) \, dt = -\pi^3 \int_C 0 \, dt = 0$

35. $\mathbf{F} = -175\mathbf{j}$

(a) $\mathbf{r}(t) = t\mathbf{i} + (50 - t)\mathbf{j}, \ \ 0 \le t \le 50$

$d\mathbf{r} = (\mathbf{i} - \mathbf{j}) \, dt$

$\int_C \mathbf{F} \cdot d\mathbf{r} = \int_0^{50} 175 \, dt = 8750 \ \text{ft} \cdot \text{lb}$

(b) $\mathbf{r}(t) = t\mathbf{i} + \dfrac{1}{50}(50 - t)^2\mathbf{j}, \ \ 0 \le t \le 50$

$d\mathbf{r} = \mathbf{i} - \dfrac{1}{25}(50 - t)\mathbf{j}$

$\int_C \mathbf{F} \cdot d\mathbf{r} = \int_0^{50} (175)\dfrac{1}{25}(50 - t) \, dt$

$= 7\left[50t - \dfrac{t^2}{2} \right]_0^{50} = 8750 \ \text{ft} \cdot \text{lb}$

37. The partial derivatives of $\mathbf{F}$ are not continuous at $(0, 0)$.

Draw an open connected region that excludes the origin.

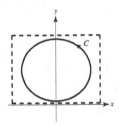

39. $F(x, y) = \left(x^2 y^2 - 3x\right)\mathbf{i} + \frac{2}{3}x^3 y\mathbf{j}$

(a) **F** is conservative; $f = \frac{x^3}{3}y^2 - \frac{3}{2}x^2$

$$\int_C \mathbf{F} \cdot d\mathbf{r} = \left[\frac{x^3}{3}y^2 - \frac{3}{2}x^2\right]_{(2, 3/2)}^{(1/2, 6)} = \frac{9}{8} - 0 = \frac{9}{8}$$

(b) $\mathbf{r}(t) = \frac{1}{t}\mathbf{i} + 3t\mathbf{j}$, $\mathbf{r}'(t) = -\frac{1}{t^2}\mathbf{i} + 3\mathbf{j}$, $\mathbf{F}(t) = \left(9 - \frac{3}{t}\right)\mathbf{i} + \frac{2}{t^2}\mathbf{j}$

$$\int_C \mathbf{F} \cdot d\mathbf{r} = \int_{1/2}^{2}\left[\left(9 - \frac{3}{t}\right)\left(-\frac{1}{t^2}\right) + \left(\frac{2}{t^2}\right)(3)\right] dt = \int_{1/2}^{2}\left(-\frac{3}{t^2} + \frac{3}{t^3}\right) dt = \left[\frac{3}{t} - \frac{3}{2}t^{-2}\right]_{1/2}^{2} = \left(\frac{3}{2} - \frac{3}{8}\right) - \left(6 - \frac{3}{2}(4)\right) = \frac{9}{8} - 0 = \frac{9}{8}$$

41. Conservative. $\int_C \mathbf{F} \cdot d\mathbf{r}$ is independent of path.

43. False, it would be true if **F** were conservative.

45. True

47. Let

$$\mathbf{F} = M\mathbf{i} + N\mathbf{j} = \frac{\partial f}{\partial y}\mathbf{i} - \frac{\partial f}{\partial x}\mathbf{j}.$$

Then $\dfrac{\partial M}{\partial y} = \dfrac{\partial}{\partial y}\left(\dfrac{\partial f}{\partial y}\right) = \dfrac{\partial^2 f}{\partial y^2}$ and $\dfrac{\partial N}{\partial x} = \dfrac{\partial}{\partial x}\left(-\dfrac{\partial f}{\partial x}\right) = -\dfrac{\partial^2 f}{\partial x^2}$. Because $\dfrac{\partial^2 f}{\partial x^2} + \dfrac{\partial^2 f}{\partial y^2} = 0$ you have $\dfrac{\partial M}{\partial y} = \dfrac{\partial N}{\partial x}$.

So, **F** is conservative. Therefore, by Theorem 15.7, you have $\displaystyle\int_C\left(\frac{\partial f}{\partial y}\, dx - \frac{\partial f}{\partial x}\, dy\right) = \int_C\left(M\, dx + N\, dy\right) = \int_C \mathbf{F} \cdot d\mathbf{r} = 0$

for every closed curve in the plane.

49. $F(x, y) = \dfrac{y}{x^2 + y^2}\mathbf{i} - \dfrac{x}{x^2 + y^2}\mathbf{j}$

(a) $M = \dfrac{y}{x^2 + y^2}$

$\dfrac{\partial M}{\partial y} = \dfrac{\left(x^2 + y^2\right)(1) - y(2y)}{\left(x^2 + y^2\right)^2} = \dfrac{x^2 - y^2}{\left(x^2 + y^2\right)^2}$

$N = -\dfrac{x}{x^2 + y^2}$

$\dfrac{\partial N}{\partial x} = \dfrac{\left(x^2 + y^2\right)(-1) + x(2x)}{\left(x^2 + y^2\right)^2} = \dfrac{x^2 - y^2}{\left(x^2 + y^2\right)^2}$

So, $\dfrac{\partial N}{\partial x} = \dfrac{\partial M}{\partial y}$.

(b) $\mathbf{r}(t) = \cos t\mathbf{i} + \sin t\mathbf{j}$, $0 \le t \le \pi$

$\mathbf{F} = \sin t\mathbf{i} - \cos t\mathbf{j}$

$d\mathbf{r} = (-\sin t\mathbf{i} + \cos t\mathbf{j})\, dt$

$\displaystyle\int_C \mathbf{F} \cdot d\mathbf{r} = \int_0^\pi \left(-\sin^2 t - \cos^2 t\right) dt$

$= [-t]_0^\pi = -\pi$

(c) $\mathbf{r}(t) = \cos t\mathbf{i} - \sin t\mathbf{j}$, $0 \le t \le \pi$

$\mathbf{F} = -\sin t\mathbf{i} - \cos t\mathbf{j}$

$d\mathbf{r} = (-\sin t\mathbf{i} - \cos t\mathbf{j})\, dt$

$\displaystyle\int_C \mathbf{F} \cdot d\mathbf{r} = \int_0^\pi \left(\sin^2 t + \cos^2 t\right) dt = [t]_0^\pi = \pi$

(d) $\mathbf{r}(t) = \cos t\mathbf{i} + \sin t\mathbf{j}$, $0 \le t \le 2\pi$

$\mathbf{F} = \sin t\mathbf{i} - \cos t\mathbf{j}$

$d\mathbf{r} = (-\sin t\mathbf{i} + \cos t\mathbf{j})\, dt$

$\displaystyle\int_C \mathbf{F} \cdot d\mathbf{r} = \int_0^{2\pi} \left(-\sin^2 t - \cos^2 t\right) dt$

$= [-t]_0^{2\pi} = -2\pi$

(e) This does not contradict Theorem 15.7 because **F** is not continuous at $(0, 0)$ in R enclosed by curve C.

(f) $\nabla\left(\arctan\dfrac{x}{y}\right) = \dfrac{1/y}{1 + (x/y)^2}\mathbf{i} + \dfrac{-x/y^2}{1 + (x/y)^2}\mathbf{j}$

$= \dfrac{y}{x^2 + y^2}\mathbf{i} - \dfrac{x}{x^2 + y^2}\mathbf{j} = \mathbf{F}$

Section 15.4   Green's Theorem

1. A curve is simple if it does not cross itself. A connected plane region is simply connected if every simple closed curve in the region encloses only points that are in the region. For example, a region with a hole is not simply connected.

3. You are working with a simple closed curve with a boundary oriented counterclockwise.

5. $\mathbf{r}(t) = \begin{cases} t\mathbf{i} + t^2\mathbf{j}, & 0 \le t \le 1 \\ (2-t)\mathbf{i} + (2-t)\mathbf{j}, & 1 \le t \le 2 \end{cases}$

$\int_C y^2\,dx + x^2\,dy = \int_0^1 \left[t^4(dt) + t^2(2t\,dt)\right] + \int_1^2 \left[(2-t)^2(-dt) + (2-t)^2(-dt)\right]$

$= \int_0^1 (t^4 + 2t^3)\,dt + \int_1^2 2(2-t)^2(-dt) = \left[\dfrac{t^5}{5} + \dfrac{t^4}{2}\right]_0^1 + \left[\dfrac{2(2-t)^3}{3}\right]_1^2 = \dfrac{7}{10} - \dfrac{2}{3} = \dfrac{1}{30}$

By Green's Theorem,

$\int_R\int \left(\dfrac{\partial N}{\partial x} - \dfrac{\partial M}{\partial y}\right)dA = \int_0^1\int_{x^2}^x (2x - 2y)\,dy\,dx = \int_0^1 \left[2xy - y^2\right]_{x^2}^x\,dx$

$= \int_0^1 (x^2 - 2x^3 + x^4)\,dx = \left[\dfrac{x^3}{3} - \dfrac{x^4}{2} + \dfrac{x^5}{5}\right]_0^1 = \dfrac{1}{30}$

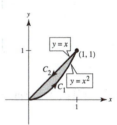

7. $\mathbf{r}(t) = \begin{cases} t\mathbf{i} & 0 \le t \le 1 \\ \mathbf{i} + (t-1)\mathbf{j} & 1 \le t \le 2 \\ (3-t)\mathbf{i} + \mathbf{j} & 2 \le t \le 3 \\ (4-t)\mathbf{j} & 3 \le t \le 4 \end{cases}$

$\int_C y^2\,dx + x^2\,dy = \int_0^1 \left[0\,dt + t^2(0)\right] + \int_1^2 \left[(t-1)^2(0) + 1\,dt\right] + \int_2^3 \left[1(-dt) + (3-t)^2(0)\right] + \int_3^4 \left[(4-t)^2(0) + 0(-dt)\right]$

$= \int_1^2 dt + \int_2^3 -dt = 1 - 1 = 0$

By Green's Theorem,

$\int_R\int \left(\dfrac{\partial N}{\partial x} - \dfrac{\partial M}{\partial y}\right)dA = \int_0^1\int_0^1 (2x - 2y)\,dy\,dx$

$= \int_0^1 \left[2xy - y^2\right]_0^1\,dx = \int_0^1 (2x - 1)\,dx = \left[x^2 - x\right]_0^1 = 0$

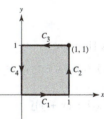

9. $C:\ x^2 + y^2 = 4$

Let $x = 2\cos t$ and $y = 2\sin t$, $0 \le t \le 2\pi$.

$\int_C xe^y\,dx + e^x\,dy = \int_0^{2\pi} \left[2\cos t\,e^{2\sin t}(-2\sin t) + e^{2\cos t}(2\cos t)\right]dt \approx 19.99$

$\int_R\int \left(\dfrac{\partial N}{\partial x} - \dfrac{\partial M}{\partial y}\right)dA = \int_{-2}^2 \int_{-\sqrt{4-x^2}}^{\sqrt{4-x^2}} (e^x - xe^y)\,dy\,dx = \int_{-2}^2 \left[2\sqrt{4-x^2}\,e^x - xe^{\sqrt{4-x^2}} + xe^{-\sqrt{4-x^2}}\right]dx \approx 19.99$

In Exercises 11–13, $\dfrac{\partial N}{\partial x} - \dfrac{\partial M}{\partial y} = 1.$

11. $\int_C (y - x)\,dx + (2x - y)\,dy = \int_0^3\int_{x^2-2x}^x dy\,dx$

$= \int_0^3 \left[x - (x^2 - 2x)\right]dx = \left[-\dfrac{x^3}{3} + \dfrac{3x^2}{2}\right]_0^3 = -9 + \dfrac{27}{2} = \dfrac{9}{2}$

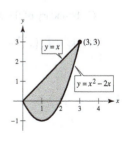

13. From the accompanying figure, we see that R is the shaded region. So, Green's Theorem yields

$$\int_C (y - x)\, dx + (2x - y)\, dy = \int_R \int 1\, dA = \text{Area of } R = 6(10) - 2(2) = 56.$$

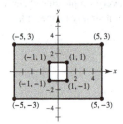

15. $\int_C 2xy\, dx + (x + y)\, dy = \int_R \int \left(\dfrac{\partial N}{\partial x} - \dfrac{\partial M}{\partial y} \right) dA$

$$= \int_{-1}^{1} \int_{0}^{1-x^2} (1 - 2x)\, dy\, dx = \int_{-1}^{1} \left[y - 2xy \right]_0^{1-x^2} dx = \int_{-1}^{1} \left[(1 - x^2) - 2x(1 - x^2) \right] dx$$

$$= \int_{-1}^{1} \left[1 - x^2 - 2x + 2x^3 \right] dx = \left[x - \frac{x^3}{3} - x^2 + \frac{x^4}{2} \right]_{-1}^{1} = \frac{1}{6} + \frac{7}{6} = \frac{4}{3}$$

17. $\int_C (x^2 - y^2)\, dx + 2xy\, dy = \int_R \int \left(\dfrac{\partial N}{\partial x} - \dfrac{\partial M}{\partial y} \right) dA = \int_{-4}^{4} \int_{-\sqrt{16-x^2}}^{\sqrt{16-x^2}} (2y + 2y)\, dy\, dx = \int_{-4}^{4} \left[2y^2 \right]_{-\sqrt{16-x^2}}^{\sqrt{16-x^2}} dx = 0$

19. Because $\dfrac{\partial M}{\partial y} = -2e^x \sin 2y = \dfrac{\partial N}{\partial x}$ you have $\int_R \int \left(\dfrac{\partial N}{\partial x} - \dfrac{\partial M}{\partial y} \right) dA = 0.$

21. By Green's Theorem,

$$\int_C \cos y\, dx + (xy - x \sin y)\, dy = \int_R \int (y - \sin y + \sin y)\, dA = \int_0^1 \int_x^{\sqrt{x}} y\, dy\, dx = \int_0^1 \left[\frac{y^2}{2} \right]_x^{\sqrt{x}} dx$$

$$= \int_0^1 \left(\frac{x}{2} - \frac{x^2}{2} \right) dx = \left[\frac{x^2}{4} - \frac{x^3}{6} \right]_0^1 = \frac{1}{4} - \frac{1}{6} = \frac{1}{12}$$

23. By Green's Theorem,

$$\int_C (x - 3y)\, dx + (x + y)\, dy = \int_R \int (1 + 3)\, dA = 4[\text{Area Large Circle} - \text{Area Small Circle}] = 4[9\pi - \pi] = 32\pi.$$

25. $\mathbf{F}(x, y) = xy\mathbf{i} + (x + y)\mathbf{j}$

$C: x^2 + y^2 = 1$

$$\text{Work} = \int_C xy\, dx + (x + y)\, dy = \int_R \int (1 - x)\, dA = \int_0^{2\pi} \int_0^1 (1 - r \cos \theta)\, r\, dr\, d\theta$$

$$= \int_0^{2\pi} \left[\frac{r^2}{2} - \frac{r^3}{3} \cos \theta \right]_0^1 d\theta = \int_0^{2\pi} \left(\frac{1}{2} - \frac{1}{3} \cos \theta \right) d\theta = \left[\frac{1}{2}\theta - \frac{1}{3} \sin \theta \right]_0^{2\pi} = \pi$$

27. $\mathbf{F}(x, y) = \left(x^{3/2} - 3y \right)\mathbf{i} + \left(6x + 5\sqrt{y} \right)\mathbf{j}$

$C:$ boundary of the triangle with vertices $(0, 0), (5, 0), (0, 5)$

$$\text{Work} = \int_C \left(x^{3/2} - 3y \right) dx + \left(6x + 5\sqrt{y} \right) dy = \int_R \int 9\, dA = 9\left(\tfrac{1}{2} \right)(5)(5) = \frac{225}{2}$$

29. Let $x = 2\cos t,\ y = 2\sin t,\ 0 \le t \le 2\pi$.

$$A = \frac{1}{2}\int_C x\,dy - y\,dx = \frac{1}{2}\int_0^{2\pi}\left[(2\cos t)(2\cos t) - 2\sin t(-2\sin t)\right]dt$$

$$= \frac{1}{2}\int_0^{2\pi} 4\left(\cos^2 t + \sin^2 t\right)dt = 2\int_0^{2\pi} dt = 2(2\pi) = 4\pi$$

Note: The region is a circle of radius 2.

31. $C_1:\ y = x^2 + 1, \quad dy = 2x\,dx$

$\quad C_2:\ y = 5x - 3, \quad dy = 5\,dx$

So, by Theorem 15.9 you have

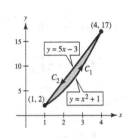

$$A = \frac{1}{2}\int_1^4\left(x(2x) - \left(x^2 + 1\right)\right)dx + \frac{1}{2}\int_4^1\left(x(5) - (5x - 3)\right)dx$$

$$= \frac{1}{2}\left[\frac{x^3}{3} - x\right]_1^4 + \frac{1}{2}[3x]_4^1 = \frac{1}{2}[18] + \frac{1}{2}[-9] = \frac{9}{2}.$$

33. For the moment about the x-axis, $M_x = \int_R\int y\,dA$. Let $N = 0$ and $M = -y^2/2$. By Green's Theorem,

$$M_x = \int_C -\frac{y^2}{2}\,dx = -\frac{1}{2}\int_C y^2\,dx \text{ and } \overline{y} = \frac{M_x}{2A} = -\frac{1}{2A}\int_C y^2\,dx.$$

For the moment about the y-axis, $M_y = \int_R\int x\,dA$. Let $N = x^2/2$ and $M = 0$. By Green's Theorem,

$$M_y = \int_C \frac{x^2}{2}\,dy = \frac{1}{2}\int_C x^2\,dy \text{ and } \overline{x} = \frac{M_y}{2A} = \frac{1}{2A}\int_C x^2\,dy.$$

35. $A = \int_{-2}^2\left(4 - x^2\right)dx = \left[4x - \frac{x^3}{3}\right]_{-2}^2 = \frac{32}{3}$

$$\overline{x} = \frac{1}{2A}\int_{C_1} x^2\,dy + \frac{1}{2A}\int_{C_2} x^2\,dy$$

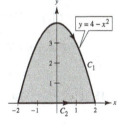

For C_1, $dy = -2x\,dx$ and for C_2, $dy = 0$. So, $\overline{x} = \frac{1}{2(32/3)}\int_2^{-2} x^2(-2x\,dx) = \left[\frac{3}{64}\left(-\frac{x^4}{2}\right)\right]_2^{-2} = 0.$

To calculate $\overline{y}$, note that $y = 0$ along C_2. So,

$$\overline{y} = \frac{-1}{2(32/3)}\int_2^{-2}\left(4 - x^2\right)^2 dx = \frac{3}{64}\int_{-2}^2\left(16 - 8x^2 + x^4\right)dx = \frac{3}{64}\left[16x - \frac{8x^3}{3} + \frac{x^5}{5}\right]_{-2}^2 = \frac{8}{5}.$$

$$(\overline{x}, \overline{y}) = \left(0, \frac{8}{5}\right)$$

37. Because $A = \int_0^1\left(x - x^3\right)dx = \left[\frac{x^2}{2} - \frac{x^4}{4}\right]_0^1 = \frac{1}{4}$, you have $\frac{1}{2A} = 2$. On C_1 you have $y = x^3$, $dy = 3x^2\,dx$ and on C_2 you

have $y = x$, $dy = dx$. So,

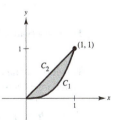

$$\overline{x} = 2\int_C x^2\,dy = 2\int_{C_1} x^2(3x^2\,dx) + 2\int_{C_2} x^2\,dx = 6\int_0^1 x^4\,dx + 2\int_1^0 x^2\,dx = \frac{6}{5} - \frac{2}{3} = \frac{8}{15}$$

$$\overline{y} = -2\int_C y^2\,dx = -2\int_0^1 x^6\,dx - 2\int_1^0 x^2\,dx = -\frac{2}{7} + \frac{2}{3} = \frac{8}{21}.$$

$$(\overline{x}, \overline{y}) = \left(\frac{8}{15}, \frac{8}{21}\right)$$

39. $r = 6(1 - \cos\theta), 0 \le \theta \le 2\pi$

$$A = \frac{1}{2}\int_c r^2\, d\theta = \frac{1}{2}\int_0^{2\pi} 36(1 - \cos\theta)^2\, d\theta = 18\int_0^{2\pi}(1 - 2\cos\theta + \cos^2\theta)\, d\theta$$

$$= 18\int_0^{2\pi}\left(1 - 2\cos\theta + \frac{1}{2} + \frac{\cos 2\theta}{2}\right) d\theta = 18\left[\frac{3\theta}{2} - 2\sin\theta + \frac{1}{4}\sin 2\theta\right]_0^{2\pi}$$

$$= 18[3\pi] = 54\pi$$

41. In this case the inner loop has domain $\dfrac{2\pi}{3} \le \theta \le \dfrac{4\pi}{3}$. So,

$$A = \frac{1}{2}\int_{2\pi/3}^{4\pi/3}(1 + 4\cos\theta + 4\cos^2\theta)\, d\theta = \frac{1}{2}\int_{2\pi/3}^{4\pi/3}(3 + 4\cos\theta + 2\cos 2\theta)\, d\theta$$

$$= \frac{1}{2}[3\theta + 4\sin\theta + \sin 2\theta]_{2\pi/3}^{4\pi/3} = \pi - \frac{3\sqrt{3}}{2}.$$

43. (a) $\displaystyle\int_{C_1} y^3\, dx + (27x - x^3)\, dy = \int_R\int\left[(27 - 3x^2) - 3y^2\right] dA$

$$= \int_0^{2\pi}\int_0^1(27 - 3r^2)\, r\, dr\, d\theta$$

$$= \int_0^{2\pi}\left[\frac{27r^2}{2} - \frac{3r^4}{4}\right]_0^1 d\theta$$

$$= \int_0^{2\pi}\frac{51}{4}\, d\theta$$

$$= \frac{51}{2}\pi$$

(b) You want to find c such that $\displaystyle\int_0^C(27 - 3r^2)r\, dr\, d\theta$ is a maximum:

$$f(c) = \frac{27c^2}{2} - \frac{3}{4}c^4$$

$$f'(c) = 27c - 3c^3 \Rightarrow c = 3$$

Maximum Value: $\displaystyle\int_0^{2\pi}\int_0^3(27 - 3r^2)r\, dr\, d\theta = \frac{243\pi}{2}$

45. $\displaystyle\int_C\left(e^{-x^2/2} - y\right) dx + \left(e^{-y^2/2} + x\right) dy = \int_R\int\left(\frac{\partial N}{\partial x} - \frac{\partial M}{\partial y}\right) dA = \int_R\int(1 - (-1))\, dA = 2(\text{area of } R)$

$$= 2(\pi r^2 - \pi ab) = 2(\pi(s^2) - \pi(2)(1)) = 46\pi$$

47. $I = \displaystyle\int_C\frac{y\, dx - x\, dy}{x^2 + y^2}$

(a) Let $\mathbf{F} = \dfrac{y}{x^2 + y^2}\mathbf{i} - \dfrac{x}{x^2 + y^2}\mathbf{j}.$

$\mathbf{F}$ is conservative because $\dfrac{\partial N}{\partial x} = \dfrac{\partial M}{\partial y} = \dfrac{x^2 - y^2}{\left(x^2 + y^2\right)^2}.$

$\mathbf{F}$ is defined and has continuous first partials everywhere except at the origin. If C is a circle (a closed path) that does not contain the origin, then

$$\int_C \mathbf{F}\cdot d\mathbf{r} = \int_C M\, dx + N\, dy = \int_R\int\left(\frac{\partial N}{\partial x} - \frac{\partial M}{\partial y}\right) dA = 0.$$

(b) Let $\mathbf{r} = a\cos t\mathbf{i} - a\sin t\mathbf{j}$, $0 \le t \le 2\pi$ be a circle C_1 oriented clockwise inside C (see figure). Introduce line segments C_2 and C_3 as illustrated in Example 6 of this section in the text. For the region inside C and outside C_1, Green's Theorem applies. Note that since C_2 and C_3 have opposite orientations, the line integrals over them cancel. So, $C_4 = C_1 + C_2 + C + C_3$ and

$$\int_{C_4} \mathbf{F} \cdot d\mathbf{r} = \int_{C_1} \mathbf{F} \cdot d\mathbf{r} + \int_C \mathbf{F} \cdot d\mathbf{r} = 0.$$

But,

$$\int_{C_1} \mathbf{F} \cdot d\mathbf{r} = \int_0^{2\pi} \left[\frac{(-a\sin t)(-a\sin t)}{a^2\cos^2 t + a^2\sin^2 t} + \frac{(-a\cos t)(-a\cos t)}{a^2\cos^2 t + a^2\sin^2 t} \right] dt$$

$$= \int_0^{2\pi} \left(\sin^2 t + \cos^2 t \right) dt = \left[t \right]_0^{2\pi} = 2\pi.$$

Finally, $\displaystyle\int_C \mathbf{F} \cdot d\mathbf{r} = -\int_{C_1} \mathbf{F} \cdot d\mathbf{r} = -2\pi.$

Note: If C were oriented clockwise, then the answer would have been 2π.

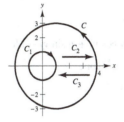

49. (a) Let C be the line segment joining (x_1, y_1) and (x_2, y_2).

$$y = \frac{y_2 - y_1}{x_2 - x_1}(x - x_1) + y_1$$

$$dy = \frac{y_2 - y_1}{x_2 - x_1} dx$$

$$\int_C -y\, dx + x\, dy = \int_{x_1}^{x_2} \left[-\frac{y_2 - y_1}{x_2 - x_1}(x - x_1) - y_1 + x\left(\frac{y_2 - y_1}{x_2 - x_1}\right) \right] dx = \int_{x_1}^{x_2} \left[x_1\left(\frac{y_2 - y_1}{x_2 - x_1}\right) - y_1 \right] dx$$

$$= \left[\left[x_1\left(\frac{y_2 - y_1}{x_2 - x_1}\right) - y_1 \right] x \right]_{x_1}^{x_2} = \left[x_1\left(\frac{y_2 - y_1}{x_2 - x_1}\right) - y_1 \right](x_2 - x_1) = x_1(y_2 - y_1) - y_1(x_2 - x_1) = x_1 y_2 - x_2 y_1$$

(b) Let C be the boundary of the region $A = \dfrac{1}{2}\displaystyle\int_C -y\, dx + x\, dy = \dfrac{1}{2}\iint_R (1 - (-1))\, dA = \iint_R dA.$

So,

$$\iint_R dA = \frac{1}{2}\left[\int_{C_1} -y\, dx + x\, dy + \int_{C_2} -y\, dx + x\, dy + \cdots + \int_{C_n} -y\, dx + x\, dy \right]$$

where C_1 is the line segment joining (x_1, y_1) and (x_2, y_2), C_2 is the line segment joining (x_2, y_2) and (x_3, y_3), ..., and C_n is the line segment joining (x_n, y_n) and (x_1, y_1). So,

$$\iint_R dA = \frac{1}{2}\left[(x_1 y_2 - x_2 y_1) + (x_2 y_3 - x_3 y_2) + \cdots + (x_{n-1} y_n - x_n y_{n-1}) + (x_n y_1 - x_1 y_n) \right].$$

51. Because $\displaystyle\int_C \mathbf{F} \cdot \mathbf{N}\, ds = \iint_R \operatorname{div} \mathbf{F}\, dA$, then

$$\int_C f D_{\mathbf{N}} g\, ds = \int_C f\nabla g \cdot \mathbf{N}\, ds = \iint_R \operatorname{div}(f\nabla g)\, dA = \iint_R \left(f\operatorname{div}(\nabla g) + \nabla f \cdot \nabla g \right) dA = \iint_R \left(f\nabla^2 g + \nabla f \cdot \nabla g \right) dA.$$

53. $\mathbf{F} = M\mathbf{i} + N\mathbf{j}$

$$\frac{\partial N}{\partial x} = \frac{\partial M}{\partial y} \;\Rightarrow\; \frac{\partial N}{\partial x} - \frac{\partial M}{\partial y} = 0$$

$$\int_C \mathbf{F} \cdot d\mathbf{r} = \int_C M\, dx + N\, dy = \iint_R \left(\frac{\partial N}{\partial x} - \frac{\partial M}{\partial y} \right) dA = \iint_R (0)\, dA = 0$$

Section 15.5 Parametric Surfaces

1. S is traced out by the position vector $\mathbf{r}(u, v)$ as the point (u, v) moves throughout the domain. To sketch the surface, it is helpful to relate x, y, and z as functions of u and v.

3. $\mathbf{r}(u, v) = u\mathbf{i} + v\mathbf{j} + uv\mathbf{k}$

   $z = xy$

   Matches (e)

4. $\mathbf{r}(u, v) = u \cos v\mathbf{i} + u \sin v\mathbf{j} + u\mathbf{k}$

   $x^2 + y^2 = z^2$, cone

   Matches (f)

5. $\mathbf{r}(u, v) = u\mathbf{i} + \frac{1}{2}(u + v)\mathbf{j} + v\mathbf{k}$

   $2y = x + z$, plane

   Matches (b)

6. $\mathbf{r}(u, v) = v\mathbf{i} + \cos u\mathbf{j} + \sin u\mathbf{k}$

   $y^2 + z^2 = \cos^2 u + \sin^2 u = 1$, cylinder

   Matches (c)

7. $\mathbf{r}(u, v) = 2 \cos v \cos u\mathbf{i} + 2 \cos v \sin u\mathbf{j} + 2 \sin v\mathbf{k}$

   $x^2 + y^2 + z^2 = 4 \cos^2 v \cos^2 u + 4 \cos^2 v \sin^2 u + 4 \sin^2 v = 4 \cos^2 v + 4 \sin^2 v = 4$, sphere

   Matches (d)

8. $\mathbf{r}(u, v) = u\mathbf{i} + \frac{1}{4}v^3\mathbf{j} + v\mathbf{k}$

   $4y = z^3$

   Matches (a)

9. $\mathbf{r}(u, v) = u\mathbf{i} + v\mathbf{j} + \dfrac{v}{2}\mathbf{k}$

   $y - 2z = 0$

   Plane

11. $\mathbf{r}(u, v) = 2 \cos u\mathbf{i} + v\mathbf{j} + 2 \sin u\mathbf{k}$

    $x^2 + z^2 = 4$

    Cylinder

13. $\mathbf{r}(u, v) = 2u \cos v\mathbf{i} + 2u \sin v\mathbf{j} + u^4\mathbf{k}$,

    $0 \le u \le 1$, $0 \le v \le 2\pi$

    $z = \dfrac{(x^2 + y^2)^2}{16}$

15. $\mathbf{r}(u, v) = (u - \sin u) \cos v\mathbf{i} + (1 - \cos u) \sin v\mathbf{j} + u\mathbf{k}$,

    $0 \le u \le \pi$, $0 \le v \le 2\pi$

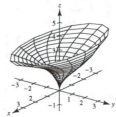

17. $z = 3y$

    $\mathbf{r}(u, v) = u\mathbf{i} + v\mathbf{j} + 3v\mathbf{k}$

19. $y = \sqrt{4x^2 + 9z^2}$

    $\mathbf{r}(x, y) = x\mathbf{i} + \sqrt{4x^2 + 9z^2}\mathbf{j} + z\mathbf{k}$

    or,

    $\mathbf{r}(u, v) = \frac{1}{2}u \cos v\mathbf{i} + u\mathbf{j} + \frac{1}{3}u \sin v\mathbf{k}$,

    $u \ge 0$, $0 \le v \le 2\pi$

21. $x^2 + y^2 = 25$

    $\mathbf{r}(u, v) = 5 \cos u\mathbf{i} + 5 \sin u\mathbf{j} + v\mathbf{k}$

23. $x = y^2 + z^2 + 7$

       $x - 7 = y^2 + z^2$

    $\mathbf{r}(u, v) = u\mathbf{i} + \sqrt{u - 7} \cos v\mathbf{j} + \sqrt{u - 7} \sin v\mathbf{k}$

    Note:

    $y^2 + z^2 = (u - 7)(\cos^2 v + \sin^2 v) = u - 7 = x - 7$

25. $z = 4$ inside $x^2 + y^2 = 9$.

$\mathbf{r}(u, v) = v \cos u\mathbf{i} + v \sin u\mathbf{j} + 4\mathbf{k}, \ 0 \le v \le 3$

27. Function: $y = \dfrac{x}{2}, \ 0 \le x \le 6$

Axis of revolution: x-axis

$x = u, \ y = \dfrac{u}{2} \cos v, \ z = \dfrac{u}{2} \sin v$

$0 \le u \le 6, \ 0 \le v \le 2\pi$

29. Function: $x = \sin z, \ 0 \le z \le \pi$

Axis of revolution: z-axis

$x = \sin u \cos v, \ y = \sin u \sin v, \ z = u$

$0 \le u \le \pi, \ 0 \le v \le 2\pi$

31. Function: $z = \cos^2 y, \dfrac{\pi}{2} \le y \le \pi$

Axis of revolution: y-axis

$x = \cos^2 u \cos v, \ y = u, \ z = \cos^2 u \sin v$

$\dfrac{\pi}{2} \le u \le \pi, 0 \le v \le 2\pi$

33. $\mathbf{r}(u, v) = 3 \cos v \cos u\mathbf{i} + 2 \cos v \sin u\mathbf{j} + 4 \sin v\mathbf{k}, \left(0, \sqrt{3}, 2\right)$

$\mathbf{r}_u(u, v) = -3 \cos v \sin u\mathbf{i} + 2 \cos v \cos u\mathbf{j}$

$\mathbf{r}_v(u, v) = -3 \sin v \cos u\mathbf{i} - 2 \sin v \sin u\mathbf{j} + 4 \cos v\mathbf{k}$

At $\left(0, \sqrt{3}, 2\right), v = \dfrac{\pi}{6}$ and $u = \dfrac{\pi}{2}$.

$\mathbf{r}_u\left(\dfrac{\pi}{2}, \dfrac{\pi}{6}\right) = \dfrac{-3\sqrt{3}}{2}\mathbf{i}$

$\mathbf{r}_v\left(\dfrac{\pi}{2}, \dfrac{\pi}{6}\right) = -\mathbf{j} + 2\sqrt{3}\mathbf{k}$

$\mathbf{N} = \mathbf{r}_u \times \mathbf{r}_v = \begin{vmatrix} \mathbf{i} & \mathbf{j} & \mathbf{k} \\ \dfrac{-3\sqrt{3}}{2} & 0 & 0 \\ 0 & -1 & 2\sqrt{3} \end{vmatrix} = 9\mathbf{j} + \dfrac{3\sqrt{3}}{2}\mathbf{k}$

Tangent plane: $0(x - 0) + 9\left(y - \sqrt{3}\right) + \dfrac{3\sqrt{3}}{2}(z - 2) = 0$

$9y + \dfrac{3\sqrt{3}}{2}z = 12\sqrt{3}$

35. $\mathbf{r}(u, v) = 2u \cos v\mathbf{i} + 3u \sin v\mathbf{j} + u^2\mathbf{k}, \quad (0, 6, 4)$

$\mathbf{r}_u(u, v) = 2 \cos v\mathbf{i} + 3 \sin v\mathbf{j} + 2u\mathbf{k}$

$\mathbf{r}_v(u, v) = -2u \sin v\mathbf{i} + 3u \cos v\mathbf{j}$

At $(0, 6, 4), u = 2$ and $v = \pi/2$.

$\mathbf{r}_u\left(2, \dfrac{\pi}{2}\right) = 3\mathbf{j} + 4\mathbf{k}, \mathbf{r}_v\left(2, \dfrac{\pi}{2}\right) = -4\mathbf{i}$

$\mathbf{N} = \mathbf{r}_u\left(2, \dfrac{\pi}{2}\right) \times \mathbf{r}_v\left(2, \dfrac{\pi}{2}\right) = \begin{vmatrix} \mathbf{i} & \mathbf{j} & \mathbf{k} \\ 0 & 3 & 4 \\ -4 & 0 & 0 \end{vmatrix} = -16\mathbf{j} + 12\mathbf{k}$

Direction numbers: $0, 4, -3$

Tangent plane: $4(y - 6) - 3(z - 4) = 0$

$4y - 3z = 12$

37. $\mathbf{r}(u, v) = 4u\mathbf{i} - v\mathbf{j} + v\mathbf{k}, \quad 0 \le u \le 2, 0 \le v \le 1$

$\mathbf{r}_u(u, v) = 4\mathbf{i}, \mathbf{r}_v(u, v) = -\mathbf{j} + \mathbf{k}$

$\mathbf{r}_u \times \mathbf{r}_v = \begin{vmatrix} \mathbf{i} & \mathbf{j} & \mathbf{k} \\ 4 & 0 & 0 \\ 0 & -1 & 1 \end{vmatrix} = -4\mathbf{j} - 4\mathbf{k}$

$\|\mathbf{r}_u \times \mathbf{r}_v\| = \sqrt{16 + 16} = 4\sqrt{2}$

$A = \int_0^1 \int_0^2 4\sqrt{2} \, du \, dv = 4\sqrt{2}(2)(1) = 8\sqrt{2}$

39. $\mathbf{r}(u, v) = au \cos v\mathbf{i} + au \sin v\mathbf{j} + u\mathbf{k}, \ 0 \le u \le b, \ 0 \le v \le 2\pi$

$\mathbf{r}_u(u, v) = a \cos v\mathbf{i} + a \sin v\mathbf{j} + \mathbf{k}$

$\mathbf{r}_v(u, v) = -au \sin v\mathbf{i} + au \cos v\mathbf{j}$

$\mathbf{r}_u \times \mathbf{r}_v = \begin{vmatrix} \mathbf{i} & \mathbf{j} & \mathbf{k} \\ a \cos v & a \sin v & 1 \\ -au \sin v & au \cos v & 0 \end{vmatrix} = -au \cos v\mathbf{i} - au \sin v\mathbf{j} + a^2 u\mathbf{k}$

$\|\mathbf{r}_u \times \mathbf{r}_v\| = au\sqrt{1 + a^2}$

$A = \int_0^{2\pi} \int_0^b a\sqrt{1 + a^2}\, u\, du\, dv = \pi ab^2 \sqrt{1 + a^2}$

41. $\mathbf{r}(u, v) = \sqrt{u} \cos v\mathbf{i} + \sqrt{u} \sin v\mathbf{j} + u\mathbf{k}, 0 \le u \le 4, 0 \le v \le 2\pi$

$\mathbf{r}_u(u, v) = \dfrac{\cos v}{2\sqrt{u}}\mathbf{i} + \dfrac{\sin v}{2\sqrt{u}}\mathbf{j} + \mathbf{k}$

$\mathbf{r}_v(u, v) = -\sqrt{u} \sin v\mathbf{i} + \sqrt{u} \cos v\mathbf{j}$

$\mathbf{r}_u \times \mathbf{r}_v = \begin{vmatrix} \mathbf{i} & \mathbf{j} & \mathbf{k} \\ \dfrac{\cos v}{2\sqrt{u}} & \dfrac{\sin v}{2\sqrt{u}} & 1 \\ -\sqrt{u} \sin v & \sqrt{u} \cos v & 0 \end{vmatrix} = -\sqrt{u} \cos v\mathbf{i} - \sqrt{u} \sin v\mathbf{j} + \dfrac{1}{2}\mathbf{k}$

$\|\mathbf{r}_u \times \mathbf{r}_v\| = \sqrt{u + \dfrac{1}{4}}$

$A = \int_0^{2\pi} \int_0^4 \sqrt{u + \dfrac{1}{4}}\, du\, dv = \dfrac{\pi}{6}\left(17\sqrt{17} - 1\right) \approx 36.177$

For Exercises 43–45, $\mathbf{r}(u, v) = u \cos v i + u \sin v j + u^2 k, 0 \le u \le 2, 0 \le v \le 2\pi.$
Eliminating the parameter yields $z = x^2 + y^2, 0 \le z \le 4.$

43. $\mathbf{s}(u, v) = u \cos v\mathbf{i} + u \sin v\mathbf{j} - u^2\mathbf{k}, \ 0 \le u \le 2, \ 0 \le v \le 2\pi$

$z = -\left(x^2 + y^2\right)$

The paraboloid is reflected (inverted) through the *xy*-plane.

45. $\mathbf{s}(u, v) = u \cos v\mathbf{i} + u \sin v\mathbf{j} + u^2\mathbf{k}, \ 0 \le u \le 3, \ 0 \le v \le 2\pi$

The height of the paraboloid is increased from 4 to 9.

47. Function: $z = x$

Axis of revolution: *z*-axis

$x = u \cos v, \ y = u \sin v, \ z = u$

$\mathbf{r}(u, v) = u \cos v\mathbf{i} + u \sin v\mathbf{j} + u\mathbf{k}$

$u \le 0, \quad 0 \le v \le 2\pi$

49. $\mathbf{r}(u, v) = a \sin^3 u \cos^3 v \mathbf{i} + a \sin^3 u \sin^3 v \mathbf{j} + a \cos^3 u \mathbf{k}$

$0 \leq u \leq \pi, \quad 0 \leq v \leq 2\pi$

$x = a \sin^3 u \cos^3 v \Rightarrow x^{2/3} = a^{2/3} \sin^2 u \cos^2 v$

$y = a \sin^3 u \sin^3 v \Rightarrow y^{2/3} = a^{2/3} \sin^2 u \sin^2 v$

$z = a \cos^3 u \Rightarrow z^{2/3} = a^{2/3} \cos^2 u$

$x^{2/3} + y^{2/3} + z^{2/3} = a^{2/3}\left[\sin^2 u \cos^2 v + \sin^2 u \sin^2 v + \cos^2 u\right] = a^{2/3}\left[\sin^2 u + \cos^2 u\right] = a^{2/3}$

51. (a) $\mathbf{r}(u, v) = (4 + \cos v) \cos u \mathbf{i} +$
$(4 + \cos v) \sin u \mathbf{j} + \sin v \mathbf{k},$
$0 \leq u \leq 2\pi, 0 \leq v \leq 2\pi$

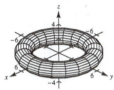

(b) $\mathbf{r}(u, v) = (4 + 2\cos v) \cos u \mathbf{i} +$
$(4 + 2\cos v) \sin u \mathbf{j} + 2\sin v \mathbf{k},$
$0 \leq u \leq 2\pi, 0 \leq v \leq 2\pi$

(c) $\mathbf{r}(u, v) = (8 + \cos v) \cos u \mathbf{i} +$
$(8 + \cos v) \sin u \mathbf{j} + \sin v \mathbf{k},$
$0 \leq u \leq 2\pi, 0 \leq v \leq 2\pi$

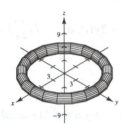

(d) $\mathbf{r}(u, v) = (8 + 3\cos v) \cos u \mathbf{i} +$
$(8 + 3\cos v) \sin u \mathbf{j} + 3\sin v \mathbf{k},$
$0 \leq u \leq 2\pi, 0 \leq v \leq 2\pi$

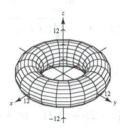

The radius of the generating circle that is revolved about the z-axis is b, and its center is a units from the axis of revolution.

53. $\mathbf{r}(u, v) = 20 \sin u \cos v \mathbf{i} + 20 \sin u \sin v \mathbf{j} + 20 \cos u \mathbf{k}, 0 \leq u \leq \pi/3, \quad 0 \leq v \leq 2\pi$

$\mathbf{r}_u = 20 \cos u \cos v \mathbf{i} + 20 \cos u \sin v \mathbf{j} - 20 \sin u \mathbf{k}$

$\mathbf{r}_v = -20 \sin u \sin v \mathbf{i} + 20 \sin u \cos v \mathbf{j}$

$$\mathbf{r}_u \times \mathbf{r}_v = \begin{vmatrix} \mathbf{i} & \mathbf{j} & \mathbf{k} \\ 20 \cos u \cos v & 20 \cos u \sin v & -20 \sin u \\ -20 \sin u \sin v & 20 \sin u \cos v & 0 \end{vmatrix}$$

$= 400 \sin^2 u \cos v \mathbf{i} + 400 \sin^2 u \sin v \mathbf{j} + 400\left(\cos u \sin u \cos^2 v + \cos u \sin u \sin^2 v\right)\mathbf{k}$

$= 400\left[\sin^2 u \cos v \mathbf{i} + \sin^2 u \sin v \mathbf{j} + \cos u \sin u \mathbf{k}\right]$

$\|\mathbf{r}_u \times \mathbf{r}_v\| = 400\sqrt{\sin^4 u \cos^2 v + \sin^4 u \sin^2 v + \cos^2 u \sin^2 u} = 400\sqrt{\sin^4 u + \cos^2 u \sin^2 u} = 400\sqrt{\sin^2 u} = 400 \sin u$

$S = \int_S \int dS = \int_0^{2\pi} \int_0^{\pi/3} 400 \sin u \, du \, dv = \int_0^{2\pi} \left[-400 \cos u\right]_0^{\pi/3} dv = \int_0^{2\pi} 200 \, dv = 400\pi \text{ m}^2$

55. $\mathbf{r}(u, v) = u \cos v\mathbf{i} + u \sin v\mathbf{j} + 2v\mathbf{k}, \quad 0 \le u \le 3, 0 \le v \le 2\pi$

$\mathbf{r}_u(u, v) = \cos v\mathbf{i} + \sin v\mathbf{j}$

$\mathbf{r}_v(u, v) = -u \sin v\mathbf{i} + u \cos v\mathbf{j} + 2\mathbf{k}$

$$\mathbf{r}_u \times \mathbf{r}_v = \begin{vmatrix} \mathbf{i} & \mathbf{j} & \mathbf{k} \\ \cos v & \sin v & 0 \\ -u \sin v & u \cos v & 2 \end{vmatrix} = 2 \sin v\mathbf{i} - 2 \cos v\mathbf{j} + u\mathbf{k}$$

$\|\mathbf{r}_u \times \mathbf{r}_v\| = \sqrt{4 + u^2}$

$$A = \int_0^{2\pi} \int_0^3 \sqrt{4 + u^2} \, du \, dv = \pi\left[3\sqrt{13} + 4 \ln\left(\frac{3 + \sqrt{13}}{2}\right)\right]$$

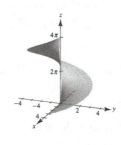

57. Answers will vary.

Section 15.6   Surface Integrals

1. First solve for y in the equation of the surface. Then use the integral

$$\iint_S f(x, g(x, z), z)\sqrt{1 + \left[g_x(x, z)\right]^2 + \left[g_z(x, z)\right]^2} \, dA.$$

3. An orientable surface has two distinct sides.

5. $S: z = 4 - x, \quad 0 \le x \le 4, \quad 0 \le y \le 3, \quad \dfrac{\partial z}{\partial x} = -1, \quad \dfrac{\partial z}{\partial y} = 0$

$$\iint_S (x - 2y + z) \, dS = \int_0^4 \int_0^3 (x - 2y + 4 - x)\sqrt{1 + (-1)^2 + 0^2} \, dy \, dx = \sqrt{2}\int_0^4 \int_0^3 (4 - 2y) \, dy \, dx = \sqrt{2}\int_0^4 3 \, dx = 12\sqrt{2}$$

7. $S: z = 2, \quad x^2 + y^2 \le 1, \quad \dfrac{\partial z}{\partial x} = \dfrac{\partial z}{\partial y} = 0$

$$\iint_S (x - 2y + z) \, dS = \int_{-1}^1 \int_{-\sqrt{1-x^2}}^{\sqrt{1-x^2}} (x - 2y + 2)\sqrt{1 + 0^2 + 0^2} \, dy \, dx = \int_0^{2\pi} \int_0^1 (r \cos \theta - 2r \sin \theta + 2) \, r \, dr \, d\theta$$

$$= \int_0^{2\pi}\left[\frac{1}{3} \cos \theta - \frac{2}{3} \sin \theta + 1\right] d\theta = \left[\frac{1}{3} \sin \theta + \frac{2}{3} \cos \theta + \theta\right]_0^{2\pi} = \frac{2}{3} + 2\pi - \frac{2}{3} = 2\pi$$

9. $S: z = 3 - x - y \quad$ (first octant), $\quad \dfrac{\partial z}{\partial x} = -1, \quad \dfrac{\partial z}{\partial y} = -1$

$$\iint_S xy \, dS = \int_0^3 \int_0^{3-x} xy\sqrt{1 + (-1)^2 + (-1)^2} \, dy \, dx = \sqrt{3}\int_0^3 \left[x\frac{y^2}{2}\right]_0^{3-x} dx$$

$$= \frac{\sqrt{3}}{2}\int_0^3 x(3 - x)^2 \, dx = \frac{\sqrt{3}}{2}\left[\frac{x^4}{4} - 2x^3 + \frac{9x^2}{2}\right]_0^3$$

$$= \frac{\sqrt{3}}{2}\left[\frac{27}{4}\right] = \frac{27\sqrt{3}}{8}$$

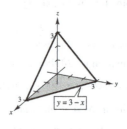

$y = 3 - x$

11. $S: z = 10 - x^2 - y^2, \, 0 \le x \le 2, \, 0 \le y \le 2$

$$\iint_S (x^2 - 2xy) \, dS = \int_0^2 \int_0^2 (x^2 - 2xy)\sqrt{1 + 4x^2 + 4y^2} \, dy \, dx \approx -11.47$$

13. $S: 2x + 3y + 6z = 12$ (first octant) $\Rightarrow z = 2 - \frac{1}{3}x - \frac{1}{2}y$

$\rho(x, y, z) = x^2 + y^2$

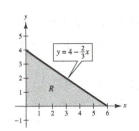

$m = \int_R \int (x^2 + y^2)\sqrt{1 + \left(-\frac{1}{3}\right)^2 + \left(-\frac{1}{2}\right)^2}\, dA = \frac{7}{6}\int_0^6 \int_0^{4-(2x/3)} (x^2 + y^2)\, dy\, dx$

$= \frac{7}{6}\int_0^6 \left[x^2\left(4 - \frac{2}{3}x\right) + \frac{1}{3}\left(4 - \frac{2}{3}x\right)^3\right] dx = \frac{7}{6}\left[\frac{4}{3}x^3 - \frac{1}{6}x^4 - \frac{1}{8}\left(4 - \frac{2}{3}x\right)^4\right]_0^6 = \frac{364}{3}$

15. $S: \mathbf{r}(u, v) = u\mathbf{i} + v\mathbf{j} + 2v\mathbf{k}$, $0 \le u \le 1$, $0 \le v \le 2$

$\mathbf{r}_u = \mathbf{i}$, $\mathbf{r}_v = \mathbf{j} + 2\mathbf{k}$

$\mathbf{r}_u \times \mathbf{r}_v = \begin{vmatrix} \mathbf{i} & \mathbf{j} & \mathbf{k} \\ 1 & 0 & 0 \\ 0 & 1 & 2 \end{vmatrix} = -2\mathbf{j} + \mathbf{k}$

$\|\mathbf{r}_u \times \mathbf{r}_v\| = \sqrt{5}$

$\int_S \int (y + 5)\, dS = \int_0^2 \int_0^1 (v + 5)\sqrt{5}\, du\, dv = \int_0^2 (v + 5)\sqrt{5}\, dv = \sqrt{5}\left[\frac{v^2}{2} + 5v\right]_0^2 = 12\sqrt{5}$

17. $S: \mathbf{r}(u, v) = \cos u\mathbf{i} + \sin u\mathbf{j} + v\mathbf{k}, 0 \le u \le \frac{\pi}{3}, 0 \le v \le 1$

$\mathbf{r}_u = -\sin u\mathbf{i} + \cos u\mathbf{j}, \mathbf{r}_v = \mathbf{k}$

$\mathbf{r}_u \times \mathbf{r}_v = \begin{vmatrix} \mathbf{i} & \mathbf{j} & \mathbf{k} \\ -\sin u & \cos u & 0 \\ 0 & 0 & 1 \end{vmatrix} = \cos u\mathbf{i} + \sin u\mathbf{j}$

$\|\mathbf{r}_u \times \mathbf{r}_v\| = \sqrt{\cos^2 u + \sin^2 u} = 1$

$\int_S \int (3y - x)\, dS = \int_0^1 \int_0^{\pi/3} (3\sin u - \cos u)\, du\, dv$

$= \int_0^1 [-3\cos u - \sin u]_0^{\pi/3}\, dv$

$= \int_0^1 \left(-\frac{3}{2} - \frac{\sqrt{3}}{2} + 3\right) dv = \int_0^1 \left(\frac{3}{2} - \frac{\sqrt{3}}{2}\right) dv$

$= \frac{3}{2} - \frac{\sqrt{3}}{2} = \frac{3 - \sqrt{3}}{2}$

19. $f(x, y, z) = x^2 + y^2 + z^2$

$S: z = x + y$, $x^2 + y^2 \le 1$, $\dfrac{\partial z}{\partial x} = \dfrac{\partial z}{\partial y} = 1$

$\int_S \int f(x, y, z)\, dS = \int_{-1}^1 \int_{-\sqrt{1-x^2}}^{\sqrt{1-x^2}} \left[x^2 + y^2 + (x + y)^2\right]\sqrt{1 + 1^2 + 1^2}\, dy\, dx$

$= \sqrt{3}\int_{-1}^1 \int_{-\sqrt{1-x^2}}^{\sqrt{1-x^2}} [2x^2 + 2y^2 + 2xy]\, dy\, dx = \sqrt{3}\int_0^{2\pi} \int_0^1 (2r^2 + 2r\cos\theta\, r\sin\theta)\, r\, dr\, d\theta$

$= 2\sqrt{3}\int_0^{2\pi} \left[\frac{r^4}{4} + \frac{r^4}{4}\cos\theta\sin\theta\right]_0^1 d\theta = \frac{\sqrt{3}}{2}\int_0^{2\pi} (1 + \cos\theta\sin\theta)\, d\theta = \frac{\sqrt{3}}{2}\left[\theta + \frac{\sin^2\theta}{2}\right]_0^{2\pi} = \sqrt{3}\pi$

21. $f(x, y, z) = \sqrt{x^2 + y^2 + z^2}$

$S: z = \sqrt{x^2 + y^2}, x^2 + y^2 \leq 4$

$$\int_S \int f(x, y, z)\, dS = \int_{-2}^{2} \int_{-\sqrt{4-x^2}}^{\sqrt{4-x^2}} \sqrt{x^2 + y^2 + \left(\sqrt{x^2+y^2}\right)^2} \sqrt{1 + \left(\frac{x}{\sqrt{x^2+y^2}}\right)^2 + \left(\frac{y}{\sqrt{x^2+y^2}}\right)^2}\, dy\, dx$$

$$= \sqrt{2}\int_{-2}^{2}\int_{-\sqrt{4-x^2}}^{\sqrt{4-x^2}} \sqrt{x^2+y^2} \sqrt{\frac{x^2+y^2+x^2+y^2}{x^2+y^2}}\, dy\, dx$$

$$= 2\int_{-2}^{2}\int_{-\sqrt{4-x^2}}^{\sqrt{4-x^2}} \sqrt{x^2+y^2}\, dy\, dx$$

$$= 2\int_0^{2\pi}\int_0^2 r^2\, dr\, d\theta$$

$$= 2\int_0^{2\pi}\left[\frac{r^3}{3}\right]_0^2 d\theta$$

$$= \left[\frac{16}{3}\theta\right]_0^{2\pi}$$

$$= \frac{32\pi}{3}$$

23. $f(x, y, z) = x^2 + y^2 + z^2$

$S: x^2 + y^2 = 9,\ 0 \leq x \leq 3,\ 0 \leq y \leq 3,\ 0 \leq z \leq 9$

Project the solid onto the *yz*-plane; $x = \sqrt{9 - y^2},\ 0 \leq y \leq 3,\ 0 \leq z \leq 9$.

$$\int_S \int f(x, y, z)\, dS = \int_0^3 \int_0^9 \left[(9 - y^2) + y^2 + z^2\right]\sqrt{1 + \left(\frac{-y}{\sqrt{9-y^2}}\right)^2 + (0)^2}\, dz\, dy$$

$$= \int_0^3 \int_0^9 (9 + z^2)\frac{3}{\sqrt{9-y^2}}\, dz\, dy = \int_0^3 \left[\frac{3}{\sqrt{9-y^2}}\left(9z + \frac{z^3}{3}\right)\right]_0^9 dy$$

$$= 324\int_0^3 \frac{3}{\sqrt{9-y^2}}\, dy = \left[972\arcsin\left(\frac{y}{3}\right)\right]_0^3 = 972\left(\frac{\pi}{2} - 0\right) = 486\pi$$

25. $\mathbf{F}(x, y, z) = 3z\mathbf{i} - 4\mathbf{j} + y\mathbf{k}$

$S: z = 1 - x - y$ (first octant)

$G(x, y, z) = x + y + z - 1$

$\nabla G(x, y, z) = \mathbf{i} + \mathbf{j} + \mathbf{k}$

$$\int_S \int \mathbf{F} \cdot \mathbf{N}\, dS = \int_R \int \mathbf{F} \cdot \nabla G\, dA = \int_0^1 \int_0^{1-x} (3z - 4 + y)\, dy\, dx$$

$$= \int_0^1 \int_0^{1-x} \left[3(1 - x - y) - 4 + y\right] dy\, dx$$

$$= \int_0^1 \int_0^{1-x} (-1 - 3x - 2y)\, dy\, dx$$

$$= \int_0^1 \left[-y - 3xy - y^2\right]_0^{1-x} dx$$

$$= -\int_0^1 \left[(1 - x) + 3x(1 - x) + (1 - x)^2\right] dx$$

$$= -\int_0^1 (2 - 2x^2)\, dx = -\frac{4}{3}$$

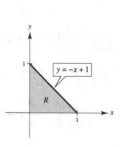

27. $\mathbf{F}(x, y, z) = x\mathbf{i} + y\mathbf{j} + z\mathbf{k}$

$S: z = 1 - x^2 - y^2, \quad z \geq 0$

$G(x, y, z) = x^2 + y^2 + z - 1$

$\nabla G(x, y, z) = 2x\mathbf{i} + 2y\mathbf{j} + \mathbf{k}$

$$\int_S \int \mathbf{F} \cdot \mathbf{N}\, dS = \int_R \int \mathbf{F} \cdot \nabla G\, dA$$

$$= \int_R \int \left(2x^2 + 2y^2 + z\right) dA$$

$$= \int_R \int \left(2x^2 + 2y^2 + \left(1 - x^2 - y^2\right)\right) dA$$

$$= \int_R \int \left(1 + x^2 + y^2\right) dA$$

$$= \int_0^{2\pi} \int_0^1 \left(r^2 + 1\right) r\, dr\, d\theta$$

$$= \int_0^{2\pi} \left[\frac{r^4}{4} + \frac{r^2}{2}\right]_0^1 d\theta = \int_0^{2\pi} \frac{3}{4}\, d\theta = \frac{3\pi}{2}$$

29. $\mathbf{F}(x, y, z) = 4\mathbf{i} - 3\mathbf{j} + 5\mathbf{k}$

$S: z = x^2 + y^2, x^2 + y^2 \leq 4$

$G(x, y, z) = -x^2 - y^2 + z$

$\nabla G(x, y, z) = -2x\mathbf{i} - 2y\mathbf{j} + \mathbf{k}$

$$\int_S \int \mathbf{F} \cdot \mathbf{N}\, dS = \int_R \int \mathbf{F} \cdot \nabla G\, dA = \int_R \int \left(-8x + 6y + 5\right) dA$$

$$= \int_0^{2\pi} \int_0^2 \left[-8r\cos\theta + 6r\sin\theta + 5\right] r\, dr\, d\theta$$

$$= \int_0^{2\pi} \left[-\tfrac{8}{3}r^3 \cos\theta + 2r^3 \sin\theta + \tfrac{5}{2}r^2\right]_0^2 d\theta$$

$$= \int_0^{2\pi} \left[-\tfrac{64}{3}\cos\theta + 16\sin\theta + 10\right] d\theta$$

$$= \left[-\tfrac{64}{3}\sin\theta - 16\cos\theta + 10\theta\right]_0^{2\pi} = 20\pi$$

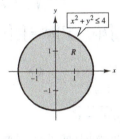

31. $\mathbf{F}(x, y, z) = (x + y)\mathbf{i} + y\mathbf{j} + z\mathbf{k}$

$S: z = 16 - x^2 - y^2, \quad z = 0$

$G(x, y, z) = z + x^2 + y^2 - 16$

$\nabla G(x, y, z) = 2x\mathbf{i} + 2y\mathbf{j} + \mathbf{k}$

$\mathbf{F} \cdot \nabla G = 2x(x + y) + 2y^2 + z = 2x^2 + 2xy + 2y^2 + 16 - x^2 - y^2 = x^2 + y^2 + 2xy + 16$

$$\int_S \int \mathbf{F} \cdot \mathbf{N}\, dS = \int_R \int \mathbf{F} \cdot \nabla G\, dA$$

$$= \int_0^{2\pi} \int_0^4 \left(r^2 + 2r^2 \cos\theta \sin\theta + 16\right) r\, dr\, d\theta$$

$$= \int_0^{2\pi} \left[\frac{r^4}{4} + \frac{r^4}{2}\cos\theta\sin\theta + 8r^2\right]_0^4 d\theta = \int_0^{2\pi} \left[192 + 128\cos\theta\sin\theta\right] d\theta = \left[192 + 64\sin^2\theta\right]_0^{2\pi} = 384\pi$$

(The flux across the bottom $z = 0$ is 0.)

33. $S: z = 16 - x^2 - y^2, z \geq 0$

$\mathbf{F}(x, y, z) = 0.5z\mathbf{k}$

$$\int_S \int \rho\mathbf{F} \cdot \mathbf{N} \, dS = \int_R \int \rho\mathbf{F} \cdot \left(-g_x(x, y)\mathbf{i} - g_y(x, y)\mathbf{j} + \mathbf{k}\right) dA = \int_R \int 0.5\rho z\mathbf{k} \cdot (2x\mathbf{i} + 2y\mathbf{j} + \mathbf{k}) \, dA$$

$$= \int_R \int 0.5\rho z \, dA = \int_R \int 0.5\rho\left(16 - x^2 - y^2\right) dA$$

$$= 0.5\rho \int_0^{2\pi} \int_0^4 \left(16 - r^2\right)r \, dr \, d\theta = 0.5\rho \int_0^{2\pi} 64 \, d\theta = 64\pi\rho$$

35. $\mathbf{E} = yz\mathbf{i} + xz\mathbf{j} + xy\mathbf{k}$

$S: z = \sqrt{1 - x^2 - y^2}$

$$\int_S \int \mathbf{E} \cdot \mathbf{N} \, dS = \int_R \int \mathbf{E} \cdot \left(-g_x(x, y)\mathbf{i} - g_y(x, y)\mathbf{j} + \mathbf{k}\right) dA$$

$$= \int_R \int (yz\mathbf{i} + xz\mathbf{j} + xy\mathbf{k}) \cdot \left(\frac{x}{\sqrt{1 - x^2 - y^2}}\mathbf{i} + \frac{y}{\sqrt{1 - x^2 - y^2}}\mathbf{j} + \mathbf{k}\right) dA$$

$$= \int_R \int \left(\frac{2xyz}{\sqrt{1 - x^2 - y^2}} + xy\right) dA = \int_R \int 3xy \, dA = \int_{-1}^1 \int_{-\sqrt{1-x^2}}^{\sqrt{1-x^2}} 3xy \, dy \, dx = 0$$

37. $z = \sqrt{x^2 + y^2}, 0 \leq z \leq a$

$$m = \int_S \int k \, dS = k\int_R \int \sqrt{1 + \left(\frac{x}{\sqrt{x^2 + y^2}}\right)^2 + \left(\frac{y}{\sqrt{x^2 + y^2}}\right)^2} \, dA = k\int_R \int \sqrt{2} \, dA = \sqrt{2}\, k\pi a^2$$

$$I_z = \int_S \int k\left(x^2 + y^2\right) dS = \int_R \int k\left(x^2 + y^2\right)\sqrt{2} \, dA = \sqrt{2}\, k\int_0^{2\pi} \int_0^a r^3 \, dr \, d\theta = \frac{\sqrt{2}\, ka^4}{4}(2\pi) = \frac{\sqrt{2}\, k\pi a^4}{2} = \frac{a^2}{2}\left(\sqrt{2}\, k\pi a^2\right) = \frac{a^2 m}{2}$$

39. $x^2 + y^2 = a^2, 0 \leq z \leq h$

$\rho(x, y, z) = 1$

$y = \pm\sqrt{a^2 - x^2}$

Project the solid onto the *xz*-plane.

$$I_z = 4\int_S \int \left(x^2 + y^2\right)(1) \, dS = 4\int_0^h \int_0^a \left[x^2 + \left(a^2 - x^2\right)\right]\sqrt{1 + \left(\frac{-x}{\sqrt{a^2 - x^2}}\right)^2 + (0)^2} \, dx \, dz$$

$$= 4a^3 \int_0^h \int_0^a \frac{1}{\sqrt{a^2 - x^2}} \, dx \, dz = 4a^3 \int_0^h \left[\arcsin\frac{x}{a}\right]_0^a \, dz = 4a^3\left(\frac{\pi}{2}\right)(h) = 2\pi a^3 h$$

41. (a) $S: z = 4 - 2x - 2y, 0 \leq x \leq 2, 0 \leq y \leq 2 - x$

$$\frac{\partial z}{\partial x} = -2, \frac{\partial z}{\partial y} = -2, \, dS = \sqrt{1 + 4 + 4} \, dA = 3 \, dA$$

$$\int_S \int (x + 2y) \, dS = \int_0^2 \int_0^{2-x} (x + 2y)3 \, dy \, dx = 12$$

(b) $S: y = 2 - x - \frac{z}{2}, 0 \leq x \leq 2, 0 \leq z \leq 4 - 2x$

$$\frac{\partial y}{\partial x} = -1, \frac{\partial y}{\partial z} = -\frac{1}{2}, \, dS = \sqrt{1 + 1 + \frac{1}{4}} = \frac{3}{2}$$

$$\int_S \int (x + 2y) \, dS = \int_S \int x + (4 - 2x - z) \, dS = \int_0^2 \int_0^{4-2x} (4 - z - x)\left(\frac{3}{2}\right) dz \, dx = 12$$

(c) $S: x = 2 - y - \dfrac{z}{2}, 0 \le y \le 2, 0 \le z \le 4 - 2y$

$$\frac{\partial x}{\partial y} = -1, \frac{\partial x}{\partial z} = -\frac{1}{2}, dS = \sqrt{1 + 1 + \frac{1}{4}} = \frac{3}{2}$$

$$\int_S\!\int (x + 2y)\, dS = \int_S\!\int \left[\left(2 - y - \frac{z}{2}\right) + 2y\right] dS$$

$$= \int_0^2 \int_0^{4-2y} \left(2 + y - \frac{z}{2}\right)\frac{3}{2}\, dz\, dy = 12$$

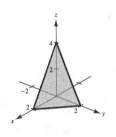

43. (a)

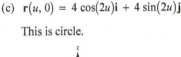

(b) If a normal vector at a point P on the surface is moved around the Möbius strip once, it will point in the opposite direction.

(c) $\mathbf{r}(u, 0) = 4\cos(2u)\mathbf{i} + 4\sin(2u)\mathbf{j}$

This is circle.

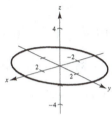

(d) Construction

(e) You obtain a strip with a double twist and twice as long as the original Möbius strip.

Section 15.7   Divergence Theorem

1. You could use the Divergence Theorem or two surface integrals. In this case, the Divergence Theorem is easier.

3. Surface Integral: There are six surfaces to the cube, each with $dS = \sqrt{1}\, dA$.

$z = 0$: $\mathbf{N} = -\mathbf{k}, \mathbf{F} \cdot \mathbf{N} = -z^2, \displaystyle\int_{S_1}\!\int 0\, dA = 0$

$z = 1$: $\mathbf{N} = \mathbf{k}, \mathbf{F} \cdot \mathbf{N} = z^2, \displaystyle\int_{S_2}\!\int 1\, dA = \int_0^1 \int_0^1 dx\, dy = 1$

$x = 0$: $\mathbf{N} = -\mathbf{i}, \mathbf{F} \cdot \mathbf{N} = -2x, \displaystyle\int_{S_3}\!\int 0\, dA = 0$

$x = 1$: $\mathbf{N} = \mathbf{i}, \mathbf{F} \cdot \mathbf{N} = 2x, \displaystyle\int_{S_4}\!\int 2\, dA = 2$

$y = 0$: $\mathbf{N} = -\mathbf{j}, \mathbf{F} \cdot \mathbf{N} = 2y, \displaystyle\int_{S_5}\!\int 0\, dA = 0$

$y = 1$: $\mathbf{N} = \mathbf{j}, \mathbf{F} \cdot \mathbf{N} = -2y, \displaystyle\int_{S_6}\!\int -2\, dA = -2$

So, $\displaystyle\int_S\!\int \mathbf{F} \cdot \mathbf{N}\, dS = 1 + 2 - 2 = 1.$

Divergence Theorem: $\text{div } \mathbf{F} = 2 - 2 + 2z = 2z$

$$\iiint_Q \text{div } \mathbf{F}\, dV = \int_0^1 \int_0^1 \int_0^1 2z\, dz\, dy\, dx = \int_0^1 \int_0^1 dy\, dx = 1$$

5. Surface Integral: There are four surfaces to this solid.

$z = 0, \quad \mathbf{N} = -\mathbf{k}, \quad \mathbf{F} \cdot \mathbf{N} = -z$

$\displaystyle\int_{S_1}\!\!\int 0 \, dS = 0$

$y = 0, \quad \mathbf{N} = -\mathbf{j}, \quad \mathbf{F} \cdot \mathbf{N} = 2y - z, \quad dS = dA = dx \, dz$

$\displaystyle\int_{S_2}\!\!\int -z \, dS = \int_0^6\!\int_0^{6-z} -z \, dx \, dz = \int_0^6 (z^2 - 6z) \, dz = -36$

$x = 0, \quad \mathbf{N} = -\mathbf{i}, \quad \mathbf{F} \cdot \mathbf{N} = y - 2x, \quad dS = dA = dz \, dy$

$\displaystyle\int_{S_3}\!\!\int y \, dS = \int_0^3\!\int_0^{6-2y} y \, dz \, dy = \int_0^3 (6y - 2y^2) \, dy = 9$

$x + 2y + z = 6, \mathbf{N} = \dfrac{\mathbf{i} + 2\mathbf{j} + \mathbf{k}}{\sqrt{6}}, \mathbf{F} \cdot \mathbf{N} = \dfrac{2x - 5y + 3z}{\sqrt{6}}, dS = \sqrt{6} \, dA$

$\displaystyle\int_{S_4}\!\!\int (2x - 5y + 3z) \, dz \, dy = \int_0^3\!\int_0^{6-2y} (18 - x - 11y) \, dx \, dy = \int_0^3 (90 - 90y + 20y^2) \, dy = 45$

So, $\displaystyle\int_S\!\!\int \mathbf{F} \cdot \mathbf{N} \, dS = 0 - 36 + 9 + 45 = 18.$

Divergence Theorem: Because div $\mathbf{F} = 1$, you have

$\displaystyle\iiint_Q dV = (\text{Volume of solid}) = \frac{1}{3}(\text{Area of base}) \times (\text{Height}) = \frac{1}{3}(9)(6) = 18.$

7. $F(x, y, z) = xz\mathbf{i} + yz\mathbf{j} + 2z^2\mathbf{k}$

Surface Integral: There are two surfaces.

Bottom: $z = 0, \quad \mathbf{N} = -\mathbf{k}, \quad \mathbf{F} \cdot \mathbf{N} = -2z^2$

$\displaystyle\int_{S_1}\!\!\int \mathbf{F} \cdot \mathbf{N} \, dS = \int_R\!\!\int -2z^2 \, dA = \int_0^1\!\int_0^1 0 \, dx \, dy = 0$

Side: Outward unit normal is

$\mathbf{N} = \dfrac{2x\mathbf{i} + 2y\mathbf{j} + \mathbf{k}}{\sqrt{4x^2 + 4y^2 + 1}}$

$\mathbf{F} \cdot \mathbf{N} = \dfrac{1}{\sqrt{4x^2 + 4y^2 + 1}} \left[2x^2z + 2y^2z + 2z^2 \right]$

$\displaystyle\int_{S_2}\!\!\int \mathbf{F} \cdot \mathbf{N} \, dS = \int_{S_2}\!\!\int \left[2(x^2 + y^2)z + 2z^2 \right] dA$

$= \int_0^{2\pi}\!\int_0^1 \left[2r^2(1 - r^2) + 2(1 - r^2)^2 \right] r \, dr \, d\theta = \int_0^{2\pi}\!\int_0^1 (2r - 2r^3) \, dr \, d\theta = \int_0^{2\pi} \frac{1}{2} \, d\theta = \pi$

Divergence Theorem: div $\mathbf{F} = z + z + 4z = 6z$

$\displaystyle\iiint_Q \text{div } \mathbf{F} \, dV = \int_0^{2\pi}\!\int_0^1\!\int_0^{1-r^2} 6z \, r \, dz \, dr \, d\theta$

$= \int_0^{2\pi}\!\int_0^1 3(1 - r^2)^2 \, r \, dr \, d\theta = \int_0^{2\pi}\!\int_0^1 (3 - 6r^2 + 3r^4) \, r \, dr \, d\theta = \int_0^{2\pi} \left[\frac{3}{2} - \frac{3}{2} + \frac{1}{2} \right] d\theta = \pi$

9. Because div $\mathbf{F} = 2x + 2y + 2z$, you have

$\displaystyle\iiint_Q \text{div } \mathbf{F} \, dV = \int_0^a\!\int_0^a\!\int_0^a (2x + 2y + 2z) \, dz \, dy \, dx$

$= \int_0^a\!\int_0^a (2ax + 2ay + a^2) \, dy \, dx = \int_0^a (2a^2x + 2a^3) \, dx = \left[a^2x^2 + 2a^3x \right]_0^a = 3a^4.$

11. Because div $\mathbf{F} = 2x - 2x + 2xyz = 2xyz$,

$$\iiint\limits_Q \text{div } \mathbf{F} \, dV = \iiint\limits_Q 2xyz \, dV = \int_0^a \int_0^{2\pi} \int_0^{\pi/2} 2(\rho \sin \phi \cos \theta)(\rho \sin \phi \sin \theta)(\rho \cos \phi)\rho^2 \sin \phi \, d\phi \, d\theta \, d\rho$$

$$= \int_0^a \int_0^{2\pi} \int_0^{\pi/2} 2\rho^5 (\sin \theta \cos \theta)(\sin^3 \phi \cos \phi) \, d\phi \, d\theta \, d\rho$$

$$= \int_0^a \int_0^{2\pi} \frac{1}{2}\rho^5 \sin \theta \cos \theta \, d\theta \, d\rho = \int_0^a \left[\left(\frac{\rho^5}{2}\right) \frac{\sin^2 \theta}{2} \right]_0^{2\pi} d\rho = 0.$$

13. Because div $\mathbf{F} = 3$, you have

$$\iiint\limits_Q 3 \, dV = 3 \, (\text{Volume of Sphere}) = 3\left[\frac{4}{3}\pi(3^3)\right] = 108\pi.$$

15. Because div $\mathbf{F} = 1 + 2y - 1 = 2y$, you have

$$\iiint\limits_Q 2y \, dV = \int_0^7 \int_{-5}^5 \int_{-\sqrt{25-y^2}}^{\sqrt{25-y^2}} 2y \, dx \, dy \, dz = \int_0^7 \int_{-5}^5 4y \sqrt{25 - y^2} \, dy \, dz = \int_0^7 \left[\frac{-4}{3}(25 - y^2)^{3/2} \right]_{-5}^5 dz = 0.$$

17. Because div $\mathbf{F} = e^z + e^z + e^z = 3e^z$, you have

$$\iiint\limits_Q 3e^z \, dV = \int_0^6 \int_0^4 \int_0^{4-y} 3e^z \, dz \, dy \, dx = \int_0^6 \int_0^4 3\left[e^{4-y} - 1\right] dy \, dx = \int_0^6 3(e^4 - 5) \, dx = 18(e^4 - 5).$$

19. $\mathbf{F}(x, y, z) = 2\mathbf{i} + y\mathbf{j} + \mathbf{k}, (2, 2, 1)$

div $\mathbf{F} = 1 > 0 \Rightarrow$ Source

21. $\mathbf{F}(x, y, z) = \sin x\mathbf{i} + \cos y\mathbf{j} + z^3 \sin y\mathbf{k}, \left(\frac{\pi}{2}, \pi, 4\right)$

div $\mathbf{F} = \cos x - \sin y + 3z^2 \sin y$

At $\left(\frac{\pi}{2}, \pi, 4\right)$, div $\mathbf{F} = 0 - 0 + 0 = 0 \Rightarrow$ Incompressible

23. $\mathbf{F}(x, y, z) = x^2 yz\mathbf{i} + x\mathbf{j} - z\mathbf{k}$

div $\mathbf{F} = 2xyz - 1 > 0 \Rightarrow xyz > \frac{1}{2}$.

Any point that satisfies $xyz > \frac{1}{2}$, for example, $(1, 1, 1)$.

25. Using the Divergence Theorem, you have $\int_S \int \text{curl } \mathbf{F} \cdot \mathbf{N} \, dS = \iiint\limits_Q \text{div (curl } \mathbf{F}) \, dV$. Let

$$\mathbf{F}(x, y, z) = M\mathbf{i} + N\mathbf{j} + P\mathbf{k}$$

$$\text{curl } \mathbf{F} = \left(\frac{\partial P}{\partial y} - \frac{\partial N}{\partial z}\right)\mathbf{i} - \left(\frac{\partial P}{\partial x} - \frac{\partial M}{\partial z}\right)\mathbf{j} + \left(\frac{\partial N}{\partial x} - \frac{\partial M}{\partial y}\right)\mathbf{k}$$

$$\text{div (curl } \mathbf{F}) = \frac{\partial^2 P}{\partial x \partial y} - \frac{\partial^2 N}{\partial x \partial z} - \frac{\partial^2 P}{\partial y \partial x} + \frac{\partial^2 M}{\partial y \partial z} + \frac{\partial^2 N}{\partial z \partial x} - \frac{\partial^2 M}{\partial z \partial y} = 0.$$

So, $\int_S \int \text{curl } \mathbf{F} \cdot \mathbf{N} \, dS = \iiint\limits_Q 0 \, dV = 0.$

27. (a) Using the triple integral to find volume, you need **F** so that

$$\text{div } \mathbf{F} = \frac{\partial M}{\partial x} + \frac{\partial N}{\partial y} + \frac{\partial P}{\partial z} = 1.$$

So, you could have $\mathbf{F} = x\mathbf{i}$, $\mathbf{F} = y\mathbf{j}$, or $\mathbf{F} = z\mathbf{k}$.

For $dA = dy\,dz$ consider $\mathbf{F} = x\mathbf{i}$, $x = f(y, z)$, then $\mathbf{N} = \dfrac{\mathbf{i} + f_y\mathbf{j} + f_z\mathbf{k}}{\sqrt{1 + f_y^2 + f_z^2}}$ and $dS = \sqrt{1 + f_y^2 + f_z^2}\,dy\,dz$.

For $dA = dz\,dx$ consider $\mathbf{F} = y\mathbf{j}$, $y = f(x, z)$, then $\mathbf{N} = \dfrac{f_x\mathbf{i} + \mathbf{j} + f_z\mathbf{k}}{\sqrt{1 + f_x^2 + f_z^2}}$ and $dS = \sqrt{1 + f_x^2 + f_z^2}\,dz\,dx$.

For $dA = dx\,dy$ consider $\mathbf{F} = z\mathbf{k}$, $z = f(x, y)$, then $\mathbf{N} = \dfrac{f_x\mathbf{i} + f_y\mathbf{j} + \mathbf{k}}{\sqrt{1 + f_x^2 + f_y^2}}$ and $dS = \sqrt{1 + f_x^2 + f_y^2}\,dx\,dy$.

Correspondingly, you then have $\mathbf{V} = \int_S\int \mathbf{F} \cdot \mathbf{N}\,dS = \int_S\int x\,dy\,dz = \int_S\int y\,dz\,dx = \int_S\int z\,dx\,dy$.

(b) $v = \int_0^a\int_0^a x\,dy\,dz = \int_0^a\int_0^a a\,dy\,dz = \int_0^a a^2\,dz = a^3$

Similarly, $\int_0^a\int_0^a y\,dz\,dx = \int_0^a\int_0^a z\,dx\,dy = a^3$.

29. If $\mathbf{F}(x, y, z) = x\mathbf{i} + y\mathbf{j} + z\mathbf{k}$, then div $\mathbf{F} = 3$.

$$\int_S\int \mathbf{F} \cdot \mathbf{N}\,dS = \iiint_Q \text{div } \mathbf{F}\,dV = \iiint_Q 3\,dV = 3V.$$

31. $\int_S\int f D_{\mathbf{N}}g\,dS = \int_S\int f\nabla g \cdot \mathbf{N}\,dS = \iiint_Q \text{div}(f\nabla g)\,dV = \iiint_Q (f\,\text{div}\nabla g + \nabla f \cdot \nabla g)\,dV = \iiint_Q (f\nabla^2 g + \nabla f \cdot \nabla g)\,dV$

Section 15.8   Stokes's Theorem

1. Stoke's Theorem allows you to evaluate a line integral using a single double integral.

3. $C: x^2 + y^2 = 9,\quad z = 0, dz = 0$

Line Integral:

$$\int_C \mathbf{F} \cdot d\mathbf{r} = \int_C -y\,dx + x\,dy$$

$x = 3\cos t$, $dx = -3\sin t\,dt$, $y = 3\sin t$, $dy = 3\cos t\,dt$

$$\int_C \mathbf{F} \cdot d\mathbf{r} = \int_0^{2\pi}\left[(-3\sin t)(-3\sin t) + (3\cos t)(3\cos t)\right]dt = \int_0^{2\pi} 9\,dt = 18\pi$$

Double Integral: $g(x, y) = 9 - x^2 - y^2$, $g_x = -2x$, $g_y = -2y$

curl $\mathbf{F} = 2\mathbf{k}$

$$\int_S\int \text{curl } \mathbf{F} \cdot \mathbf{N}\,dS = \int_R\int 2\,dA = 2(\text{area circle}) = 18\pi$$

5. Line Integral:

From the figure you see that

C_1: $z = 0$, $dz = 0$

C_2: $x = 0$, $dx = 0$

C_3: $y = 0$, $dy = 0$

$$\int_C \mathbf{F} \cdot d\mathbf{r} = \int_C xyz \, dx + y \, dy + z \, dz = \int_{C_1} y \, dy + \int_{C_2} y \, dy + z \, dz + \int_{C_3} z \, dz = \int_0^2 y \, dy + \int_2^0 y \, dy + \int_0^{12} z \, dz + \int_{12}^0 z \, dz = 0$$

Double Integral: $\text{curl } \mathbf{F} = xy\mathbf{j} - xz\mathbf{k}$

Letting $z = 12 - 6x - 6y = g(x, y)$, $g_x = -6 = g_y$.

$$\int_S \int (\text{curl } \mathbf{F}) \cdot \mathbf{N} \, dS = \int_R \int (\text{curl } \mathbf{F}) \cdot [6\mathbf{i} + 6\mathbf{j} + \mathbf{k}] \, dA = \int_R \int (6xy - xz) \, dA$$

$$= \int_0^2 \int_0^{2-x} [6xy - x(12 - 6x - 6y)] \, dy \, dx = \int_0^2 \int_0^{2-x} (12xy - 12x + 6x^2) \, dy \, dx$$

$$= \int_0^2 [6xy^2 - 12xy + 6x^2y]_0^{2-x} \, dx = 0$$

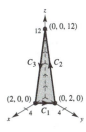

7. These three points have the equation $x + y + z = 2$.

Normal vector: $\mathbf{N} = \mathbf{i} + \mathbf{j} + \mathbf{k}$

$$\text{curl } \mathbf{F} = \begin{vmatrix} \mathbf{i} & \mathbf{j} & \mathbf{k} \\ \dfrac{\partial}{\partial x} & \dfrac{\partial}{\partial y} & \dfrac{\partial}{\partial z} \\ 2y & 3z & x \end{vmatrix} = -3\mathbf{i} - \mathbf{j} - 2\mathbf{k}$$

$$\int_S \int \text{curl } \mathbf{F} \cdot \mathbf{N} \, dS = \int_R \int (-6) \, dA = -6(\text{area of triangle in } xy\text{-plane}) = -6(2) = -12$$

9. $\mathbf{F}(x, y, z) = z^2 \mathbf{i} + 2x\mathbf{j} + y^2\mathbf{k}$, $S: z = 1 - x^2 - y^2$, $z > 0$

$$\text{curl } \mathbf{F} = \begin{vmatrix} \mathbf{i} & \mathbf{j} & \mathbf{k} \\ \dfrac{\partial}{\partial x} & \dfrac{\partial}{\partial y} & \dfrac{\partial}{\partial z} \\ z^2 & 2x & y^2 \end{vmatrix} = 2y\mathbf{i} + 2z\mathbf{j} + 2\mathbf{k}$$

$z = G(x, y) = 1 - x^2 - y^2$, $G_x = -2x$, $G_y = -2y$

$$\int_S \int \text{curl } \mathbf{F} \cdot \mathbf{N} \, dS = \int_R \int (2y\mathbf{i} + 2z\mathbf{j} + 2\mathbf{k}) \cdot (2x\mathbf{i} + 2y\mathbf{j} + \mathbf{k}) \, dA = \int_R \int [4xy + 4y(1 - x^2 - y^2) + 2] \, dA$$

$$= \int_{-1}^1 \int_{-\sqrt{1-x^2}}^{\sqrt{1-x^2}} [4xy + 4y - 4x^2y - 4y^3 + 2] \, dy \, dx$$

$$= \int_{-1}^1 4\sqrt{1 - x^2} \, dx = 2 \left[\arcsin x + x\sqrt{1 - x^2} \right]_{-1}^1 = 2\pi$$

11. $\mathbf{F}(x, y, z) = z^2\mathbf{i} + y\mathbf{j} + z\mathbf{k}, \ S: z = \sqrt{4 - x^2 - y^2}$

$$\text{curl } \mathbf{F} = \begin{vmatrix} \mathbf{i} & \mathbf{j} & \mathbf{k} \\ \dfrac{\partial}{\partial x} & \dfrac{\partial}{\partial y} & \dfrac{\partial}{\partial z} \\ z^2 & y & z \end{vmatrix} = 2z\mathbf{j}$$

$$z = G(x, y) = \sqrt{4 - x^2 - y^2}, G_x = \frac{-x}{\sqrt{4 - x^2 - y^2}}, G_y = \frac{-y}{\sqrt{4 - x^2 - y^2}}$$

$$\int_S \int \text{curl } \mathbf{F} \cdot \mathbf{N} = \int_R \int (2z\mathbf{j}) \cdot \left(\frac{x}{\sqrt{4 - x^2 - y^2}}\mathbf{i} + \frac{y}{\sqrt{4 - x^2 - y^2}}\mathbf{j} + \mathbf{k} \right) dA$$

$$= \int_R \int \frac{2yz}{\sqrt{4 - x^2 - y^2}} \, dA = \int_R \int \frac{2y\sqrt{4 - x^2 - y^2}}{\sqrt{4 - x^2 - y^2}} \, dA = \int_{-2}^{2} \int_{-\sqrt{4-x^2}}^{\sqrt{4-x^2}} 2y \, dy \, dx = 0$$

13. $\mathbf{F}(x, y, z) = -\ln\sqrt{x^2 + y^2}\,\mathbf{i} + \arctan\dfrac{x}{y}\mathbf{j} + \mathbf{k}$

$$\text{curl } \mathbf{F} = \begin{vmatrix} \mathbf{i} & \mathbf{j} & \mathbf{k} \\ \dfrac{\partial}{\partial x} & \dfrac{\partial}{\partial y} & \dfrac{\partial}{\partial z} \\ -1/2 \ln(x^2 + y^2) & \arctan x/y & 1 \end{vmatrix} = \left[\frac{(1/y)}{1 + (x^2/y^2)} + \frac{y}{x^2 + y^2} \right]\mathbf{k} = \left[\frac{2y}{x^2 + y^2} \right]\mathbf{k}$$

$S: z = 9 - 2x - 3y$ over one petal of $r = 2 \sin 2\theta$ in the first octant.

$G(x, y, z) = 2x + 3y + z - 9$

$\nabla G(x, y, z) = 2\mathbf{i} + 3\mathbf{j} + \mathbf{k}$

$$\int_S \int (\text{curl } \mathbf{F}) \cdot \mathbf{N} \, dS = \int_R \int \frac{2y}{x^2 + y^2} \, dA = \int_0^{\pi/2} \int_0^{2 \sin 2\theta} \frac{2r \sin \theta}{r^2} r \, dr \, d\theta$$

$$= \int_0^{\pi/2} \int_0^{4 \sin \theta \cos \theta} 2 \sin \theta \, dr \, d\theta = \int_0^{\pi/2} 8 \sin^2 \theta \cos \theta \, d\theta = \left[\frac{8 \sin^3 \theta}{3} \right]_0^{\pi/2} = \frac{8}{3}$$

15. $\mathbf{F}(x, y, z) = xyz\mathbf{i} + y\mathbf{j} + z\mathbf{k}, \ S: z = x^2, 0 \le x \le a, 0 \le y \le a$

$$\text{curl } \mathbf{F} = \begin{vmatrix} \mathbf{i} & \mathbf{j} & \mathbf{j} \\ \dfrac{\partial}{\partial x} & \dfrac{\partial}{\partial y} & \dfrac{\partial}{\partial z} \\ xyz & y & z \end{vmatrix} = xy\mathbf{j} - xz\mathbf{k}$$

$z = G(x, y) = x^2, G_x = 2x, G_y = 0, \mathbf{N} = -2x\mathbf{i} + \mathbf{k}$

$$\int_S \int \text{curl } \mathbf{F} \cdot \mathbf{N} \, dS = \int_R \int (-xz) \, dA = \int_R \int -x(x^2) \, dA$$

$$= \int_0^a \int_0^a (-x^3) \, dy \, dx = -\frac{a^5}{4}$$

17. $\mathbf{F}(x, y, z) = -\frac{1}{6}y^3\mathbf{i} + \frac{1}{6}x^3\mathbf{j} + 5\mathbf{k}$

$$\text{curl } \mathbf{F} = \begin{vmatrix} \mathbf{i} & \mathbf{j} & \mathbf{k} \\ \dfrac{\partial}{\partial x} & \dfrac{\partial}{\partial y} & \dfrac{\partial}{\partial z} \\ -\dfrac{1}{6}y^3 & \dfrac{1}{6}x^3 & 5 \end{vmatrix} = \left(\frac{1}{2}x^2 + \frac{1}{2}y^2\right)\mathbf{k}$$

$\mathbf{N} = \mathbf{k}$

$\text{curl } \mathbf{F} \cdot \mathbf{N} = \frac{1}{2}x^2 + \frac{1}{2}y^2 = \frac{1}{2}r^2$

$$\int_S\!\!\int \text{curl } \mathbf{F} \cdot \mathbf{N} \, dS = \int_0^{2\pi}\!\!\int_0^3 \frac{1}{2}r^2 \, r \, dr \, d\theta = \int_0^{2\pi} \left[\frac{r^4}{8}\right]_0^3 d\theta = \frac{81}{8}(2\pi) = \frac{81\pi}{4}$$

19. Yes. Let $\mathbf{K} = a\mathbf{i} + b\mathbf{j} + c\mathbf{k}$, then $\frac{1}{2}\int_C(\mathbf{K} \times \mathbf{r}) \cdot d\mathbf{r} = \frac{1}{2}\int_S\!\!\int\text{curl}(\mathbf{K} \times \mathbf{r}) \cdot \mathbf{N} \, dS = \frac{1}{2}\int_S\!\!\int 2\mathbf{K} \cdot \mathbf{N} \, dS = \int_S\!\!\int \mathbf{K} \cdot \mathbf{N} \, dS$

because $\mathbf{K} \times \mathbf{r} = \begin{vmatrix} \mathbf{i} & \mathbf{j} & \mathbf{k} \\ a & b & c \\ x & y & z \end{vmatrix} = (bz - cy)\mathbf{i} - (az - cx)\mathbf{j} + (ay - bx)\mathbf{k}$

and $\text{curl}(\mathbf{K} \times \mathbf{r}) = \begin{vmatrix} \mathbf{i} & \mathbf{j} & \mathbf{k} \\ \dfrac{\partial}{\partial x} & \dfrac{\partial}{\partial y} & \dfrac{\partial}{\partial z} \\ bz - cy & cx - az & ay - bx \end{vmatrix} = 2(a\mathbf{i} + b\mathbf{j} + c\mathbf{k}) = 2\mathbf{K}.$

21. Let S be the upper portion of the ellipsoid

$x^2 + 4y^2 + z^2 = 4, z \geq 0$

Let $C: \mathbf{r}(t) = \langle 2\cos t, \sin t, 0 \rangle, 0 \leq t \leq 2\pi$, be the boundary of S.

If $\mathbf{F} = \langle M, N, P \rangle$ exists, then

$\begin{aligned} 0 &= \int_S\!\!\int (\text{curl } \mathbf{F}) \cdot \mathbf{N} \, dS && \text{(by (i))} \\ &= \int_C \mathbf{F} \cdot d\mathbf{r} && \text{(Stokes's Theorem)} \\ &= \int_C \mathbf{G} \cdot d\mathbf{r} && \text{(by (iii))} \\ &= \int_0^{2\pi} \left\langle \frac{-\sin t}{4}, \frac{2\cos t}{4}, 0 \right\rangle \cdot \langle -2\sin t, \cos t, 0 \rangle \, dt = \frac{1}{4}\int_0^{2\pi}(2\sin^2 t + 2\cos^2 t)\, dt = \pi \end{aligned}$

So, there is no such $\mathbf{F}$.

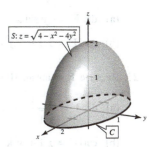

Review Exercises for Chapter 15

1. $\mathbf{F}(x, y, z) = x\mathbf{i} + \mathbf{j} + 2\mathbf{k}$

$\|\mathbf{F}\| = \sqrt{x^2 + 1^2 + 2^2} = \sqrt{x^2 + 5}$

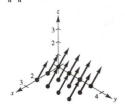

3. $f(x, y) = \sin xy - y^2$

$\mathbf{F}(x, y) = \nabla f = y\cos xy\mathbf{i} + (x\cos xy - 2y)\mathbf{j}$

5. $f(x, y, z) = 2x^2 + xy + z^2$

$\mathbf{F}(x, y, z) = \nabla f = (4x + y)\mathbf{i} + x\mathbf{j} + 2z\mathbf{k}$

7. $\mathbf{F}(x, y) = \cosh y\mathbf{i} + x\sinh x\mathbf{j}$

$\dfrac{\partial N}{\partial x} = \sinh x + x\cosh x \neq \dfrac{\partial M}{\partial y} = \sinh y$

Not conservative

9. $F(x, y, z) = y^2 \mathbf{i} + 2xy\mathbf{j} + \cos z\mathbf{k}$

$$\text{curl } \mathbf{F} = \begin{vmatrix} \mathbf{i} & \mathbf{j} & \mathbf{k} \\ \dfrac{\partial}{\partial x} & \dfrac{\partial}{\partial y} & \dfrac{\partial}{\partial z} \\ y^2 & 2xy & \cos z \end{vmatrix} = (2y - 2y)\mathbf{k} = \mathbf{0}$$

Conservative

11. Because $\partial M/\partial y = -1/x^2 = \partial N/\partial x$, $\mathbf{F}$ is conservative.

From $M = \partial U/\partial x = -y/x^2$ and $N = \partial U/\partial y = 1/x$,

partial integration yields $U = (y/x) + h(y)$ and

$U = (y/x) + g(x)$ which suggests that

$U(x, y) = (y/x) + C$.

13. Because $\dfrac{\partial M}{\partial y} = 2xy$ and $\dfrac{\partial N}{\partial x} = 2xy$, $\mathbf{F}$ is conservative.

From $M = \dfrac{\partial U}{\partial x} = xy^2 - x^2$ and $N = \dfrac{\partial U}{\partial y} = x^2 y + y^2$,

partial integration yields

$U = \dfrac{1}{2}x^2 y^2 - \dfrac{x^3}{3} + h(y)$ and

$U = \dfrac{1}{2}x^2 y^2 + \dfrac{y^3}{3} + g(x)$.

So, $h(y) = y^3/3$ and $g(x) = -x^3/3$. So,

$U(x, y) = \dfrac{1}{2}x^2 y^2 - \dfrac{x^3}{3} + \dfrac{y^3}{3} + C$.

15. Because $\dfrac{\partial M}{\partial y} = 8xy$ and $\dfrac{\partial N}{\partial x} = 4x$, $\dfrac{\partial M}{\partial y} \neq \dfrac{\partial N}{\partial x}$, so $\mathbf{F}$ is

not conservative.

17. Because

$$\dfrac{\partial M}{\partial y} = \dfrac{-1}{y^2 z} = \dfrac{\partial N}{\partial x}, \dfrac{\partial M}{\partial z} = \dfrac{-1}{yz^2} = \dfrac{\partial P}{\partial x},$$

$$\dfrac{\partial N}{\partial z} = \dfrac{x}{y^2 z^2} = \dfrac{\partial P}{\partial y},$$

$\mathbf{F}$ is conservative. From

$$M = \dfrac{\partial U}{\partial x} = \dfrac{1}{yz}, \ N = \dfrac{\partial U}{\partial y} = \dfrac{-x}{y^2 z}, \ P = \dfrac{\partial U}{\partial z} = \dfrac{-x}{yz^2}$$

you obtain

$$U = \dfrac{x}{yz} + f(y, z), \ U = \dfrac{x}{yz} + g(x, z),$$

$$U = \dfrac{x}{yz} + h(x, y) \Rightarrow f(x, y\ z) = \dfrac{x}{yz} + K.$$

19. Because $\mathbf{F}(x, y, z) = x^2\, \mathbf{i} + xy^2\, \mathbf{j} + x^2 z\, \mathbf{k}$:

(a) $\text{div } \mathbf{F} = 2x + 2xy + x^2$

(b) $\text{curl } \mathbf{F} = \begin{vmatrix} \mathbf{i} & \mathbf{j} & \mathbf{k} \\ \dfrac{\partial}{\partial x} & \dfrac{\partial}{\partial y} & \dfrac{\partial}{\partial z} \\ x^2 & xy^2 & x^2 z \end{vmatrix} = -(2xz)\mathbf{j} + y^2\mathbf{k}$

21. Because $\mathbf{F} = (\cos y + y \cos x)\mathbf{i} + (\sin x - x \sin y)\mathbf{j} + xyz\,\mathbf{k}$:

(a) $\text{div } \mathbf{F} = -y \sin x - x \cos y + xy$

(b) $\text{curl } \mathbf{F} = xz\mathbf{i} - yz\mathbf{j} + (\cos x - \sin y + \sin y - \cos x)\mathbf{k} = xz\mathbf{i} - yz\mathbf{j}$

23. Because $\mathbf{F} = \arcsin x\mathbf{i} + xy^2\mathbf{j} + yz^2\mathbf{k}$:

(a) $\text{div } \mathbf{F} = \dfrac{1}{\sqrt{1 - x^2}} + 2xy + 2yz$

(b) $\text{curl } \mathbf{F} = z^2\mathbf{i} + y^2\mathbf{k}$

25. Because $\mathbf{F} = \ln(x^2 + y^2)\mathbf{i} + \ln(x^2 + y^2)\mathbf{j} + z\mathbf{k}$:

(a) $\text{div } \mathbf{F} = \dfrac{2x}{x^2 + y^2} + \dfrac{2y}{x^2 + y^2} + 1 = \dfrac{2x + 2y}{x^2 + y^2} + 1$

(b) $\text{curl } \mathbf{F} = \dfrac{2x - 2y}{x^2 + y^2}\mathbf{k}$

27. (a) Let $x = 3t, y = 4t, \quad 0 \le t \le 1$,

then $ds = \sqrt{9 + 16}\, dt = 5\, dt$.

$$\int_C (x^2 + y^2)\, ds = \int_0^1 (9t^2 + 16t^2)5\, dt$$

$$= \left[125\dfrac{t^3}{3}\right]_0^1 = \dfrac{125}{3}$$

(b) Let $x = \cos t, y = \sin t, \quad 0 \le t \le 2\pi$,

then $ds = \sqrt{(-\sin t)^2 + (\cos t)^2} = 1$.

$$\int_C (x^2 + y^2)\, ds = \int_0^{2\pi} dt = 2\pi$$

29. $x = 1 - \sin t, y = 1 - \cos t, 0 \le t \le 2\pi$

$\dfrac{dx}{dt} = -\cos t, \dfrac{dy}{dt} = \sin t, ds = \sqrt{(-\cos t)^2 + (\sin t)^2}\, dt = dt$

$\displaystyle\int_C (x^2 + y^2)\, ds = \int_0^{2\pi}\left[(1 - \sin t)^2 + (1 - \cos t)^2\right] dt = \int_0^{2\pi}\left[1 - 2\sin t + \sin^2 t + 1 - 2\cos t + \cos^2 t\right] dt$

$\qquad = \displaystyle\int_0^{2\pi}\left[3 - 2\sin t - 2\cos t\right] dt = \left[3t + 2\cos t - 2\sin t\right]_0^{2\pi} = 6\pi$

31. $\displaystyle\int_C (2x + y)\, ds, \mathbf{r}(t) = a\cos^3 t\,\mathbf{i} + a\sin^3 t\,\mathbf{j}, 0 \le t \le \dfrac{\pi}{2}$

$x'(t) = -3a \cdot \cos^2 t \sin t$

$y'(t) = 3a \cdot \sin^2 t \cos t$

$\displaystyle\int_C (2x + y)\, ds = \int_0^{\pi/2}\left(2(a \cdot \cos^3 t) + a \cdot \sin^3 t\right)\sqrt{x'(t)^2 + y'(t)^2}\, dt = \dfrac{9a^2}{5}$

33. $\mathbf{r}(t) = 3\cos t\,\mathbf{i} + 3\sin t\,\mathbf{j}, 0 \le t \le \pi$

$\mathbf{r}'(t) = -3\sin t\,\mathbf{i} + 3\cos t\,\mathbf{j}$

$\|\mathbf{r}'(t)\| = \sqrt{9\sin^2 t + 9\cos^2 t} = 3$

Mass $= \displaystyle\int_C (1 + x)\, ds$

$\qquad = \displaystyle\int_0^{\pi} (1 + 3\cos t)\, 3\, dt$

$\qquad = 3\left[t + 3\sin t\right]_0^{\pi}$

$\qquad = 3\pi$

35. $\mathbf{F}(x, y) = xy\mathbf{i} + 2xy\mathbf{j}$

$\mathbf{r}(t) = t^2\mathbf{i} + t^2\mathbf{j}, \; 0 \le t \le 1$

$\mathbf{r}'(t) = 2t\mathbf{i} + 2t\mathbf{j}$

$\displaystyle\int_C \mathbf{F} \cdot d\mathbf{r} = \int_0^1\left[t^2(t^2)(2t) + 2(t^2)(t^2)(2t)\right] dt$

$\qquad = \displaystyle\int_0^1 6t^5\, dt = t^6\Big]_0^1 = 1$

37. $d\mathbf{r} = \left[(-2\sin t)\mathbf{i} + (2\cos t)\mathbf{j} + \mathbf{k}\right] dt$

$\mathbf{F} = (2\cos t)\mathbf{i} + (2\sin t)\mathbf{j} + t\mathbf{k}, 0 \le t \le 2\pi$

$\displaystyle\int_C \mathbf{F} \cdot d\mathbf{r} = \int_0^{2\pi} t\, dt = 2\pi^2$

39. $\mathbf{F} = x\mathbf{i} - \sqrt{y}\,\mathbf{j}$

$\dfrac{\partial N}{\partial x} = 0 = \dfrac{\partial M}{\partial y} \Rightarrow$ Conservative

$f(x, y) = \dfrac{1}{2}x^2 - \dfrac{2}{3}y^{3/2}$

Work $= \left[\dfrac{1}{2}x^2 - \dfrac{2}{3}y^{3/2}\right]_{(0,0)}^{(4,8)} = \dfrac{1}{2}(16) - \left(\dfrac{2}{3}\right)8^{3/2} = \dfrac{8}{3}\left(3 - 4\sqrt{2}\right)$

41. $C_1: y = -2x, 0 \le x \le 2, dy = -2\, dx$

$C_2: y = -4, 2 \le x \le 4, dy = 0\, dx$

$\displaystyle\int_{C_1} (y - x)\, dx + (2x + 5y)\, dy + \int_{C_2} (y - x)\, dx + (2x + 5y)\, dy$

$= \displaystyle\int_0^2 (-2x - x)\, dx + (2x - 10x)(-2\, dx) + \int_2^4 (-4 - x)\, dx + 0$

$= \displaystyle\int_0^2 13x\, dx + \int_2^4 (-4 - x)\, dx$

$= \left[\dfrac{13x}{2}\right]_0^2 + \left[-4x - \dfrac{x^2}{2}\right]_2^4$

$= 26 - 14 = 12$

43. $f(x, y) = 3 + \sin(x + y)$

$C: y = 2x$ from $(0, 0)$ to $(2, 4)$

$\mathbf{r}(t) = t\mathbf{i} + 2t\mathbf{j}, \ 0 \leq t \leq 2$

$\mathbf{r}'(t) = \mathbf{i} + 2\mathbf{j}$

$\|\mathbf{r}'(t)\| = \sqrt{5}$

Lateral surface area:

$$\int_C f(x, y) \, ds = \int_0^2 \left[3 + \sin(t + 2t)\right]\sqrt{5} \, dt = \sqrt{5}\int_0^2 [3 + \sin 3t] \, dt = \sqrt{5}\left[3t - \frac{1}{3}\cos 3t\right]_0^2$$

$$= \sqrt{5}\left[6 - \frac{1}{3}\cos 6 + \frac{1}{3}\right] = \frac{\sqrt{5}}{3}(19 - \cos 6) \approx 13.446$$

45. $\mathbf{F}(x, y) = (3x + 4)\mathbf{i} + y^3\mathbf{j}$

(a) $\dfrac{\partial N}{\partial x} = 0 = \dfrac{\partial M}{\partial y} \Rightarrow$ Conservative

(b) (i) $\mathbf{r}(t) = t\mathbf{i} + t\mathbf{j}, \ 0 \leq t \leq 4$

$\mathbf{r}'(t) = \mathbf{i} + \mathbf{j}$

$\mathbf{F}(t) = (3t + 4)\mathbf{i} + t^3\mathbf{j}$

$$\int_{C_1} \mathbf{F} \cdot d\mathbf{r} = \int_0^4 \left[(3t + 4) + t^3\right] dt = \left[\frac{3t^2}{2} + 4t + \frac{t^4}{4}\right]_0^4 = 24 + 16 + 64 = 104$$

(ii) $\mathbf{r}(w) = w^2\mathbf{i} + w^2\mathbf{j}, \ 0 \leq w \leq 2$

$\mathbf{r}'(w) = 2w\mathbf{i} + 2w\mathbf{j}$

$\mathbf{F}(w) = (3w^2 + 4)\mathbf{i} + w^6\mathbf{j}$

$$\int_{C_2} \mathbf{F} \cdot d\mathbf{r} = \int_0^2 \left[(3w^2 + 4)(2w) + w^6(2w)\right] dw = \int_0^2 (8w + 6w^3 + 2w^7) \, dw$$

$$= \left[4w^2 + \frac{3}{2}w^4 + \frac{1}{4}w^8\right]_0^2 = 16 + 24 + 64 = 104$$

47. $\mathbf{F} = e^{2x}\mathbf{i} + e^{2y}\mathbf{j}$

$\dfrac{\partial N}{\partial x} = 0 = \dfrac{\partial M}{\partial y} \Rightarrow$ Conservative

$f(x, y) = \dfrac{1}{2}e^{2x} + \dfrac{1}{2}e^{2y}$

$\displaystyle\int_C \mathbf{F} \cdot d\mathbf{r} = \left[\frac{1}{2}e^{2x} + \frac{1}{2}e^{2y}\right]_{(-1,-1)}^{(0, 0)}$

$= \left(\dfrac{1}{2} + \dfrac{1}{2}\right) - \left(\dfrac{1}{2}e^{-2} + \dfrac{1}{2}e^{-2}\right)$

$= 1 - e^{-2} = 1 - \dfrac{1}{e^2}$

49. $\mathbf{F}(x, y, z) = 2xyz\mathbf{i} + x^2z\mathbf{j} + x^2y\mathbf{k}$

$$\text{curl } \mathbf{F} = \begin{vmatrix} \mathbf{i} & \mathbf{j} & \mathbf{k} \\ \dfrac{\partial}{\partial x} & \dfrac{\partial}{\partial y} & \dfrac{\partial}{\partial z} \\ 2xyz & x^2z & x^2y \end{vmatrix} = \mathbf{0} \Rightarrow \text{Conservative}$$

$f(x, y, z) = x^2yz$

$\displaystyle\int_C \mathbf{F} \cdot d\mathbf{r} = \left[x^2yz\right]_{(0, 0, 0)}^{(1, 3, 2)} = 6$

51. $\mathbf{F}(x, y) = (1 - 3xy^2)\mathbf{i} - 3x^2y\mathbf{j}$

(a) $\dfrac{\partial N}{\partial x} = -6xy = \dfrac{\partial M}{\partial y} \Rightarrow$ Conservative

(b) $f(x, y) = x - \dfrac{3}{2}x^2y^2$

$$W = \int_C \mathbf{F} \cdot d\mathbf{r} = \left[x - \dfrac{3}{2}x^2y^2 \right]_{(4, 2)}^{(0, 1)}$$

$$= 0 - \left(4 - \dfrac{3}{2}64 \right) = 92$$

53. $\displaystyle\int_C y\, dx + 2x\, dy = \int_0^1 \int_0^1 \left(\dfrac{\partial N}{\partial x} - \dfrac{\partial M}{\partial y} \right) dy\, dx$

$$= \int_0^1 \int_0^1 (2 - 1)\, dy\, dx = 1$$

55. $\displaystyle\int_C xy^2\, dx + x^2y\, dy = \int_R \int \left(\dfrac{\partial N}{\partial x} - \dfrac{\partial M}{\partial y} \right) dA$

$$= \int_R \int (2xy - 2xy)\, dA = 0$$

57. $\displaystyle\int_C xy\, dx + x^2\, dy = \int_R \int \left(\dfrac{\partial N}{\partial x} - \dfrac{\partial M}{\partial y} \right) dA$

$$= \int_{-1}^1 \int_{x^2}^1 (2x - x)\, dy\, dx$$

$$= \int_{-1}^1 [xy]_{x^2}^1 = dx$$

$$= \int_{-1}^1 (x - x^3)\, dx$$

$$= \left[\dfrac{x^2}{2} - \dfrac{x^4}{4} \right]_{-1}^1 = 0$$

59. $\mathbf{F}(x, y) = y^2\mathbf{i} + 2xy\mathbf{j}$

$$\int_C M\, dx + N\, dy = \int_R \int (2y - 2y)\, dA = 0$$

61. $A = \dfrac{1}{2}\displaystyle\int_C x\, dy - y\, dx$

$$= \dfrac{1}{2}\int_{C_1} x\, dy - y\, dx + \dfrac{1}{2}\int_{C_2} x\, dy - y\, dx + \dfrac{1}{2}\int_{C_3} x\, dy - y\, dx$$

$$= \dfrac{1}{2}\int_0^4 x\left(\dfrac{1}{2}\, dx \right) - \dfrac{1}{2}x\, dx + \dfrac{1}{2}\int_4^3 x(-dx) - (6 - x)\, dx + \dfrac{1}{2}\int_3^0 x\, dx - x\, dx$$

$$= 0 + \dfrac{1}{2}\int_4^3 -6\, dx + 0 = 3$$

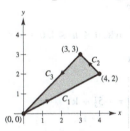

63. $\mathbf{r}(u, v) = 3u\cos v\mathbf{i} + 3u\sin v\mathbf{j} + 18u^2\mathbf{k}$

$x^2 + y^2 = 9u^2\cos^2 v + 9u^2\sin^2 v = 9u^2$

$2(x^2 + y^2) = z$

Paraboloid

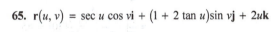

65. $\mathbf{r}(u, v) = \sec u\cos v\mathbf{i} + (1 + 2\tan u)\sin v\mathbf{j} + 2u\mathbf{k}$

$$0 \le u \le \dfrac{\pi}{3}, \quad 0 \le v \le 2\pi$$

67. $\dfrac{x^2}{1} + \dfrac{y^2}{8} + \dfrac{z^2}{9} = 1$

$\mathbf{r}(u, v) = \cos v \cos u\mathbf{i} + 2\sqrt{2}\cos v \sin u\mathbf{j} + 3\sin v\mathbf{k}$

Check: $x^2 + \dfrac{y^2}{8} + \dfrac{z^2}{9} = \cos^2 v \cos^2 u + \dfrac{8\cos^2 v \sin^2 u}{8} + \dfrac{9\sin^2 v}{9}$

$\qquad\qquad = \cos^2 v + \sin^2 v = 1$

69. $y = 2x^3, 0 \le x \le 2$, x-axis

$x = u, y = 2u^3 \cos v, z = 2u^3 \sin v, 0 \le u \le 2, 0 \le v \le 2\pi$

Check: $y^2 + z^2 = \left(2u^3\right)^2$

71. $\mathbf{r}(u, v) = 4u\mathbf{i} + (3u - v)\mathbf{j} + v\mathbf{k}, 0 \le u \le 3, 0 \le v \le 1$

$\mathbf{r}_u = 4\mathbf{i} + 3\mathbf{j}, \mathbf{r}_v = -\mathbf{j} + \mathbf{k}$

$\mathbf{r}_u \times \mathbf{r}_v = \begin{vmatrix} \mathbf{i} & \mathbf{j} & \mathbf{k} \\ 4 & 3 & 0 \\ 0 & -1 & 1 \end{vmatrix} = 3\mathbf{i} - 4\mathbf{j} - 4\mathbf{k}$

$\|\mathbf{r}_u \times \mathbf{r}_v\| = \sqrt{9 + 16 + 16} = \sqrt{41}$

$A = \int_0^3 \int_0^1 \sqrt{41}\, dv\, du = 3\sqrt{41}$

73. $S: z = x + \dfrac{y}{2}, 0 \le x \le 2, 0 \le y \le 5$

$g(x, y) = x + \dfrac{y}{2}, g_x = 1, g_y = \dfrac{1}{2}$

$\sqrt{1 + (g_x)^2 + (g_y)^2} = \sqrt{1 + 1 + \dfrac{1}{4}} = \dfrac{3}{2}$

$\int_S\!\!\int (5x + y - 2z)\, dS = \int_R\!\!\int \left[5x + y - 2\left(x + \dfrac{y}{2}\right)\right]\dfrac{3}{2}\, dA$

$\qquad = \dfrac{3}{2}\int_0^2 \int_0^5 3x\, dy\, dx$

$\qquad = \dfrac{3}{2}\int_0^2 15x\, dx$

$\qquad = \dfrac{3}{2}\left[\dfrac{15x^2}{2}\right]_0^2 = \dfrac{3}{2}(30) = 45$

75. $S: 2y + 6x + z = 18$, first octant, $\rho(x, y, z) = 2x$

$g(x, y) = 18 - 2y - 6x, g_x = -6, g_y = -2$

$\sqrt{1 + (g_x)^2 + (g_y)^2} = \sqrt{1 + 36 + 4} = \sqrt{41}$

Mass $= \int_S\!\!\int 2x\, dS = \int_0^3 \int_0^{9-3x} 2x\sqrt{41}\, dy\, dx$

$\qquad = \sqrt{41}\int_0^3 \left(18x - 6x^2\right) dx = 27\sqrt{41}$

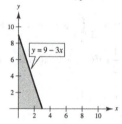

77. $\mathbf{r}(u, v) = u\mathbf{i} + v\mathbf{j} + 5v\mathbf{k}, 0 \le u \le 1, 0 \le v \le 3$

$\mathbf{r}_u = \mathbf{i}, \mathbf{r}_v = \mathbf{j} + 5\mathbf{k}$

$\mathbf{r}_u \times \mathbf{r}_v = \begin{vmatrix} \mathbf{i} & \mathbf{j} & \mathbf{k} \\ 1 & 0 & 0 \\ 0 & 1 & 5 \end{vmatrix} = -5\mathbf{j} + \mathbf{k}$

$\|\mathbf{r}_u \times \mathbf{r}_v\| = \sqrt{25 + 1} = \sqrt{26}$

$\int_S\!\!\int (x + y)\, dS = \int_0^3 \int_0^1 (u + v)\sqrt{26}\, du\, dv$

$\qquad = \sqrt{26}\int_0^3 \left(u + \dfrac{1}{2}\right) du = 6\sqrt{26}$

79. $\mathbf{F}(x, y, z) = -2\mathbf{i} - 2\mathbf{j} + \mathbf{k}$

$S: z = 25 - x^2 - y^2, z \ge 0$

$z = g(x, y) = 25 - x^2 - y^2$

$g_x = -2x, g_y = -2y$

$\int_S\!\!\int \mathbf{F} \cdot \mathbf{N}\, dS = \int_R\!\!\int (-2\mathbf{i} - 2\mathbf{j} + \mathbf{k}) \cdot (2x\mathbf{i} + 2y\mathbf{j} + \mathbf{k})\, dA = \int_R\!\!\int (-4x - 4y + 1)\, dA$

$\qquad = \int_0^{2\pi} \int_0^5 (-4r\cos\theta - 4r\sin\theta + 1)r\, dr\, d\theta = 25\pi$

81. $\mathbf{F}(x, y, z) = x^2\mathbf{i} + xy\mathbf{j} + z\mathbf{k}$

Q: solid region bounded by the coordinates planes and the plane $2x + 3y + 4z = 12$

$$\int_{S_4}\!\!\int \mathbf{F} \cdot \mathbf{N} \, dS = \frac{1}{4}\int_R\!\!\int \left(2x^2 + 3xy + 4z\right) dA$$

$$= \frac{1}{4}\int_0^6 \int_0^{4-(2x/3)} \left(2x^2 + 3xy + 12 - 2x - 3y\right) dy \, dx$$

$$= \frac{1}{4}\int_0^6 \left[2x^2\left(\frac{12-2x}{3}\right) + \frac{3x}{2}\left(\frac{12-2x}{3}\right)^2 + 12\left(\frac{12-2x}{3}\right) - 2x\left(\frac{12-2x}{3}\right) - \frac{3}{2}\left(\frac{12-2x}{3}\right)^2\right] dx$$

$$= \frac{1}{6}\int_0^6 \left(-x^3 + x^2 + 24x + 36\right) dx$$

$$= \frac{1}{6}\left[-\frac{x^4}{4} + \frac{x^3}{3} + 12x^2 + 36x\right]_0^6 = 66$$

Divergence Theorem: Because div $\mathbf{F} = 2x + x + 1 = 3x + 1$, Divergence Theorem yields

$$\iiint_Q \text{div } \mathbf{F} \, dV = \int_0^6 \int_0^{(12-2x)/3} \int_0^{(12-2x-3y)/4} \left(3x + 1\right) dz \, dy \, dx$$

$$= \int_0^6 \int_0^{(12-2x)/3} (3x + 1)\left(\frac{12 - 2x - 3y}{4}\right) dy \, dx$$

$$= \frac{1}{4}\int_0^6 (3x + 1)\left(12y - 2xy - \frac{3}{2}y^2\right)\Big|_0^{(12-2x)/3} dx$$

$$= \frac{1}{4}\int_0^6 (3x + 1)\left[4(12 - 2x) - 2x\left(\frac{12-2x}{3}\right) - \frac{3}{2}\left(\frac{12-2x}{3}\right)^2\right] dx$$

$$= \frac{1}{4}\int_0^6 \frac{2}{3}\left(3x^3 - 35x^2 + 96x + 36\right) dx$$

$$= \frac{1}{6}\left[\frac{3x^4}{4} - \frac{35x^3}{3} + 48x^2 + 36x\right] = 66.$$

83. $\mathbf{F}(x, y, z) = \left(\cos y + y \cos x\right)\mathbf{i} + \left(\sin x - x \sin y\right)\mathbf{j} + xyz\mathbf{k}$

S: portion of $z = y^2$ over the square in the xy-plane with vertices $(0, 0), (a, 0), (a, a), (0, a)$

$$\int_C \mathbf{F} \cdot d\mathbf{r} = \int_C (\cos y + y \cos x) \, dx + (\sin x - x \sin y) \, dy + xyz \, dz$$

$$= \int_{C_1} dx + \int_{C_2} 0 + \int_{C_3} (\cos a + a \cos x) \, dx + \int_{C_4} (\sin a - a \sin y) \, dy + ay^3(2y \, dy)$$

$$= \int_0^a dx + \int_a^0 (\cos a + a \cos x) \, dx + \int_0^a (\sin a - a \sin y) \, dy + \int_0^a 2ay^4 \, dy$$

$$= a + \left[x \cos a + a\sin x\right]_a^0 + \left[y \sin a + a \cos y\right]_0^a + \left[2a\frac{y^5}{5}\right]_0^a$$

$$= a - a \cos a - a \sin a + a \sin a + a \cos a - a + \frac{2a^6}{5} = \frac{2a^6}{5}$$

Double Integral: Considering $f(x, y, z) = z - y^2$, you have:

$$\mathbf{N} = \frac{\nabla f}{\|\nabla f\|} = \frac{-2y\mathbf{j} + \mathbf{k}}{\sqrt{1 + 4y^2}}, \quad dS = \sqrt{1 + 4y^2}\, dA, \text{ and curl } \mathbf{F} = xz\mathbf{i} - yz\mathbf{j}.$$

So, $\displaystyle\int_S\!\!\int (\text{curl } \mathbf{F}) \cdot \mathbf{N} \, dS = \int_0^a \int_0^a 2y^2z \, dy \, dx = \int_0^a \int_0^a 2y^4z \, dy \, dx = \int_0^a \frac{2a^5}{5} \, dx = \frac{2a^6}{5}.$

85. $\mathbf{F}(x, y, z) = \mathbf{i} + x\mathbf{j} - \mathbf{k}$

$$\text{curl } \mathbf{F} = \begin{vmatrix} \mathbf{i} & \mathbf{j} & \mathbf{k} \\ \dfrac{\partial}{\partial x} & \dfrac{\partial}{\partial y} & \dfrac{\partial}{\partial z} \\ 1 & x & -1 \end{vmatrix} = \mathbf{k}$$

$\mathbf{N} = \mathbf{k}$

$$\int_S\!\!\int (\text{curl } \mathbf{F}) \cdot \mathbf{N} \, dS = \int_R\!\!\int dA = \int_0^{2\pi}\!\int_0^4 r \, dr \, d\theta = 16\pi$$

Problem Solving for Chapter 15

1. (a) $\nabla T = \dfrac{-25}{\left(x^2 + y^2 + z^2\right)^{3/2}}[x\mathbf{i} + y\mathbf{j} + z\mathbf{k}]$

$\mathbf{N} = x\mathbf{i} + \sqrt{1 - x^2}\,\mathbf{k}$

$dS = \dfrac{1}{\sqrt{1 - x^2}}\, dA$

$$\text{Flux} = \int_S\!\!\int -k\nabla T \cdot \mathbf{N} \, dS = 25k \int_R\!\!\int \left[\frac{x^2}{\left(x^2 + y^2 + z^2\right)^{3/2}\left(1 - x^2\right)^{1/2}} + \frac{z}{\left(x^2 + y^2 + z^2\right)^{3/2}} \right] dA$$

$$= 25k \int_{-1/2}^{1/2}\!\int_0^1 \left[\frac{x^2}{\left(x^2 + y^2 + z^2\right)^{3/2}\left(1 - x^2\right)^{1/2}} + \frac{1 - x^2}{\left(x^2 + y^2 + z^2\right)^{3/2}\left(1 - x^2\right)^{1/2}} \right] dy \, dx$$

$$= 25k \int_{-1/2}^{1/2}\!\int_0^1 \frac{1}{\left(1 + y^2\right)^{3/2}\left(1 - x^2\right)^{1/2}} \, dy \, dx = 25k \int_0^1 \frac{1}{\left(1 + y^2\right)^{3/2}} \, dy \int_{-1/2}^{1/2} \frac{1}{\left(1 - x^2\right)^{1/2}} \, dx = 25k\left(\frac{\sqrt{2}}{2}\right)\left(\frac{\pi}{3}\right) = 25k\frac{\sqrt{2}\pi}{6}$$

(b) $\mathbf{r}(u,v) = \langle \cos u, v, \sin u \rangle$

$\mathbf{r}_u = \langle -\sin u, 0, \cos u \rangle, \; \mathbf{r}_v = \langle 0, 1, 0 \rangle$

$\mathbf{r}_u \times \mathbf{r}_v = \langle -\cos u, 0, \sin u \rangle$

$\nabla T = \dfrac{-25}{\left(x^2 + y^2 + z^2\right)^{3/2}}[x\mathbf{i} + y\mathbf{j} + z\mathbf{k}] = \dfrac{-25}{\left(v^2 + 1\right)^{3/2}}[\cos u\mathbf{i} + v\mathbf{j} + \sin u\mathbf{k}]$

$\nabla T \cdot (\mathbf{r}_u \times \mathbf{r}_v) = \dfrac{-25}{\left(v^2 + 1\right)^{3/2}}\left(-\cos^2 u - \sin^2 u\right) = \dfrac{25}{\left(v^2 + 1\right)^{3/2}}$

$\text{Flux} = \displaystyle\int_0^1\!\int_{\pi/3}^{2\pi/3} \dfrac{25}{\left(v^2 + 1\right)^{3/2}} \, du \, dv = 25k\dfrac{\sqrt{2}\pi}{6}$

3. $\mathbf{r}(t) = \langle 3\cos t, 3\sin t, 2t \rangle$

$\mathbf{r}'(t) = \langle -3\sin t, 3\cos t, 2t \rangle, \|\mathbf{r}'(t)\| = \sqrt{13}$

$I_x = \displaystyle\int_C \left(y^2 + z^2\right)\rho \, ds = \int_0^{2\pi} \left(9\sin^2 t + 4t^2\right)\sqrt{13}\, dt = \frac{1}{3}\sqrt{13}\pi\left(32\pi^2 + 27\right)$

$I_y = \displaystyle\int_C \left(x^2 + z^2\right)\rho \, ds = \int_0^{2\pi} \left(9\cos^2 t + 4t^2\right)\sqrt{13}\, dt = \frac{1}{3}\sqrt{13}\pi\left(32\pi^2 + 27\right)$

$I_z = \displaystyle\int_C \left(x^2 + y^2\right)\rho \, ds = \int_0^{2\pi} \left(9\cos^2 t + 9\sin^2 t\right)\sqrt{13}\, dt = 18\pi\sqrt{13}$

5. (a) $\ln f = \frac{1}{2}\ln(x^2 + y^2 + z^2)$

$\nabla(\ln f) = \frac{x}{x^2 + y^2 + z^2}\mathbf{i} + \frac{y}{x^2 + y^2 + z^2}\mathbf{j} + \frac{z}{x^2 + y^2 + z^2}\mathbf{k} = \frac{x\mathbf{i} + y\mathbf{j} + z\mathbf{k}}{x^2 + y^2 + z^2} = \frac{\mathbf{F}}{f^2}$

(b) $\dfrac{1}{f} = \dfrac{1}{\sqrt{x^2 + y^2 + z^2}}$

$\nabla\left(\frac{1}{f}\right) = \frac{-x}{(x^2 + y^2 + z^2)^{3/2}}\mathbf{i} + \frac{-y}{(x^2 + y^2 + z^2)^{3/2}}\mathbf{j} + \frac{-z}{(x^2 + y^2 + z^2)^{3/2}}\mathbf{k} = \frac{-(x\mathbf{i} + y\mathbf{j} + z\mathbf{k})}{\left(\sqrt{x^2 + y^2 + z^2}\right)^3} = \frac{\mathbf{F}}{f^3}$

(c) $f^n = \left(\sqrt{x^2 + y^2 + z^2}\right)^n$

$\nabla f^n = n\left(\sqrt{x^2 + y^2 + z^2}\right)^{n-1}\frac{x}{\sqrt{x^2 + y^2 + z^2}}\mathbf{i} + n\left(\sqrt{x^2 + y^2 + z^2}\right)^{n-1}\frac{y}{\sqrt{x^2 + y^2 + z^2}}\mathbf{j}$

$\qquad + n\left(\sqrt{x^2 + y^2 + z^2}\right)^{n-1}\frac{z}{\sqrt{x^2 + y^2 + z^2}}\mathbf{k}$

$\qquad = n\left(\sqrt{x^2 + y^2 + z^2}\right)^{n-2}(x\mathbf{i} + y\mathbf{j} + z\mathbf{k})$

$\qquad = nf^{n-2}\mathbf{F}$

(d) $w = \dfrac{1}{f} = \dfrac{1}{\sqrt{x^2 + y^2 + z^2}}$ $\qquad \dfrac{\partial^2 w}{dx^2} = \dfrac{2x^2 - y^2 - z^2}{(x^2 + y^2 + z^2)^{5/2}}$

$\dfrac{\partial w}{dx} = -\dfrac{x}{(x^2 + y^2 + z^2)^{3/2}}$ $\qquad \dfrac{\partial^2 w}{dy^2} = \dfrac{2y^2 - x^2 - z^2}{(x^2 + y^2 + z^2)^{5/2}}$

$\dfrac{\partial w}{dy} = -\dfrac{y}{(x^2 + y^2 + z^2)^{3/2}}$ $\qquad \dfrac{\partial^2 w}{dz^2} = \dfrac{2z^2 - x^2 - y^2}{(x^2 + y^2 + z^2)^{5/2}}$

$\dfrac{\partial w}{dz} = -\dfrac{z}{(x^2 + y^2 + z^2)^{3/2}}$ $\qquad \nabla^2 w = \dfrac{\partial^2 w}{dx^2} + \dfrac{\partial^2 w}{dy^2} + \dfrac{\partial^2 w}{dz^2} = 0$

Therefore $w = \dfrac{1}{f}$ is harmonic.

7. $\frac{1}{2}\displaystyle\int_C x\,dy - y\,dx = \frac{1}{2}\int_0^{2\pi}\left[a(\theta - \sin\theta)(a\sin\theta)\,d\theta - a(1 - \cos\theta)(a(1 - \cos\theta))\,d\theta\right]$

$\qquad = \frac{1}{2}a^2\int_0^{2\pi}\left[\theta\sin\theta = \sin^2\theta - 1 + 2\cos\theta - \cos^2\theta\right]d\theta$

$\qquad = \frac{1}{2}a^2\int_0^{2\pi}(\theta\sin\theta + 2\cos\theta - 2)\,d\theta$

$\qquad = -3\pi a^2$

So, the area is $3\pi a^2$.

9. (a) $\mathbf{r}(t) = t\,\mathbf{j}, 0 \le t \le 1$

   $\mathbf{r}'(t) = \mathbf{j}$

   $W = \int_C \mathbf{F} \cdot d\mathbf{r} = \int_0^1 (t\,\mathbf{i} + \mathbf{j}) \cdot \mathbf{j}\,dt = \int_0^1 dt = 1$

(b) $\mathbf{r}(t) = (t - t^2)\mathbf{i} + t\,\mathbf{j}, 0 \le t \le 1$

   $\mathbf{r}'(t) = (1 - 2t)\mathbf{i} + \mathbf{j}$

   $W = \mathbf{F} \cdot d\mathbf{r} = \int_0^1 \left((2t - t^2)\mathbf{i} + \left[(t - t^2)^2 + 1 \right]\mathbf{j} \right) \cdot ((1 - 2t)\mathbf{i} + \mathbf{j})\,dt$

   $\quad = \int_0^1 \left[(1 - 2t)(2t - t^2) + (t^4 - 2t^3 + t^2 + 1) \right] dt$

   $\quad = \int_0^1 (t^4 - 4t^2 + 2t + 1)\,dt$

   $\quad = \dfrac{13}{15}$

(c) $\mathbf{r}(t) = c(t - t^2)\mathbf{i} + t\,\mathbf{j}, 0 \le t \le 1$

   $\mathbf{r}'(t) = c(1 - 2t)\mathbf{i} + \mathbf{j}$

   $\mathbf{F} \cdot d\mathbf{r} = \left(c(t - t^2) + t \right)\left(c(1 - 2t) \right) + \left(c^2(t - t^2)^2 + 1 \right)(1)$

   $\quad = c^2 t^4 - 2c^2 t^2 + c^2 t - 2ct^2 + ct + 1$

   $W = \int_C \mathbf{F} \cdot d\mathbf{r} = \dfrac{1}{30}c^2 - \dfrac{1}{6}c + 1$

   $\dfrac{dW}{dc} = \dfrac{1}{15}c - \dfrac{1}{6} = 0 \Rightarrow c = \dfrac{5}{2}$

   $\dfrac{d^2W}{dc^2} = \dfrac{1}{15} > 0 \quad c = \dfrac{5}{2}$ minimum.

11. Area $= \pi ab$

   $\mathbf{r}(t) = a \cos t\,\mathbf{i} + b \sin t\,\mathbf{j}, 0 \le t \le 2\pi$

   $\mathbf{r}'(t) = -a \sin t\,\mathbf{i} + b \cos t\,\mathbf{j}$

   $\mathbf{F} = -\dfrac{1}{2}b \sin t\,\mathbf{i} + \dfrac{1}{2}a \cos t\,\mathbf{j}$

   $\mathbf{F} \cdot d\mathbf{r} = \left[\dfrac{1}{2}ab \sin^2 t + \dfrac{1}{2}ab \cos^2 t \right] dt = \dfrac{1}{2}ab$

   $W = \int_0^{2\pi} \mathbf{F} \cdot d\mathbf{r} = \dfrac{1}{2}ab(2\pi) = \pi ab$

   Same as area.

CHAPTER 16
Additional Topics in Differential Equations

C H A P T E R   16
Additional Topics in Differential Equations

Section 16.1 Exact First-Order Equations

1. The equation $M(x, y)\,dx + N(x, y)\,dy = 0$ is exact if there exists a function $f(x, y)$ such that $f_x(x, y) = M(x, y)$ and $f_y = N(x, y)$. f must have continuous first partial derivatives.

To test for exactness, determine whether $M_y = N_x$.

3. $(2x + xy^2)\,dx + (3 + x^2y)\,dy = 0$

$$\frac{\partial M}{\partial y} = 2xy$$

$$\frac{\partial N}{\partial x} = 2xy$$

$$\frac{\partial M}{\partial y} = \frac{\partial N}{\partial x} \quad \text{Exact}$$

5. $x \sin y\,dx + x \cos y\,dy = 0$

$$\frac{\partial M}{\partial y} = x \cos y$$

$$\frac{\partial N}{\partial x} = \cos y$$

$$\frac{\partial M}{\partial y} \neq \frac{\partial N}{\partial x} \quad \text{Not exact}$$

7. $(2x - 3y)\,dx + (2y - 3x)\,dy = 0$

$$\frac{\partial M}{\partial y} = -3 = \frac{\partial N}{\partial x} \quad \text{Exact}$$

$$f(x, y) = \int M(x, y)\,dx$$

$$= \int (2x - 3y)\,dx$$

$$= x^2 - 3xy + g(y)$$

$$f_y(x, y) = -3x + g'(y)$$

$$= 2y - 3x \Rightarrow g'(y) = 2y$$

$$\Rightarrow g(y) = y^2 + C_1$$

$$f(x, y) = x^2 - 3xy + y^2 + C_1$$

$$x^2 - 3xy + y^2 = C$$

9. $(3y^2 + 10xy^2)\,dx + (6xy - 2 + 10x^2y) = 0$

$$\frac{\partial M}{\partial y} = 6y + 20xy = \frac{\partial N}{\partial x} \quad \text{Exact}$$

$$f(x, y) = \int M(x, y)\,dx = \int (3y^2 + 10xy^2)\,dx$$

$$= 3xy^2 + 5x^2y^2 + g(y)$$

$$f_y(x, y) = 6xy + 10x^2y + g'(y) = 6xy - 2 + 10x^2y$$

$$\Rightarrow g'(y) = -2 \Rightarrow g(y) = -2y + C_1$$

$$f(x, y) = 3xy^2 + 5x^2y^2 - 2y + C_1$$

$$3xy^2 + 5x^2y^2 - 2y = C$$

11. $\dfrac{-y}{x^2 + y^2}\,dx + \dfrac{x}{x^2 + y^2}\,dy = 0$

$$\frac{\partial M}{\partial y} = \frac{y^2 - x^2}{(x^2 + y^2)^2} = \frac{\partial N}{\partial x} \quad \text{Exact}$$

$$f(x, y) = \int M(x, y)\,dx = -\arctan\left(\frac{x}{y}\right) + g(y)$$

$$f_y(x, y) = \frac{x}{x^2 + y^2} + g'(y)$$

$$= \frac{x}{x^2 + y^2} \Rightarrow g'(y) = 0 \Rightarrow g(y) = C_1$$

$$f(x, y) = -\arctan\left(\frac{x}{y}\right) + C_1$$

$$\arctan\left(\frac{x}{y}\right) = C$$

13. $\dfrac{x}{y^2}\,dx - \dfrac{x^2}{y^3}\,dy = 0$

$$\frac{\partial M}{\partial y} = -2xy^{-3} = \frac{\partial N}{\partial x} \quad \text{Exact}$$

$$f(x, y) = \int N(x, y)\,dy = \int -x^2y^{-3}\,dy = \frac{x^2}{2y^2} + g(x)$$

$$f_x(x, y) = \frac{x}{y^2} + g'(x) \Rightarrow g'(x) = 0$$

$$f(x, y) = \frac{x^2}{2y^2} + C_1$$

$$\frac{x^2}{2y^2} = C$$

15. (a) and (c)

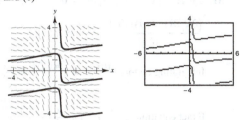

(b) $(2x \tan y + 5) \, dx + (x^2 \sec^2 y) \, dy = 0, \, y\left(\frac{1}{2}\right) = \frac{\pi}{4}$

$\dfrac{\partial M}{\partial y} = 2x \sec^2 y = \dfrac{\partial N}{\partial x} \Rightarrow$ Exact

$f(x, y) = \int M(x, y) \, dx = \int (2x \tan y + 5) \, dx = x^2 \tan y + 5x + g(y)$

$f_y(x, y) = x^2 \sec^2 y + g'(y) = x^2 \sec^2 y \Rightarrow g'(y) = 0 \Rightarrow g(y) = C$

$f(x, y) = x^2 \tan y + 5x = C$

$f\left(\dfrac{1}{2}, \dfrac{\pi}{4}\right) = \dfrac{1}{4} + \dfrac{5}{2} = \dfrac{11}{4} = C$

Answer: $x^2 \tan y + 5x = \dfrac{11}{4}$

17. $(2xy - 9x^2) \, dx + (2y + x^2 + 1) \, dy = 0$

$\dfrac{\partial M}{\partial y} = 2x = \dfrac{\partial N}{\partial x}$ Exact

$f(x, y) = \int M(x, y) \, dx = \int (2xy - 9x^2) \, dx = x^2 y - 3x^3 + g(y)$

$f_y(x, y) = x^2 + g'(y) = 2y + x^2 + 1 \Rightarrow g'(y) = 2y + 1 \Rightarrow g(y) = y^2 + y + C_1$

$f(x, y) = x^2 y - 3x^3 + y^2 + y + C_1$

$x^2 y - 3x^3 + y^2 + y = C$

$y(0) = -3 \colon 9 - 3 = 6 = C$

Solution: $x^2 y - 3x^3 + y^2 + y = 6$

19. $(e^{3x} \sin 3y) \, dx + (e^{3x} \cos 3y) \, dy = 0$

$\dfrac{\partial M}{\partial y} = 3e^{3x} \cos 3y = \dfrac{\partial N}{\partial x}$ Exact

$f(x, y) = \int M(x, y) \, dx$

$\quad = \int e^{3x} \sin 3y \, dx = \dfrac{1}{3} e^{3x} \sin 3y + g(y)$

$f_y(x, y) = e^{3x} \cos 3y + g'(y)$

$\quad \Rightarrow g'(y) = 0 \Rightarrow g(y) = C_1$

$f(x, y) = \dfrac{1}{3} e^{3x} \sin 3y + C_1$

$e^{3x} \sin 3y = C$

$y(0) = \pi \colon C = 0$

Solution: $e^{3x} \sin 3y = 0$

21. $\dfrac{y}{x - 1} \, dx + \left[\ln(x - 1) + 2y\right] dy = 0$

$\dfrac{\partial M}{\partial y} = \dfrac{1}{x - 1} = \dfrac{\partial N}{\partial x}$ Exact

$f(x, y) = \int M(x, y) \, dx = y \ln(x - 1) + g(y)$

$f_y(x, y) = \ln(x - 1) + g'(y)$

$\quad \Rightarrow g'(y) = 2y \Rightarrow g(y) = y^2 + C_1$

$f(x, y) = y \ln(x - 1) + y^2 + C_1$

$y \ln(x - 1) + y^2 = C$

$y(2) = 4 \colon 4 \ln(2 - 1) + 16 = C \Rightarrow C = 16$

Solution: $y \ln(x - 1) + y^2 = 16$

23. $y^2\,dx + 5xy\,dy = 0$

$\dfrac{(\partial N/\partial x) - (\partial M/\partial y)}{M} = \dfrac{5y - 2y}{y^2} = \dfrac{3}{y}$

Integrating factor: $e^{\int k(y)\,dy} = e^{3\ln y} = y^3$

Exact equation: $y^5 + 5xy^4 = 0$

$f(x, y) = xy^5 + g(y)$

$f_y(x, y) = 5xy^4 + g'(y) \Rightarrow g'(y) = C_1$

$xy^5 = C$

25. $y\,dx - \left(x + 6y^2\right) dy = 0$

$\dfrac{(\partial N/\partial x) - (\partial M/\partial y)}{M} = -\dfrac{2}{y} = k(y)$

Integrating factor: $e^{\int k(y)\,dy} = e^{\ln y^{-2}} = \dfrac{1}{y^2}$

Exact equation: $\dfrac{1}{y}\,dx - \left(\dfrac{x}{y^2} + 6\right) dy = 0$

$f(x, y) = \dfrac{x}{y} + g(y)$

$g'(y) = -6$

$g(y) = -6y + C_1$

$\dfrac{x}{y} - 6y = C$

27. $(x + y)\,dx + (\tan x)\,dy = 0$

$\dfrac{(\partial M/\partial y) - (\partial N/\partial x)}{N} = -\tan x = h(x)$

Integrating factor: $e^{\int h(x)\,dx} = e^{\ln \cos x} = \cos x$

Exact equation: $(x + y)\cos x\,dx + \sin x\,dy = 0$

$f(x, y) = x\sin x + \cos x + y\sin x + g(y)$

$g'(y) = 0$

$g(y) = C_1$

$x\sin x + \cos x + y\sin x = C$

29. $y^2\,dx + (xy - 1)\,dy = 0$

$\dfrac{(\partial N/\partial x) - (\partial M/\partial y)}{M} = -\dfrac{1}{y} = k(y)$

Integrating factor: $e^{\int k(y)\,dy} = e^{\ln(1/y)} = \dfrac{1}{y}$

Exact equation: $y\,dx + \left(x - \dfrac{1}{y}\right) dy = 0$

$f(x, y) = xy + g(y)$

$g'(y) = -\dfrac{1}{y}$

$g(y) = -\ln|y| + C_1$

$xy - \ln|y| = C$

31. $2y\,dx + \left(x - \sin\sqrt{y}\right) dy = 0$

$\dfrac{(\partial N/\partial x) - (\partial M/\partial y)}{M} = \dfrac{-1}{2y} = k(y)$

Integrating factor: $e^{\int k(y)\,dy} = e^{\ln(1/\sqrt{y})} = \dfrac{1}{\sqrt{y}}$

Exact equation: $2\sqrt{y}\,dy + \left(\dfrac{x}{\sqrt{y}} - \dfrac{\sin\sqrt{y}}{\sqrt{y}}\right) dy = 0$

$f(x, y) = 2\sqrt{y}\,x + g(y)$

$g'(y) = -\dfrac{\sin\sqrt{y}}{\sqrt{y}}$

$g(y) = 2\cos\sqrt{y} + C_1$

$\sqrt{y}\,x + \cos\sqrt{y} = C$

33. $\left(4x^2 y + 2y^2\right) dx + \left(3x^3 + 4xy\right) dy = 0$

Integrating factor: xy^2

Exact equation:

$\left(4x^3 y^3 + 2xy^4\right) dy + \left(3x^4 y^2 + 4x^2 y^3\right) dy = 0$

$f(x, y) = x^4 y^3 + x^2 y^4 + g(y)$

$g'(y) = 0$

$g(y) = C_1$

$x^4 y^3 + x^2 y^4 = C$

35. $\left(-y^5 + x^2 y\right) dx + \left(2xy^4 - 2x^3\right) dy = 0$

Integrating factor: $x^{-2} y^{-3}$

Exact equation:

$\left(-\dfrac{y^2}{x^2} + \dfrac{1}{y^2}\right) dx + \left(2\dfrac{y}{x} - 2\dfrac{x}{y^3}\right) dy = 0$

$f(x, y) = \dfrac{y^2}{x} + \dfrac{x}{y^2} + g(y)$

$g'(y) = 0$

$g(y) = C_1$

$\dfrac{y^2}{x} + \dfrac{x}{y^2} = C$

37. $y\,dx - x\,dy = 0$

(a) $\dfrac{1}{x^2}, \dfrac{y}{x^2}\,dx - \dfrac{1}{x}\,dy = 0, \dfrac{\partial M}{\partial y} = \dfrac{1}{x^2} = \dfrac{\partial N}{\partial x}$

(b) $\dfrac{1}{y^2}, \dfrac{1}{y}\,dx - \dfrac{x}{y^2}\,dy = 0, \dfrac{\partial M}{\partial y} = \dfrac{-1}{y^2} = \dfrac{\partial N}{\partial x}$

(c) $\dfrac{1}{xy}, \dfrac{1}{x}\,dx - \dfrac{1}{y}\,dy = 0, \dfrac{\partial M}{\partial y} = 0 = \dfrac{\partial N}{\partial x}$

(d) $\dfrac{1}{x^2 + y^2}, \dfrac{y}{x^2 + y^2}\,dx - \dfrac{x}{x^2 + y^2}\,dy = 0,$

$\dfrac{\partial M}{\partial y} = \dfrac{x^2 - y^2}{\left(x^2 + y^2\right)^2} = \dfrac{\partial N}{\partial x}$

39. $\mathbf{F}(x, y) = \dfrac{y}{\sqrt{x^2 + y^2}}\mathbf{i} - \dfrac{x}{\sqrt{x^2 + y^2}}\mathbf{j}$

$\dfrac{dy}{dx} = -\dfrac{x}{y}$

$y\,dy + x\,dx = 0$

$y^2 + x^2 = C$

Family of circles

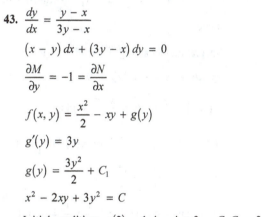

41. $\mathbf{F}(x, y) = 4x^2 y\mathbf{i} - \left(2xy^2 + \dfrac{x}{y^2}\right)\mathbf{j}$

$\dfrac{dy}{dx} = \dfrac{-y}{2x} - \dfrac{1}{4xy^3}$

$\dfrac{8y^3}{2y^4 + 1}\,dy = -\dfrac{2}{x}\,dx$

$\ln\left(2y^4 + 1\right) = \ln\left(\dfrac{1}{x^2}\right) + \ln C$

$2y^4 + 1 = \dfrac{C}{x^2}$

$2x^2 y^4 + x^2 = C$

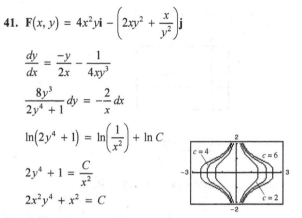

43. $\dfrac{dy}{dx} = \dfrac{y - x}{3y - x}$

$(x - y)\,dx + (3y - x)\,dy = 0$

$\dfrac{\partial M}{\partial y} = -1 = \dfrac{\partial N}{\partial x}$

$f(x, y) = \dfrac{x^2}{2} - xy + g(y)$

$g'(y) = 3y$

$g(y) = \dfrac{3y^2}{2} + C_1$

$x^2 - 2xy + 3y^2 = C$

Initial condition: $y(2) = 1, 4 - 4 + 3 = C, C = 3$

Particular solution: $x^2 - 2xy + 3y^2 = 3$

45. $E(x) = \dfrac{20x - y}{2y - 10x} = \dfrac{x\,dy}{y\,dx}$

$\left(20xy - y^2\right)dx + \left(10x^2 - 2xy\right)dy = 0$

$\dfrac{\partial M}{\partial y} = 20x - 2y = \dfrac{\partial N}{\partial x}$

$f(x, y) = 10x^2 y - xy^2 + g(y)$

$g'(y) = 0$

$g(y) = C_1$

$10x^2 y - xy^2 = K$

Initial condition: $C(100) = 500, 100 \le x, K = 25{,}000{,}000$

$$10x^2 y - xy^2 = 25{,}000{,}000$$

$xy^2 - 10x^2 y + 25{,}000{,}000 = 0$ Quadratic Formula

$$y = \dfrac{10x^2 + \sqrt{100x^4 - 4x(25{,}000{,}000)}}{2x} = \dfrac{5\left(x^2 + \sqrt{x^4 - 1{,}000{,}000x}\right)}{x}$$

47. (a) $y(4) \approx 0.5231$

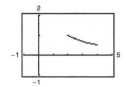

(b) $\dfrac{dy}{dx} = \dfrac{-xy}{x^2 + y^2}$

$xy\, dx + \left(x^2 + y^2\right) dy = 0$

$\dfrac{1}{M}\left[N_x - M_y\right] = \dfrac{1}{xy}[2x - x] = \dfrac{1}{y}$ function of y alone.

Integrating factor: $e^{\int (1/y)\, dy} = e^{\ln y} = y$

$xy^2\, dx + \left(x^2 y + y^3\right) dy = 0$

$f(x, y) = \int xy^2\, dx = \dfrac{x^2 y^2}{2} + g(y)$

$f_y(x, y) = x^2 y + g'(y) \Rightarrow g(y) = \dfrac{y^4}{4} + C_1$

$f(x, y) = \dfrac{x^2 y^2}{2} + \dfrac{y^4}{4} = C$

Initial condition: $y(2) = 1, \dfrac{4}{2} + \dfrac{1}{4} = \dfrac{9}{4} = C$

Particular solution: $\dfrac{x^2 y^2}{2} + \dfrac{y^4}{4} = \dfrac{9}{4}$ or

$2x^2 y^2 + y^4 = 9.$

For $x = 4$, $32y^2 + y^4 = 9 \Rightarrow y(4) = 0.528$

(c)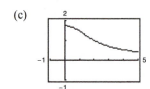

49. (a) $y(4) \approx 0.408$

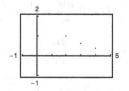

(b) $\dfrac{dy}{dx} = \dfrac{-xy}{x^2 + y^2}$

$xy\, dx + \left(x^2 + y^2\right) dy = 0$

$\dfrac{1}{M}\left[N_x - M_y\right] = \dfrac{1}{xy}[2x - x]$

$= \dfrac{1}{y}$ function of y alone.

Integrating factor: $e^{\int 1/y\, dy} = e^{\ln y} = y$

$xy^2\, dx + \left(x^2 y + y^3\right) dy = 0$

$f(x, y) = \int xy^2\, dx = \dfrac{x^2 y^2}{2} + g(y)$

$f_y(x, y) = x^2 y + g'(y) \Rightarrow g(y) = \dfrac{y^4}{4} + C_1$

$f(x, y) = \dfrac{x^2 y^2}{2} + \dfrac{y^4}{4} = C$

Initial condition: $y(2) = 1, \dfrac{4}{2} + \dfrac{1}{4} = \dfrac{9}{4} = C$

Particular solution: $\dfrac{x^2 y^2}{2} + \dfrac{y^4}{4} = \dfrac{9}{4}$ or

$2x^2 y^2 + y^4 = 9.$

For $x = 4$, $32y^2 + y^4 = 9 \Rightarrow y(4) = 0.528$

(c)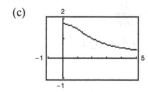

The solution is less accurate. For Exercise 47, Euler's Method gives $y(4) \approx 0.523$, whereas in Exercise 49, you obtain $y(4) \approx 0.408$. The errors are $0.528 - 0.523 = 0.005$ and $0.528 - 0.408 = 0.120$.

51. $M = xy^2 + kx^2 y + x^3$, $N = x^3 + x^2 y + y^2$

$\dfrac{\partial M}{\partial y} = 2xy + kx^2$, $\dfrac{\partial N}{\partial x} = 3x^2 + 2xy$

$\dfrac{\partial M}{\partial y} = \dfrac{\partial N}{\partial x} \Rightarrow k = 3$

53. $M = g(y) \sin x,\ N = y^2 f(x)$

$$\frac{\partial M}{\partial y} = g'(y) \sin x,\ \frac{\partial N}{\partial x} = y^2 f'(x)$$

$$\frac{\partial M}{\partial y} = \frac{\partial N}{\partial x}:\ g'(y) \sin x = f'(x) y^2$$

$$g'(y) = y^2 \Rightarrow g(y) = \frac{y^3}{3} + C_1$$

55. True

$$M(x)\ dx + N(y)\ dy = 0 \text{ is exact.}$$

57. True

$$\frac{\partial}{\partial y}\big[f(x) + M\big] = \frac{\partial M}{\partial y} \text{ and } \frac{\partial}{\partial x}\big[g(y) + N\big] = \frac{\partial N}{\partial x}$$

Section 16.2   Second-Order Homogeneous Linear Equations

1. (a) Order 5; homogenous

   (b) $y'' + 3e^x y = -2x$

   Order 2; nonhomogeneous

3. (a) $y = C_1 e^{-x} + C_2 e^{3x}$

   (b) $y = C_1 e^{2x} + C_2 x e^{2x}$

5.
$$y = C_1 e^{-3x} + C_2 x e^{-3x}$$
$$y' = -3C_1 e^{-3x} + C_2 e^{-3x} - 3C_2 x e^{-3x}$$
$$y'' = 9C_1 e^{-3x} - 6C_2 e^{-3x} + 9C_2 x e^{-3x}$$

$$y'' + 6y' + 9y = \big(9C_1 e^{-3x} - 6C_2 e^{-3x} + 9C_2 x e^{-3x}\big) + \big(-18C_1 e^{-3x} + 6C_2 e^{-3x} - 18C_2 x e^{-3x}\big) + \big(9C_1 e^{-3x} + 9C_2 x e^{-3x}\big) = 0$$

y approaches zero as $x \to \infty$.

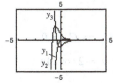

7.
$$y = C_1 \cos 2x + C_2 \sin 2x$$
$$y' = -2C_1 \sin 2x + 2C_2 \cos 2x$$
$$y'' = -4C_1 \cos 2x - 4C_2 \sin 2x = -4y$$
$$y'' + 4y = -4y + 4y = 0$$

The graphs are basically the same shape, with left and right shifts and varying ranges.

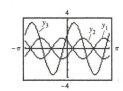

9. $y'' - y' = 0$

Characteristic equation: $m^2 - m = 0$

Roots: $m = 0, 1$

$y = C_1 + C_2 e^x$

11. $y'' - y' - 6y = 0$

Characteristic equation: $m^2 - m - 6 = 0$

Roots: $m = 3, -2$

$y = C_1 e^{3x} + C_2 e^{-2x}$

13. $2y'' + 3y' - 2y = 0$

Characteristic equation: $2m^2 + 3m - 2 = 0$

Roots: $m = \frac{1}{2}, -2$

$y = C_1 e^{(1/2)x} + C_2 e^{-2x}$

15. $y'' + 6y' + 9y = 0$

Characteristic equation: $m^2 + 6m + 9 = 0$

Roots: $m = -3, -3$

$y = C_1 e^{-3x} + C_2 x e^{-3x}$

17. $16y'' - 8y' + y = 0$

Characteristic equation: $16m^2 - 8m + 1 = 0$

Roots: $m = \frac{1}{4}, \frac{1}{4}$

$y = C_1 e^{(1/4)x} + C_2 x e^{(1/4)x}$

19. $y'' + y = 0$

Characteristic equation: $m^2 + 1 = 0$

Roots: $m = -i, i$

$y = C_1 \cos x + C_2 \sin x$

21. $4y'' - 5y = 0$

Characteristic equation: $4m^2 - 5 = 0$

Roots: $m = \pm\frac{\sqrt{5}}{2}$

$y = C_1 e^{\sqrt{5}/2x} + C_2{}^{-\sqrt{5}/2x}$

23. $y'' - 2y' + 4y = 0$

Characteristic equation: $m^2 - 2m + 4 = 0$

Roots: $m = 1 - \sqrt{3}i, 1 + \sqrt{3}i$

$y = e^x\left(C_1 \cos \sqrt{3}x + C_2 \sin \sqrt{3}x\right)$

25. $y'' - 3y' + y = 0$

Characteristic equation: $m^2 - 3m + 1 = 0$

Roots: $m = \frac{3 - \sqrt{5}}{2}, \frac{3 + \sqrt{5}}{2}$

$y = C_1 e^{\left[(3+\sqrt{5})/2\right]x} + C_2 e^{\left[3-\sqrt{5}/2\right]x}$

27. $9y'' - 12y' + 11y = 0$

Characteristic equation: $9m^2 - 12m + 11 = 0$

Roots: $m = \frac{2 + \sqrt{7}i}{3}, \frac{2 - \sqrt{7}i}{3}$

$y = e^{(2/3)x}\left[C_1 \cos\left(\frac{\sqrt{7}}{3}x\right) + C_2 \sin\left(\frac{\sqrt{7}}{3}x\right)\right]$

29. $y^{(4)} - y = 0$

Characteristic equation: $m^4 - 1 = 0$

Roots: $m = -1, 1, -i, i$

$y = C_1 e^x + C_2 e^{-x} + C_3 \cos x + C_4 \sin x$

31. $y''' - 6y'' + 11y' - 6y = 0$

Characteristic equation: $m^3 - 6m^2 + 11m - 6 = 0$

Roots: $m = 1, 2, 3$

$y = C_1 e^x + C_2 e^{2x} + C_3 e^{3x}$

33. $y''' - 3y'' + 7y' - 5y = 0$

Characteristic equation: $m^3 - 3m^2 + 7m - 5 = 0$

Roots: $m = 1, 1 - 2i, 1 + 2i$

$y = C_1 e^x + e^x\left(C_2 \cos 2x + C_3 \sin 2x\right)$

35. $y^{(4)} - 2y'' + y = 0$

Characteristic equation:

$m^4 - 2m^2 + 1 = 0 \Rightarrow \left(m^2 - 1\right)^2 = 0$

Roots: $m = 1, 1, -1, -1$

$y = C_1 e^{-x} + C_2 x e^{-x} + C_3 e^x + C_4 x e^x$

37. $y'' + 100y = 0$

$y = C_1 \cos 10x + C_2 \sin 10x$

$y' = -10C_1 \sin 10x + 10C_2 \cos 10x$

(a) $y(0) = 2$: $2 = C_1$

$y'(0) = 0$: $0 = 10C_2 \Rightarrow C_2 = 0$

Particular solution: $y = 2 \cos 10x$

(b) $y(0) = 0$: $0 = C_1$

$y'(0) = 2$: $2 = 10C_2 \Rightarrow C_2 = \frac{1}{5}$

Particular solution: $y = \frac{1}{5} \sin 10x$

(c) $y(0) = -1$: $-1 = C_1$

$y'(0) = 3$: $3 = 10C_2 \Rightarrow C_2 = \frac{3}{10}$

Particular solution: $y = -\cos 10x + \frac{3}{10} \sin 10x$

39. $y'' - y' - 30y = 0$, $y(0) = 1$, $y'(0) = -4$

Characteristic equation: $m^2 - m - 30 = 0$

Roots: $m = 6, -5$

$y = C_1 e^{6x} + C_2 e^{-5x}$, $y' = 6C_1 e^{6x} - 5C_2 e^{-5x}$

Initial conditions: $y(0) = 1$, $y'(0) = -4$, $1 = C_1 + C_2$,

$-4 = 6C_1 - 5C_2$

Solving simultaneously: $C_1 = \frac{1}{11}, C_2 = \frac{10}{11}$

Particular solution: $y = \frac{1}{11}\left(e^{6x} + 10e^{-5x}\right)$

41. $y'' + 16y = 0$, $y(0) = 0$, $y'(0) = 2$

Characteristic equation: $m^2 + 16 = 0$

Roots: $m = \pm 4i$

$y = C_1 \cos 4x + C_2 \sin 4x$

$y' = -4C_1 \sin 4x + 4C_2 \cos 4x$

Initial conditions: $y(0) = 0 = C_1$

$$y'(0) = 2 = 4C_2 \Rightarrow C_2 = \tfrac{1}{2}$$

Particular solution: $y = \tfrac{1}{2} \sin 4x$

43. $9y'' - 6y' + y = 0$, $y(0) = 2$, $y'(0) = 1$

Characteristic equation: $9m^2 - 6m + 1 = 0$

Roots: $m = \tfrac{1}{3}, \tfrac{1}{3}$

$y = C_1 e^{(1/3)x} + C_2 x e^{(1/3)x}$

$y' = \tfrac{1}{3} C_1 e^{(1/3)x} + \tfrac{1}{3} C_2 x e^{(1/3)x} + C_2 e^{(1/3)x}$

Initial conditions: $y(0) = 2$, $y'(0) = 1$

$\left.\begin{array}{l} C_1 = 2 \\ \tfrac{1}{3} C_1 + C_2 = 1 \end{array}\right\} \Rightarrow C_1 = 2, C_2 = \tfrac{1}{3}$

Particular solution: $y = 2e^{x/3} + \tfrac{1}{3} x e^{x/3}$

45. $y'' - 4y' + 3y = 0$, $y(0) = 1$, $y(1) = 3$

Characteristic equation: $m^2 - 4m + 3 = 0$

Roots: $m = 1, 3$

$y = C_1 e^x + C_2 e^{3x}$

$y(0) = 1$: $C_1 + C_2 = 1$

$y(1) = 3$: $C_1 e + C_2 e^3 = 3$

Solving simultaneously, $C_1 = \dfrac{e^3 - 3}{e^3 - e}, C_2 = \dfrac{3 - e}{e^3 - e}$

Solution: $y = \dfrac{e^3 - 3}{e^3 - e} e^x + \dfrac{3 - e}{e^3 - e} e^{3x}$

47. $y'' + 9y = 0$, $y(0) = 3$, $y\left(\dfrac{\pi}{2}\right) = 4$

Characteristic equation: $m^2 + 9m = 0$

Roots: $m = \pm 3i$

$y = C_1 \cos 3x + C_2 \sin 3x$

$y(0) = 3$: $C_1 = 3$

$y\left(\dfrac{\pi}{2}\right) = 4$: $3 \cos \dfrac{3\pi}{2} + C_2 \sin \dfrac{3\pi}{2} = C_2 \Rightarrow C_2 = -4$

$y = 3 \cos 3x - 4 \sin 3x$

49. $4y'' - 28y' + 49y = 0$, $y(0) = 2$, $y(1) = -1$

Characteristic equation: $4m^2 - 28m + 49 = 0$

Roots: $m = \dfrac{7}{2}, \dfrac{7}{2}$

$y = C_1 e^{(7/2)x} + C_2 x e^{(7/2)x}$

$y(0) = 2$: $C_1 = 2$

$y(1) = -1$: $C_1 e^{7/2} + C_2 e^{7/2} = -1 \Rightarrow C_2 = \dfrac{-1 - 2e^{7/2}}{e^{7/2}}$

Solution: $y = 2e^{(7/2)x} + \left(\dfrac{-1 - 2e^{7/2}}{e^{7/2}}\right) x e^{(7/2)x}$

51. $y = C_1 \sinh x + C_2 \cosh x$

$y' = C_1 \cosh x + C_2 \sinh x$

$y'' = C_1 \sinh x + C_2 \cosh x$

$y'' - y = 0$

$y = C_1 \left(\dfrac{e^x - e^{-x}}{2}\right) + C_2 \left(\dfrac{e^x + e^{-x}}{2}\right)$

$ = C_3 e^x + C_4 e^{-x}$

53. The motion of a spring in a shock absorber is damped.

55. $y'' + 9y = 0$

Undamped vibration

Period: $\dfrac{2\pi}{3}$

Matches (b)

56. $y'' + 25y = 0$

Undamped vibration

Period: $\dfrac{2\pi}{5}$

Matches (d)

57. $y'' + 2y' + 10y = 0$

Damped vibration

Matches (c)

58. $y'' + y' + \tfrac{37}{4}y = 0$

Damped vibration

Matches (a)

59. By Hooke's Law, $F = kx$

$$k = \dfrac{F}{x} = \dfrac{32}{2/3} = 48.$$

Also, $F = ma$, and $m = \dfrac{F}{a} = \dfrac{32}{32} = 1.$

So, $y = \dfrac{1}{2} \cos\left(4\sqrt{3}t\right)$

61. $y = C_1 \cos\left(\sqrt{k/m}\,t\right) + C_2 \sin\left(\sqrt{k/m}\,t\right)$, $\sqrt{k/m} = \sqrt{48} = 4\sqrt{3}$

Initial conditions: $y(0) = -\dfrac{2}{3}$, $y'(0) = \dfrac{1}{2}$

$y = C_1 \cos\left(4\sqrt{3}t\right) + C_2 \sin\left(4\sqrt{3}t\right)$

$y(0) = C_1 = -\dfrac{2}{3}$

$y'(t) = -4\sqrt{3}\,C_1 \sin\left(4\sqrt{3}t\right) + 4\sqrt{3}\,C_2 \cos\left(4\sqrt{3}t\right)$

$y'(0) = 4\sqrt{3}\,C_2 = \dfrac{1}{2} \Rightarrow C_2 = \dfrac{1}{8\sqrt{3}} = \dfrac{\sqrt{3}}{24}$

$y(t) = -\dfrac{2}{3}\cos\left(4\sqrt{3}t\right) + \dfrac{\sqrt{3}}{24}\sin\left(4\sqrt{3}t\right)$

63. By Hooke's Law, $32 = k(2/3)$, so $k = 48$. Moreover, because the weight w is given by mg, it follows that $m = w/g = 32/32 = 1$. Also, the damping force is given by $(-1/8)(dy/dt)$. So, the differential equation for the oscillations of the weight is

$$m\left(\frac{d^2y}{dt^2}\right) = -\frac{1}{8}\left(\frac{dy}{dt}\right) - 48y$$

$$m\left(\frac{d^2y}{dt^2}\right) + \frac{1}{8}\left(\frac{dy}{dt}\right) + 48y = 0.$$

In this case the characteristic equation is $8m^2 + m + 384 = 0$ with complex roots $m = (-1/16) \pm \left(\sqrt{12{,}287}/16\right)i$.

So, the general solution is $y(t) = e^{-t/16}\left(C_1 \cos\dfrac{\sqrt{12{,}287}t}{16} + C_2 \sin\dfrac{\sqrt{12{,}287}t}{16}\right)$.

Using the initial conditions, you have $y(0) = C_1 = \dfrac{1}{2}$

$$y'(t) = e^{-t/16}\left[\left(-\frac{\sqrt{12{,}287}}{16}C_1 - \frac{C_2}{16}\right)\sin\frac{\sqrt{12{,}287}t}{16} + \left(\frac{\sqrt{12{,}287}}{16}C_2 - \frac{C_1}{16}\right)\cos\frac{\sqrt{12{,}287}t}{16}\right]$$

$$y'(0) = \frac{\sqrt{12{,}287}}{16}C_2 - \frac{C_1}{16} = 0 \Rightarrow C_2 = \frac{\sqrt{12{,}287}}{24{,}574}$$

and the particular solution is

$$y(t) = \frac{e^{-t/16}}{2}\left(\cos\frac{\sqrt{12{,}287}t}{16} + \frac{\sqrt{12{,}287}}{12{,}287}\sin\frac{\sqrt{12{,}287}t}{16}\right).$$

65. Because $m = -a/2$ is a double root of the characteristic equation, you have

$$\left(m + \frac{a}{2}\right)^2 = m^2 + am + \frac{a^2}{4} = 0$$

and the differential equation is $y'' + ay' + \left(a^2/4\right)y = 0$. The solution is

$$y = (C_1 + C_2 x)e^{-(a/2)x}$$

$$y' = \left(-\frac{C_1 a}{2} + C_2 - \frac{C_2 a}{2}x\right)e^{-(a/2)x}$$

$$y'' = \left(\frac{C_1 a^2}{4} - aC_2 + \frac{C_2 a^2}{4}x\right)e^{-(a/2)x}$$

$$y'' + ay' + \frac{a^2}{4}y = e^{-(a/2)x}\left[\left(\frac{C_1 a^2}{4} - C_2 a + \frac{C_2 a^2}{4}x\right) + \left(-\frac{C_1 a^2}{2} + C_2 a - \frac{C_2 a^2}{2}x\right) + \left(\frac{C_1 a^2}{4} + \frac{C_2 a^2}{4}x\right)\right] = 0.$$

67. True

69. True

71. $y_1 = e^{ax}$, $y_2 = e^{bx}$, $a \neq b$

$$W(y_1, y_2) = \begin{vmatrix} e^{ax} & e^{bx} \\ ae^{ax} & be^{bx} \end{vmatrix} = (b - a)e^{ax+bx} \neq 0 \text{ for any value of } x.$$

73. $y_1 = e^{ax}\sin bx$, $y_2 = e^{ax}\cos bx$, $b \neq 0$

$$W(y_1, y_2) = \begin{vmatrix} e^{ax}\sin bx & e^{ax}\cos bx \\ ae^{ax}\sin bx + be^{ax}\cos bx & ae^{ax}\cos bx - be^{ax}\sin bx \end{vmatrix}$$

$$= -be^{2ax}\sin^2 bx - be^{2ax}\cos^2 bx = -be^{2ax} \neq 0 \text{ for any value of } x.$$

Section 16.3   Second-Order Nonhomogeneous Linear Equations

1. The general form is the sum of the general solution of the corresponding homogeneous equation and a particular solution of the nonhomogeneous equation: $y = y_h + y_p$.

3. $y'' + 7y' + 12y = 3x + 1$

$y'' + 7y' + 12y = 0$

$m^2 - 7m + 12 = (m - 3)(m - 4) = 0$ when $m = 3, 4$

$y_h = C_1 e^{3x} + C_2 e^{4x}$

$y_p = A_0 + A_1 x$

$y_p' = A_1, \; y_p'' = 0$

$y_p'' + 7y_p' + 12y_p = 7A_1 + 12(A_0 + A_1 x) = 3x + 1$

$\left. \begin{array}{l} 12A_1 = 3 \\ 7A_1 + 12A_0 = 1 \end{array} \right\} \Rightarrow A_1 = \frac{1}{4}, A_0 = -\frac{1}{16}$

$y_p = -\frac{1}{16} + \frac{1}{4}x$

Solution: $y = y_h + y_p = C_1 e^{-3x} + C_2 e^{-4x} - \dfrac{1}{16} + \dfrac{1}{4}x$

5. $y'' - 8y' + 16y = e^{3x}$

$y'' - 8y' + 16y = 0$

$m^2 - 8m + 16 = (m - 4)^2 = 0$ when $m = 4$

$y_h = C_1 e^{4x} + C_2 x e^{4x}$

$y_p = Ae^{3x}, \; y_p' = 3Ae^{3x}, \; y_p'' = 9Ae^{3x}$

$y_p'' - 8y_p' + 16y_p = 9Ae^{3x} - 8(3Ae^{3x}) + 16(Ae^{3x})$

$\qquad\qquad = e^{3x}$

$9A - 24A + 16A = 1 \Rightarrow A = 1$

$y_p = e^{3x}$

Solution: $y = y_h + y_p = C_1 e^{4x} + C_2 x e^{4x} + e^{3x}$

7. $y'' + y' = 4x + 6$

The initial guess of $y_p = A + Bx$ must be modified because y_h already has a constant term.

Use $y_p = Ax + Bx^2$.

9. $3y'' + 6y' = 4 + \sin x$

The initial guess of $y_p = A + B \sin x + C \cos x$ must be modified because y_h already has a constant term.

Use $y_p = Ax + B \sin x + C \cos x$.

11. $y'' + 2y' = e^{-2x}$

$y'' + 2y' = 0$

$m^2 + 2m = m(m + 2) = 0 \Rightarrow m = 0, -2$

$y_h = C_1 + C_2 e^{-2x}$

$y_p = Axe^{-2x}, \ y_p' = Ae^{-2x} - 2Axe^{-2x}$

$y_p'' = -2Ae^{-2x} - 2Ae^{-2x} + 4Axe^{-2x} = -4Ae^{-2x} + 4Axe^{-2x}$

$y_p' + 2y_p' = \left(-4Ae^{-2x} + 4Axe^{-2x}\right) + 2\left(Ae^{-2x} - 2Axe^{-2x}\right) = e^{-2x}$

$\Rightarrow -2Ae^{-2x} = e^{-2x} \Rightarrow A = -\frac{1}{2}$

$y_p = -\frac{1}{2}xe^{-2x}$

$y = y_h + y_p = C_1 + C_2 e^{-2x} - \frac{1}{2}xe^{-2x}$

13. $y'' + 9y = \sin 3x$

$y'' + 9y = 0$

$m^2 + 9 = 0$ when $m = -3i, 3i$.

$y_h = C_1 \cos 3x + C_2 \sin 3x$

$y_p = Ax \cos 3x + Bx \sin 3x$

$y_p' = -3Ax \sin 3x + A \cos 3x + 3Bx \cos 3x + B \sin 3x$

$y_p'' = -9Ax \cos 3x - 3A \sin 3x - 3A \sin 3x - 9B \sin 3x + 3B \cos 3x + 3B \cos 3x$

$y_p'' + 9y_p = -6A \sin 3x + 6B \cos 3x = \sin 3x$

$A = -\frac{1}{6}, \ B = 0$

$y = y_h + y_p = \left(C_1 - \frac{1}{6}x\right) \cos 3x + C_2 \sin 3x$

15. $y''' - 3y'' + 4y = 2 + e^{2x}$

$y''' - 3y'' + 4y = 0$

$m^3 - 3m^2 + 4 = (m + 1)(m - 2)^2 \Rightarrow m = -1, 2, 2$

$y_h = C_1 e^{-x} + C_2 e^{2x} + C_3 x e^{2x}$

$y_p = A + Bx^2 e^{2x}$

$y_p' = B(2x^2 + 2x)e^{2x}$

$y_p'' = B(4x^2 + 8x + 2)e^{2x}$

$y_p''' = B(8x^2 + 24x + 12)e^{2x}$

$y_p''' - 3y_p'' + 4y = B(8x^2 + 24x + 12)e^{2x} - 3B(4x^2 + 8x + 2)e^{2x} + 4(A + Bx^2 e^{2x}) = 2 + e^{2x}$

$4A = 2 \Rightarrow A = \dfrac{1}{2}$

$(12 - 6)B = 1 \Rightarrow B = \dfrac{1}{6}$

$y_p = \dfrac{1}{2} + \dfrac{1}{6}x^2 e^{2x}$

$y = y_h + y_p = C_1 e^{-x} + C_2 e^{2x} + C_3 x e^{2x} + \dfrac{1}{2} + \dfrac{1}{6}x^2 e^{2x}$

17. $y'' + y = x^3, \, y(0) = 1, \, y'(0) = 0$

$y'' + y = 0$

$m^2 + 1 = 0$ when $m = i, -i.$

$y_h = C_1 \cos x + C_2 \sin x$

$y_p = A_0 + A_1 x + A_2 x^2 + A_3 x^3$

$y_p' = A_1 + 2A_2 x + 3A_3 x^2$

$y_p'' = 2A_2 + 6A_3 x$

$y_p'' + y_p = A_3 x^3 + A_2 x^2 + (A_1 + 6A_3)x + (A_0 + 2A_2)$

$\qquad = x^3$

or $A_3 = 1, \, A_2 = 0, \, A_1 = -6, \, A_0 = 0$

$y = C_1 \cos x + C_2 \sin x + x^3 - 6x$

$y' = -C_1 \sin x + C_2 \cos x + 3x^2 - 6$

Initial conditions:

$y(0) = 1, \, y'(0) = 0, 1 = C_1, 0 = C_2 - 6, C_2 = 6$

Particular solution: $y = \cos x + 6 \sin x + x^3 - 6x$

19. $y'' + y' = 2 \sin x, \, y(0) = 0, \, y'(0) = -3$

$y'' + y' = 0$

$m^2 + m = 0$ when $m = 0, -1.$

$y_h = C_1 + C_2 e^{-x}$

$y_p = A \cos x + B \sin x$

$y_p' = -A \sin x + B \cos x$

$y_p'' = -A \cos x - B \sin x$

$y_p'' + y_p' = (-A + B) \cos x + (-A - B) \sin x = 2 \sin x$

$\left.\begin{array}{r} -A + B = 0 \\ -A - B = 2 \end{array}\right\} A = -1, B = -1$

$y = C_1 + C_2 e^x - (\cos x + \sin x)$

$y' = -C_2 e^{-x} - (-\sin x + \cos x)$

Initial conditions: $y(0) = 0, \, y'(0) = -3,$

$\qquad 0 = C_1 + C_2 - 1, -3 = -C_2 - 1,$

$\qquad C_2 = 2, C_1 = -1$

Particular solution: $y = -1 + 2e^{-x} - (\cos x + \sin x)$

21. $y' - 4y = xe^x - xe^{4x}$, $y(0) = \frac{1}{3}$

$y' - 4y = 0$

$m - 4 = 0$ when $m = 4$.

$y_h = Ce^{4x}$

$y_p = (A_0 + A_1 x)e^x + (A_2 x + A_3 x^2)e^{4x}$

$y_p' = (A_0 + A_1 x)e^x + A_1 e^x$
$\qquad + 4(A_2 x + A_3 x^2)e^{4x} + (A_2 + 2A_3 x)e^{4x}$

$y_p' - 4y_p = (-3A_0 - 3A_1 x)e^x + A_1 e^x + A_2 e^{4x}$
$\qquad + 2A_3 xe^{4x} = xe^x - xe^{4x}$

$A_0 = -\frac{1}{9}$, $A_1 = -\frac{1}{3}$, $A_2 = 0$, $A_3 = -\frac{1}{2}$

$y = \left(C - \frac{1}{2}x^2\right)e^{4x} - \frac{1}{9}(1 + 3x)e^x$

Initial conditions: $y(0) = \frac{1}{3}$, $\frac{1}{3} = C - \frac{1}{9}$, $C = \frac{4}{9}$

Particular solution: $y = \left(\frac{4}{9} - \frac{1}{2}x^2\right)e^{4x} - \frac{1}{9}(1 + 3x)e^x$

23. $y'' + y = \sec x$

$y'' + y = 0$

$m^2 + 1 = 0$ when $m = -i, i$.

$y_h = C_1 \cos x + C_2 \sin x$

$y_p = u_1 \cos x + u_2 \sin x$

$u_1' \cos x + u_2' \sin x = 0$

$u_1'(-\sin x) + u_2'(\cos x) = \sec x$

$u_1' = \dfrac{\begin{vmatrix} 0 & \sin x \\ \sec x & \cos x \end{vmatrix}}{\begin{vmatrix} \cos x & \sin x \\ -\sin x & \cos x \end{vmatrix}} = -\tan x$

$u_1 = \int -\tan x \, dx = \ln|\cos x|$

$u_2' = \dfrac{\begin{vmatrix} \cos x & 0 \\ -\sin x & \sec x \end{vmatrix}}{\begin{vmatrix} \cos x & \sin x \\ -\sin x & \cos x \end{vmatrix}} = 1$

$u_2 = \int dx = x$

$y = (C_1 + \ln|\cos x|)\cos x + (C_2 + x)\sin x$

25. $y'' + 4y = \csc 2x$

$y'' + 4y = 0$

$m^2 + 4 = 0$ when $m = -2i, 2i$.

$y_h = C_1 \cos 2x + C_2 \sin 2x$

$y_p = u_1 \cos 2x + u_2 \sin 2x = 0$

$u_1' \cos 2x + u_2' \sin 2x = 0$

$u_1'(-2 \sin 2x) + u_2'(2 \cos 2x) = \csc 2x$

$u_1' = \dfrac{\begin{vmatrix} 0 & \sin 2x \\ \csc 2x & 2\cos 2x \end{vmatrix}}{\begin{vmatrix} \cos 2x & \sin 2x \\ -2\sin 2x & 2\cos 2x \end{vmatrix}} = -\frac{1}{2}$

$u_1 = \int -\frac{1}{2} \, dx = -\frac{1}{2}x$

$u_2' = \dfrac{\begin{vmatrix} \cos 2x & 0 \\ -2\sin 2x & \csc 2x \end{vmatrix}}{\begin{vmatrix} \cos 2x & \sin 2x \\ -2\sin 2x & 2\cos 2x \end{vmatrix}} = \frac{1}{2}\cot 2x$

$u_2 = \int \frac{1}{2}\cot 2x \, dx = \frac{1}{4}\ln|\sin 2x|$

$y = \left(C_1 - \frac{1}{2}x\right)\cos 2x + \left(C_2 + \frac{1}{4}\ln|\sin 2x|\right)\sin 2x$

27. $y'' - 2y' + y = e^x \ln x$

$y'' - 2y' + y = 0$

$m^2 - 2m + 1 = 0$ when $m = 1, 1$.

$y_h = (C_1 + C_2 x)e^x$

$y_p = (u_1 + u_2 x)e^x$

$u_1' e^x + u_2' x e^x = 0$

$u_1' e^x + u_2'(x + 1)e^x = e^x \ln x$

$u_1' = -x \ln x$

$u_1 = \int -x \ln x \, dx = -\frac{x^2}{2}\ln x + \frac{x^2}{4}$

$u_2' = \ln x$

$u_2 = \int \ln x \, dx = x \ln x - x$

$y = (C_1 + C_2 x)e^x + \frac{x^2 e^x}{4}(\ln x^2 - 3)$

29. $q'' + 10q' + 25q = 6 \sin 5t, q(0) = 0, q'(0) = 0$

$m^2 + 10m + 25 = 0$ when $m = -5, -5$.

$q_h = (C_1 + C_2 t)e^{-5t}$

$q_p = A \cos 5t + B \sin 5t$

$q_p' = -5A \sin 5t + 5B \cos 5t$

$q_p'' = -25A \cos 5t - 25B \sin 5t$

$q_p'' + 10q_p' + 25q_p = 50B \cos 5t - 50A \sin 5t = 6 \sin 5t, A = -\frac{3}{25}, B = 0$

$q = (C_1 + C_2 t)e^{-5t} - \frac{3}{25} \cos 5t$

Initial conditions: $q(0) = 0, q'(0) = 0, C_1 - \frac{3}{25} = 0, -5C_1 + C_2 = 0, C_1 = \frac{3}{25}, C_2 = \frac{3}{5}$

Particular solution: $q = \frac{3}{25}\left(e^{-5t} + 5te^{-5t} - \cos 5t\right)$

31. $\frac{24}{32}y'' + 48y = \frac{24}{32}(48 \sin 4t), y(0) = \frac{1}{4}, y'(0) = 0$

$\frac{24}{32}m^2 + 48 = 0$ when $m = \pm 8i$.

$y_h = C_1 \cos 8t + C_2 \sin 8t$

$y_p = A \sin 4t + B \cos 4t$

$y_p' = 4A \cos 4t - 4B \sin 4t$

$y_p'' = -16A \sin 4t - 16B \cos 4t$

$\frac{24}{32}y_p'' + 48y_p = 36A \sin 4t + 36B \cos 4t$

$\quad = \frac{24}{32}(48 \sin 4t), B = 0, A = 1$

$y = y_h + y_p = C_1 \cos 8t + C_2 \sin 8t + \sin 4t$

Initial conditions: $y(0) = \frac{1}{4}, y'(0) = 0, \frac{1}{4} = C_1,$

$\qquad 0 = 8C_2 + 4 \Rightarrow C_2 = -\frac{1}{2}$

Particular solution: $y = \frac{1}{4} \cos 8t - \frac{1}{2} \sin 8t + \sin 4t$

33. $\frac{2}{32}y'' + y' + 4y = \frac{2}{32}(4 \sin 8t), y(0) = \frac{1}{4}, y'(0) = -3$

$\frac{1}{16}m^2 + m + 4 = 0$

when $m = -8, -8$.

$y_h = (C_1 + C_2 t)e^{-8t}$

$y_p = A \sin 8t + B \cos 8t$

$y_p' = 8A \cos 8t - 8B \sin 8t$

$y_p'' = -64A \sin 8t - 64B \cos 8t$

$\frac{2}{32}y_p'' + y_p' + 4y_p = -8B \sin 8t + 8A \cos 8t$

$\quad = \frac{2}{32}(4 \sin 8t) - 8B$

$\quad = \frac{1}{4} \Rightarrow B = -\frac{1}{32}, 8A = 0 \Rightarrow A = 0$

Initial conditions:

$y(0) = \frac{1}{4}, y'(0) = -3, \frac{1}{4} = C_1 - \frac{1}{32} \Rightarrow C_1 = \frac{9}{32},$

$\qquad -3 = -8C_1 + C_2 \Rightarrow C_2 = -\frac{3}{4}$

Particular solution: $y = \left(\frac{9}{32} - \frac{3}{4}t\right)e^{-8t} - \frac{1}{32} \cos 8t$

35. In Exercise 31,

$$y_h = \frac{1}{4} \cos 8t - \frac{1}{2} \sin 8t - \frac{\sqrt{5}}{4} \sin\left[8t + \pi + \arctan\left(-\frac{1}{2}\right)\right] = \frac{\sqrt{5}}{4} \sin\left(8t + \pi - \arctan\frac{1}{2}\right) \approx \frac{\sqrt{5}}{4} \sin(8t + 2.6779).$$

37. (a) $\frac{4}{32}y'' + \frac{25}{2}y = 0$

$y = C_1 \cos 10x + C_2 \sin 10x$

$y(0) = \frac{1}{2}: \frac{1}{2} = C_1$

$y'(0) = -4: -4 = 10C_2 \Rightarrow C_2 = -\frac{2}{5}$

$y = \frac{1}{2} \cos 10x - \frac{2}{5} \sin 10x$

The motion is undamped.

(b) If $b > 0$, the motion is damped.

(c) If $b > \frac{5}{2}$, the solution to the differential equation is of the form $y = C_1 e^{m_1 x} + C_2 e^{m_2 x}$.

There would be no oscillations in this case.

39. True. $y_p = -e^{2x} \cos e^{-x}$

$$y_p' = e^{2x} \sin e^{-x}(-e^{-x}) - 2e^{2x} \cos e^{-x} = -e^x \sin e^{-x} - 2e^{2x} \cos e^{-x}$$

$$y_p'' = \left[-e^x \cos e^{-x}(-e^{-x}) - e^x \sin e^{-x}\right] + \left[2e^{2x} \sin e^{-x}(-e^{-x}) - 4e^{2x} \cos e^{-x}\right]$$

So, $y_p'' - 3y_p' + 2y_p = \left[\cos e^{-x} - e^x \sin e^{-x} - 2e^x \sin e^{-x} - 4e^{2x} \cos e^{-x}\right] - 3\left[-e^x \sin e^{-x} - 2e^{2x} \cos e^{-x}\right] - 2e^{2x} \cos e^{-x}$

$$= \left[-e^x - 2e^x + 3e^x\right] \sin e^{-x} + \left[1 - 4e^{2x} + 6e^{2x} - 2e^{2x}\right] \cos e^{-x} = \cos e^{-x}.$$

41. $y'' - 2y' + y = 2e^x$

$m^2 - 2m + 1 = 0 \Rightarrow m = 1, 1$

$y_h = C_1 e^x + C_2 x e^x$, $y_p = x^2 e^x$, particular solution

General solution: $f(x) = (C_1 + C_2 x)e^x + x^2 e^x = (C_1 + C_2 x + x^2)e^x$

$f'(x) = (C_2 + 2x + C_1 + C_2 x + x^2)e^x = (x^2 + (C_2 + 2)x + (C_1 + C_2))e^x$

(a) No. If $f(x) > 0$ for all x, then $x^2 + C_2 x + C_1 > 0 \Leftrightarrow C_2^2 - 4C_1 < 0$ for all x.

So, let $C_1 = C_2 = 1$. Then $f'(x) = (x^2 + 3x + 2)e^x$ and $f'(-\frac{3}{2}) = -\frac{1}{4} < 0$.

(b) Yes. If $f'(x) > 0$ for all x, then

$$(C_2 + 2)^2 - 4(C_1 + C_2) < 0$$
$$\Rightarrow C_2^2 - 4C_1 + 4 < 0$$
$$C_2^2 - 4C_1 < -4$$
$$C_2^2 - 4C_1 < 0$$
$$\Rightarrow f(x) > 0 \text{ for all } x.$$

Section 16.4   Series Solutions of Differential Equations

1. Given a differential equation, assume that the solution is of the form $y = \sum\limits_{n=0}^{\infty} a_n x^n$. Then substitute y and its derivatives into the differential equation. You should then be able to determine the coefficients $a_0, a_1, \ldots$.

3. $5y' + y = 0$. Letting $y = \sum\limits_{n=0}^{\infty} a_n x^n$:

$$y' = \sum_{n=1}^{\infty} n a_n x^{n-1}$$

$$5y' + y = 5 \sum_{n=1}^{\infty} n a_n x^{n-1} + \sum_{n=0}^{\infty} a_n x^n = 5 \sum_{n=0}^{\infty} (n+1) a_{n+1} x^n + \sum_{n=0}^{\infty} a_n x^n = 0$$

$$5(n+1)a_{n+1} + a_n = 0$$

$$a_{n+1} = -\frac{a_n}{5(n+1)}$$

$$a_1 = -\frac{a_0}{5}, \, a_2 = -\frac{a_1}{5(2)} = \frac{a_0}{5^2(2)}$$

$$a_3 = -\frac{a_2}{5(3)} = -\frac{a_0}{5^3(3)(2)}, \cdots, a_n = \frac{(-1)^n a_0}{5^n n!}$$

$$y = \sum_{n=0}^{\infty} a_n x^n = a_0 \sum_{n=0}^{\infty} \left(-\frac{x}{5}\right)^n \frac{1}{n!}$$

Note: $y = a_0 e^{-x/5}$

5. $y' + 3xy = 0$. Letting $y = \sum\limits_{n=0}^{\infty} a_n x^n$:

$$y' + 3xy = \sum_{n=1}^{\infty} na_n x^{n-1} + \sum_{n=0}^{\infty} 3a_n x^{n+1} = 0$$

$$\sum_{n=-1}^{\infty} (n+2)a_{n+2}x^{n+1} = \sum_{n=0}^{\infty} -3a_n x^{n+1} \Rightarrow a_1 = 0 \text{ and } a_{n+2} = \frac{-3a_n}{n+2}$$

$a_0 = a_0$ \qquad\qquad\qquad\qquad $a_1 = 0$

$a_2 = -\dfrac{3a_0}{2}$ \qquad\qquad\qquad $a_3 = -\dfrac{3a_1}{3} = 0$

$a_4 = -\dfrac{3}{4}\left(-\dfrac{3a_0}{2}\right) = \dfrac{3^2}{2^3}a_0$ \qquad $a_5 = -\dfrac{3}{5}\left(-\dfrac{3a_1}{3}\right) = 0$

$a_6 = -\dfrac{3}{6}\left(-\dfrac{3^2}{2^3}a_0\right) = -\dfrac{3^3 a_0}{2^3(3 \cdot 2)}$ \qquad $a_7 = -\dfrac{3}{7}\left(\dfrac{3^2 a_1}{3 \cdot 5}\right) = 0$

$a_8 = -\dfrac{3}{8}\left(-\dfrac{3^3 a_0}{2^3(3 \cdot 2)}\right) = \dfrac{3^4 a_0}{2^4(4 \cdot 3 \cdot 2)}$ \qquad $a_9 = -\dfrac{3}{9}\left(-\dfrac{3^3 a_1}{3 \cdot 5 \cdot 7}\right) = 0$

$$y = a_0 \sum_{n=0}^{\infty} \frac{(-3)^n x^{2n}}{2^n n!}$$

7. $(x^2 + 4)y'' + y = 0$. Letting $y = \sum\limits_{n=0}^{\infty} a_n x^n$:

$$(x^2 + 4)y'' + y = \sum_{n=2}^{\infty} n(n-1)a_n x^n + 4\sum_{n=2}^{\infty} n(n-1)a_n x^{n-2} + \sum_{n=0}^{\infty} a_n x^n$$

$$= \sum_{n=0}^{\infty} (n^2 - n + 1)a_n x^n + \sum_{n=0}^{\infty} 4(n+2)(n+1)a_{n+2}x^n = 0$$

$$a_{n+2} = -\frac{(n^2 - n + 1)a_n}{4(n+2)(n+1)}$$

$a_0 = a_0$ \qquad\qquad\qquad\qquad\qquad $a_1 = a_1$

$a_2 = -\dfrac{a_0}{4(2)(1)} = -\dfrac{a_0}{8}$ \qquad\qquad $a_3 = -\dfrac{a_1}{4(3)(2)} = -\dfrac{a_1}{24}$

$a_4 = -\dfrac{3a_2}{4(4)(3)} = \dfrac{a_0}{128}$ \qquad\qquad $a_5 = -\dfrac{7a_3}{4(5)(4)} = \dfrac{7a_1}{1920}$

$$y = a_0\left(1 - \frac{x^2}{8} + \frac{x^4}{128} - \cdots\right) + a_1\left(x - \frac{x^3}{24} + \frac{7x^5}{1920} - \cdots\right)$$

9. $y' + (2x - 1)y = 0$, $y(0) = 2$

$y' = (1 - 2x)y$ \qquad\qquad $y'(0) = 0$

$y'' = (1 - 2x)y' - 2y$ \qquad $y''(0) = -2$

$y''' = (1 - 2x)y'' - 4y'$ \qquad $y'''(0) = -10$

$y^4 = (1 - 2x)y''' - 6y''$ \qquad $y^{(4)}(0) = 2$

$\vdots$ \qquad\qquad\qquad\qquad $\vdots$

$$y(x) = 2 + \frac{2}{1!}x - \frac{2}{2!}x^2 - \frac{10}{3!}x^3 + \frac{2}{4!}x^4 + \cdots$$

Using the first five terms of the series, $y\left(\dfrac{1}{2}\right) = \dfrac{163}{64} \approx 2.547$.

11. $y'' - 2xy = 0$, $y(0) = 1$, $y'(0) = -3$

$$y'' = 2xy \qquad\qquad y''(0) = 0$$
$$y''' = 2(xy' + y) \qquad\qquad y'''(0) = 2$$
$$y^{(4)} = 2(xy'' + 2y') \qquad\qquad y^{(4)}(0) = -12$$
$$y^{(5)} = 2(xy''' + 3y'') \qquad\qquad y^{(5)}(0) = 0$$
$$y^{(6)} = 2(xy^{(4)} + 4y''') \qquad\qquad y^{(6)}(0) = 16$$
$$y^{(7)} = 2(xy^{(5)} + 5y^{(4)}) \qquad\qquad y^{(7)}(0) = -120$$
$$\vdots \qquad\qquad\qquad \vdots$$

$$y \approx 1 - \frac{3}{1!}x + \frac{2}{3!}x^3 - \frac{12}{4!}x^4 + \frac{16}{6!}x^6 - \frac{120}{7!}x^7$$

Using the first six terms of the series, $y\left(\dfrac{1}{4}\right) \approx 0.253$.

13. $y'' + x^2 y' - (\cos x)y = 0$, $y(0) = 3$, $y'(0) = 2$

$$y'' = -x^2 y' + (\cos x)y \qquad\qquad\qquad y''(0) = 3$$
$$y''' = -2x^2 y' - x^2 y'' - (\sin x)y + (\cos x)y' \qquad\qquad y'''(0) = 2$$

$$y \approx 3 + \frac{2}{1!}x + \frac{3}{2!}x^2 + \frac{2}{3!}x^3$$

Using the first four terms of the series, $y\left(\dfrac{1}{3}\right) \approx 3.846$.

15. $y' - ky = 0$. Letting $y = \displaystyle\sum_{n=0}^{\infty} a_n x^n$:

$$y' - ky = \sum_{n=1}^{\infty} na_n x^{n-1} - k\sum_{n=0}^{\infty} a_n x^n = \sum_{n=0}^{\infty} (n+1)a_{n+1}x^n - \sum_{n=0}^{\infty} ka_n x^n = 0$$

$$(n + 1)a_{n+1} = ka_n$$

$$a_{n+1} = \frac{ka_n}{n+1}$$

$$a_1 = ka_0, \; a_2 = \frac{ka_1}{2} = \frac{k^2 a_0}{2}, \; a_3 = \frac{ka_2}{3} = \frac{k^3 a_0}{1 \cdot 2 \cdot 3}, \dots, \; a_n = \frac{k^n}{n!}a_0$$

$$y = \sum_{n=0}^{\infty} \frac{k^n}{n!}a_0 x^n = a_0 \sum_{n=0}^{\infty} \frac{(kx)^n}{n!} = a_0 e^{kx}$$

Check: By separation of variables, you have:

$$\int \frac{dy}{y} = \int k \, dx$$

$$\ln y = kx + C_1$$

$$y = Ce^{kx}$$

17. $y'' - k^2 y = 0.$ Letting $y = \sum_{n=0}^{\infty} a_n x^n$:

$$y'' - k^2 y = \sum_{n=2}^{\infty} n(n-1)a_n x^{n-2} - k^2 \sum_{n=0}^{\infty} a_n x^n = \sum_{n=0}^{\infty} (n+2)(n+1)a_{n+2} x^n - \sum_{n=0}^{\infty} k^2 a_n x^n = 0$$

$$(n+2)(n+1)a_{n+2} = k^2 a_n$$

$$a_{n+2} = \frac{k^2 a_n}{(n+2)(n+1)}$$

$a_0 = a_0$ $\qquad\qquad\qquad\qquad a_1 = a_1$

$a_2 = \dfrac{k^2 a_0}{2}$ $\qquad\qquad\qquad a_3 = \dfrac{k^2 a_1}{3 \cdot 2}$

$a_4 = \dfrac{k^2 a_2}{4 \cdot 3} = \dfrac{k^4 a_0}{4 \cdot 3 \cdot 2 \cdot 1}$ $\qquad a_5 = \dfrac{k^2 a_3}{5 \cdot 4} = \dfrac{k^4 a_1}{5 \cdot 4 \cdot 3 \cdot 2 \cdot 1}$

$\qquad\qquad \vdots$ $\qquad\qquad\qquad\qquad\qquad \vdots$

$a_{2n} = \dfrac{k^{2n} a_n}{(2n)!}$ $\qquad\qquad\qquad a_{2n+1} = \dfrac{k^{2n} a_1}{(2n+1)!}$

$$y = \sum_{n=0}^{\infty} \frac{k^{2n} a_0}{(2n)!} x^{2n} + \sum_{n=0}^{\infty} \frac{k^{2n} a_1}{(2n+1)!} x^{2n+1} = a_0 \sum_{n=0}^{\infty} \frac{(kx)^{2n}}{(2n)!} + \frac{a_1}{k} \sum_{n=0}^{\infty} \frac{(kx)^{2n+1}}{(2n+1)!} = C_0 \sum_{n=0}^{\infty} \frac{(kx)^n}{n!} + C_1 \sum_{n=0}^{\infty} \frac{(-kx)^n}{n!}$$

$$= C_0 e^{kx} + C_1 e^{-kx}, \text{ where } C_0 + C_1 = a_0 \text{ and } C_0 - C_1 = \frac{a_1}{k}$$

Check: $y'' - k^2 y = 0$ is a second-order homogeneous linear equation.

$m^2 - k^2 = 0 \Rightarrow m_1 = k$ and $m_2 = -k$

$y = C_1 e^{kx} + C_2 e^{-kx}$

19. (a) $y'' - xy' = 0.$ Letting $y = \sum_{n=0}^{\infty} a_n x^n$:

$$y'' - xy' = \sum_{n=2}^{\infty} n(n-1)a_n x^{n-2} - x \sum_{n=1}^{\infty} n a_n x^{n-1} = 0$$

$$\sum_{n=2}^{\infty} n(n-1)a_n x^{n-2} = \sum_{n=0}^{\infty} n a_n x^n$$

$$\sum_{n=0}^{\infty} (n+2)(n+1)a_{n+2} x^n = \sum_{n=0}^{\infty} n a_n x^n$$

$$a_{n+2} = \frac{n a_n}{(n+2)(n+1)}$$

$a_0 = a_0$ $\qquad\qquad\qquad\qquad a_1 = a_1$

$a_2 = 0$ $\qquad\qquad\qquad\qquad a_3 = \dfrac{a_1}{3 \cdot 2}$

There are no even powered terms. $a_5 = \dfrac{3a_3}{5 \cdot 4} = \dfrac{3a_1}{5!}$

$\qquad\qquad\qquad\qquad\qquad a_7 = \dfrac{5a_5}{7 \cdot 6} = \dfrac{5 \cdot 3a_1}{7!}$

$$y = a_0 + a_1 \sum_{n=0}^{\infty} \frac{1 \cdot 3 \cdot 5 \cdot 7 \cdots (2n-1)x^{2n+1}}{(2n+1)!} = a_0 + a_1 \sum_{n=0}^{\infty} \frac{(2n)! x^{2n+1}}{2^n n!(2n+1)!} = a_0 + a_1 \sum_{n=0}^{\infty} \frac{x^{2n+1}}{2^n n!(2n+1)}$$

$y(0) = 0 \Rightarrow a_0 = 0$

$$y' = a_1 \sum_{n=0}^{\infty} \frac{(2n+1)x^{2n}}{2^2 n!(2n+1)} = a_1 \sum_{n=0}^{\infty} \frac{x^{2n}}{2^n n!}$$

$y'(0) = 2 = a_1$

$$y = 2 \sum_{n=0}^{\infty} \frac{x^{2n+1}}{2^n n!(2n+1)}$$

(b) $P_3(x) = 2\left[x + \dfrac{x^3}{2 \cdot 3}\right] = 2x + \dfrac{x^3}{3}$

$P_5(x) = 2x + \dfrac{x^3}{3} + 2\dfrac{x^5}{4 \cdot 2 \cdot 5} = 2x + \dfrac{x^3}{3} + \dfrac{x^5}{20}$

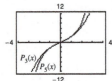

(c) The solution is symmetric about the origin.

21. $f(x) = e^x, f'(x) = e^x, y' - y = 0$. Assume $y = \displaystyle\sum_{n=0}^{\infty} a_n x^n$, then:

$$y' = \sum_{n=1}^{\infty} n a_n x^{n-1}$$

$$\sum_{n=1}^{\infty} n a_n x^{n-1} = \sum_{n=0}^{\infty} a_n x^n$$

$$\sum_{n=0}^{\infty} (n+1) a_{n+1} x^n = \sum_{n=0}^{\infty} a_n x^n$$

$$a_{n+1} = \frac{a_n}{n+1}, n \geq 0$$

$n = 0, \qquad a_1 = a_0$

$n = 1, \qquad a_2 = \dfrac{a_1}{2} = \dfrac{a_0}{2}$

$n = 2, \qquad a_3 = \dfrac{a_2}{3} = \dfrac{a_0}{2(3)}$

$n = 3, \qquad a_4 = \dfrac{a_3}{4} = \dfrac{a_0}{2(3)(4)}$

$n = 4, \qquad a_5 = \dfrac{a_4}{5} = \dfrac{a_0}{2(3)(4)(5)}$

$$\vdots$$

$$a_{n+1} = \frac{a_0}{(n+1)!} \Rightarrow a_n = \frac{a_0}{n!}$$

$y = a_0 \displaystyle\sum_{n=0}^{\infty} \dfrac{x^n}{n!}$ which converges on $(-\infty, \infty)$. When $a_0 = 1$, you have the Maclaurin Series for $f(x) = e^x$.

23.
$$f(x) = \arctan x$$

$$f'(x) = \frac{1}{1 + x^2}$$

$$f''(x) = \frac{-2x}{\left(1 + x^2\right)^2}$$

$$y'' = \frac{-2x}{1 + x^2}y'$$

$$\left(1 + x^2\right)y'' + 2xy' = 0$$

Assume $y = \displaystyle\sum_{n=0}^{\infty} a_n x^n$, then:

$$y' = \sum_{n=1}^{\infty} n a_n x^{n-1}$$

$$y'' = \sum_{n=2}^{\infty} n(n-1)a_n x^{n-2}$$

$$\left(1 + x^2\right)y'' + 2xy' = \sum_{n=2}^{\infty} n(n-1)a_n x^{n-2} + \sum_{n=0}^{\infty} n(n-1)a_n x^n + \sum_{n=0}^{\infty} 2n a_n x^n = 0$$

$$\sum_{n=2}^{\infty} n(n-1)a_n x^{n-2} = -\sum_{n=0}^{\infty} n(n-1)a_n x^n - \sum_{n=0}^{\infty} 2n a_n x^n$$

$$\sum_{n=0}^{\infty} (n+2)(n+1)a_{n+2} x^n = -\sum_{n=0}^{\infty} n(n+1)a_n x^n$$

$$(n+2)(n+1)a_{n+2} = -n(n+1)a_n$$

$$a_{n+2} = -\frac{n}{n+2}a_n, n \geq 0$$

$n = 0 \Rightarrow a_2 = 0 \Rightarrow$ all the even-powered terms have a coefficient of 0.

$n = 1$, $a_3 = -\dfrac{1}{3}a_1$

$n = 3$, $a_5 = -\dfrac{3}{5}a_3 = \dfrac{1}{5}a_1$

$n = 5$, $a_7 = -\dfrac{5}{7}a_5 = -\dfrac{1}{7}a_1$

$n = 7$, $a_9 = -\dfrac{7}{9}a_7 = \dfrac{1}{9}a_1$

$$\vdots$$

$$a_{2n+1} = \frac{(-1)^n a_1}{2n+1}$$

$y = a_1 \displaystyle\sum_{n=0}^{\infty} \frac{(-1)^n x^{2n+1}}{2n+1}$ which converges on $(-1, 1)$. When $a_1 = 1$, you have the Maclaurin Series for $f(x) = \arctan x$.

25. $y'' - xy = 0$. Let $y = \sum_{n=0}^{\infty} a_n x^n$.

$$y'' - xy = \sum_{n=2}^{\infty} n(n-1)a_n x^{n-2} - x\sum_{n=0}^{\infty} a_n x^n = \sum_{n=-1}^{\infty} (n+3)(n+2)a_{n+3}x^{n+1} - \sum_{n=0}^{\infty} a_n x^{n+1} = 0$$

$$2a_2 + \sum_{n=0}^{\infty} \left[(n+3)(n+2)a_{n+3} - a_n \right] x^{n+1} = 0$$

So, $a_2 = 0$ and $a_{n+3} = \dfrac{a_n}{(n+3)(n+2)}$ for $n = 0, 1, 2, \ldots$

The constants a_0 and a_1 are arbitrary.

$a_0 = a_0$ $a_1 = a_1$

$a_3 = \dfrac{a_0}{3 \cdot 2}$ $a_4 = \dfrac{a_1}{4 \cdot 3}$

$a_6 = \dfrac{a_3}{6 \cdot 5} = \dfrac{a_0}{6 \cdot 5 \cdot 3 \cdot 2}$ $a_7 = \dfrac{a_4}{7 \cdot 6} = \dfrac{a_1}{7 \cdot 6 \cdot 4 \cdot 3}$

So, $y = a_0 + a_1 x + \dfrac{a_0}{6}x^3 + \dfrac{a_1}{12}x^4 + \dfrac{a_0}{180}x^6 + \dfrac{a_1}{504}x^7$.

Review Exercises for Chapter 16

1. $\left(y + x^3 + xy^2 \right) dx - x\, dy = 0$

$\dfrac{\partial M}{\partial y} = 1 + 2xy \neq \dfrac{\partial N}{\partial x} = -1$

Not exact

3. $(10x + 8y + 2)\, dx + (8x + 5y + 2)\, dy = 0$

Exact: $\dfrac{\partial M}{\partial y} = 8 = \dfrac{\partial N}{\partial x}$

$f(x, y) = \int (10x + 8y + 2)\, dx = 5x^2 + 8xy + 2x + g(y)$

$f_y(x, y) = 8x + g'(y) = 8x + 5y + 2$

$g'(y) = 5y + 2$

$g(y) = \dfrac{5}{2}y^2 + 2y + C_1$

$f(x, y) = 5x^2 + 8xy + 2x + \dfrac{5}{2}y^2 + 2y + C_1$

$5x^2 + 8xy + 2x + \dfrac{5}{2}y^2 + 2y = C$

5. $(x - y - 5)\, dx - (x + 3y - z)\, dy = 0$

$\dfrac{\partial M}{\partial y} = -1 = \dfrac{\partial N}{\partial x}$ Exact

$f(x, y) = \int (x - y - 5)\, dx = \dfrac{x^2}{2} - xy - 5x + g(y)$

$f_y(x, y) = -x + g'(y) = -x - 3y + 2$

$g'(y) = -3y + 2$

$g(y) = \dfrac{-3}{2}y^2 + 2y + C_1$

$\dfrac{x^2}{2} - xy - 5x - \dfrac{3}{2}y^2 + 2y + C_1 = 0$

$x^2 - 2xy - 10x - 3y^2 + 4y = C$

7. (a)

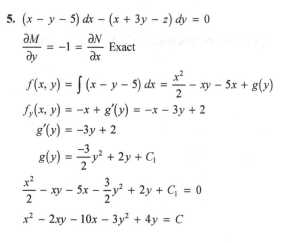

(b) $(2x - y)\, dx + (2y - x)\, dy = 0$

$\dfrac{\partial M}{\partial y} = -1 = \dfrac{\partial N}{\partial x}$ Exact

$f(x, y) = \int (2x - y)\, dx = x^2 - xy + g(y)$

$f_y(x, y) = -x + g'(y) = 2y - x$

$g'(y) = 2y$

$g(y) = y^2 + C_1$

$x^2 - xy + y^2 = C$

$y(2) = 2 : 4 - 4 + 4 = 4 = C$

Particular solution: $x^2 - xy + y^2 = 4$

(c)

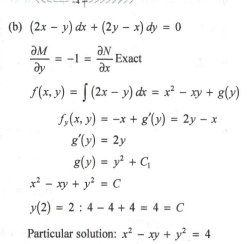

9. $(2x + y - 3)\, dx + (x - 3y + 1)\, dy = 0$

Exact: $\dfrac{\partial M}{\partial y} = 1 = \dfrac{\partial N}{\partial x}$

$$f(x, y) = \int (2x + y - 3)\, dx$$
$$= x^2 + xy - 3x + g(y)$$
$$f_y(x, y) = x + g'(y)$$
$$= x - 3y + 1$$
$$g'(y) = -3y + 1$$
$$g(y) = -\frac{3}{2}y^2 + y + C_1$$
$$f(x, y) = x^2 + xy - 3x$$
$$\qquad -\frac{3}{2}y^2 + y + C_1$$

$2x^2 + 2xy - 6x - 3y^2 + 2y = C$

Initial condition:
$y(2) = 0$
$8 + 0 - 12 - 0 + 0 = C \Rightarrow C = -4$

Particular solution: $2x^2 + 2xy - 6x - 3y^2 + 2y = -4$

11. $-\cos 2y\, dx + 2x \sin 2y\, dy = 0$

Exact: $\dfrac{\partial N}{\partial x} = 2 \sin 2y = \dfrac{\partial M}{\partial y}$

$f(x, y) = -x \cos 2y + g(y)$

$f_y(x, y) = 2x \sin 2y + g'(y) \Rightarrow g'(y) = 0$

$f(x, y) = -x \cos 2y + C_1$

$x \cos 2y = C$

Initial condition: $y(3) = \pi \Rightarrow 3 \cos 2\pi = C = 3$

Particular solution: $x \cos 2y = 3$

13. $\left(3x^2 - y^2\right) dx + 2xy\, dy = 0$

$\dfrac{(\partial M / \partial y) - (\partial N / \partial x)}{N} = \dfrac{-2y - 2y}{2xy} = -\dfrac{2}{x} = h(x)$

Integrating factor: $e^{\int h(x)\, dx} = e^{\ln x^{-2}} = \dfrac{1}{x^2}$

Exact equation: $\left(3 - \dfrac{y^2}{x^2}\right) dx + \dfrac{2y}{x}\, dy = 0$

$f(x, y) = \int \left(3 - \dfrac{y^2}{x^2}\right) dx = 3x + \dfrac{y^2}{x} + g(y)$

$f_y(x, y) = \dfrac{2y}{x} + g'(y) = \dfrac{2y}{x}$

$g'(y) = 0 \Rightarrow g(y) = C_1$

$3x + \dfrac{y^2}{x} = C$

15. $dx + \left(3x - e^{-2y}\right) dy = 0$

$\dfrac{(\partial N / \partial x) - (\partial M / \partial y)}{M} = \dfrac{3 - 0}{1} = 3 = k(y)$

Integrating factor: $e^{\int k(y)\, dy} = e^{3y}$

Exact equation: $e^{3y}\, dx + \left(3xe^{3y} - e^y\right) dy = 0$

$f(x, y) = \int e^{3y}\, dx = xe^{3y} + g(y)$

$f_y(x, y) = 3xe^{3y} + g'(y) = 3xe^{3y} - e^y$

$g'(y) = -e^y$

$g(y) = -e^y + C_1$

$xe^{3y} - e^y = C$

17. $y = C_1 e^{-3x} + C_2 e^{2x}$

$y' = -3C_1 e^{-3x} + 2C_2 e^{2x}$

$y'' = 9C_1 e^{-3x} + 4C_2 e^{2x}$

$y'' + y' - 6y = \left(9C_1 e^{-3x} + 4C_2 e^{2x}\right) + \left(-3C_1 e^{-3x} + 2C_2 e^{2x}\right) - 6\left(C_1 e^{-3x} + C_2 e^{2x}\right) = 0$

19. $2y'' + 5y' + 3y = 0$

$2m^2 + 5m + 3 = 0$

$(m + 1)(2m + 3) = 0 \Rightarrow m = -1, -\frac{3}{2}$

$y = C_1 e^{-x} + C_2 e^{(-3/2)x}$

21. $y'' - 6y' = 0$

$m^2 - 6m = m(m - 6) = 0 \Rightarrow m = 0, 6$

$y = C_1 + C_2 e^{6x}$

23. $y'' + 8y = 0$

$m^2 + 8 = 0 \Rightarrow m = \pm 2\sqrt{2}\,i$

$y = C_1 \cos\left(2\sqrt{2}x\right) + C_2 \sin\left(2\sqrt{2}x\right)$

25. $y''' - 2y'' - 3y' = 0$

$m^3 - 2m^2 - 3m = m(m^2 - 2m - 3) = m(m - 3)(m + 1) = 0$

$$m = 0, -1, 3$$

27. $y^{(4)} - 5y'' = 0$

$m^4 - 5m^2 = m^2(m^2 - 5) = 0 \Rightarrow m = 0, 0, \pm\sqrt{5}$

$y = C_1 + C_2 x + C_3 e^{\sqrt{5}x} + C_4 e^{-\sqrt{5}x}$

29. $y'' - y' - 2y = 0$

$m^2 - m - 2 = (m - 2)(m + 1) = 0 \Rightarrow m = 2, -1$

$y = C_1 e^{2x} + C_2 e^{-x}$

$y' = 2C_1 e^{2x} - C_2 e^{-x}$

$y(0) = 0 = C_1 + C_2$

$y'(0) = 3 = 2C_1 - C_2$

Adding these equations, $3 = 3C_1 \Rightarrow C_1 = 1$ and $C_2 = -1$.

$y = e^{2x} - e^{-x}$

31. $y'' + 2y' - 3y = 0$

$m^2 + 2m - 3 = (m + 3)(m - 1) = 0 \Rightarrow m = -3, 1$

$y = C_1 e^{-3x} + C_2 e^{x}$

$y' = -3C_1 e^{-3x} + C_2 e^{x}$

$y(0) = 2 = C_1 + C_2$

$y'(0) = 0 = -3C_1 + C_2$

Subtracting these equations, $2 = 4C_1 \Rightarrow C_1 = \frac{1}{2}$ and $C_2 = \frac{3}{2}$.

$y = \frac{3}{2}e^{x} + \frac{1}{2}e^{-3x}$

33. $y'' + 2y' + 5y = 0$

$m^2 + 2m + 5 = 0 \Rightarrow m = \dfrac{-2 \pm \sqrt{4 - 20}}{2} = -1 \pm 2i$

$y = e^{-x}(C_1 \cos 2x + C_2 \sin 2x)$

$y(1) = 4 = e^{-1}(C_1 \cos 2 + C_2 \sin 2)$

$y(2) = 0 = e^{-2}(C_1 \cos 4 + C_2 \sin 4)$

Solving this system, you obtain $C_1 = -9.0496$, $C_2 = 7.8161$.

$y = e^{-x}(-9.0496 \cos 2x + 7.8161 \sin 2x)$

35. By Hooke's Law, $F = kx, k = F/x = 64/(4/3) = 48$.

Also, $F = ma$ and $m = F/a = 64/32 = 2$.

So, $\dfrac{d^2 y}{dt^2} + \left(\dfrac{48}{2}\right) y = 0$

$y = C_2 \cos(2\sqrt{6}t) + C_2 \sin(2\sqrt{6}t)$.

Because $y(0) = \frac{1}{2}$ you have $C_1 = \frac{1}{2}$ and $y'(0) = 0$

yields $C_2 = 0$. So, $y = \frac{1}{2}\cos(2\sqrt{6}t)$.

37. $y'' + y = x^3 + x$

$m^2 + 1 = 0$ when $m = -i, i$.

$y_h = C_1 \cos x + C_2 \sin x$

$y_p = A_0 + A_1 x + A_2 x^2 + A_3 x^3$

$y_p' = A_1 + 2A_2 x + 3A_3 x^2$

$y_p'' = 2A_2 + 6A_3 x$

$y_p'' + y_p = (A_0 + 2A_2) + (A_1 + 6A_3)x + A_2 x^2 + A_3 x^3$

$= x^3 + x$

$A_0 = 0, A_1 = -5, A_2 = 0, A_3 = 1$

$y = C_1 \cos x + C_2 \sin x - 5x + x^3$

39. $y'' - 8y' - 9y = 9x - 10$

$y'' - 8y' - 9y = 0$

$m^2 - 8m - 9 = (m - 9)(m + 1) = 0 \Rightarrow m = 9, -1$

$y_p = Ax + B, y_p' = A, y_p'' = 0$

$y_p'' - 8y_p' - 9y_p = -8A - 9(Ax + B) = 9x - 10$

$\Rightarrow -9Ax - 8A - 9B = 9x - 10$

$\Rightarrow A = -1, B = 2$

$y_h = C_1 e^{-x} + C_2 e^{9x}$

$y_p = -x + 2$

$y = C_1 e^{-x} + C_2 e^{9x} - x + 2$

41. $y'' - 4y' + 3y = e^x + 8e^{3x}$

Because e^x and e^{3x} are solutions to the homogeneous equation, use

$y_p = Axe^x + Bxe^{3x}$.

43. $y'' + y = 2 \cos x$

$m^2 + 1 = 0$ when $m = -i, i$.

$$y_h = C_1 \cos x + C_2 \sin x$$
$$y_p = Ax \cos x + Bx \sin x$$
$$y_p' = (Bx + A) \cos x + (B - Ax) \sin x$$
$$y_p'' = (2B - Ax) \cos x + (-Bx - 2A) \sin x$$
$$y_p'' + y_p = 2B \cos x - 2A \sin x = 2 \cos x$$
$$A = 0, B = 1$$
$$y = C_1 \cos x + (C_2 + x) \sin x$$

45. $y'' + y' - 6y = 54, y(0) = 2, y'(0) = 0$

$$m^2 - m - 6 = 0$$
$$(m - 3)(m + 2) = 0$$
$$m_1 = 3, m_2 = -2$$
$$y_h = C_1 e^{3x} + C_2 e^{-2x}$$
$$y_p = -9 \text{ by inspection}$$
$$y = y_h + y_p = C_1 e^{3x} + C_2 e^{-2x} - 9$$

Initial conditions: $y(0) = 2$: $2 = C_1 + C_2 - 9 \Rightarrow C_1 + C_2 = 11$

$$y'(0) = 0: 0 = 3C_1 - 2C_2 \Rightarrow C_1 = \tfrac{22}{5}, C_2 = \tfrac{33}{5}$$

$$y = \tfrac{11}{5}\left(2e^{3x} + 3e^{-2x}\right) - 9$$

47. $y'' + 4y = \cos x$

$$m^2 + 4 = 0 \Rightarrow m = \pm 2i$$
$$y_h = C_1 \cos 2x + C_2 \sin 2x$$
$$y_p = A \cos x + B \sin x$$
$$y_p' = -A \sin x + B \cos x$$
$$y_p'' = -A \cos x - B \sin x$$
$$y_p'' + 4y_p = (-A \cos x - B \sin x) + 4(A \cos x + B \sin x) = \cos x$$
$$3A \cos x + 3B \sin x = \cos x \Rightarrow A = \tfrac{1}{3} \text{ and } B = 0$$

$$y_p = \tfrac{1}{3} \cos x$$
$$y = y_h + y_p = C_1 \cos 2x + C_2 \sin 2x + \tfrac{1}{3} \cos x$$

Initial conditions: $y(0) = 6: 6 = C_1 + \tfrac{1}{3} \Rightarrow C_1 = \tfrac{17}{3}$

$$y'(0) = -6: -6 = 2C_2 \Rightarrow C_2 = -3$$

Particular solution: $y = \tfrac{17}{3} \cos 2x - 3 \sin 2x + \tfrac{1}{3} \cos x$

49. $y'' - y' - 2y = 1 + xe^{-x}$, $y(0) = 1$, $y'(0) = 3$

$m^2 - m - 2 = (m - 2)(m + 1) = 0 \Rightarrow m = 2, -1$

$y_h = C_1 e^{2x} + C_2 e^{-x}$

$y_p = A + (Bx + Cx^2)e^{-x}$

$y_p' = -(Bx + Cx^2)e^{-x} + (B + 2Cx)e^{-x} = (B + (2C - B)x - Cx^2)e^{-x}$

$y_p'' = -(B + (2C - B)x - Cx^2)e^{-x} + (2C - B - 2Cx)e^{-x} = (Cx^2 + (B - 4C)x + 2C - 2B)e^{-x}$

$y_p'' - y_p' - 2y_p = (2C - 2B + (-4C + B)x + Cx^2)e^{-x} - (B + (2C - B)x - Cx^2)e^{-x} - 2(A + (Bx + Cx^2)e^{-x})$

$\qquad = -2A + (-6Cx + 2C - 3B)e^{-x} = 1 + xe^{-x} \Rightarrow A = -\dfrac{1}{2}, -6C = 1$ and $2C - 3B = 0$.

So, $C = -\dfrac{1}{6}$ and $B = -\dfrac{1}{9}$.

$y = y_h + y_p = C_1 e^{2x} + C_2 e^{-x} - \dfrac{1}{2} + \left(-\dfrac{1}{9}x - \dfrac{1}{6}x^2\right)e^{-x}$

Initial conditions: $y(0) = 1 = C_1 + C_2 - \dfrac{1}{2} \Rightarrow C_1 + C_2 = \dfrac{3}{2}$

$\qquad\qquad\qquad y'(0) = 3 = 2C_1 - C_2 - \dfrac{1}{9} \Rightarrow 2C_1 - C_2 = \dfrac{28}{9}$

Adding, $3C_1 = \dfrac{83}{18} \Rightarrow C_1 = \dfrac{83}{54}$.

So, $C_2 = -\dfrac{1}{27}$.

Particular solution: $y = \dfrac{83}{54}e^{2x} - \dfrac{1}{27}e^{-x} - \dfrac{1}{2} - \left(\dfrac{1}{9} + \dfrac{1}{6}x\right)xe^{-x}$

51. $y'' + 9y = \csc 3x$

$m^2 + 9 = 0 \Rightarrow m = \pm 3i$

$y_h = C_1 \cos 3x + C_2 \sin 3x$

$y_p = u_1 y_1 + u_2 y_2 = u_1 \cos 3x + u_2 \sin 3x$

$\qquad u_1' \cos 3x + u_2' \sin 3x = 0 \qquad (1)$

$3u_1'(-\sin 3x) + 3u_2' \cos 3x = \csc 3x \quad (2)$

Multiply (1) by $3 \sin 3x$ and (2) by $\cos 3x$:

$3u_1' \cos 3x \sin 3x + 3u_2' \sin 3x \sin 3x = 0$

$-3u_1' \sin 3x \cos 3x + 3u_2' \cos 3x \cos 3x = (\csc 3x)\cos 3x$

Adding these equations, $3u_2' = \csc 3x \cos 3x = \cot 3x$.

Then $3u_1' \cos 3x \sin 3x + \cot 3x \sin^2 3x = 0$

$\Rightarrow u_1' \cos 3x \sin 3x = -\dfrac{1}{3} \cos 3x \sin 3x \Rightarrow u_1' = -\dfrac{1}{3}$

Integrating, $u_1 = -\dfrac{1}{3}x$, $u_2 = \dfrac{1}{9} \ln |\sin 3x|$

$y = y_h + y_p = C_1 \cos 3x + C_2 \sin 3x - \dfrac{1}{3}x \cos 3x + \dfrac{1}{9} \ln |\sin 3x| \sin 3x$

53. $y'' - 2y' + y = 2xe^x$

$m^2 - 2m + 1 = 0$ when $m = 1, 1$.

$y_h = (C_1 + C_2x)e^x$

$y_p = (u_1 + u_2x)e^x$

$u_1'e^x + u_2'xe^x = 0$

$u_1'e^x + u_2'(x + 1)e^x = 2xe^x$

$u_1' = -2x^2$

$u_1 = \int -2x^2 \, dx = -\frac{2}{3}x^3$

$u_2' = 2x$

$u_2 = \int 2x \, dx = x^2$

$y = \left(C_1 + C_2x + \frac{1}{3}x^3\right)e^x$

55. $\dfrac{d^2q}{dt^2} + 4\dfrac{dq}{dt} + 8q = 3 \sin 4t$

$m^2 + 4m + 8 = 0 \Rightarrow m = \dfrac{-4 \pm \sqrt{16 - 32}}{2} = -2 \pm 2i$

$q_h = C_1e^{-2t} \cos 2t + C_2e^{-2t} \sin 2t$

$q_p = A \sin 4t + B \cos 4t$

$q_p' = 4A \cos 4t - 4B \sin 4t$

$q_p'' = -16A \sin 4t - 16B \cos 4t$

$q_p'' + 4q_p' + 8q_p = (-16A \sin 4t - 16B \cos 4t) + 4(4A \cos 4t - 4B \sin 4t) + 8(A \sin 4t + B \cos 4t) = 3 \sin 4t$

$-16A - 16B + 8A = 3$

$-16B + 16A + 8B = 0$

Solving for A and B, $A = -\frac{3}{40}$ and $B = -\frac{3}{20}$.

$y = C_1e^{-2t} \cos 2t + C_2e^{-2t} \sin 2t - \frac{3}{40} \sin 4t - \frac{3}{20} \cos 4t$

Initial conditions: $q(0) = 0 \Rightarrow C_1 - \frac{3}{20} = 0 \Rightarrow C_1 = \frac{3}{20}$

$q'(0) = 0 \Rightarrow C_2 = \frac{3}{10}$

$q(t) = \frac{3}{10}e^{-2t} \sin 2t + \frac{3}{20}e^{-2t} \cos 2t - \frac{3}{40} \sin 4t - \frac{3}{20} \cos 4t$

57. (a) $y_p'' = -A \sin x$ and $3y_p = 3A \sin x$.

So, $y_p'' + 3y_p = -A \sin x + 3A \sin x = 2A \sin x = 12 \sin x$

(b) $y_p = \dfrac{5}{2} \cos x$

(c) If $y_p = A \cos x + B \sin x$, then $y_p'' = -A \cos x - B \sin x$, and solving for A and B would be more difficult.

59. $(x - 4)y' + y = 0$. Letting $y = \sum\limits_{n=0}^{\infty} a_n x^n$:

$$xy' - 4y' + y = \sum_{n=0}^{\infty} na_n x^n - 4\sum_{n=1}^{\infty} na_n x^{n-1} + \sum_{n=0}^{\infty} a_n x^n$$

$$= \sum_{n=0}^{\infty} (n+1)a_n x^n - \sum_{n=1}^{\infty} 4na_n x^{n-1} = \sum_{n=0}^{\infty} (n+1)a_n x^n - \sum_{n=-1}^{\infty} 4(n+1)a_{n+1} x^n = 0$$

$$(n+1)a_n = 4(n+1)a_{n+1}$$

$$a_{n+1} = \frac{1}{4}a_n$$

$$a_0 = a_0, a_1 = \frac{1}{4}a_0, a_2 = \frac{1}{4}a_1 = \frac{1}{4^2}a_0, \ldots, a_n = \frac{1}{4^n}a_0$$

$$y = a_0 \sum_{n=0}^{\infty} \frac{x^n}{4^n}$$

61. $y'' + y' - e^x y = 0$, $y(0) = 2$, $y'(0) = 0$

$y'' = -y' + e^x y$ $\qquad\qquad\qquad y''(0) = 2$

$y''' = -y'' + e^x(y + y')$ $\qquad\qquad y'''(0) = -2 + 2 = 0$

$y^{(4)} = -y''' + e^x(y + 2y' + y'')$ $\qquad y^{(4)}(0) = 4$

$y^{(5)} = -y^{(4)} + e^x(y + 3y' + 3y'' + y''')$ $\qquad y^{(5)}(0) = -4 + 8 = 4$

$$y \approx y(0) + y'(0)x + \frac{y''(0)}{2!}x^2 + \frac{y'''(0)}{3!}x^3 + \frac{y^{(4)}(0)}{4!}x^4 + \frac{y^{(5)}(0)}{5!}x^5 = 2 + x^2 + \frac{1}{6}x^4 + \frac{1}{30}x^5$$

Using the first four terms of the series, $y\left(\frac{1}{4}\right) \approx 2.063$.

Problem Solving for Chapter 16

1. $(3x^2 + kxy^2)\,dx - (5x^2y + ky^2)\,dy = 0$

$$\frac{\partial M}{\partial y} = 2kxy$$

$$\frac{\partial N}{\partial x} = -10xy$$

$$\frac{\partial M}{\partial y} = \frac{\partial N}{\partial x} \Rightarrow k = -5$$

$(3x^2 - 5xy^2)\,dx - (5x^2y - 5y^2)\,dy = 0$ Exact

$$f(x, y) = \int (3x^2 - 5xy^2)\,dx = x^3 - \frac{5}{2}x^2y^2 + g(y)$$

$$f_y(x, y) = -5x^2y + g'(y) = -5x^2y + 5y^2$$

$$g'(y) = 5y^2 \Rightarrow g(y) = \frac{5}{3}y^3 + C_1$$

$$x^3 - \frac{5}{2}x^2y^2 + \frac{5}{3}y^3 = C_2$$

$$6x^3 - 15x^2y^2 + 10y^3 = C$$

3. $y'' - a^2y = 0$, $y > 0$

$$m^2 - a^2 = (m + a)(m - a) = 0 \Rightarrow m = \pm a$$

$$y = B_1 e^{ax} + B_2 e^{-ax} = \frac{C_1 + C_2}{2}e^{ax} + \frac{C_1 - C_2}{2}e^{-ax}$$

$$= C_1\left(\frac{e^{ax} + e^{-ax}}{2}\right) + C_2\left(\frac{e^{ax} - e^{-ax}}{2}\right)$$

$$= C_1 \cosh ax + C_2 \sinh ax$$

5. The general solution to $y'' + ay' + by = 0$ is

$$y = B_1 e^{(r+s)x} + B_2 e^{(r-s)x}.$$

Let $C_1 = B_1 + B_2$ and $C_2 = B_1 - B_2$.

Then $B_1 = \dfrac{C_1 + C_2}{2}$ and $B_2 = \dfrac{C_1 - C_2}{2}$.

So $y = \left(\dfrac{C_1 + C_2}{2}\right)e^{(r+s)x} + \left(\dfrac{C_1 - C_2}{2}\right)e^{(r-s)x}$

$$= e^{rx}\left[C_1\left(\frac{e^{sx} + e^{-sx}}{2}\right) + C_2\left(\frac{e^{sx} - e^{-sx}}{2}\right)\right]$$

$$= e^{rx}[C_1 \cosh sx + C_2 \sinh sx].$$

7. $y'' + ay = 0$, $y(0) = y(L) = 0$

(a) If $a = 0$, $y'' = 0 \Rightarrow y = cx + d$. $y(0) = 0 = d$ and $y(L) = 0 = cL \Rightarrow c = 0$. So $y = 0$ is the solution.

(b) If $a < 0$, $y'' + ay = 0$ has characteristic equation $m^2 + a = 0 \Rightarrow m = \pm\sqrt{-a}$.

$$y = C_1 e^{\sqrt{-a}\,x} + C_2 e^{-\sqrt{-a}\,x}$$

$$y(0) = 0 = C_1 + C_2 \Rightarrow -C_1 = C_2$$

$$y(L) = 0 = C_1 e^{\sqrt{-a}L} + C_2 e^{-\sqrt{-a}L} = C_1 e^{\sqrt{-a}L} - C_1 e^{-\sqrt{-a}L} = 2C_1\left(\frac{e^{\sqrt{-a}L} - e^{-\sqrt{-a}L}}{2}\right) = 2C_1 \sinh\left(\sqrt{-a}\,L\right) \Rightarrow C_1 = 0 = C_2$$

So, $y = 0$ is the only solution.

(c) For $a > 0$:

$$m^2 + a = 0 \Rightarrow m = \pm\sqrt{a}\,i$$

$$y = C_1 \cos\left(\sqrt{a}\,x\right) + C_2 \sin\left(\sqrt{a}\,x\right).$$

$$y(0) = 0 = C_1$$

$$y = C_2 \sin\left(\sqrt{a}\,x\right)$$

$$y(L) = 0 = C_2 \sin\left(\sqrt{a}\,L\right)$$

So $\sqrt{a}\,L = n\pi$

$$a = \left(\frac{n\pi}{L}\right)^2, \ n \text{ an integer.}$$

9. $\dfrac{d^2\theta}{dt^2} + \dfrac{g}{L}\theta = 0$, $\dfrac{g}{L} > 0$

(a) $\theta(t) = C_1 \sin\left(\sqrt{\dfrac{g}{L}}\,t\right) + C_2 \cos\left(\sqrt{\dfrac{g}{L}}\,t\right)$

Let ϕ be given by $\tan\left(\sqrt{\dfrac{g}{L}}\,\phi\right) = -\dfrac{C_1}{C_2}$, $-\dfrac{\pi}{2} < \phi < \dfrac{\pi}{2}$.

Then $C_2 \sin\left(\sqrt{\dfrac{g}{L}}\,\phi\right) = -C_1 \cos\left(\sqrt{\dfrac{g}{L}}\,\phi\right)$.

Let $A = \dfrac{C_2}{\cos\left(\sqrt{\dfrac{g}{L}}\,\phi\right)} = -\dfrac{C_1}{\sin\left(\sqrt{\dfrac{g}{L}}\,\phi\right)}$

$\theta(t) = C_1 \sin\left(\sqrt{\dfrac{g}{L}}\,t\right) + C_2 \cos\left(\sqrt{\dfrac{g}{L}}\,t\right)$

$= -A \sin\left(\sqrt{\dfrac{g}{L}}\,\phi\right)\sin\left(\sqrt{\dfrac{g}{L}}\,t\right) + A\cos\left(\sqrt{\dfrac{g}{L}}\,\phi\right)\cos\left(\sqrt{\dfrac{g}{L}}\,t\right)$

$= A \cos\left[\sqrt{\dfrac{g}{L}}(t + \phi)\right]$

(b) $\theta(t) = A \cos\left[\sqrt{\dfrac{g}{L}}(t + \phi)\right], g = 9.8, L = 0.25$

  $\theta(0) = A \cos\left[\sqrt{39.2}\ \phi\right] = 0.1$

  $\theta'(t) = -A\sqrt{\dfrac{g}{L}}\ \sin\left[\sqrt{\dfrac{g}{4}}(t + \phi)\right]$

  $\theta'(0) = -A\sqrt{39.2}\ \sin\left[\sqrt{39.2}\ \phi\right] = 0.5$

  Dividing, $\tan\left[\sqrt{39.2}\ \phi\right] = \dfrac{-5}{\sqrt{39.2}} \Rightarrow \phi \approx -0.1076 \Rightarrow A \approx 0.128.$

  $\theta(t) = 0.128 \cos\left[\sqrt{39.2}\,(t - 0.108)\right]$

(c) Period $= \dfrac{2\pi}{\sqrt{39.2}} \approx 1 \sec$

(d) Maximum is 0.128.

(e) $\theta(t) = 0$ at $t \approx 0.359$ sec, and at $t \approx 0.860$ sec.

(f) $a = \left(\dfrac{n\pi}{L}\right)^2, n$

11. $y'' + 8y' + 16y = 0, y(0) = 1, y'(0) = 1$

(a) $\lambda = 4, \omega = 4, \lambda^2 - \omega^2 = 0$, critically damped

(b) $m_1 = m_2 = -4$

  $y = (C_1 + C_2 t)e^{-4t},$
  $y' = -4(C_1 + C_2 t)e^{-4t} + C_2 e^{-4t}$

  $y(0) = 1 = C_1$
  $y'(0) = 1 = -4 + C_2 \Rightarrow C_2 = 5$
  $y = (1 + 5t)e^{-4t}$

(c)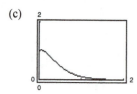

The solution tends to zero quickly.

13. $y'' + 20y' + 64y = 0, y(0) = 2, y'(0) = -20$

(a) $\lambda = 10, \omega = 8, \lambda^2 - \omega^2 = 36 > 0$, overdamped

(b) $m_1 = -10 + 6 = -4, m_2 = -10 - 6 = -16$
  $y = C_1 e^{-4t} + C_2 e^{-16t}$
  $y(0) = 2 = C_1 + C_2$
  $y'(t) = -4C_1 e^{-4t} - 16C_2 e^{-16t}$
  $y'(0) = -20 = -4C_1 - 16C_2$

  $\left.\begin{array}{r} C_1 + C_2 = 2 \\ -C_1 - 4C_2 = -5 \end{array}\right\} C_1 = 1, C_2 = 1$

  $y = e^{-4t} + e^{-16t}$

(c)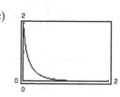

The solution tends to zero quickly.

15. Airy's Equation: $y'' - xy = 0$

$$y'' - xy + y - y = y'' - (x - 1)y - y = 0$$

Let $y = \sum_{n=0}^{\infty} a_n(x - 1)^n$, $y' = \sum_{n=1}^{\infty} na_n(x - 1)^{n-1}$, $y'' = \sum_{n=2}^{\infty} n(n - 1)a_n(x - 1)^{n-2}$.

$$y'' - (x - 1)y - y = 0$$

$$\sum_{n=2}^{\infty} n(n - 1)a_n(x - 1)^{n-2} - (x - 1)\sum_{n=0}^{\infty} a_n(x - 1)^n - \sum_{n=0}^{\infty} a_n(x - 1)^n = 0$$

$$\sum_{n=-1}^{\infty} (n + 3)(n + 2)a_{n+3}(x - 1)^{n+1} - \sum_{n=0}^{\infty} a_n(x - 1)^{n+1} - \sum_{n=-1}^{\infty} a_{n+1}(x - 1)^{n+1} = 0$$

$$(2a_2 - a_0) + \sum_{n=0}^{\infty} \left[(n + 3)(n + 2)a_{n+3} - a_n - a_{n+1}\right](x - 1)^{n+1} = 0$$

$$2a_2 - a_0 = 0 \Rightarrow a_2 = \frac{1}{2}a_0; a_0, a_1 \text{ arbitrary}$$

In general, $a_{n+3} = \dfrac{a_n + a_{n+1}}{(n + 3)(n + 2)}$.

$$a_3 = \frac{a_0 + a_1}{6}$$

$$a_4 = \frac{a_1 + a_2}{12} = \frac{a_1 + \left(\frac{1}{2}a_0\right)}{12} = \frac{2a_1 + a_0}{24}$$

$$a_5 = \frac{a_2 + a_3}{20} = \frac{\frac{1}{2}a_0 + \frac{a_0 + a_1}{6}}{20} = \frac{4a_0 + a_1}{120}$$

$$a_6 = \frac{a_3 + a_4}{30} = \frac{\left(\frac{a_0 + a_1}{6}\right) + \left(\frac{2a_1 + a_0}{24}\right)}{30} = \frac{5a_0 + 6a_1}{720}$$

$$a_7 = \frac{a_4 + a_5}{42} = \frac{\left(\frac{2a_1 + a_0}{24}\right) + \left(\frac{4a_0 + a_1}{120}\right)}{42} = \frac{9a_0 + 11a_1}{5040}$$

So, the first eight terms are

$$y = a_0 + a_1(x - 1) + \frac{a_0}{2}(x - 1)^2 + \frac{a_0 + a_1}{6}(x - 1)^3 + \frac{2a_1 + a_0}{24}(x - 1)^4 + \frac{4a_0 + a_1}{120}(x - 1)^5$$

$$+ \frac{5a_0 + 6a_1}{720}(x - 1)^6 + \frac{9a_0 + 11a_1}{5040}(x - 1)^7.$$

17. $x^2y'' + xy' + x^2y = 0$ Bessell equation of order zero

(a) Let $y = \displaystyle\sum_{n=0}^{\infty} a_n x^n$, $y' = \displaystyle\sum_{n=1}^{\infty} na_n x^{n-1}$, $y'' = \displaystyle\sum_{n=2}^{\infty} n(n-1)a_n x^{n-2}$.

$x^2 y'' + xy' + x^2 y = 0$

$x^2 \displaystyle\sum_{n=2}^{\infty} n(n-1)a_n x^{n-2} + x\displaystyle\sum_{n=1}^{\infty} na_n x^{n-1} + x^2\displaystyle\sum_{n=0}^{\infty} a_n x^n = 0$

$\displaystyle\sum_{n=2}^{\infty} n(n-1)a_n x^{n} + \displaystyle\sum_{n=1}^{\infty} na_n x^{n} + \displaystyle\sum_{n=0}^{\infty} a_n x^{n+2} = 0$

$\displaystyle\sum_{n=0}^{\infty} (n+2)(n+1)a_{n+2} x^{n+2} + \displaystyle\sum_{n=-1}^{\infty} (n+2)a_{n+2} x^{n+2} + \displaystyle\sum_{n=0}^{\infty} a_n x^{n+2} = 0$

$a_1 x + \displaystyle\sum_{n=0}^{\infty} \left[(n+2)(n+1)a_{n+2} + (n+2)a_{n+2} + a_n \right] x^{n+2} = 0$

$a_1 = 0$ and $a_{n+2} = \dfrac{-a_n}{(n+2)^2}$.

All odd terms a_i are 0.

$a_2 = \dfrac{-a_0}{2^2}$

$a_4 = \dfrac{-a_2}{4^2} = a_0\dfrac{1}{2^2 \cdot 4^2} = \dfrac{a_0}{2^4(1 \cdot 2)^2}$

$a_6 = \dfrac{-a_4}{6^2} = -a_0\dfrac{1}{2^2 \cdot 4^2 \cdot 6^2} = \dfrac{-a_0}{2^6(3!)^2}$

$y = a_0\displaystyle\sum_{n=0}^{\infty} \dfrac{(-1)^n x^{2n}}{2^{2n}(n!)^2}$

(b) This is the same function (assuming $a_0 = 1$).

19. (a) Let $y = \displaystyle\sum_{n=0}^{\infty} a_n x^n$, $y' = \displaystyle\sum_{n=1}^{\infty} na_n x^{n-1}$,

$y'' = \displaystyle\sum_{n=2}^{\infty} n(n-1)a_n x^{n-2}$.

$y'' - 2xy' + 8y = 0$

$\displaystyle\sum_{n=2}^{\infty} n(n-1)a_n x^{n-2} - 2x\displaystyle\sum_{n=1}^{\infty} na_n x^{n-1} + 8\displaystyle\sum_{n=0}^{\infty} a_n x^n = 0$

$\displaystyle\sum_{n=0}^{\infty} (n+2)(n+1)a_{n+2} x^{n} - \displaystyle\sum_{n=0}^{\infty} 2na_n x^{n} + \displaystyle\sum_{n=0}^{\infty} 8a_n x^{n} = 0$

$\displaystyle\sum_{n=0}^{\infty} \left[(n+2)(n+1)a_{n+2} - 2na_n + 8a_n \right] x^n = 0$

$a_{n+2} = \dfrac{2(n-4)}{(n+2)(n+1)} a_n$

$a_4 = 16 = \dfrac{2(-2)}{4(3)} a_2 = -\dfrac{1}{3}a_2 \Rightarrow a_2 = -48$

$a_2 = -48 = \dfrac{2(-4)}{2} a_0 = -4a_0 \Rightarrow a_0 = 12$

$H_4(x) = 16x^4 - 48x^2 + 12$

(b) $H_0(x) = \dfrac{(2x)^0}{0!} = 1$

$H_1(x) = \dfrac{(2x)^1}{1!} = 2x$

$H_2(x) = \displaystyle\sum_{n=0}^{1} \dfrac{(-1)^n 2!(2x)^{2-2n}}{n!(2-2n)!}$

$= \dfrac{2(2x)^2}{2!} - \dfrac{2}{1} = 4x^2 - 2$

$H_3(x) = \displaystyle\sum_{n=0}^{1} \dfrac{(-1)^n 3!(2x)^{3-2n}}{n!(3-2n)!}$

$= \dfrac{3!(2x)^3}{3!} - \dfrac{3!(2x)^1}{1}$

$= 8x^3 - 12x$

$H_4(x) = \displaystyle\sum_{n=0}^{2} \dfrac{(-1)^n 4!(2x)^{4-2n}}{n!(4-2n)!}$

$= \dfrac{4!(2x)^4}{4!} - \dfrac{4!(2x)^2}{2!} + \dfrac{4!}{2!}$

$= 16x^4 - 48x^2 + 12$